W9-BGN-203

Mike Holt's Illustrated Guide to

Understanding the *National Electrical Code* Volume 2

Based on the 2005 *NEC*©

www.*NEC*code.com
1.888.NEC®code

Mike Holt Enterprises, Inc.
1.888.NEC.CODE • NECcode.com • Info@NECcode.com

NOTICE TO THE READER

Publisher does not warrant or guarantee any of the products described herein or perform any independent analysis in connection with any of the product information contained herein. Publisher does not assume, and expressly disclaims, any obligation to obtain and include information other than that provided to it by the manufacturer.

The reader is expressly warned to consider and adopt all safety precautions that might be indicated by the activities herein and to avoid all potential hazards. By following the instructions contained herein, the reader willingly assumes all risks in connection with such instructions.

The publisher makes no representation or warranties of any kind, including but not limited to, the warranties of fitness for particular purpose or merchantability, nor are any such representations implied with respect to the material set forth herein, and the publisher takes no responsibility with respect to such material. The publisher shall not be liable for any special, consequential, or exemplary damages resulting, in whole or part, from the reader's use of, or reliance upon, this material.

Mike Holt's Illustrated Guide to
Understanding the *National Electrical Code*, Volume 2
5th Edition
Graphic Illustrations: Mike Culbreath
Cover Design: Tracy Jette
Layout Design and Typesetting: Cathleen Kwas

COPYRIGHT © 2005 Charles Michael Holt Sr.
ISBN: 1-932685-17-0

For more information, call 1-888-NEC CODE, or email info@MikeHolt.com.

All rights reserved. No part of this work covered by the copyright hereon may be reproduced or used in any form or by any means graphic, electronic, or mechanical, including photocopying, recording, taping, or information storage and retrieval systems without the written permission of the publisher. You can request permission to use material from this text, phone 1.888.NEC.CODE, Sales@NECcode.com, or www.NECcode.com.

NEC, *NFPA*, and *National Electrical Code* are registered trademarks of the National Fire Protection Association.

This logo is a registered trademark of Mike Holt Enterprises, Inc.

To request examination copies of this or other Mike Holt Publications, call:
Phone: 1-888-NEC CODE • Fax: 1-954-720-7944
or E-mail: info@MikeHolt.com
or visit Mike Holt Online: www.NECcode.com

You can download a sample PDF of all our publications by visiting www.NECcode.com

I dedicate this book to the
Lord Jesus Christ,
my mentor and teacher.

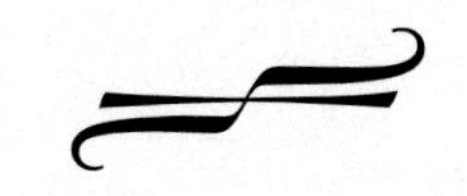

To understand the NEC®, this series is for you. Mike explains, in great detail, the history of the rule, the reason for the rule, and how to apply the Code® for everyday use. This program explains the 2002 or the 2005 Code, how to use the NEC, general installation requirements, branch circuits, feeders, services and overcurrent protection, grounding versus bonding, conductors, cables and raceways, boxes, panels, motors and transformers, and more. Understanding the National Electrical Code was written to provide insight into, and an understanding of, many of the technical rules of the NEC. This series contains clear, color graphics and an easy-to-follow format.

Call us today at 1.888.NEC.Code, or visit us online at www.NECcode.com, for the latest information and pricing.

ONE TEAM

To Our Instructors and Students:

We are committed to providing you the finest product with the fewest errors, but we are realistic and know that there will be errors found and reported after the printing of this book. The last thing we want is for you to have problems finding, communicating, or accessing this information. It is unacceptable to us for there to be even one error in our textbooks or answer keys. For this reason, we are asking you to work together with us as **One Team**.

Students: Please report any errors that you may find to your instructor.
Instructors: Please communicate these errors to us.

Our Commitment:

We will continue to list all of the corrections that come through for all of our textbooks and answer keys on our Website. We will always have the most up-to-date answer keys available to instructors to download from our instructor Website. The last thing that we want is for you to have problems finding this updated information, so we're outlining where to go for all of this below:

To view textbook and answer key corrections: Students and instructors go to our Website, www.MikeHolt.com, click on "Books" in the sidebar of links, and then click on "Corrections."

To download the most up-to-date answer keys: Instructors go to our Website, www.MikeHolt.com, click on "Instructors" in the sidebar of links and then click on "Answer Keys." On this page you will find instructions for how to access and download these answer keys.

If you are not registered as an instructor you will need to register. Your registration will be sent to our educational director who in turn reviews and approves your registration. In your approval E-mail will be the login and password so you can have access to all of the answer keys. If you have a situation that needs immediate attention, please contact the office directly at 1-888-NEC-CODE.

1.888.NEC.Code or visit us online at www.NECcode.com

Table of Contents

Introduction

This edition of *Mike Holt's Illustrated Guide to Understanding the National Electrical Code, Volume 2* textbook is intended to provide you with the tools necessary to understand the technical requirements of the *National Electrical Code (NEC)*®. The writing style of this textbook, and in all of Mike Holt's products, is meant to be informative, practical, useful, informal, easy to read, and applicable for today's electrical professional. Also, just like all of Mike Holt's textbooks, it contains hundreds of full-color illustrations to help you see the safety requirements of the *NEC* in practical use, as they apply to today's electrical installations.

This illustrated textbook contains advice, cautions about possible conflicts or confusing *Code* requirements, tips on proper electrical installations, and warnings of dangers related to improper electrical installations. This textbook cannot eliminate confusing, conflicting, or controversial *NEC* requirements, but it does try to put these requirements into sharper focus to help you understand their intended purpose. Sometimes a requirement is so confusing that nobody really understands its actual application. When this occurs, the textbook will be upfront and straightforward to point this out.

The *NEC* is updated every three years to accommodate new electrical products and material, emerging and dynamically changing technologies, as well as improved installation techniques. Fortunately, the *Code* allows the authority having jurisdiction, typically called the "Electrical Inspector," the flexibility to waive specific *NEC* requirements, and to permit alternative wiring methods contrary to the *Code* requirements, when he/she is assured that the completed electrical installation is equivalent in establishing and maintaining effective safety [90.4].

Keeping up with the *NEC* should be the goal of all those who are involved in electrical safety as it relates to electrical installations. This includes the electrical installer, contractor, owner, inspector, architect, engineer, instructor, and others concerned with safety as it relates to electrical installations.

Scope of *Understanding the NEC, Volume 2*

This textbook, *Understanding the National Electrical Code, Volume 2*, covers the requirements for wiring in special occupancies, special equipment, under special conditions, as well as communications systems that Mike considers to be of critical importance. This textbook contains the following stipulations:

- Power Systems and Voltage. All power-supply systems are assumed to be solidly grounded and of any of the following voltages: 120V single-phase, 120/240V single-phase, 120/208V three-phase, 120/240V three-phase, or 277/480V three-phase, unless identified otherwise.
- Electrical Calculations. Unless the question or example specifies three-phase, the questions and examples are based on a single-phase power supply. In addition, all ampere calculations are rounded to the nearest ampere in accordance with 220.5(B).
- Conductor Material. All conductors are considered copper, unless aluminum is identified or specified.
- Conductor Sizing. All conductors are sized based on a THHN copper conductor terminating on a 75°C terminal in accordance with 110.14(C), unless the question or example identifies otherwise.
- Protection Device. The term "circuit protection device" refers to a molded case circuit breaker, unless identified otherwise. Where a fuse is identified, it is the single-element type, also known as a "one-time fuse," unless identified otherwise.

Workbook to Accompany Understanding the NEC, Volume 2

The *Workbook To Accompany Understanding the NEC, Volume 2* (2005 Edition) contains 670 *NEC* practice questions plus three 50-question exams which will test your knowledge and comprehension of the covered material.

Understanding the NEC, Volume 1

To understand the entire *National Electrical Code*, you need to also study Mike's *Understanding the NEC, Volume 1* textbook, which covers general installation requirements, branch circuits, feeders, services and overcurrent protection, grounding versus bonding, conductors, cables and raceways, boxes, panels, motors and transformers, and more, in Articles 90 through 460 (*NEC* Chapters 1 through 4).

Companion Video or DVDs

There are three companion videos or DVDs (a total of 13.5 hours) available for *Understanding the National Electric Code, Volume 2* (Articles 500–830). There are seven companion videos or DVDs (a total of 28 hours) available for *Understanding the National Electric Code, Volume 1* (Articles 90–460). Visit www.NECcode.com for details.

Page Numbering Sequence

As a follow up, this textbook includes coverage of Articles 90 and Chapter 1 from the *Understanding the NEC, Volume 1* textbook (pages 1-50). Beginning with Chapter 5, the page numbering sequence continues from Chapter 4 of *Volume 1*.

NEC Library

The *NEC* Library includes the Understanding the *NEC* Volumes 1 & 2 textbooks, the *NEC* Practice Questions book and ten videos or DVD's (a total of 41.5 hours). This option allows you to learn at the most cost effective price.

How to Use This Textbook

This textbook is to be used with the *NEC*, not as a replacement for the *NEC,* so be sure to have a copy of the 2005 *National Electrical Code* handy. Compare what Mike has explained in the text to your *Code* book, and discuss those topics that you find difficult to understand with others. As you read through this textbook, be sure to take the time to review the text with the outstanding graphics and examples.

Cross-References

The textbook contains thousands of *NEC* cross-references to other related *Code* requirements to help you develop a better understanding of how the *NEC* rules relate to one another. These cross-references are identified by a *Code* Section number in brackets, such as "90.4," which would look like "[90.4]."

Author's Comments

This textbook contains hundreds of "Author's Comments." These sections were written by Mike to help you better understand the *NEC* material, and to bring to your attention things he believes you should be aware of. To help you find them more easily, they are printed differently than the rest of the material. Mike's first Author's Comment is contained in 90.1(B).

Textbook Format

This textbook follows the *NEC* format, but it doesn't cover every *Code* requirement. For example, it doesn't cover every *NEC* Article, Section, Subsection, Exception, or Fine Print Note. So don't be concerned if you see that the textbook contains Exception No. 1 and Exception No. 3, but not Exception No. 2. In addition, at times, the title of an Article, Section, or Subsection might be rephrased differently.

Difficult Concepts

As you progress through this textbook, you might find that you don't understand every explanation, example, calculation, or comment. Don't get frustrated, and don't get down on yourself. Remember, this is the *National Electrical Code* and sometimes the best attempt to explain a concept isn't enough to make it perfectly clear. When this happens to you, just make it a point to highlight the section that is causing you difficulty. If you can, take the textbook to someone you feel can provide additional insight, possibly your boss, the electrical inspector, a co-worker, your instructor, etc.

Not an *NEC* Replacement

This textbook is intended to explain the requirements of the *NEC*. It isn't intended to be a replacement for the *NEC*, so be sure you have a copy of the current *Code* book, and always compare Mike's explanation, comments, and graphics to the actual language contained in the *NEC*.

You'll sometimes notice that the titles of a few Articles and Sections are different than they appear in the actual *Code*. This only occurs when Mike feels it's easier to understand the content of the rule, so please keep this in mind when comparing the two documents. For example, 250.8 in the *NEC* is titled "Connection of Grounding and Bonding Equipment," but Mike used the title "Termination of Grounding (Earthing) and Bonding Conductors," because he felt his title better reflects the content of the rule.

Textbook Errors and Corrections

Humans develop the text, graphics, and layout of this textbook, and since currently none of us is perfect, there may be a few errors. This could occur because the *NEC* is dramatically changed each *Code* cycle; new Articles are added, some deleted, some relocated, and many renumbered. In addition, this textbook must be written within a very narrow window of opportunity; after the *NEC* has been published (September), yet before it's enforceable (January).

You can be sure we work a tremendous number of hours and use all of our available resources to produce the finest product with the fewest errors. We take great care in researching the *Code* requirements to ensure this textbook is correct. If you feel there's an error of any type in this textbook (typo, grammar, or technical), no matter how insignificant, please let us know.

Any errors found after printing are listed on our Website, so if you find an error, first check to see if it has already been corrected. Go to www.MikeHolt.com, click on the "Books" link, and then the "Corrections" link (www.MikeHolt.com/bookcorrections.htm).

If you do not find the error listed on the Website, contact us by E-mailing corrections@MikeHolt.com, calling 1.888.NEC.CODE (1.888.632.2633), or faxing 954.720.7944. Be sure to include the book title, page number, and any other pertinent information.

Internet

Today as never before, you can get your technical questions answered by posting them to Mike Holt's *Code* Forum. Just visit www.MikeHolt.com and click on the "*Code* Forum" link.

Different Interpretations

Some electricians, contractors, instructors, inspectors, engineers, and others enjoy the challenge of discussing the *Code* requirements, hopefully in a positive and a productive manner. This action of challenging each other is important to the process of better understanding the *NEC*'s requirements and its intended application. However, if you're going to get into an *NEC* discussion, please do not spout out what you think without having the actual *Code* in your hand. The professional way of discussing an *NEC* requirement is by referring to a specific section, rather than by talking in vague generalities.

The *National* *Electrical Code*

The *National Electrical Code (NEC)* is written for persons who understand electrical terms, theory, safety procedures, and electrical trade practices. These individuals include electricians, electrical contractors, electrical inspectors, electrical engineers, designers, and other qualified persons. The *Code* was not written to serve as an instructive or teaching manual for untrained individuals [90.1(C)].

Learning to use the *NEC* is somewhat like learning to play the game of chess; it's a great game if you enjoy mental warfare. You must first learn the names of the game pieces, how the pieces are placed on the board, and how each piece moves.

In the electrical world, this is equivalent to completing a comprehensive course on basic electrical theory, such as:

- What electricity is and how is it produced
- Dangers of electrical potential: fire, arc blast, arc fault, and electric shock
- Direct current
- Series and parallel circuits
- Electrical formulas
- Alternating current
- Induction, motors, generators, and transformers

Once you understand the fundamentals of the game of chess, you're ready to start playing the game. Unfortunately, at this point all you can do is make crude moves, because you really do not understand how all the information works together. To play chess well, you will need to learn how to use your knowledge by working on subtle strategies before you can work your way up to the more intriguing and complicated moves.

Again, back to the electrical world, this is equivalent to completing a course on the basics of electrical theory. You have the foundation upon which to build, but now you need to take it to the next level, which you can do by reading this textbook, watching the companion video or DVD, and answering the *NEC* practice questions in the *Workbook to Accompany Understanding the National Electric Code, Volume 2.*

Not a Game

Electrical work isn't a game, and it must be taken very seriously. Learning the basics of electricity, important terms and concepts, as well as the basic layout of the *NEC* gives you just enough knowledge to be dangerous. There are thousands of specific and unique applications of electrical installations, and the *Code* doesn't cover every one of them. To safely apply the *NEC*, you must understand the purpose of a rule and how it affects the safety aspects of the installation.

NEC Terms and Concepts

The *NEC* contains many technical terms, so it's crucial that *Code* users understand their meanings and their applications. If you do not understand a term used in a *Code* rule, it will be impossible to properly apply the *NEC* requirement. Be sure you understand that Article 100 defines the terms that apply to *two or more* Articles. For example, the term "Dwelling Unit" applies to many Articles. If you do not know what a Dwelling Unit is, how can you possibly apply the *Code* requirements for it?

In addition, many Articles have terms that are unique for that specific Article. This means that the definition of those terms is only applicable for that given Article. For example, Article 250 Grounding and Bonding has the definitions of a few terms that are only to be used within Article 250.

Small Words, Grammar, and Punctuation

It's not only the technical words that require close attention, because even the simplest of words can make a big difference to the intent of a rule. The word "or" can imply alternate choices for equipment wiring methods, while "and" can mean an additional requirement. Let's not forget about grammar and punctuation. The location of a comma "," can dramatically change the requirement of a rule.

Slang Terms or Technical Jargon

Electricians, engineers, and other trade-related professionals use slang terms or technical jargon that isn't shared by all. This makes it very difficult to communicate because not everybody understands the intent or application of those slang terms. So where possible, be sure you use the proper word, and do not use a word if you do not understand its definition and application. For example, lots of electricians use the term "pigtail" when describing the short conductor for the connection of a receptacle, switch, luminaire, or equipment. Although they may understand it, not everyone does. **Figure 1**

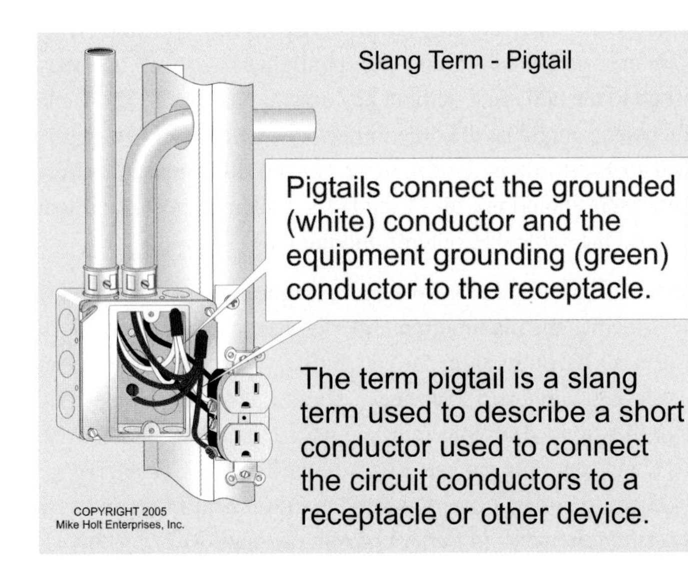

Slang Term - Pigtail
Pigtails connect the grounded (white) conductor and the equipment grounding (green) conductor to the receptacle.
The term pigtail is a slang term used to describe a short conductor used to connect the circuit conductors to a receptacle or other device.

COPYRIGHT 2005
Mike Holt Enterprises, Inc.

Figure 1

NEC Style and Layout

Before we get into the details of the *NEC*, we need to take a few moments to understand its style and layout. Understanding the structure and writing style of the *Code* is very important before it can be used effectively. If you think about it, how can you use something if you don't know how it works? Okay, let's get started. The *National Electrical Code* is organized into nine components.

- Table of Contents
- Chapters 1 through 9 (major categories)
- Articles 90 through 830 (individual subjects)
- Parts (divisions of an Article)
- Sections and Tables (*Code* requirements)
- Exceptions (*Code* permissions)
- Fine Print Notes (explanatory material)
- Index
- Annexes (information)

1. Table of Contents. The Table of Contents displays the layout of the Chapters, Articles, and Parts as well as the page numbers. It's an excellent resource and should be referred to periodically to observe the interrelationship of the various *NEC* components. When attempting to locate the rules for a particular situation, knowledgeable *Code* users often go first to the Table of Contents to quickly find the specific *NEC* section that applies.

2. Chapters. There are nine Chapters, each of which is divided into Articles. The Articles fall into one of four groupings: General Requirements (Chapters 1 through 4), Specific Requirements (Chapters 5 through 7), Communications Systems (Chapter 8), and Tables (Chapter 9).

- Chapter 1 General
- Chapter 2 Wiring and Protection
- Chapter 3 Wiring Methods and Materials
- Chapter 4 Equipment for General Use
- Chapter 5 Special Occupancies
- Chapter 6 Special Equipment
- Chapter 7 Special Conditions
- Chapter 8 Communications Systems (Telephone, Data, Satellite, and Cable TV)
- Chapter 9 Tables—Conductor and Raceway Specifications

3. Articles. The *NEC* contains approximately 140 Articles, each of which covers a specific subject. For example:

- Article 110 General Requirements
- Article 250 Grounding
- Article 300 Wiring Methods
- Article 430 Motors
- Article 500 Hazardous (Classified) Locations
- Article 680 Swimming Pools, Spas, Hot Tubs, and Fountains
- Article 725 Remote-Control, Signaling, and Power-Limited Circuits
- Article 800 Communications Systems

4. Parts. Larger Articles are subdivided into Parts. For example, Article 110 has been divided into multiple parts:

- Part I. General (Sections 110.1—110.23)
- Part II. 600 Volts, Nominal, or Less (110.26—110.27)
- Part III. Over 600 Volts, Nominal (110.30—110.59)

Note: Because the Parts of a *Code* Article aren't included in the Section numbers, we have a tendency to forget what "Part" the *NEC* rule is relating to. For example, Table 110.34(A) contains the working space clearances for electrical equipment. If we aren't careful, we might think this table applies to all electrical installations, but Table 110.34(A) is located in Part III, which contains the requirements for Over 600 Volts, Nominal installations. The rules for working clearances for electrical equipment for systems 600V or less are contained in Table 110.26(A)(1), which is located in Part II. 600 Volts, Nominal, or Less.

5. Sections and Tables.

Sections: Each *NEC* rule is called a *Code* Section. A *Code* Section may be broken down into subsections by letters in parentheses (A), (B), etc. Numbers in parentheses (1), (2), etc., may further break down a subsection, and lower-case letters (a), (b), etc., further break the rule down to the third level. For example, the rule requiring all receptacles in a dwelling unit bathroom to be GFCI protected is contained in Section 210.8(A)(1). Section 210.8(A)(1) is located in Chapter 2, Article 210, Section 8, subsection (A), sub-subsection (1).

Many in the industry incorrectly use the term "Article" when referring to a *Code* Section. For example, they say "Article 210.8," when they should say "Section 210.8."

Tables: Many *Code* requirements are contained within Tables, which are lists of *NEC* requirements placed in a systematic arrangement. The titles of the Tables are extremely important; they must be carefully read in order to understand the contents, applications, limitations, etc., of each Table in the *Code*. Many times notes are provided in a table; be sure to read them as well, since they are also part of the requirement. For example, Note 1 for Table 300.5 explains how to measure the cover when burying cables and raceways, and Note 5 explains what to do if solid rock is encountered.

6. Exceptions. Exceptions are *Code* requirements that provide an alternative method to a specific requirement. There are two types of exceptions—mandatory and permissive. When a rule has several exceptions, those exceptions with mandatory requirements are listed before the permissive exceptions.

Mandatory Exception: A mandatory exception uses the words "shall" or "shall not." The word "shall" in an exception means that if you're using the exception, you're required to do it in a particular way. The term "shall not" means it isn't permitted.

Permissive Exception: A permissive exception uses words such as "is permitted," which means that it's acceptable to do it in this way.

7. Fine Print Note (FPN). A Fine Print Note contains explanatory material intended to clarify a rule or give assistance, but it isn't a *Code* requirement.

8. Index. The Index contained in the *NEC* is excellent and is helpful in locating a specific rule.

9. Annexes. Annexes aren't a part of the *NEC* requirements, and are included in the *Code* for informational purposes only.

- Annex A. Product Safety Standards
- Annex B. Application Information for Ampacity Calculation
- Annex C. Conduit and Tubing Fill Tables for Conductors and Fixture Wires of the Same Size
- Annex D. Examples
- Annex E. Types of Construction
- Annex F. Cross-Reference Tables (1999, 2002, and 2005 *NEC*)
- Annex G. Administration and Enforcement

Note: Changes to the *NEC*, since the previous edition(s) are identified in the margins by a vertical line (|), but rules that have been relocated aren't identified as a change. In addition, the location from which the *Code* rule was removed has no identifier.

How to Locate a Specific Requirement

How to go about finding what you're looking for in the *Code* depends, to some degree, on your experience with the *NEC*. *Code* experts typically know the requirements so well that they just go to the *NEC* rule without any outside assistance. The Table of Contents might be the only thing very experienced *Code* users need to locate their requirement. On the other hand, average *Code* users should use all of the tools at their disposal, and that includes the Table of Contents and the Index.

Table of Contents: Let's work out a simple example: What *NEC* rule specifies the maximum number of disconnects permitted for a service? If you're an experienced *Code* user, you'll know that Article 230 applies to "Services," and because this Article is so large, it's divided up into multiple parts (actually 8 parts). With this knowledge, you can quickly go to the Table of Contents (page 70-2) and see that it lists the Service Equipment Disconnecting Means requirements in Part VI, starting at page 70-77.

Note: The number 70 precedes all page numbers because the *NEC* is standard number 70 within the collection of *NFPA* standards.

Index: If you used the Index, which lists subjects in alphabetical order, to look up the term "service disconnect," you would see that there's no listing. If you tried "disconnecting means," then "services," you would find the Index specifies that the rule is located at 230, Part VI. Because the *NEC* doesn't give a page number in the Index, you'll need to use the Table of Contents to get the page number, or flip through the *Code* to Article 230, then continue to flip until you find Part VI.

As you can see, although the index is very comprehensive, it's not that easy to use if you do not understand how the index works. But if you answer the over 670 *NEC* practice questions or three 50-question exams contained in the *Workbook to Accompany Understanding the National Electric Code, Volume 2,* you'll become a master at finding things in the *Code* quickly.

Many people complain that the *NEC* only confuses them by taking them in circles. As you gain experience in using the *Code* and deepen your understanding of words, terms, principles, and practices, you will find the *NEC* much easier to understand and use than you originally thought.

Customizing Your *Code* Book

One way to increase your comfort level with the *Code* is to customize it to meet your needs. You can do this by highlighting and underlining important *NEC* requirements, and by attaching tabs to important pages.

Highlighting: As you read through this textbook and answer the questions in the workbook, be sure you highlight those requirements in the *Code* that are most important to you. Use yellow for general interest and orange for important requirements you want to find quickly. Be sure to highlight terms in the Index and Table of Contents as you use them.

Because of the size of the 2005 *NEC*, I recommend you highlight in green the Parts of Articles that are important for your applications, particularly:

Article 230 Services
Article 250 Grounding
Article 430 Motors

Underlining: Underline or circle key words and phrases in the *NEC* with a red pen (not a lead pencil) and use a 6-in. ruler to keep lines straight and neat. This is a very handy way to make important requirements stand out. A small 6-in. ruler also comes in handy for locating specific information in the many *Code* tables.

Tabbing the *NEC*: Placing tabs on important *Code* Articles, Sections, and Tables will make it very easy to access important *NEC* requirements. However, too many tabs will defeat the purpose. You can order a custom set of *Code* tabs, designed by Mike Holt, online at www.MikeHolt.com, or by calling us at 1.888.NEC.Code (1.888.632.2633).

Mike Holt Enterprises Team

About the Author

Mike Holt worked his way up through the electrical trade from an apprentice electrician to become one of the most recognized experts in the world as it relates to electrical power installation. He was a Journeyman Electrician, Master Electrician, and Electrical Contractor. Mike came from the real world, and he has a unique understanding of how the *NEC* relates to electrical installations from a practical standpoint. You will find his writing style to be simple, nontechnical, and practical.

Did you know that Mike didn't finish high school? So if you struggled in high school or if you didn't finish it at all, don't let this get you down, you're in good company. As a matter of fact, Mike Culbreath, Master Electrician, who produces the finest electrical graphics in the history of the electrical industry, didn't finish high school either! So two high school dropouts produced the text and graphics in this textbook! However, realizing that success depends on one's continuing pursuit of education, Mike immediately attained his GED (as did Mike Culbreath) and ultimately attended the University of Miami's Graduate School for a Master's degree in Business Administration (MBA).

Mike Holt resides in Central Florida, is the father of seven children, and has many outside interests and activities. He is a former National Barefoot Waterskiing Champion (1988 and 1999), who set five barefoot water-ski records, and he continues to train year-round at a national competition level [www.barefootcentral.com].

Mike enjoys motocross racing, but at the age of 52 decided to retire from that activity (way too many broken bones, concussions, collapsed lung, etc., but what a rush). Mike also enjoys snow skiing and spending time with his family. What sets Mike apart from some is his commitment to living a balanced lifestyle; he places God first, then family, career, and self.

Educational Director

Sarina Snow was born and raised in "The Bronx" (Yankee Stadium Area), then moved to New Jersey with her husband Freddy where they raised their three children. They moved to Florida in 1979 and have totally loved the move to a warmer climate. Sarina has worked with Mike Holt for over twenty years and has literally learned the business from the ground up. She remembers typing Mike's books using carbon paper BC (Before Computers). She has developed a strong relationship with the industry by attending Mike's classes and seminars, and by accompanying Mike to many trade shows in various states to get a deeper understanding of the trade firsthand. Her love and devotion to Mike and Company (actually the original name of Mike Holt Enterprises) has given her the ability to be the "Mom" in the industry. "She Cares," sums it up.

Graphic Illustrator

Mike Culbreath devoted his career to the electrical industry and worked his way up from an apprentice electrician to master electrician. While working as a journeyman electrician, he suffered a serious on-the-job knee injury. With a keen interest in continuing education for electricians, and as part of his rehabilitation program, he completed courses at Mike Holt Enterprises, Inc. and then passed the exam to receive his Master Electrician's license.

In 1986, after attending classes at Mike Holt Enterprises, he joined the staff to update material and later studied computer graphics and began illustrating Mike Holt's textbooks and magazine articles. He's worked with Mike Holt Enterprises for over 15 years and, as Mike Holt has proudly acknowledged, has helped to transform his words and visions into lifelike graphics.

Mike Culbreath resides in northern Michigan with his wife Toni, and two children: Dawn and Mac. He is helping Toni fulfill her dream by helping her develop and build a quality horse boarding, training, and teaching facility. Mike enjoys working with children by volunteering as a leader for a 4-H archery club and assisting with the local 4-H horse club. He also enjoys fishing, gardening, and cooking.

Editorial

Toni Culbreath completed high school graduation requirements by the end of the first semester of her senior year. She went on to complete courses for computer programming at a trade school by March of that year, and then returned to participate in graduation ceremonies with her high school class.

Toni became associated with Mike Holt Enterprises in 1994 in the area of software support and training and now enjoys the challenges of editing Mike Holt's superb material. She is certified as a therapeutic riding instructor and is extensively involved in Michigan's 4-H horse programs at both the county and state level.

Barbara Parks has been working for Mike Holt Enterprises for the last several years as a Writer's Assistant. She has edited most of Mike Holt's books and various projects over this period of time. She is a retired lady, working part time at home and thoroughly enjoys "keeping busy."

Technical Editorial Director

Steve Arne has been involved in the electrical industry since 1974 working in various positions from electrician to full-time instructor and department chair in technical post secondary education. Steve has developed curriculum for many electrical training courses and has developed university business and leadership courses. Currently, Steve offers occasional exam prep and Continuing Education *Code* classes.

Steve believes that as a teacher he understands the joy of helping others as they learn and experience new insights. His goal is to help others understand more of the technological marvels that surround us. Steve thanks God for the wonders of His creation and for the opportunity to share it with others.

Steve and his lovely wife Deb live in Rapid City, South Dakota where they are both active in their church and community. They have two grown children and five grandchildren.

Cover Design

Tracy Jette has enjoyed working in the field of Graphic Design for over 10 years. She loves all aspects of design, and finds that spending time outdoors camping and hiking with her family and friends is a great inspiration. Tracy is very happy to have recently joined Mike Holt Enterprises and has found that working from home brings a harmony to her life with her 3 boys (10-year-old twins and a 7-year-old), her husband of 16 years, Mario, and her work life.

Layout Design and Production

Cathleen Kwas has been in the publishing industry for over 26 years. She's worn many hats–copy editor, desktop publisher, prepress manager, project coordinator, communications director, book designer, and graphic artist.

Cathleen is very happily married to Michael and lives in beautiful Lake Mary, Florida with their adorable Maltese-ShihTzu puppy, Bosco.

Acknowledgments

Special Acknowledgments

First I want to thank God for my life. I want to thank Him for even the most difficult of times, because this has helped me become a man that I hope honors Him in my actions. My loving Godly wife is always by my side, and there's no question that I could not have achieved any of my success without her continued support. She is a selfless mother and wife. She made the difficult decision to stay at home and support her family; the successes of her husband and her children are God's reward for her sacrifice. To my wonderful children, Belynda, Melissa, Autumn, Steven, Michael, Meghan, and Brittney—I love every moment we shared together (well, most moments). Thank you for loving me and knowing God.

I would like to thank all the people in my life that believed in me, and those who spurred me on. Thanks to the Electrical Construction & Maintenance *(EC&M)* magazine for my first "big break" in 1980, and Joe McPartland who helped and encouraged me from 1980 to 1992. Joe, I'll never forget to help others as you've helped me. I would also like to thank Joe Salimando, the former publisher of the Electrical Contractor magazine produced by the National Electrical Contractors Association (*NEC*A) for my second "big break" in 1995.

A special thank you must be sent to the staff at the National Fire Protection Association (NFPA), publishers of the *NEC*—in particular Jeff Sargent for his assistance in answering my many *NEC* questions. Jeff, you're a "first class" guy, and I admire your dedication and commitment to helping others, including me, to understand the *Code*. Other former NFPA staff members I would like to thank include John Caloggero, Joe Ross, and Dick Murray for their help in the past.

Phil Simmons, former Executive Director of the International Association of Electrical Inspectors (IAEI)—you're truly a Godly man whom I admire, and I do want to thank you for your help, especially in grounding. Other people who have been important in my personal and technical development include James Stallcup, Dick Loyd, Mark Ode, DJ Clements, Morris Trimmer, Tony Silvestri, and the infamous Marvin Weiss.

A personal thank you goes to Sarina, my long-time friend and office manager. Thank you for covering the office for me while I spend so much time writing textbooks, conducting seminars, and producing videos and DVDs. Your love and concern for the customer has contributed significantly to the success of Mike Holt Enterprises, Inc., and it has been wonderful working side-by-side and nurturing this company's growth from its small beginnings. Also thank you for loving my family and me and for being there during those many difficult times.

Mike Holt Enterprises Team Acknowledgments

There are many people who played a role in the development and production of this textbook. I would like to start with Mike Culbreath, Master Electrician, who has been with me for over 15 years, helping me transform my words and thoughts into lifelike graphics.

Also, a thank you goes to Cathleen Kwas for the outstanding electronic layout of this textbook, and Tracy Jette for the amazing front and back cover.

Finally, I would like to thank the following individuals who worked tirelessly to proofread and edit the final stages of this publication: Toni Culbreath and Barbara Parks. Their attention to detail and dedication to this project is greatly appreciated.

Special Acknowledgement

I would like to thank my assistant, Tara Martin, for her outstanding work in coordinating this book. In addition to her excellent organizational skills, she has displayed a great team attitude and demonstrated calm patience when juggling many projects at the same time—especially when we're on deadline!

Advisory Committee

Thanks are also in order for the following individuals who reviewed the manuscript and offered invaluable suggestions and feedback.

Victor M. Ammons, P.E.
Director of Electrical Engineering,
The Prisco Group,
Hopewell, New Jersey

Steve Arne
Technical Director, Mike Holt Enterprises, Inc.,
Rapid City, South Dakota

Mike Culbreath
Graphic Designer, Mike Holt Enterprises, Inc.
Alden, Michigan

Leo W. Moritz, P.E.
Senior Electrical Engineer,
Chicago, Illinois

Jerry Peck
Inspector and Instructor,
Inspection Services Associates, Inc.,
Pembroke Pines, Florida

Terry Schneider
Electrical Field Inspection Supervisor,
Colorado Springs, Colorado

Brooke Stauffer
Executive Director of Standards and Safety,
National Electrical Contractors Association,
Bethesda, Maryland

James Thomas
Electrical/Electronics Instructor,
James Sprunt Community College,
Kenansville, North Carolina

J. Kevin Vogel, P.E.
Design and Quotations, Crescent Electric Supply,
Coeur d'Alene, Idaho

Joseph Wages Jr.
Instructor
Siloam Springs, Arkansas

A Very Special Thank You

To my beautiful wife, Linda, and my seven children:
Belynda, Melissa, Autumn, Steven, Michael,
Meghan, and Brittney—
thank you for loving me so much.

Video Team Members

Larry Abernathy
Master Electrician/Instructor
Ford Motor Company—Research and Engineering
Dearborn, Michigan

Larry Abernathy entered the U.S. Army after graduating from high school and received training in communications and electronics. He served in Korea, Germany, and three tours in Vietnam, where he was wounded twice. Larry was trained as an Army instructor and taught communications and electronics at the U.S. Army Southeastern Signal School at Fort Gordon, Georgia. He then served an IBEW apprenticeship and became licensed as a journeyman in 1976 and as a master in 1979. Larry was employed in construction, industrial, commercial, and residential work until becoming the Electrical Inspector for the City of Ypsilanti, Michigan in 1986 with a promotion to Supervisor of Building Inspection in 1988. In 1995, he became employed by the Ford Motor Company and works in the Research and Engineering Center in Dearborn, Michigan. Larry has also taught *National Electrical Code* update classes.

Larry is married with three grown sons, one of whom is an IBEW journeyman electrician.

Steve Arne
Technical Editorial Director,
Mike Holt Enterprises, Inc.
Electrical Instructor, Arne Electro Tech,
Rapid City, South Dakota
http://electricalmaster.com

Steve Arne has been involved in the electrical industry since 1974, working in various positions from electrician to full-time instructor and department chair in technical post secondary education. He has a Bachelor's Degree in Technical Education and a Master's Degree in Administrative Studies with a human resources emphasis. Licenses held by Steve include Electrical Master, Electrical Inspector, Electrical Contractor, and Real Estate Home Inspector. He is a board member of the Black Hills Chapter of the SD Electrical Council and a member of the SD Real Estate Task Force on Home Inspection. He also enjoys developing his own Websites.

Steve and his lovely wife Deb have celebrated over 32 years of marriage in Rapid City, South Dakota where they are both active in their church and community. They have two grown children and five grandchildren.

Mike Culbreath

Graphic Illustrator,
Mike Holt Enterprises, Inc.
Alden, Michigan

Mike Culbreath devoted his career to the electrical industry and worked his way up from an apprentice electrician to master electrician. While working as a journeyman electrician, he suffered a serious on-the-job knee injury. With a keen interest in continuing education for electricians, and as part of his rehabilitation program, he completed courses at Mike Holt Enterprises, Inc. and then passed the exam to receive his Master Electrician's license.

In 1986, after attending classes at Mike Holt Enterprises, he joined the staff to update material and later studied computer graphics and began illustrating Mike Holt's textbooks and magazine articles. He's worked with Mike Holt Enterprises for over 15 years and, as Mike Holt has proudly acknowledged, has helped to transform his words and visions into lifelike graphics.

Mike resides in northern Michigan with his wife Toni, and two children: Dawn and Mac. Mike enjoys working with children by volunteering as a leader for a 4-H archery club, assisting with the local 4-H horse club, fishing, gardening, and cooking.

Ryan Jackson

Inspector/Instructor
www.RyanJacksonElectricalTraining.com
City of Draper, Utah

Ryan Jackson is a combination inspector for Draper City, Utah. Ryan is certified by the International Code Council as an electrical, plumbing, mechanical, and building inspector for both residential and commercial structures. He is also certified as an Electrical Plans Examiner and a Building Plans Examiner. Ryan is the senior electrical inspector for Draper City.

Ryan has taught continuing education seminars as well as classes for the Utah Chapter of the International Code Council, the Utah Chapter of the International Association of Electrical Inspectors, the Independent Electrical Contractors of Utah, and the International Brotherhood of Electrical Workers. He also teaches several classes for electrical contracting companies in Utah. Ryan is a board member of the Utah Chapter of the IAEI, and is currently the 2nd Vice President of the chapter.

Ryan enjoys reading, going to college football games, and spending time with his wife Sharie and their two children, Kaitlynn and Aaron. He also enjoys helping people over the Internet on various *NEC* discussion groups. On Mike Holt's Web site, Ryan has contributed to nearly 2,500 topics.

Chuck Williams

Master Electrician/Instructor
North Idaho College
Post Falls, Idaho

After serving with the U.S. Navy, he worked as a civilian on U.S. Navy aircraft and associated systems for 17 years before relocating to Idaho. He entered the Electrical Apprenticeship program in 1991 at the age of 41. He has worked in Idaho and Washington as a journeyman electrician and electrical administrator, as well as an electrical contractor.

He is in his seventh year as an electrical apprenticeship instructor at the Workforce Training Center, North Idaho College, and also teaches several courses in continuing education for electricians and contractors, including tutoring and electrical exam preparation.

Chuck began teaching electrical students and instantly fell in love with this aspect of his vocation. He considers the rewards of teaching the most significant experience of his life. "We all contribute to our industry in different ways. Teaching is only one of the many opportunities afforded electricians, as ours is a trade that is constantly evolving and moving forward to utilize new technologies, methods, and equipment. To help others see the opportunities available is a joy and I will never tire of it."

Chuck Williams resides in Hauser, Idaho with his wife of 33 years.

David A. Williams

Inspector/Instructor
Lansing Community College
Delta Township
Lansing, Michigan

Dave Williams was born and raised in Lake Station, Indiana. In 1974, he started his electrical apprenticeship. Dave relocated to Lansing, Michigan in 1985, working as an electrician at General Motors. He became an Electrical Inspector for the State of Michigan in 1989. Dave left the State of Michigan to inspect for Delta Township over ten years ago in order to be able to teach and to provide better service as an inspector. He has taught *National Electrical Code* classes at Lansing Community College since 1994 and has provided *Code* update classes for various groups including the Lansing JATC.

He is a member of the International Association of Electrical Inspectors and on the Michigan Chapter Board of Directors since 1993. He is currently the Michigan Chapter President and the Western Section Representative. Dave designed and operates the Web site for the Michigan Chapter at http://IAEI-Michigan.org.

Dave and his wife Marie have two adopted children Aaron 2 and Christina 4. In his spare time, he enjoys spending time with his family, camping, and golfing.

ARTICLE 90 Introduction to the *National Electrical Code*

Introduction

Many *NEC* violations and misunderstandings wouldn't occur if people doing the work simply understood Article 90. For example, many people see *Code* requirements as performance standards. In fact, *NEC* requirements are the bare minimum for safety. This is exactly the stance electrical inspectors, insurance companies, and courts will take when making a decision regarding electrical design or installation.

Article 90 opens by saying the *NEC* isn't intended as a design specification or instruction manual. The *National Electrical Code* has one purpose only. That is "the practical safeguarding of persons and property from hazards arising from the use of electricity."

Article 90 then describes the scope and arrangement of the *Code.* A person who says, "I can't find anything in the *Code,*" is really saying, "I never took the time to review Article 90." The balance of Article 90 provides the reader with information essential to understanding those items you do find in the *NEC.*

Typically, electrical work requires you to understand the first four Chapters of the *NEC,* plus have a working knowledge of the Chapter 9 tables. Chapters 5, 6, 7, and 8 make up a large portion of the *NEC,* but they apply to special situations. They build on, and extend, what you must know in the first four chapters. That knowledge begins with Chapter 1.

90.1 Purpose of the *NEC.*

(A) Practical Safeguarding. The purpose of the *NEC* is to ensure that electrical systems are installed in a manner that protects people and property by minimizing the risks associated with the use of electricity.

(B) Adequacy. The *Code* contains requirements that are considered necessary for a safe electrical installation. When an electrical installation is installed in compliance with the *NEC*, it will be essentially free from electrical hazards. The *NEC* is a safety standard, not a design guide.

The *NEC* requirements aren't intended to ensure that the electrical installation will be efficient, convenient, adequate for good service, or suitable for future expansion. Specific items of concern, such as electrical energy management, maintenance, and power quality issues aren't within the scope of the *NEC.* **Figure 90–1**

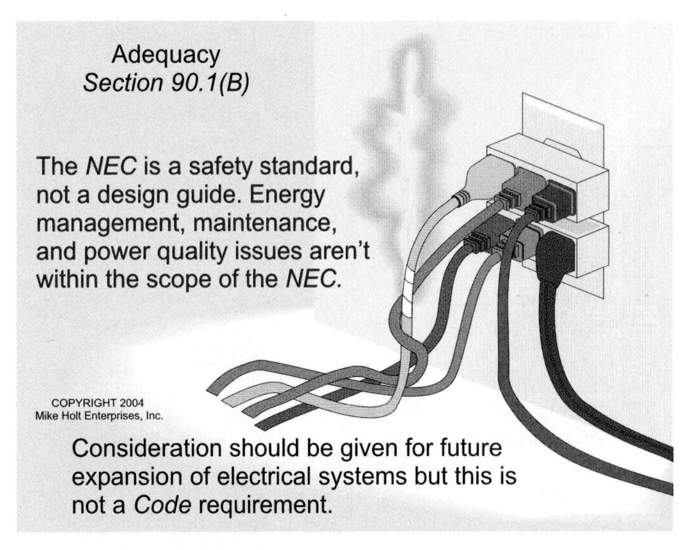

Adequacy
Section 90.1(B)
The NEC is a safety standard, not a design guide. Energy management, maintenance, and power quality issues aren't within the scope of the NEC.
COPYRIGHT 2004
Mike Holt Enterprises, Inc.
Consideration should be given for future expansion of electrical systems but this is not a Code requirement.

Figure 90–1

FPN: Hazards in electrical systems often occur because circuits are overloaded or not properly installed in accordance with the *NEC*. The initial wiring often did not provide reasonable provisions for system changes or for the increase in the use of electricity.

Author's Comments:

- See Article 100 for the definition of "Overload."
- The *NEC* does not require electrical systems to be designed or installed to accommodate future loads. However, the electrical designer, typically an electrical engineer, is concerned with not only ensuring electrical safety (*Code* compliance), but also ensuring that the system meets the customers' needs, both of today and in the near future. To satisfy customers' needs, electrical systems must be designed and installed above the minimum requirements contained in the *NEC*.

(C) Intention. The *Code* is to be used by those skilled and knowledgeable in electrical theory, electrical systems, construction, and the installation and operation of electrical equipment. It isn't a design specification standard or instruction manual for the untrained and unqualified.

(D) Relation to International Standards. The requirements of the *NEC* address the fundamental safety principles contained in International Electrotechnical Commission standards, including protection against electric shock, adverse thermal effects, overcurrent, fault currents, and overvoltage. **Figure 90–2**

Author's Comments:

- See Article 100 for the definition of "Overcurrent."
- The *NEC* is used in Chile, Ecuador, Peru, and the Philippines. It's also the *Electrical Code* for Colombia, Costa Rica, Mexico, Panama, Puerto Rico, and Venezuela. Because of these adoptions, the *NEC* is available in Spanish from the National Fire Protection Association, 1.617.770.3000.

NEC Relation to International Standards
Section 90.1(D) and FPN
The *NEC* addresses the safety principles contained in the IEC such as:
• Protection against electric shock
• Thermal effects
• Overcurrent
• Fault currents
• Overvoltage
NEC 2005
COPYRIGHT 2004
Mike Holt Enterprises, Inc.

Figure 90–2

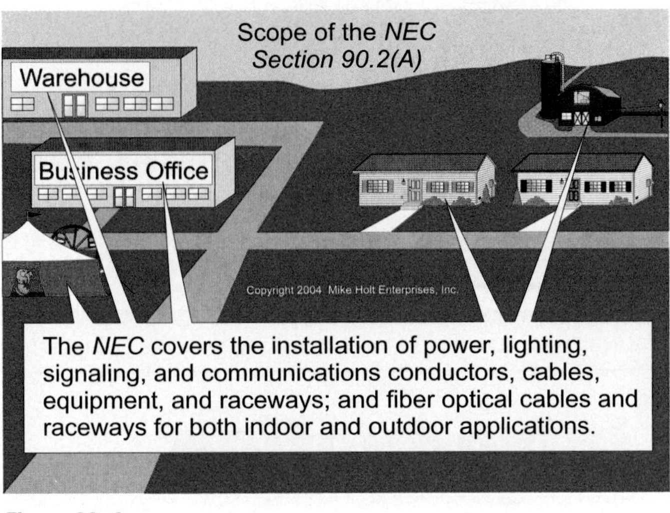

Scope of the *NEC*
Section 90.2(A)
Warehouse
Business Office
Copyright 2004 Mike Holt Enterprises, Inc.
The *NEC* covers the installation of power, lighting, signaling, and communications conductors, cables, equipment, and raceways; and fiber optical cables and raceways for both indoor and outdoor applications.

Figure 90–3

90.2 Scope of the *NEC*.

(A) What is Covered. The *NEC* contains requirements necessary for the proper electrical installation of electrical conductors, equipment, and raceways; signaling and communications conductors, equipment, and raceways; as well as fiber optic cables and raceways for the following locations: **Figure 90–3**

(1) Public and private premises, including buildings or structures, mobile homes, recreational vehicles, and floating buildings.

(2) Yards, lots, parking lots, carnivals, and industrial substations.

(3) Conductors and equipment that connect to the utility supply.

(4) Installations used by an electric utility, such as office buildings, warehouses, garages, machine shops, recreational buildings, and other electric utility buildings that are not an integral part of a utility's generating plant, substation, or control center. **Figure 90–4**

(B) What isn't Covered. The *National Electrical Code* doesn't apply to the following applications:

(1) Transportation Vehicles. Installations in cars, trucks, boats, ships and watercraft, planes, electric trains, or underground mines.

(2) Mining Equipment. Installations underground in mines and self-propelled mobile surface mining machinery and its attendant electrical trailing cables.

(3) Railways. Railway power, signaling, and communications wiring.

(4) Communications Utilities. The installation requirements of the *NEC* do not apply to communications (telephone), CATV, or network-powered broadband utility equipment located in

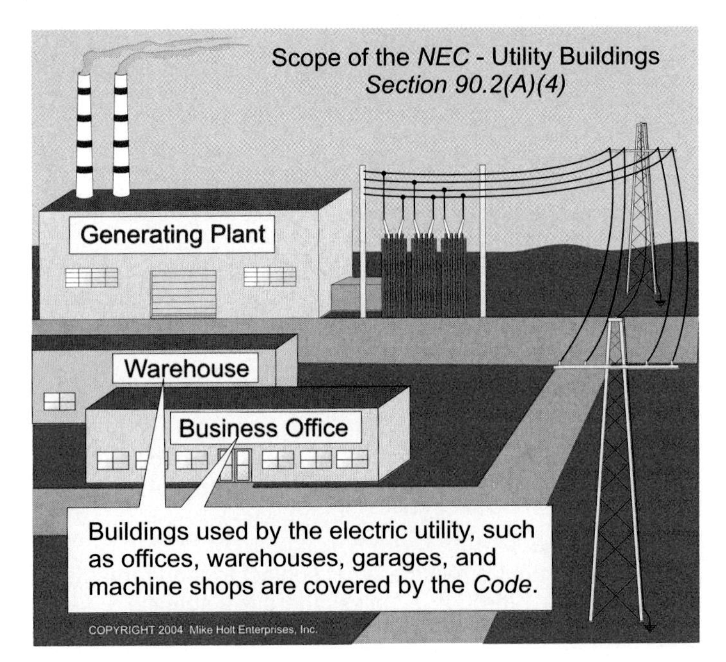

Figure 90–4

building spaces used exclusively for such use or outdoors, if the installation is under the exclusive control of the communications utility. **Figure 90–5**

Author's Comment: Interior wiring for communications systems, not in building spaces used exclusively for such use, must be installed in accordance with the following Chapter 8 requirements: **Figure 90–6**

- Phone and Data, Article 800
- CATV, Article 820
- Network-Powered Broadband, Article 830

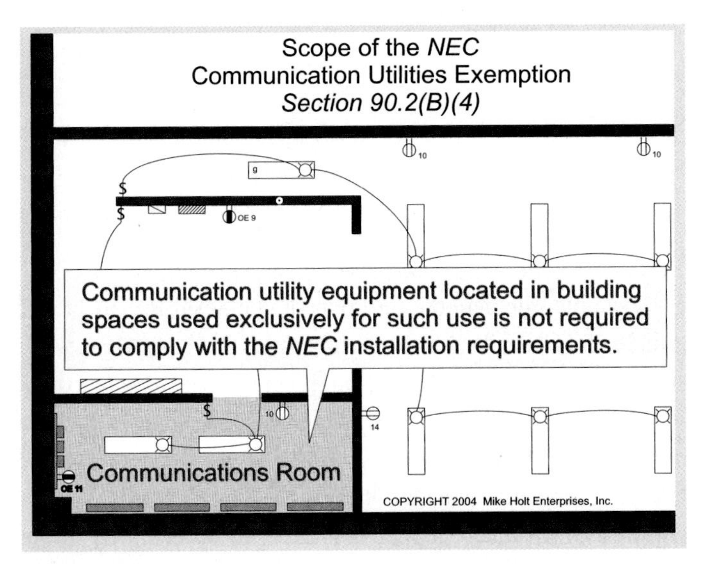

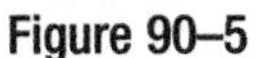
Figure 90–5

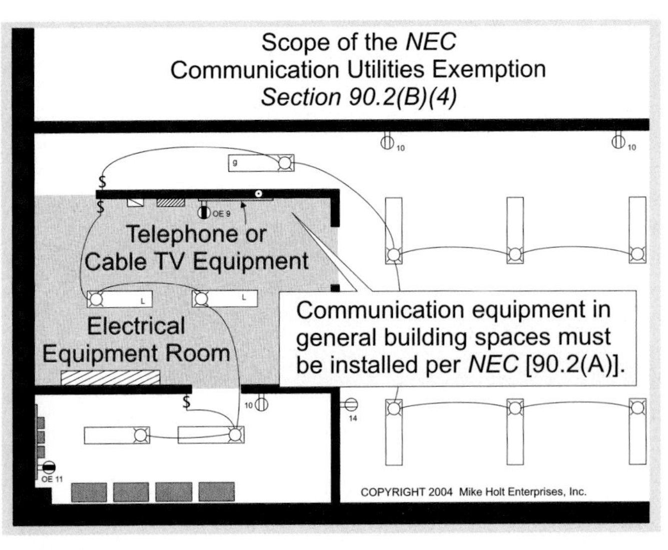

Figure 90–6

(5) Electric Utilities. The *NEC* doesn't apply to electric installations under the exclusive control of an electric utility where such installations:

a. Consist of service drops or service laterals and associated metering. **Figure 90–07**

b. Are located on legally established easements, rights-of-way, or by other agreements recognized by public/utility regulatory agencies, or property owned or leased by the electric utility. **Figure 90–8**

c. Are on property owned or leased by the electric utility for the purpose of generation, transformation, transmission, distribution, or metering of electric energy. **See Figure 90–8.**

Author's Comment: Luminaires (lighting fixtures) located in legally established easements, or rights-of-way, such as at poles supporting transmission or distribution lines, are exempt from the requirements of the *NEC*. However, if the electric utility provides

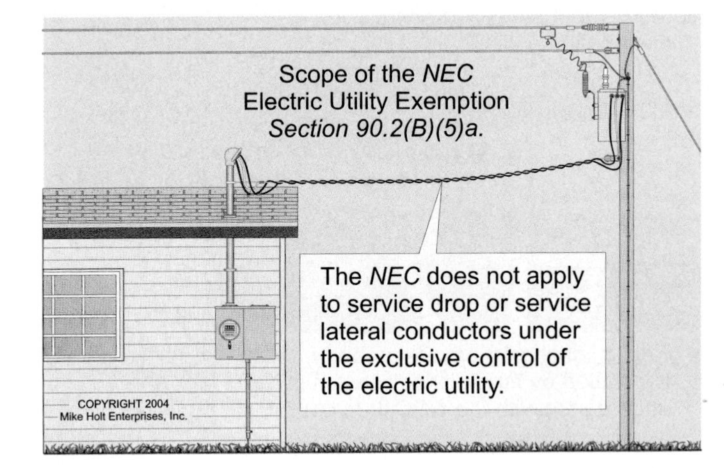

Scope of the NEC
Electric Utility Exemption
Section 90.2(B)(5)a.
The NEC does not apply to service drop or service lateral conductors under the exclusive control of the electric utility.
COPYRIGHT 2004 Mike Holt Enterprises, Inc.

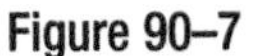
Figure 90–7

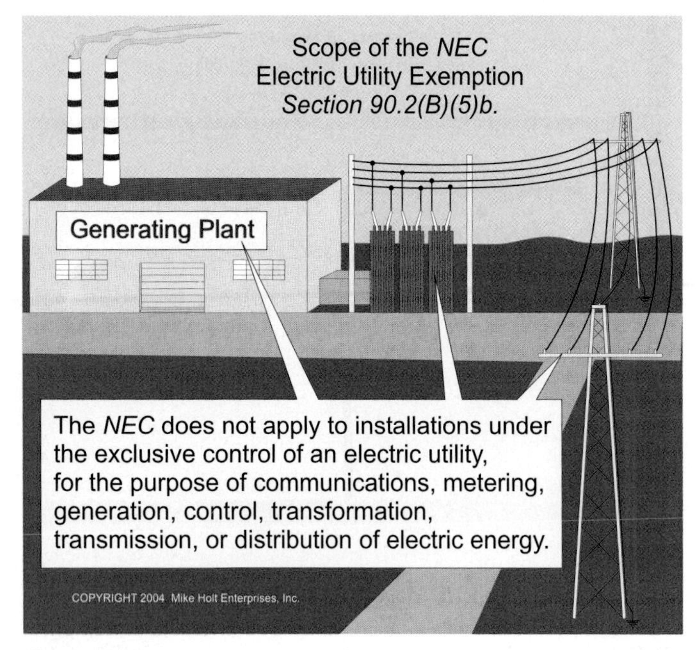

Figure 90–8

site and public lighting on private property, then the installation must comply with the *NEC* [90.2(A)(4)]. **Figure 90–9**

FPN to 90.2(B)(4) and (5): Utilities include entities that install, operate, and maintain communications systems (telephone, CATV, Internet, satellite, or data services) or electric supply systems (generation, transmission, or distribution systems) and are designated or recognized by governmental law or regulation by public service/utility commissions. Utilities may be subject to compliance with codes and standards covering their regulated activities as adopted under governmental law or regulation.

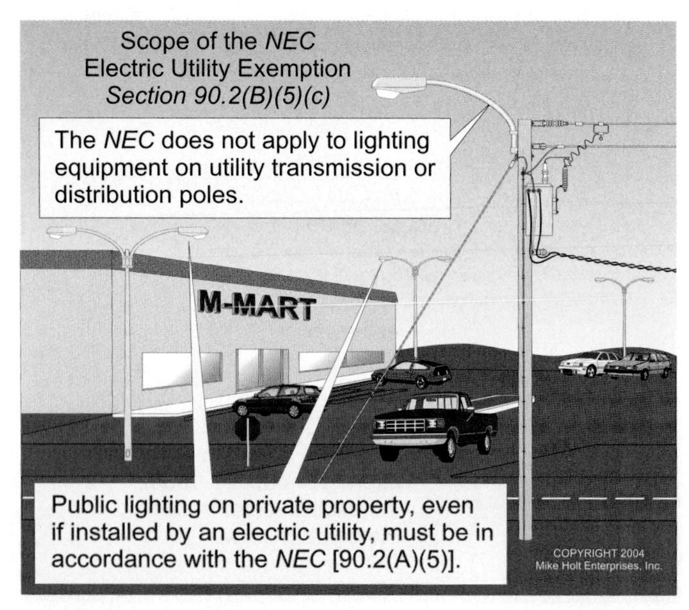

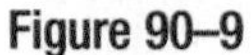

Figure 90–9

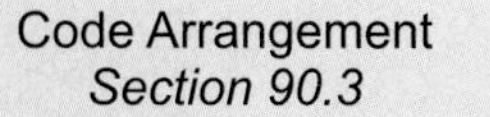

Code Arrangement
Section 90.3
General Requirements
• Chapter 1 - General
• Chapter 2 - Wiring and Protection
• Chapter 3 - Wiring Methods and Materials
• Chapter 4 - Equipment for General Use
Chapters 1 through 4 apply to all applications.
Special Requirements
• Chapter 5 - Special Occupancies
• Chapter 6 - Special Equipment
• Chapter 7 - Special Conditions
Chapters 5 through 7 can supplement or modify the general requirements of Chapters 1 through 4.
• Chapter 8 - Communications Systems
Chapter 8 requirements are not subject to requirements in Chapters 1 through 7, unless there is a specific reference in Chapter 8 to a rule in Chapters 1 through 7.
• Chapter 9 - Tables
Chapter 9 tables are used for calculating raceway sizes, conductor fill, and voltage drop.
• Annex A through F
Annexes are for information only and not enforceable.
COPYRIGHT 2004 Mike Holt Enterprises, Inc.

Figure 90–10

90.3 *Code* Arrangement.

The *Code* is divided into an Introduction and nine chapters. **Figure 90–10**

General Requirements. The requirements contained in Chapters 1, 2, 3, and 4 apply to all installations.

> **Author's Comment:** The scope of this textbook includes *NEC* Chapters 1 through 4.

Special Requirements. The requirements contained in Chapters 5, 6, and 7 apply to special occupancies, special equipment, or other special conditions. They can supplement or modify the requirements in Chapters 1 through 4.

For example, the general requirement contained in 250.118 of Article 250 Grounding and Bonding states that a metal raceway, such as Electrical Metallic Tubing, is considered suitable to provide a low-impedance path to the power supply for ground-fault current. However, 517.13(B) of Article 517 Health Care Facilities doesn't consider the raceway to be sufficient. It requires an insulated copper conductor to be installed in the raceway for this purpose.

Communications Systems. Chapter 8 contains the requirements for communications systems, such as telephone, antenna wiring, CATV, and network-powered broadband systems. Communications systems aren't subject to the general requirements of Chapters 1 through 4, or the special requirements of Chapters 5 through 7, unless there's a specific reference in Chapter 8 to a rule in Chapters 1 through 7.

> **Author's Comment:** Mike Holt's *Understanding the NEC, Volume 2 [Articles 500 through 830]*, explains the wiring requirements of special occupancies, special equipment, and special conditions, as well as communications systems.

Table. Chapter 9 consists of tables necessary to calculate raceway sizing, conductor fill, and voltage drop.

Annexes. Annexes aren't part of the *Code*, but are included for informational purposes. They are:

- Annex A. Product Safety Standards
- Annex B. Conductor Ampacity Under Engineering Supervision
- Annex C. Raceway Size Tables
- Annex D. Examples
- Annex E. Types of Construction
- Annex F. Cross-Reference Tables

90.4 Enforcement. This *Code* is intended to be suitable for enforcement by governmental bodies that exercise legal jurisdiction over electrical installations for power, lighting, signaling circuits, and communications systems, such as: **Figure 90–11**

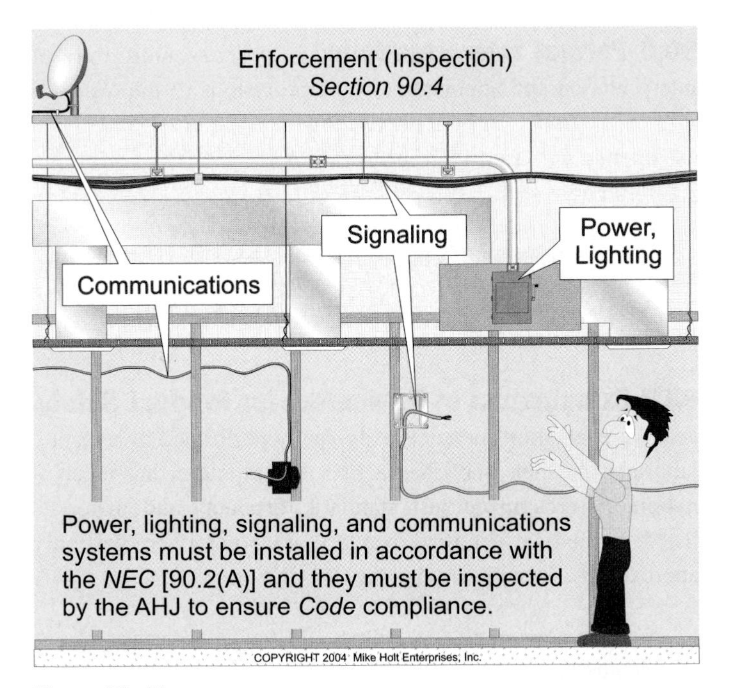

Enforcement (Inspection)
Section 90.4
Signaling
Power, Lighting
Communications
Power, lighting, signaling, and communications systems must be installed in accordance with the NEC [90.2(A)] and they must be inspected by the AHJ to ensure Code compliance.
COPYRIGHT 2004 Mike Holt Enterprises, Inc.

Figure 90–11

Signaling circuits, which include:

- Article 725 Class 1, Class 2, and Class 3 Remote-Control, Signaling, and Power-Limited Circuits
- Article 760 Fire Alarm Systems
- Article 770 Optical Fiber Cables and Raceways

Communications circuits, which include:

- Article 800 Communications Circuits (twisted-pair conductors)
- Article 810 Radio and Television Equipment (satellite dish and antenna)
- Article 820 Community Antenna Television and Radio Distribution Systems (coaxial cable)
- Article 830 Network-Powered Broadband Communications Systems

> **Author's Comment:** The installation requirements for signaling circuits and communications circuits are covered in Mike's *Understanding the NEC, Volume 2* textbook.

The enforcement of the *NEC* is the responsibility of the authority having jurisdiction (AHJ), who is responsible for interpreting requirements, approving equipment and materials, waiving *Code* requirements, and ensuring that equipment is installed in accordance with listing instructions.

> **Author's Comment:** See Article 100 for the definition of "Authority Having Jurisdiction."

Interpretation of the Requirements. The authority having jurisdiction is responsible for interpreting the *NEC*, but his or her decisions must be based on a specific *Code* requirement. If an installation is rejected, the authority having jurisdiction is legally responsible for informing the installer which specific *NEC* rule was violated.

> **Author's Comment:** The art of getting along with the authority having jurisdiction consists of doing good work and knowing what the *Code* actually says (as opposed to what you only think it says). It's also useful to know how to choose your battles when the inevitable disagreement does occur.

Approval of Equipment and Materials. Only the authority having jurisdiction has authority to approve the installation of equipment and materials. Typically, the authority having jurisdiction will approve equipment listed by a product testing organization, such as Underwriters Laboratories Inc. (UL), but the *NEC* doesn't require all equipment to be listed. See 90.7, 110.2, 110.3, and the definitions in Article 100 for Approved, Identified, Labeled, and Listed. **Figure 90–12**

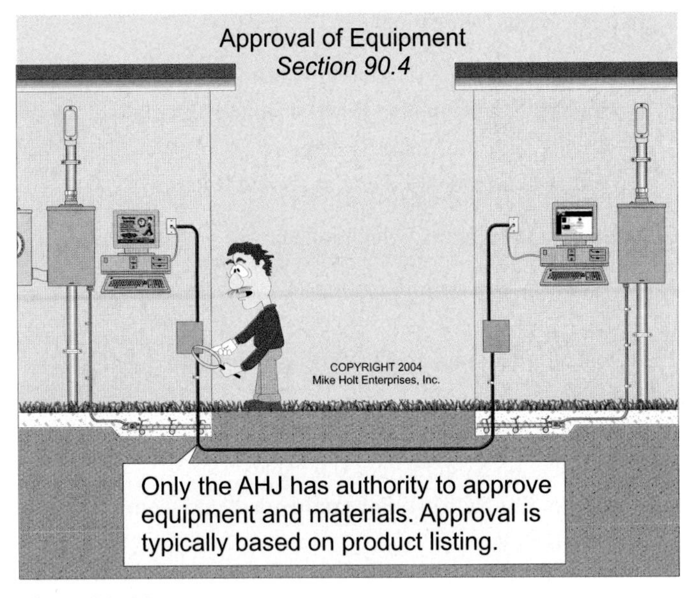

Approval of Equipment
Section 90.4
COPYRIGHT 2004
Mike Holt Enterprises, Inc.
Only the AHJ has authority to approve equipment and materials. Approval is typically based on product listing.

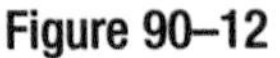
Figure 90–12

Author's Comment: According to the *NEC*, the authority having jurisdiction determines the approval of equipment. This means that he/she can reject an installation of listed equipment and he/she can approve the use of unlisted equipment. Given our highly litigious society, approval of unlisted equipment is becoming increasingly difficult to obtain.

Waiver of Requirements. By special permission, the authority having jurisdiction can waive specific requirements in this *Code* or permit alternative methods where it's assured that equivalent safety can be achieved and maintained.

Author's Comment: Special permission is defined in Article 100 as the written consent of the authority having jurisdiction.

Waiver of New Product Requirements. If the 2005 *NEC* requires products that aren't yet available at the time the *Code* is adopted, the authority having jurisdiction can allow products that were acceptable in the previous *Code* to continue to be used.

Author's Comment: Sometimes it takes years before testing laboratories establish product standards for new *NEC* product requirements, and then it takes time before manufacturers can design, manufacture, and distribute these products to the marketplace.

Compliance with Listing Instructions. It's the authority having jurisdiction's responsibility to ensure that electrical equipment is installed in accordance with equipment listing and/or labeling instructions [110.3(B)]. In addition, the authority having jurisdiction can reject the installation of equipment modified in the field [90.7].

Author's Comment: The *NEC* doesn't address the maintenance of electrical equipment (NFPA 70B does), because the *Code* is an installation standard, not a maintenance standard.

90.5 Mandatory Requirements and Explanatory Material.

(A) Mandatory Requirements. In the *NEC* the words "shall" or "shall not," indicate a mandatory requirement.

Author's Comment: For the ease of reading this textbook, the word "shall" has been replaced with the word "must," and the words "shall not" have been replaced with the word "cannot."

(B) Permissive Requirements. When the *Code* uses "shall be permitted" it means the identified actions are allowed but not required, and the authority having jurisdiction is not to restrict an installation from being done in that manner. A permissive rule is often an exception to the general requirement.

Author's Comment: For ease of reading, the phrase "shall be permitted" as used in the *Code*, has been replaced in this textbook with the words "is permitted."

(C) Explanatory Material. References to other standards or sections of the *NEC*, or information related to a *Code* rule, are included in the form of Fine Print Notes (FPN). Fine Print Notes are for information only and aren't intended to be enforceable.

For example, Fine Print Note No. 4 in 210.19(A)(1) recommends that the circuit voltage drop not exceed three percent. This isn't a requirement; it's just a recommendation.

90.6 Formal Interpretations.

To promote uniformity of interpretation and application of the provisions of the *National Electrical Code,* formal interpretation procedures have been established and are found in the NFPA Regulations Governing Committee Projects.

Author's Comment: This is rarely done because it's a very time-consuming process, and formal interpretations from the *NFPA* are not binding on the authority having jurisdiction!

90.7 Examination of Equipment for Product Safety.

Product evaluation for safety is typically performed by a testing laboratory, which publishes a list of equipment that meets a nationally recognized test standard. Products and materials listed, labeled, or identified by a testing laboratory are generally approved by the authority having jurisdiction.

Author's Comment: See Article 100 for the definition of "Approved."

Listed, factory-installed, internal wiring and construction of equipment need not be inspected at the time of installation, except to detect alterations or damage [300.1(B)]. **Figure 90–13**

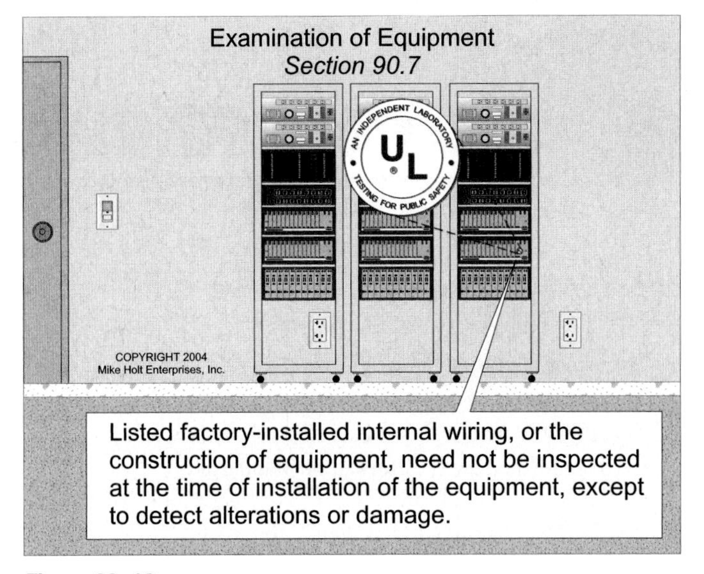

Figure 90–13

90.9 Units of Measurement.

(B) Dual Systems of Units. Both the metric and inch-pound measurement systems are shown in the *NEC*, with the metric units appearing first and the inch-pound system immediately following in parentheses.

> **Author's Comment:** This is a normal practice in all *NFPA* standards, even though the U.S. construction industry uses inch-pound units of measurement.

(D) Compliance. Installing electrical systems in accordance with the metric system or the inch-pound system is considered to comply with the *Code*.

> **Author's Comment:** Since compliance with either the metric or the inch-pound system of measurement constitutes compliance with the *NEC*, this textbook uses only inch-pound units.

Article 90 Questions

(• Indicates that 75% or fewer exam takers get the question correct.)

1. The *NEC* is _____.

   (a) intended to be a design manual
   (b) meant to be used as an instruction guide for untrained persons
   (c) for the practical safeguarding of persons and property
   (d) published by the Bureau of Standards

2. •The *Code* applies to the installation of _____.

   (a) electrical conductors and equipment within or on public and private buildings
   (b) outside conductors and equipment on the premises
   (c) optical fiber cable
   (d) all of these

3. Service laterals installed by an electrical contractor must be installed in accordance with the *NEC*.

   (a) True (b) False

4. The requirements in "Annexes" must be complied with.

   (a) True (b) False

5. Explanatory material, such as references to other standards, references to related sections of the *NEC*, or information related to a *Code* rule, are included in the form of Fine Print Notes (FPNs).

   (a) True (b) False

CHAPTER 1

General

Introduction

Many people skip Chapter 1 of the *NEC* because they want something prescriptive—they want something that tells them what to do, cookbook style. But electricity isn't a simple topic you can jump right into. You cannot just follow a few simple steps to get a safe installation. You need a foundation from which you can apply the *Code*.

Consider Ohm's law. Would Ohm's Law make sense to you if you did not know what an ohm was? Similarly, you must become familiar with a few basic rules, concepts, definitions, and requirements that apply to the rest of the *NEC*, and you must maintain that familiarity as you continue to apply the *Code*.

Chapter 1 consists of two main parts. Article 100 provides definitions so people can understand one another when trying to communicate on *Code* related matters. Article 110 provides general requirements that you need to know so you can correctly apply the rest of the *NEC*.

Time spent learning this general material is a great investment. After understanding Chapter 1, some of the *Code* requirements that seem confusing to other people—those who do not understand Chapter 1—will become increasingly straight forward to you. That is, they will strike you as being "common sense," because you'll have the foundation from which to understand and apply them. Because you'll understand the principles upon which many *NEC* requirements in later Chapters are based, you'll read those requirements and not be surprised at all. You'll read them and feel like you already knew them.

Article 100—Definitions. Part I of Article 100 contains the definitions of terms used throughout the *Code* for systems that operate at 600V or less. The definitions of terms in Part II apply to systems that operate at over 600V.

Author's Comment: The requirements covered in this textbook apply to systems that operate at 600V or less.

Definitions of standard terms, such as volt, voltage drop, ampere, impedance, and resistance, aren't listed in Article 100. If the *NEC* doesn't define a term, then a dictionary suitable to the authority having jurisdiction should be consulted. A building code glossary might provide a better definition than a dictionary found at your home or school.

Definitions at the beginning of an article apply only to that specific article. For example, the definition of a "Swimming Pool" is contained in 680.2, because this term applies only to the requirements contained in Article 680 Swimming Pools.

Article 110—Requirements for Electrical Installations. This article contains the general requirements for electrical installations for the following:

- **PART I. GENERAL**
- **PART II. 600V, NOMINAL, OR LESS**
- **PART III. OVER 600V, NOMINAL**
- **PART IV. TUNNEL INSTALLATIONS OVER 600V, NOMINAL**
- **PART V. MANHOLES AND OTHER ELECTRIC ENCLOSURES INTENDED FOR PERSONNEL ENTRY**

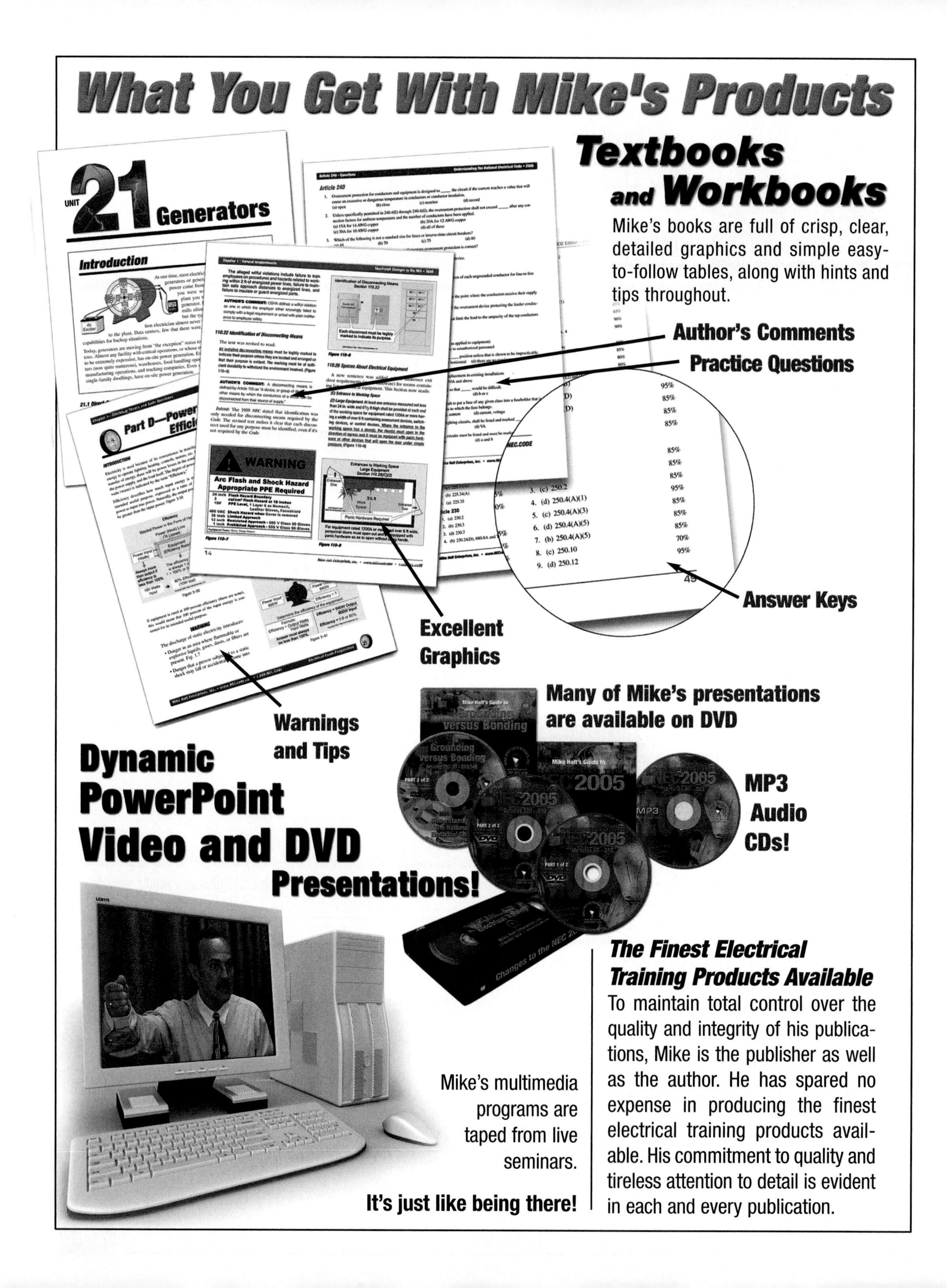
What You Get With Mike's Products
Textbooks
and Workbooks
Mike's books are full of crisp, clear, detailed graphics and simple easy-to-follow tables, along with hints and tips throughout.
Author's Comments
Practice Questions
Answer Keys
Excellent
Graphics
Warnings
and Tips
Dynamic
PowerPoint
Video and DVD
Presentations!
Many of Mike's presentations
are available on DVD
MP3
Audio
CDs!
The Finest Electrical
Training Products Available
To maintain total control over the quality and integrity of his publications, Mike is the publisher as well as the author. He has spared no expense in producing the finest electrical training products available. His commitment to quality and tireless attention to detail is evident in each and every publication.
Mike's multimedia
programs are
taped from live
seminars.
It's just like being there!

Definitions

Introduction

Have you ever had a conversation with someone, only to discover what you said and what he/she heard were completely different? This happens when one or more of the people in a conversation do not understand the definitions of the words being used, and that's why the definitions of key terms are located right up in the front of the *NEC,* in Article 100.

If we can all agree on important definitions, then we speak the same language and avoid misunderstandings. Because the *Code* exists to protect people and property, we can agree it's very important to know the definitions presented in Article 100.

Now, here are a couple of things you may not know about Article 100:

- Article 100 contains the definitions of many, but not all, of the terms used throughout the *NEC.* In general, only those terms used in two or more articles are defined in Article 100.
- Part I of Article 100 contains the definitions of terms used throughout the *Code.*
- Part II of Article 100 contains only terms that apply to systems that operate at over 600V.

How can you possibly learn all these definitions? There seem to be so many. Here are a few tips:

- Break the task down. Study a few words at a time, rather than trying to learn them all at one sitting.
- Review the graphics in the textbook. These will help you see how a term is applied.
- Relate them to your work. As you read a word, think of how it applies to the work you're doing. This will provide a natural reinforcement of the learning process.

Definitions.

Accessible as it Applies to Equipment. Admitting close approach and not guarded by locked doors, elevation, or other effective means.

Accessible as it Applies to Wiring Methods. Not permanently closed in by the building structure or finish and capable of being removed or exposed without damaging the building structure or finish. **Figure 100–1**

Author's Comments:

- Conductors in a concealed raceway are considered concealed, even though they may become accessible by withdrawing them. See the definition of "Concealed" in this article.
- Raceways, cables, and enclosures installed above a suspended ceiling or within a raised floor are considered accessible, because the wiring methods can be accessed without damaging the building structure. See "Concealed" and "Exposed."

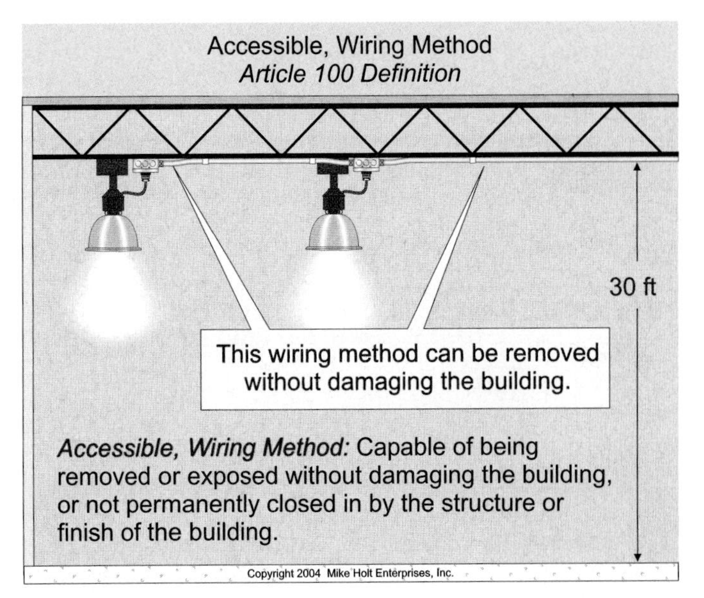

Accessible, Wiring Method
Article 100 Definition
30 ft
This wiring method can be removed without damaging the building.
Accessible, Wiring Method: Capable of being removed or exposed without damaging the building, or not permanently closed in by the structure or finish of the building.

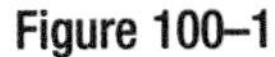
Copyright 2004 Mike Holt Enterprises, Inc.

Figure 100–1

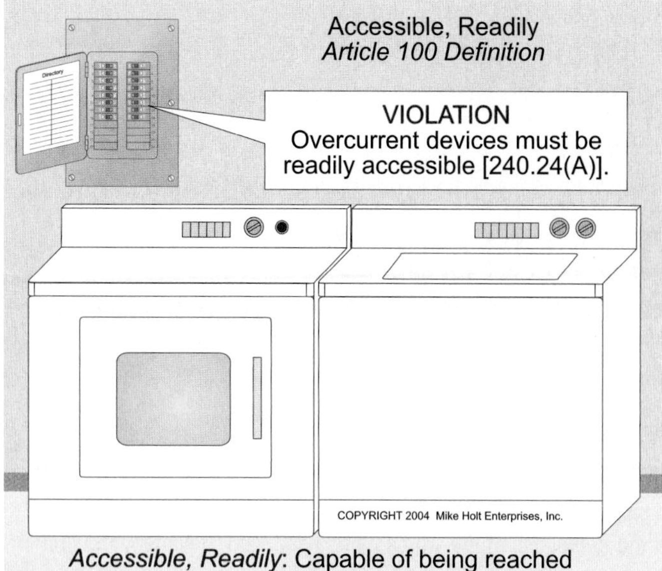

Accessible, Readily
Article 100 Definition
VIOLATION
Overcurrent devices must be readily accessible [240.24(A)].
COPYRIGHT 2004 Mike Holt Enterprises, Inc.
Accessible, Readily: Capable of being reached without having to climb over or remove obstacles, or without having to use portable ladders.

Figure 100–2

Accessible, Readily (Readily Accessible). Capable of being reached quickly without having to climb over or remove obstacles or resort to portable ladders. **Figures 100–2** and **100–3**

Ampacity. The current in amperes a conductor can carry continuously, where the temperature will not be raised in excess of the conductor's insulation temperature rating. See 310.10 and 310.15 for details and examples. **Figure 100–4**

Appliance [Article 424]. Electrical equipment, other than industrial equipment, built in standardized sizes, such as ranges, ovens, cooktops, refrigerators, drinking water coolers, or beverage dispensers.

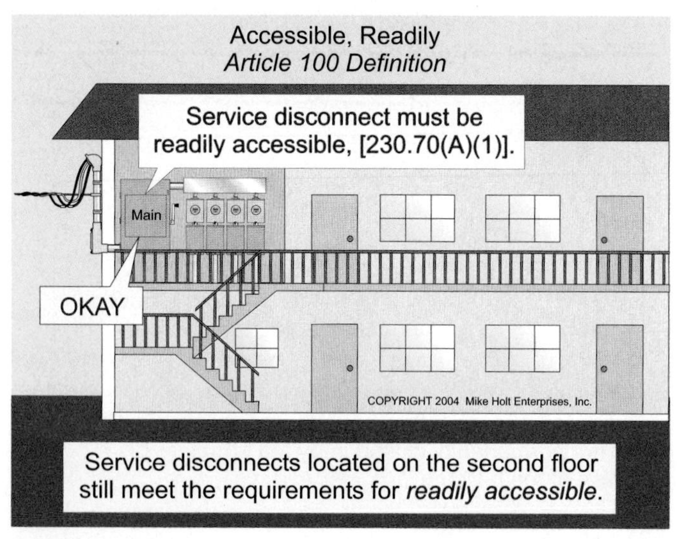

Accessible, Readily
Article 100 Definition
Service disconnect must be readily accessible, [230.70(A)(1)].
Main
OKAY
COPYRIGHT 2004 Mike Holt Enterprises, Inc.
Service disconnects located on the second floor still meet the requirements for readily accessible.

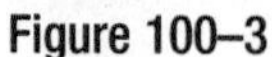
Figure 100–3

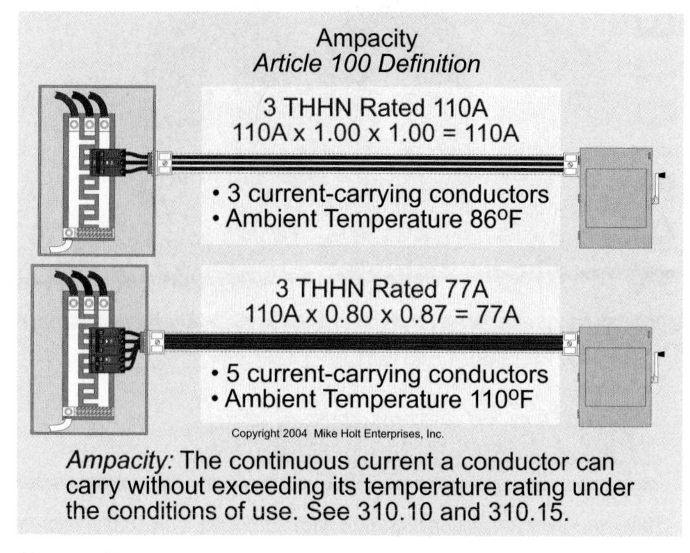

Figure 100–4

Approved. Acceptable to the authority having jurisdiction, usually the electrical inspector.

Author's Comment: Product listing doesn't mean that the product is approved, but it's a basis for approval. See 90.4, 90.7, 110.2, and the definitions in Article 100 for Authority Having Jurisdiction, Identified, Labeled, and Listed.

Attachment Plug (Plug Cap)(Plug). A wiring device at the end of flexible cord intended to be inserted into a receptacle. **Figure 100–5**

Author's Comment: The use of a cord with an attachment plug is limited by 210.50(A), 400.7, 410.14, 410.30, 422.33, 590.4, and 645.5.

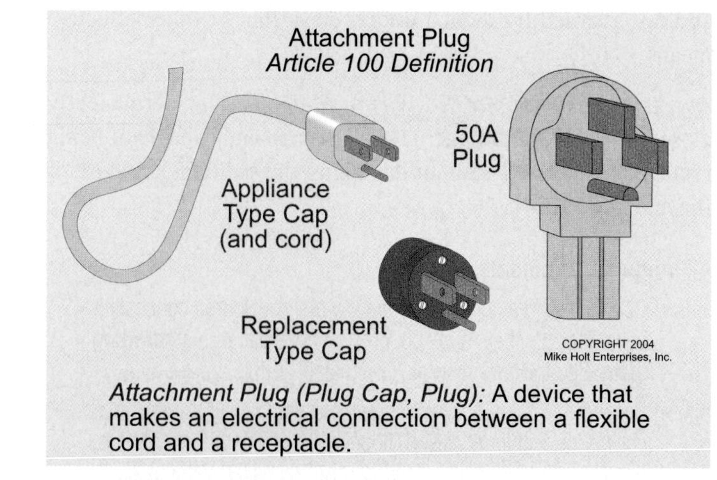

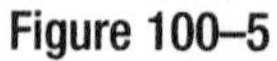
Figure 100–5

Authority Having Jurisdiction (AHJ). The organization, office, or individual that is responsible for approving equipment, materials, an installation, or a procedure. See 90.4, 90.7, and 110.2 for more information.

> **FPN:** The authority having jurisdiction may be a federal, state, or local government, or an individual such as a fire chief, fire marshal, chief of a fire prevention bureau or labor department or health department, a building official or electrical inspector, or others having statutory authority. In some circumstances, the property owner or his/her agent assumes the role, and at government installations, the commanding officer, or departmental official may be the authority having jurisdiction.

Author's Comments:

- Typically, the authority having jurisdiction will be the electrical inspector who has legal statutory authority. In the absence of federal, state, or local regulations, the operator of the facility or his/her agent, such as an architect or engineer of the facility, can assume the role.
- Many feel that the authority having jurisdiction should have a strong background in the electrical field, such as having studied electrical engineering or having obtained an electrical contractor's license, and in a few states this is a legal requirement. Memberships, certifications, and active participation in electrical organizations, such as the IAEI (www.IAEI.org), speak to an individual's qualifications.

Bathroom. A bathroom is an area that includes a basin with a toilet, tub, or shower. **Figure 100–6**

> **Author's Comment:** All 15A and 20A, 125V receptacles located in bathrooms must be GFCI protected [210.8].

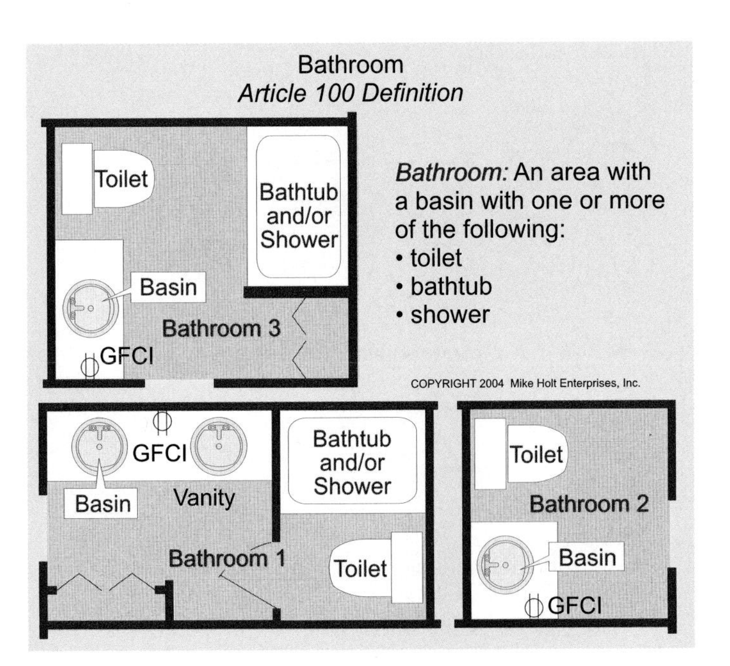

Figure 100–6

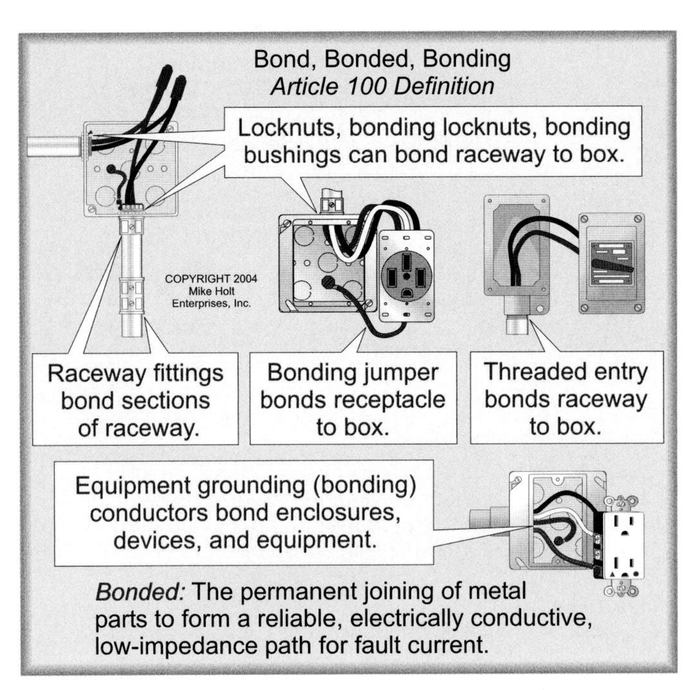

Figure 100–7

Bonding (Bond) (Bonded). The permanent joining of metallic parts together to form an electrically conductive path. Such a path must have the capacity to conduct safely any fault current likely to be imposed on it. **Figure 100–7**

> **Author's Comment:** Bonding is accomplished by the use of bonding conductors, metallic raceways and cables, connectors, couplings, or other devices listed for this purpose [250.8, 250.118, and 300.10].

Bonding Jumper. A reliable conductor that is properly sized in accordance with Article 250, to ensure electrical conductivity between metal parts of the electrical installation. **Figure 100–8**

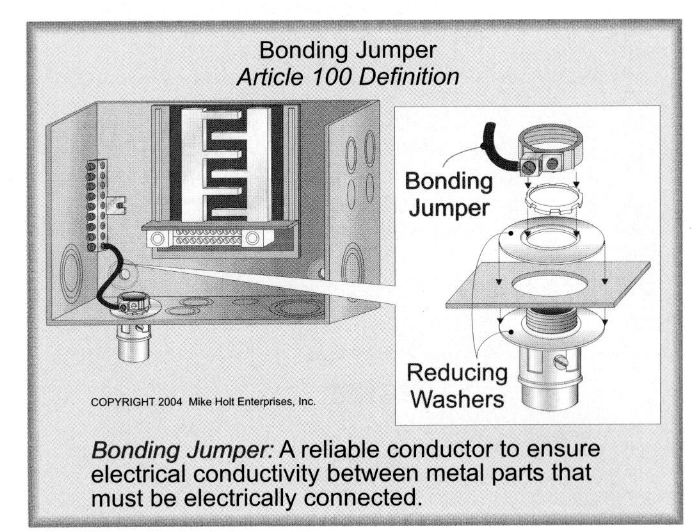

Bonding Jumper
Article 100 Definition
Bonding Jumper
Reducing Washers
COPYRIGHT 2004 Mike Holt Enterprises, Inc.
Bonding Jumper: A reliable conductor to ensure electrical conductivity between metal parts that must be electrically connected.

Figure 100–8

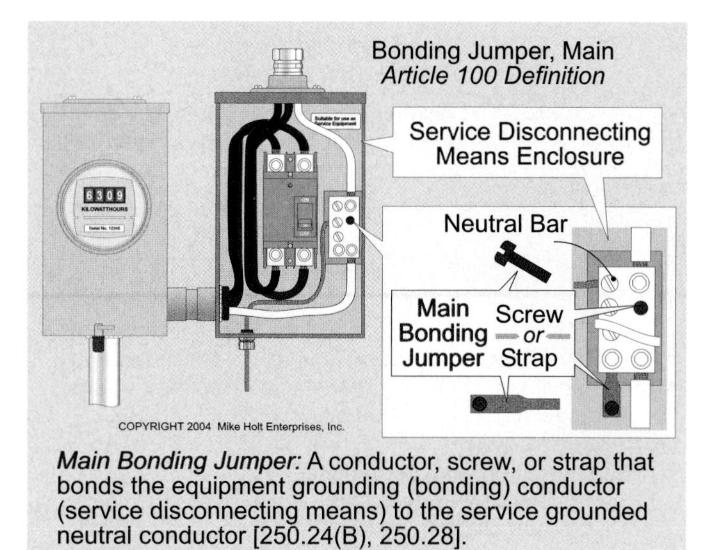
Bonding Jumper, Main
Article 100 Definition
Service Disconnecting Means Enclosure
Neutral Bar
Main Bonding Jumper
Screw or Strap
COPYRIGHT 2004 Mike Holt Enterprises, Inc.
Main Bonding Jumper: A conductor, screw, or strap that bonds the equipment grounding (bonding) conductor (service disconnecting means) to the service grounded neutral conductor [250.24(B), 250.28].

Figure 100–9

Bonding Jumper, Main. A conductor, screw, or strap that bonds the equipment grounding (bonding) conductor (service disconnecting means) to the grounded neutral conductor in accordance with 250.24(B). For more details, see 250.24(A)(4), 250.28, and 408.3(C). **Figure 100–9**

Bonding Jumper, System. The conductor, screw, or strap that bonds the metal parts of a separately derived system to a system winding in accordance with 250.30(A)(1). **Figure 100–10**

> **Author's Comment:** The system bonding jumper provides the low-impedance fault-current path to the power source to facilitate the clearing of a ground fault by opening the circuit protection device. For more information, see 250.4(A)(5), 250.28, and 250.30(A)(1).

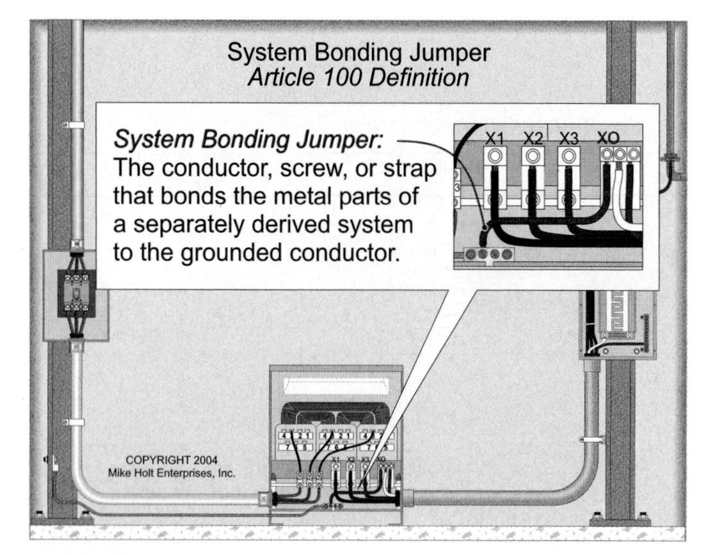
System Bonding Jumper
Article 100 Definition
System Bonding Jumper:
The conductor, screw, or strap that bonds the metal parts of a separately derived system to the grounded conductor.
X1 X2 X3 XO
COPYRIGHT 2004 Mike Holt Enterprises, Inc.

Figure 100–10

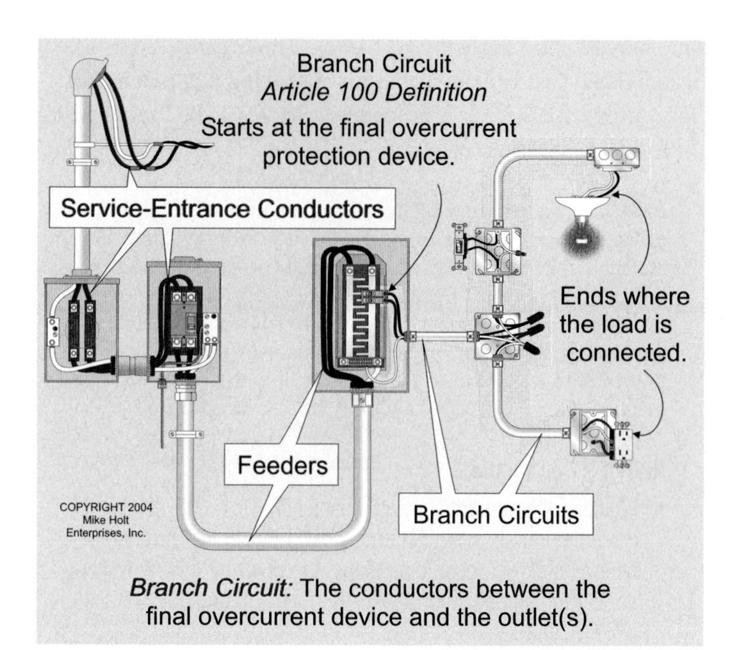
Branch Circuit
Article 100 Definition
Starts at the final overcurrent protection device.
Service-Entrance Conductors
Ends where the load is connected.
Feeders
Branch Circuits
COPYRIGHT 2004 Mike Holt Enterprises, Inc.
Branch Circuit: The conductors between the final overcurrent device and the outlet(s).

Figure 100–11

Branch Circuit [Article 210]. The conductors between the final overcurrent device and the receptacle outlets, lighting outlets, or other outlets as defined in Article 100. **Figure 100–11**

Branch Circuit, Multiwire. A branch circuit that consists of two or more ungrounded circuit conductors with a common grounded neutral conductor. There must be a voltage potential between the ungrounded conductors and an equal voltage potential from each ungrounded conductor to the grounded neutral conductor. **Figure 100-12**

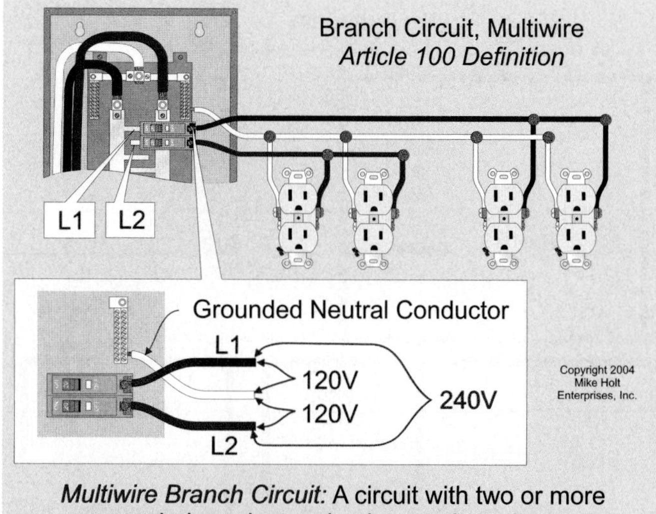
Branch Circuit, Multiwire
Article 100 Definition
L1
L2
Grounded Neutral Conductor
L1
120V
120V
240V
L2
Copyright 2004 Mike Holt Enterprises, Inc.
Multiwire Branch Circuit: A circuit with two or more ungrounded conductors having a voltage between them, with equal voltage between the ungrounded conductors and the grounded neutral conductor.

Figure 100–12

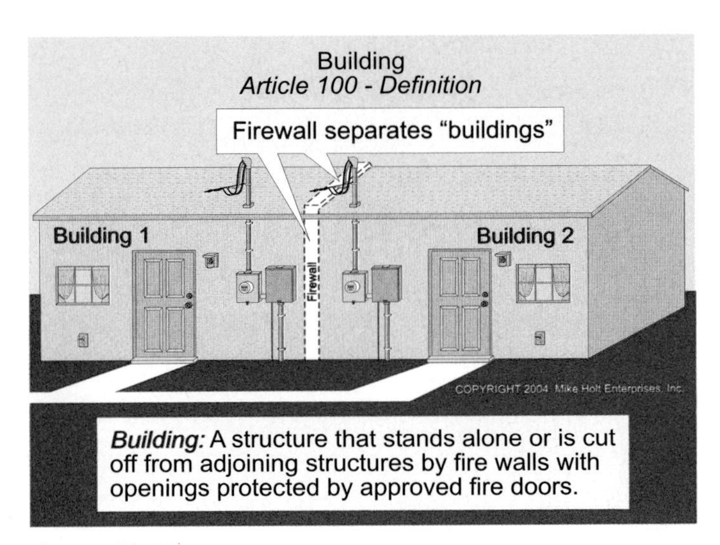

Building
Article 100 - Definition
Firewall separates "buildings"
Building 1
Building 2
Firewall
COPYRIGHT 2004 Mike Holt Enterprises, Inc.
Building: A structure that stands alone or is cut off from adjoining structures by fire walls with openings protected by approved fire doors.

Figure 100–13

Author's Comment: Multiwire branch circuits offer the advantage of fewer conductors in a raceway, smaller raceway sizing, and a reduction of material and labor costs. In addition, multiwire branch circuits can reduce circuit voltage drop by as much as 50 percent. However, because of the dangers associated with multiwire branch circuits, the *NEC* contains additional requirements to ensure a safe installation. See 210.4, 300.13(B), and 408.40 for additional details.

Building. A structure that stands alone or is cut off from other structures by firewalls with all openings protected by fire doors that are approved by the authority having jurisdiction. **Figure 100–13**

Cabinet [Article 312]. An enclosure for either surface mounting or flush mounting provided with a frame in which a door can be hung. **Figure 100–14**

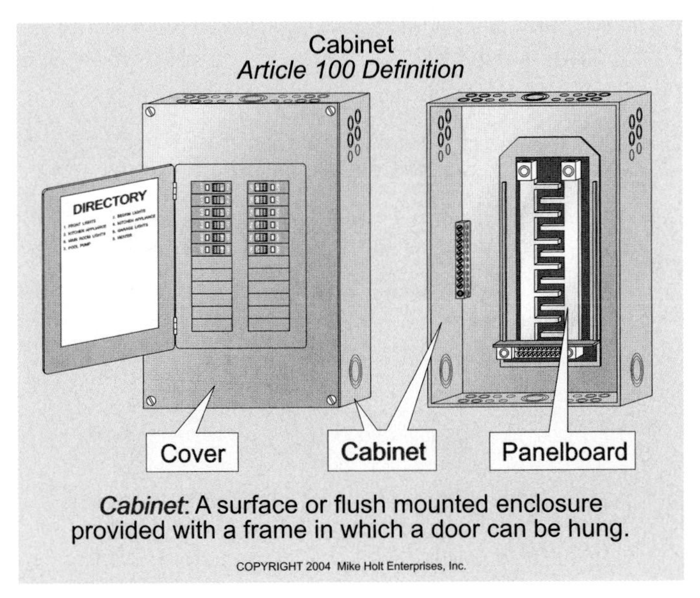

Figure 100–14

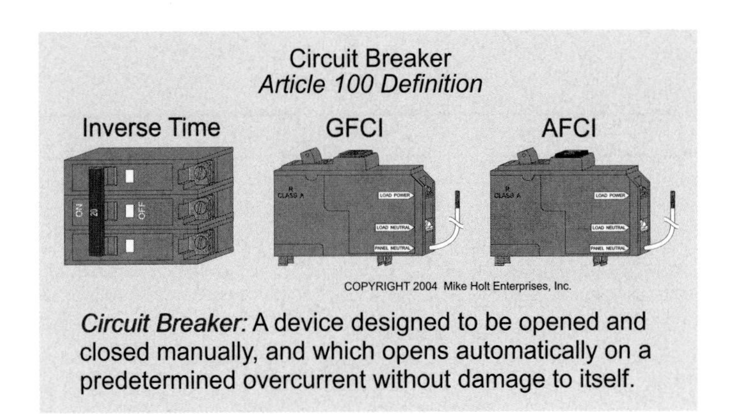

Figure 100–15

Circuit Breaker. A device designed to be opened and closed manually, and which opens automatically on a predetermined overcurrent without damage to itself. Circuit breakers are available in different configurations, such as inverse time molded case, adjustable (electronically controlled), and instantaneous trip/motor circuit protectors. **Figure 100–15**

- *Inverse Time:* Inverse-time breakers operate on the principle that as the current increases, the time it takes for the devices to open decreases. This type of breaker provides overcurrent protection (overload, short circuit, and ground fault).

- *Adjustable Trip:* Adjustable-trip breakers permit the thermal trip setting to be adjusted. The adjustment is often necessary to coordinate the operation of the circuit breakers with other overcurrent protection devices.

Author's Comment: Coordination means that the devices with the lowest ratings, closest to the fault, operate and isolate the fault and disruption, if possible, so that the rest of the system can remain energized and functional.

- *Instantaneous Trip:* Instantaneous-trip breakers operate on the principle of electromagnetism only and are used for motors; sometimes these devices are called motor short-circuit protectors (MCPs). This type of protection device doesn't provide overload protection. It only provides short-circuit and ground-fault protection; overload protection must be provided separately.

Author's Comment: Instantaneous-trip circuit breakers have no intentional time delay and are sensitive to current inrush, and to vibration and shock. Consequently, they should not be used where these factors are known to exist.

Concealed. Rendered inaccessible by the structure or finish of the building. Conductors in a concealed raceway are considered concealed, even though they may become accessible by withdrawing them. **Figure 100–16**

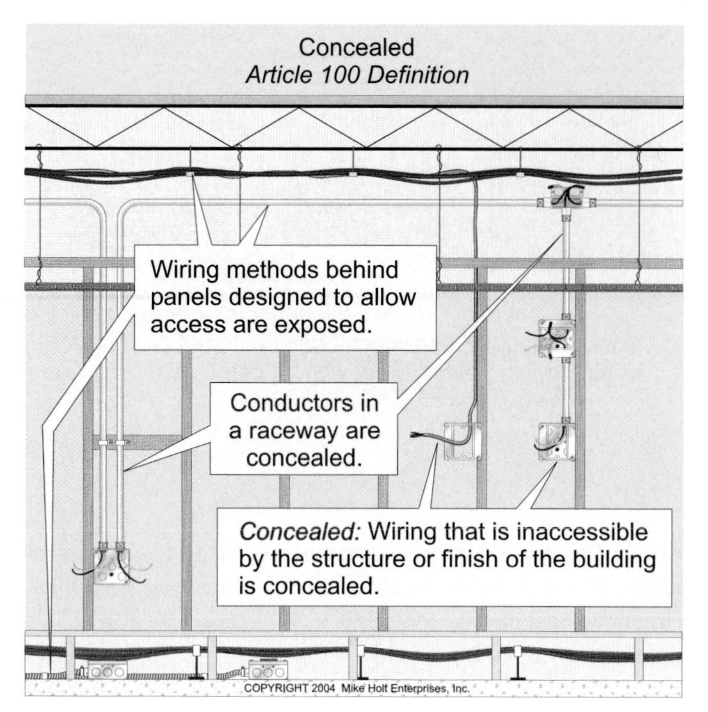

Figure 100–16

Author's Comment: Wiring behind panels that are designed to allow access is considered exposed.

Conduit Body. A fitting that provides access to conductors through a removable cover. **Figure 100–17**

Connector, Pressure (Solderless). A device that establishes a conductive connection between conductors and a terminal by the means of mechanical pressure.

Continuous Load. A load where the current is expected to exist for three hours or more, such as store or parking lot lighting.

Controller. A device that controls, in some predetermined manner, the electric power delivered to electrical equipment. This includes time clocks, lighting contactors, photocells, etc. **Figure 100–18**

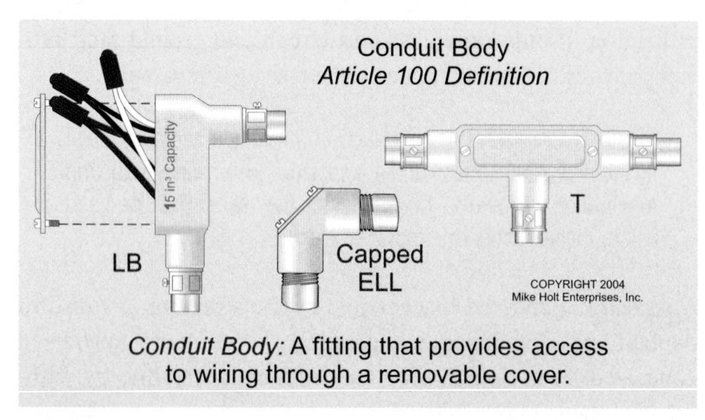

Figure 100–17

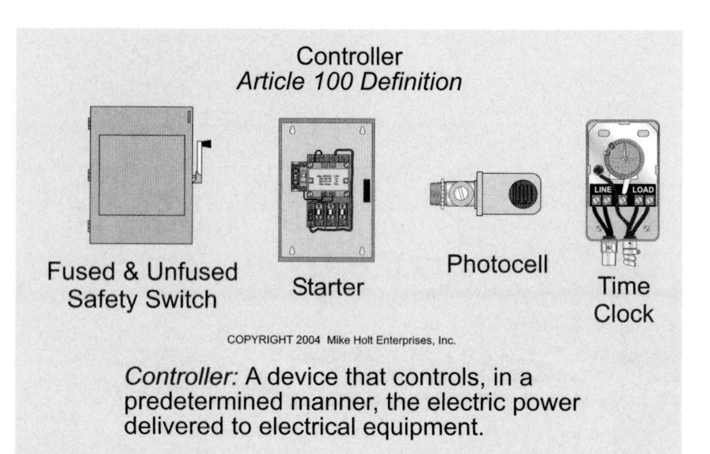

Controller
Article 100 Definition
Fused & Unfused Safety Switch
Starter
Photocell
LINE
LOAD
Time Clock
COPYRIGHT 2004 Mike Holt Enterprises, Inc.
Controller: A device that controls, in a predetermined manner, the electric power delivered to electrical equipment.

Figure 100–18

Coordination (Selective). Localization of an overcurrent condition to restrict outages to the circuit or equipment affected, accomplished by the choice of overcurrent protective devices.

Author's Comment: Selective coordination is required for:

- Orderly Shutdown, 240.12
- Motors, 430.52(C)(3)
- Elevators, 620.62
- Fire Pumps, 695.5(C)(2)
- Emergency Power Systems, 700.27
- Legally Required Standby Power Systems, 701.18

Selective coordination means the circuit protection scheme confines the interruption to a particular area rather than to the whole system. For example, if someone plugs in a space heater and raises total demand on a 20A circuit to 25A, or if a short circuit or ground fault occurs with selective coordination, the only breaker/fuse that will open is the one protecting just that branch circuit. Without selective coordination, an entire floor of a building could go dark!

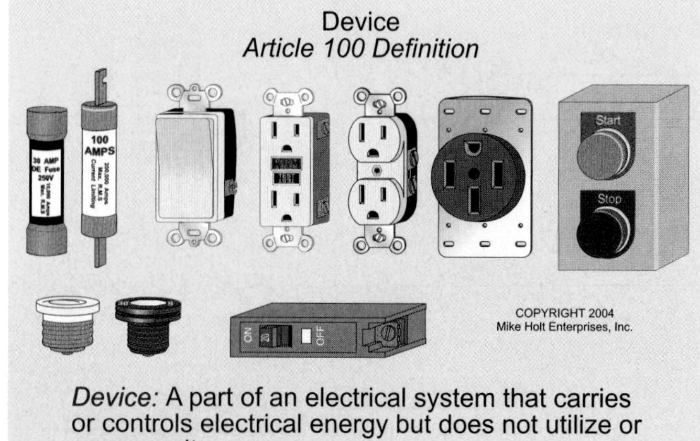

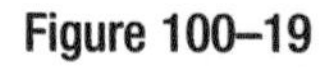
Figure 100–19

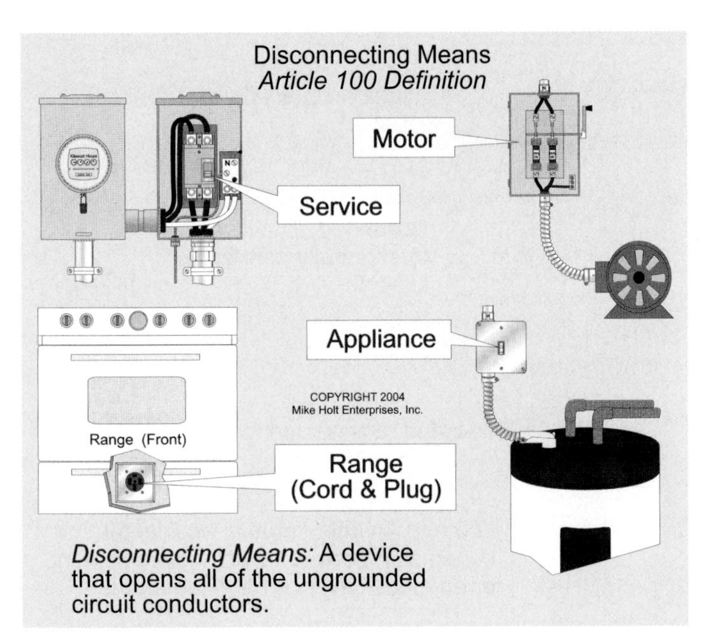
Disconnecting Means
Article 100 Definition
Motor
Service
Appliance
COPYRIGHT 2004
Mike Holt Enterprises, Inc.
Range (Front)
Range
(Cord & Plug)
Disconnecting Means: A device that opens all of the ungrounded circuit conductors.

Figure 100–20

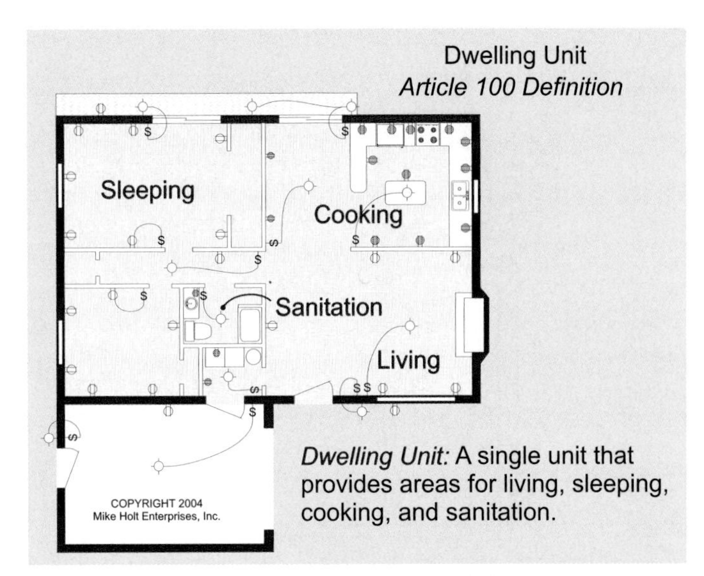
Dwelling Unit
Article 100 Definition
Sleeping
Cooking
Sanitation
Living
COPYRIGHT 2004
Mike Holt Enterprises, Inc.
Dwelling Unit: A single unit that provides areas for living, sleeping, cooking, and sanitation.

Figure 100–21

Cutout Box. Cutout boxes are designed for surface mounting with a swinging door.

Device. A component of an electrical installation that is intended to carry or control, but not consume electrical energy.

> **Author's Comment:** Devices include receptacles, switches, circuit breakers, fuses, time clocks, controllers, etc., but not locknuts or other mechanical fittings. **Figure 100–19**

Disconnecting Means. A device that opens all of the ungrounded circuit conductors from their power source. These include switches, attachment plugs and receptacles, and circuit breakers. **Figure 100–20**

> **Author's Comment:** Review the following for the specific requirements for equipment disconnecting means:
>
> - Air Conditioning and Refrigeration, 440.14
> - Appliances, Article 422, Part III
> - Building supplied by a feeder, Article 225, Part II
> - Electric space heating, 424.19
> - Electric duct heaters, 424.65
> - Motor control conductors, 430.74
> - Motor controllers, 430.102(A)
> - Motors, 430.102(B)
> - Refrigeration equipment, 440.14
> - Services, Article 230, Part VI
> - Swimming pool, spa, hot tub, and fountain equipment, 680.12

Dwelling Unit. A single unit that provides independent living facilities for persons, including permanent provisions for living, sleeping, cooking, and sanitation. **Figure 100–21**

Dwelling, Multifamily. A building that contains three or more dwelling units.

Energized. Electrically connected to, or is, a source of voltage.

Exposed (Wiring Methods). On, or attached to the surface of a building, or behind panels designed to allow access.

> **Author's Comment:** An example is wiring located in the space above a suspended ceiling or below a raised floor. **Figure 100–22**

Feeder. The conductors between the service equipment, the source of a separately derived system, or other power source and the final branch-circuit overcurrent device. **Figure 100–23**

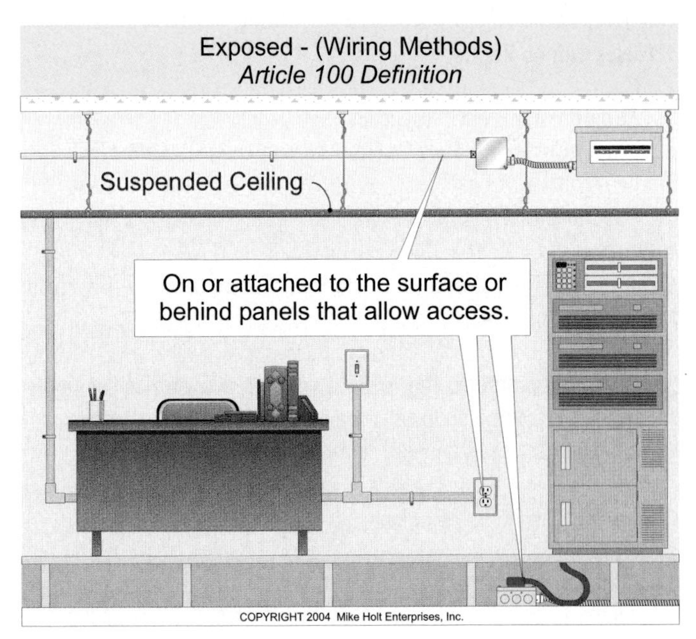
Exposed - (Wiring Methods)
Article 100 Definition
Suspended Ceiling
On or attached to the surface or behind panels that allow access.
COPYRIGHT 2004 Mike Holt Enterprises, Inc.

Figure 100–22

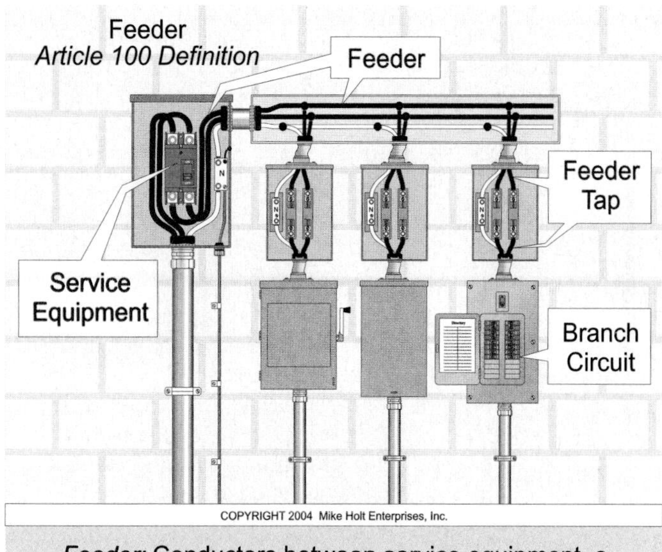
Feeder
Article 100 Definition
Feeder
Feeder Tap
Service Equipment
Branch Circuit
COPYRIGHT 2004 Mike Holt Enterprises, Inc.
Feeder: Conductors between service equipment, a separately derived system, or other power supply, and the final branch-circuit overcurrent device.

Figure 100–23

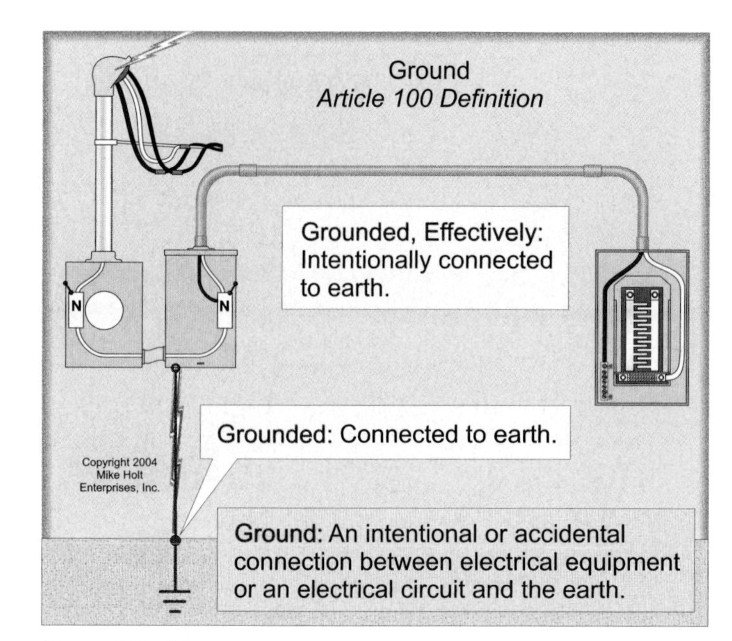
Ground
Article 100 Definition
Grounded, Effectively: Intentionally connected to earth.
N
N
Grounded: Connected to earth.
Copyright 2004 Mike Holt Enterprises, Inc.
Ground: An intentional or accidental connection between electrical equipment or an electrical circuit and the earth.

Figure 100–24

Author's Comments:

- An "other power source" would include a solar energy system (photovoltaic or PV).
- To have a better understanding of what a feeder is, be sure to review the definitions of service equipment and separately derived systems.

Fitting. An accessory, such as a locknut, that is intended to perform a mechanical function.

Garage. A building or portion of a building where self-propelled vehicles can be kept.

Author's Comment: Receptacles can be installed at any height in a dwelling unit garage, but no less than 18 in. above the floor for a commercial garage [511.3(A)(5)], unless they are listed as explosionproof.

Ground. An intentional or accidental connection to the earth. **Figure 100–24**

Author's Comment: The *NEC* also defines this term as "connection to some conducting body that serves in place of the earth," which leads to much confusion.

Grounded. Connected to earth. **See Figure 100–24.**

Author's Comment: The *NEC* also defines this term as "connected to some conducting body that serves in place of the earth." However, nobody really knows what this means.

Effectively Grounded. Intentional connection to earth through a conductor of sufficiently low impedance.

Grounded, Solidly. The intentional electrical connection of one system terminal to ground.

Author's Comment: In reality, solidly grounded means the bonding of the system to the metal case of the derived system in accordance with 250.30(A)(1). **Figure 100–25**

Grounded Neutral Conductor. The conductor that is intentionally grounded to the earth. See Article 200 in this textbook for additional details.

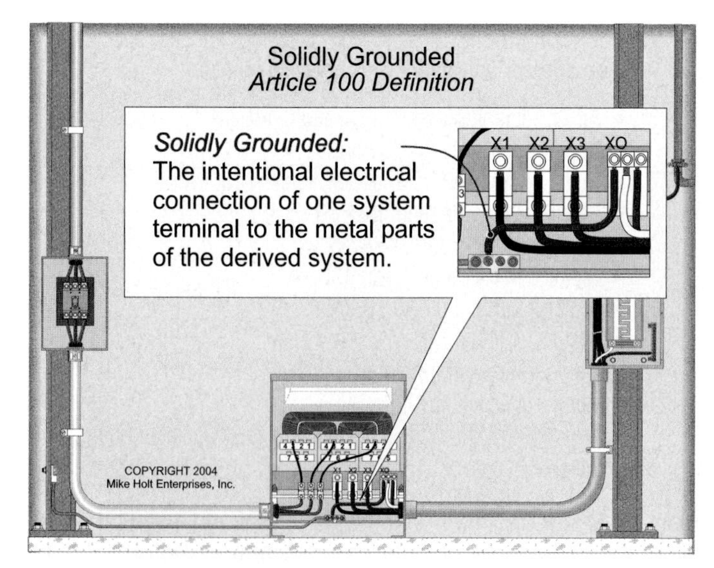
Solidly Grounded
Article 100 Definition
Solidly Grounded:
The intentional electrical connection of one system terminal to the metal parts of the derived system.
X1 X2 X3 XO
COPYRIGHT 2004 Mike Holt Enterprises, Inc.

Figure 100–25

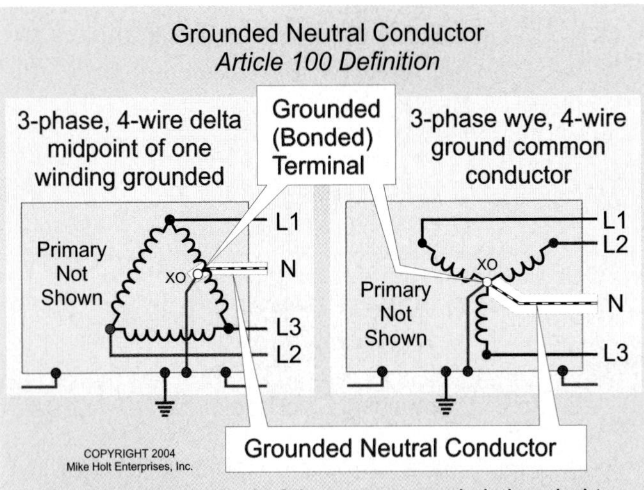

Figure 100–26

Author's Comment: In reality, one output terminal of a power supply is bonded to the case of the power supply (system bonding jumper). The conductor that is connected to this grounded terminal is called a "grounded conductor." **Figure 100–26**

Grounding (Earthing) Conductor. A conductor used to connect equipment to a grounding (earthing) electrode.

Author's Comment: A grounding (earthing) conductor is often used to connect the metal parts of electrical equipment to a supplementary grounding electrode [250.54]. This actually serves no performance purpose, but some equipment manufacturers require this connection in their installation instructions. **Figure 100–27**

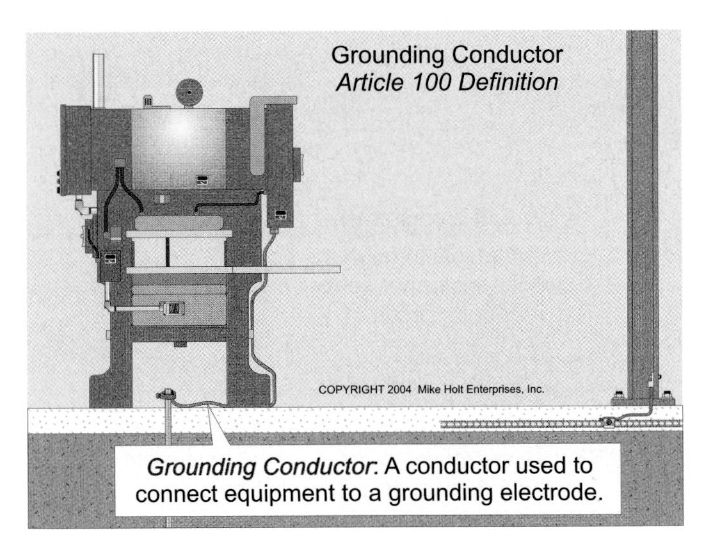

Figure 100–27

Grounding (Bonding) Conductor, Equipment. The low-impedance fault-current path used to connect the noncurrent-carrying metal parts of equipment, raceways, and other metal enclosures to the grounded neutral conductor at service equipment or at the source of a separately derived system.

Author's Comments:

- The equipment grounding (bonding) conductor actually serves as the "effective ground-fault current path" as defined in 250.2. Its purpose is to provide the low-impedance fault-current path necessary to facilitate the operation of overcurrent protection devices, and to remove dangerous voltage potentials between conductive parts of building components and electrical systems [250.4(A)(3)]. Because this is actually a bonding conductor, not a grounding (earthing) conductor, we will identify this conductor in this textbook as the equipment grounding (bonding) conductor.
- According to 250.118, the equipment grounding (bonding) conductor must be one or a combination of the following: **Figure 100–28**
  - A bare or insulated conductor
  - Rigid Metal Conduit
  - Intermediate Metal Conduit
  - Electrical Metallic Tubing
  - Listed Flexible Metal Conduit as limited by 250.118(5)
  - Listed Liquidtight Flexible Metal Conduit as limited by 250.118(6)
  - Type AC Armored Cable
  - Copper metal sheath of Mineral Insulated Cable
  - Metal Clad Cable as limited by 250.118(10)

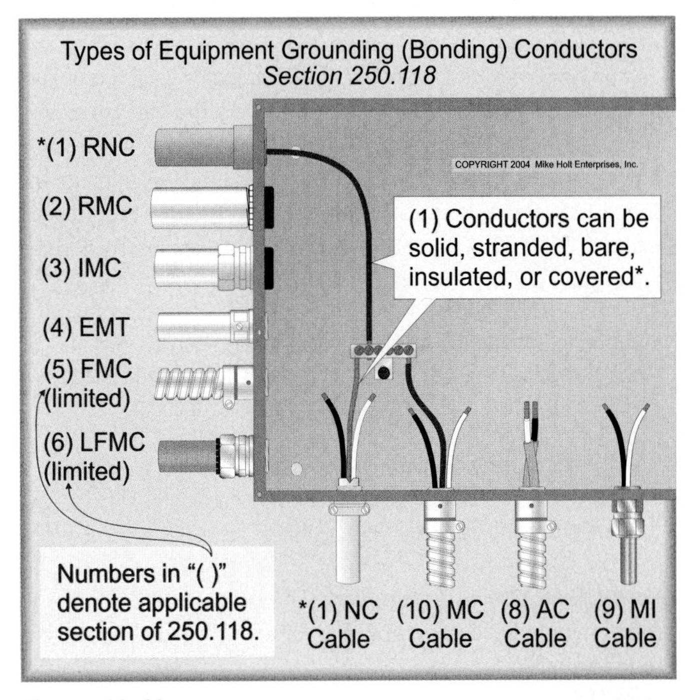

Figure 100–28

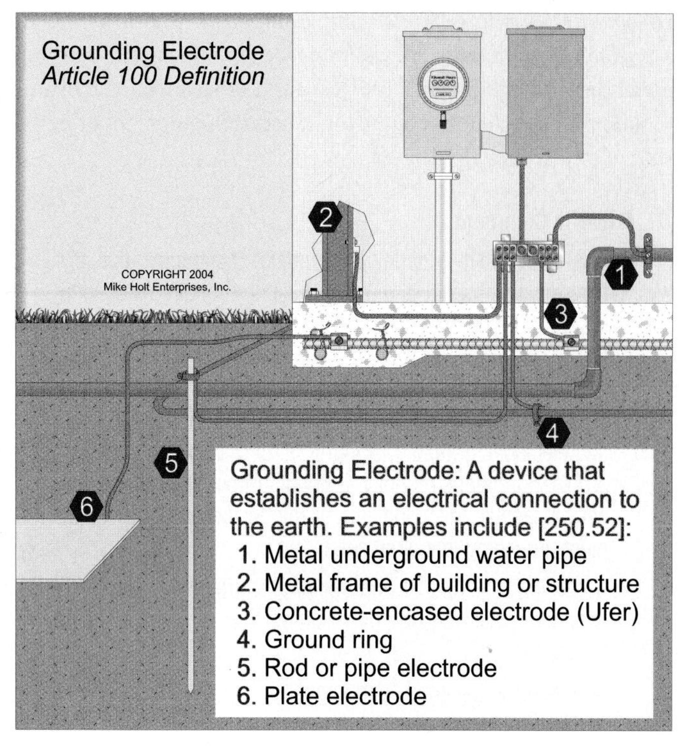
Grounding Electrode
Article 100 Definition
COPYRIGHT 2004
Mike Holt Enterprises, Inc.
1
2
3
4
5
6
Grounding Electrode: A device that establishes an electrical connection to the earth. Examples include [250.52]:
1. Metal underground water pipe
2. Metal frame of building or structure
3. Concrete-encased electrode (Ufer)
4. Ground ring
5. Rod or pipe electrode
6. Plate electrode

Figure 100–29

- Metallic cable trays as limited by 250.118(11) and 392.7
- Electrically continuous metal raceways listed for grounding
- Surface Metal Raceways listed for grounding.

Grounding (Earthing) Electrode. A device that establishes an electrical connection to the earth. See 250.50 through 250.70. **Figure 100–29**

Grounding Electrode Conductor. The conductor used to connect the grounded neutral conductor or the equipment grounding (bonding) conductor (metal parts of the disconnecting means), or both, to the grounding electrode (earthing) system at the service [250.24(A)], at each building or structure supplied by feeder(s) [250.32(A)], or the source of a separately derived system [250.30(A)]. **Figure 100–30**

> **Author's Comment:** At a service or separately derived system, the grounding electrode conductor connects the grounded neutral conductor and the equipment grounding (bonding) conductor to the grounding electrode (earthing) system [250.24(D) and 250.30]. At a separate building, the grounding electrode conductor connects the metal parts of the building disconnect to the grounding electrode [250.32(A)].

Ground-Fault Circuit Interrupter (GFCI). A device intended to protect people by de-energizing a circuit when the current-to-ground exceeds the value established for a "Class A" device. **Figure 100–31**

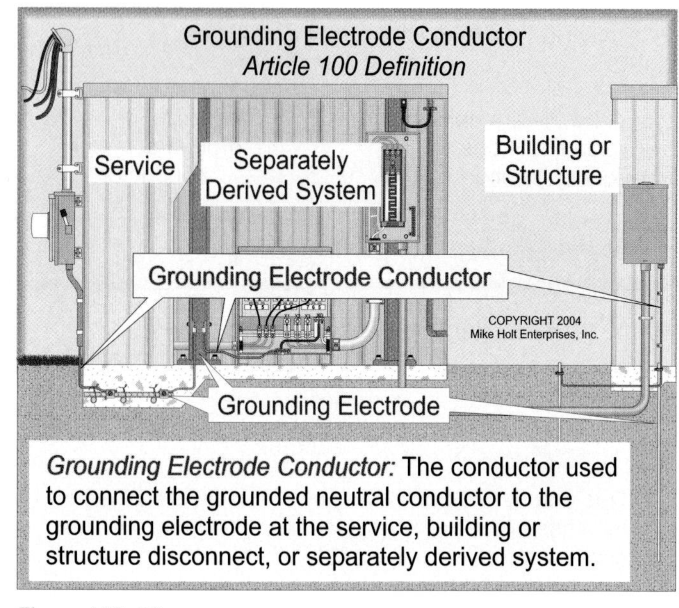
Grounding Electrode Conductor
Article 100 Definition
Service
Separately Derived System
Building or Structure
Grounding Electrode Conductor
COPYRIGHT 2004
Mike Holt Enterprises, Inc.
Grounding Electrode
Grounding Electrode Conductor: The conductor used to connect the grounded neutral conductor to the grounding electrode at the service, building or structure disconnect, or separately derived system.

Figure 100–30

> **FPN:** A "Class A" ground-fault circuit interrupter opens the circuit when the current-to-ground has a value between 4 mA and 6 mA.

> **Author's Comment:** A GFCI operates on the principle of monitoring the unbalanced current between the ungrounded and grounded neutral conductor. GFCI protective devices are commercially available in receptacles, circuit breakers, cord sets, and other types of devices. **Figure 100–32**

Ground-Fault Protection of Equipment. A system intended to provide protection of equipment from damaging ground-fault currents by opening all ungrounded conductors of the faulted circuit.

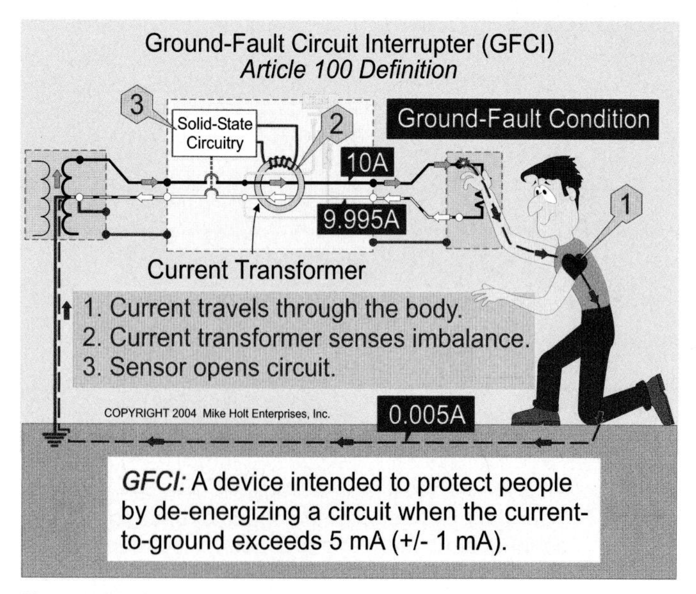
Ground-Fault Circuit Interrupter (GFCI)
Article 100 Definition
3
Solid-State Circuitry
2
Ground-Fault Condition
10A
9.995A
1
Current Transformer
1. Current travels through the body.
2. Current transformer senses imbalance.
3. Sensor opens circuit.
COPYRIGHT 2004 Mike Holt Enterprises, Inc.
0.005A
GFCI: A device intended to protect people by de-energizing a circuit when the current-to-ground exceeds 5 mA (+/- 1 mA).

Figure 100–31

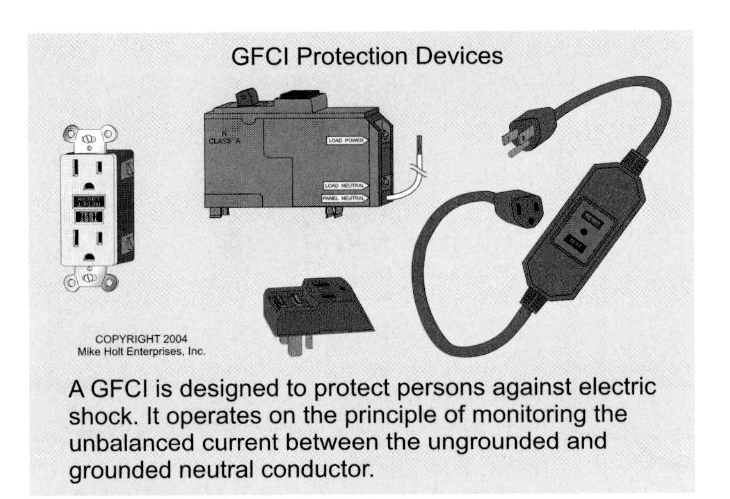

Figure 100–32

This protection is provided at current levels less than those required to protect conductors from damage through the operation of a supply circuit overcurrent device. See 215.10, 230.95, and 240.13.

> **Author's Comment:** This type of protective device isn't intended to protect people, only connected utilization equipment.

Guest Room. An accommodation that combines living, sleeping, sanitary, and storage facilities. **Figure 100–33**

Guest Suite. An accommodation with two or more contiguous rooms comprising a compartment, with or without doors between such rooms, that provides living, sleeping, sanitary, and storage facilities.

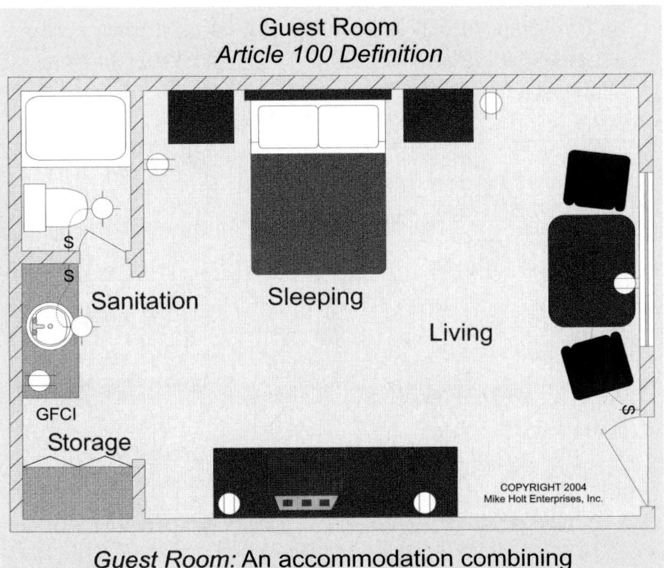

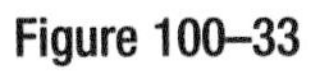
Figure 100–33

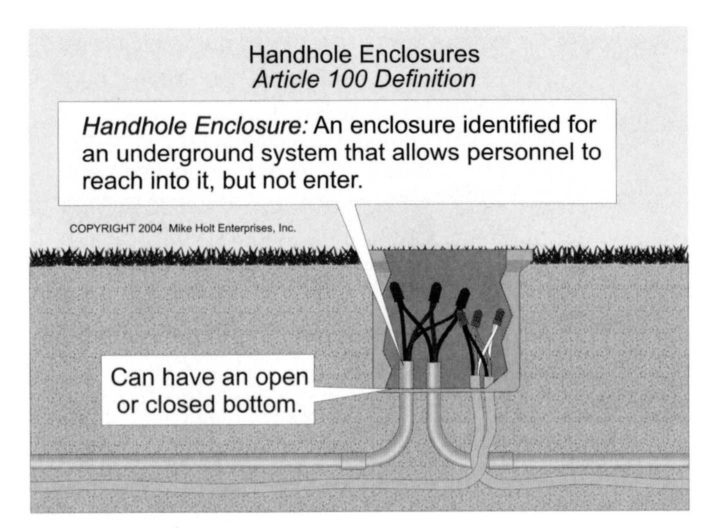

Figure 100–34

Handhole Enclosure. An enclosure identified for underground system use, provided with an open or closed bottom, and sized to allow personnel to reach into, but not enter, for the purpose of installing or maintaining equipment or wiring. **Figure 100–34**

> **Author's Comment:** See 314.30 for the installation requirements for handhole enclosures.

Identified Equipment. Recognized as suitable for a specific purpose, function, or environment by listing and labeling. See 90.4, 90.7, 110.3(A)(1), and the definitions in Article 100 for Approved, Labeled, and Listed.

In Sight From (Within Sight). Visible and not more than 50 ft distant from the equipment. **Figure 100–35**

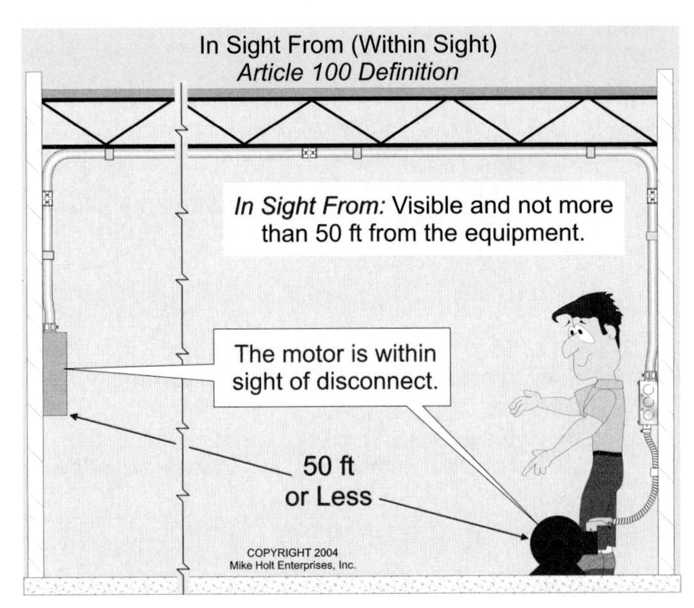

Figure 100–35

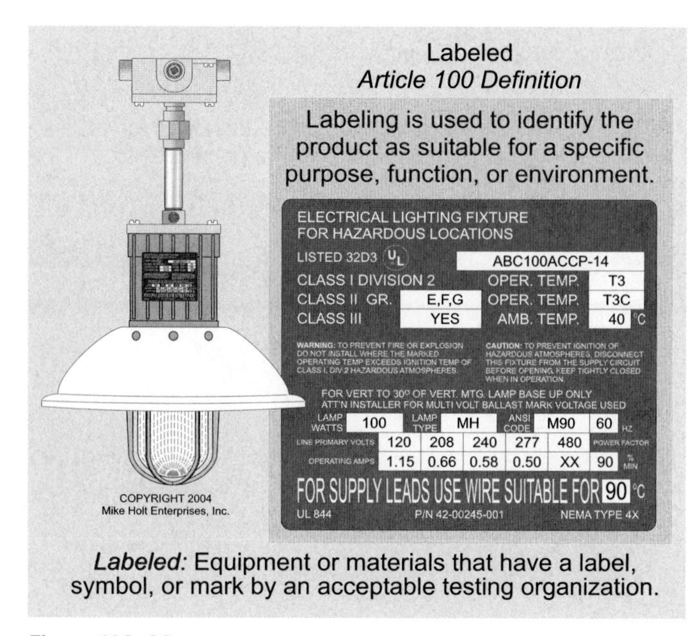

Labeled: Equipment or materials that have a label, symbol, or mark by an acceptable testing organization.

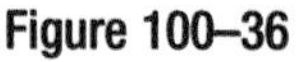
Figure 100–36

Interrupting Rating. The highest short-circuit current at rated voltage that the device can safely interrupt. For more information, see 110.9 in this textbook.

Labeled. Equipment or materials that have a label, symbol, or other identifying mark in the form of a sticker, decal, or printed label, or molded or stamped into the product by a testing laboratory acceptable to the authority having jurisdiction. See Identified and Listed. **Figure 100–36**

> **Author's Comment:** Labeling and listing of equipment typically provides the basis for equipment approval by the authority having jurisdiction. See 90.4, 90.7, 110.2, and 110.3 for more information.

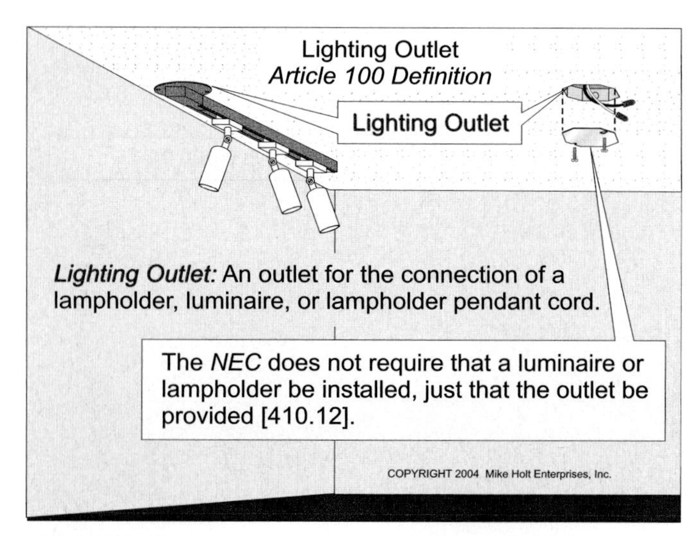

Lighting Outlet
Article 100 Definition
Lighting Outlet
The NEC does not require that a luminaire or lampholder be installed, just that the outlet be provided [410.12].
COPYRIGHT 2004 Mike Holt Enterprises, Inc.

Lighting Outlet: An outlet for the connection of a lampholder, luminaire, or lampholder pendant cord.

Figure 100–37

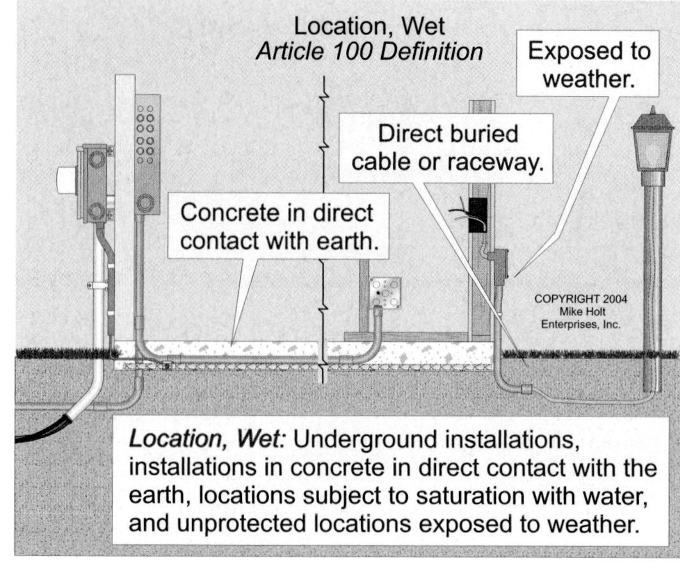

Location, Wet: Underground installations, installations in concrete in direct contact with the earth, locations subject to saturation with water, and unprotected locations exposed to weather.

Figure 100–38

Lighting Outlet. An outlet for the connection of a lampholder, luminaire, or lampholder pendant cord. **Figure 100–37**

Listed. Equipment or materials included in a list published by a testing laboratory that is acceptable to the authority having jurisdiction. The listing organization must periodically inspect the production of listed equipment or material to ensure that the equipment or material meets appropriate designated standards and is suitable for a specified purpose. See Identified and Labeled.

> **Author's Comment:** The *NEC* doesn't require all electrical equipment to be listed, but some *Code* requirements do specifically require product listing. Increasingly, organizations such as OSHA require that listed equipment be used when such equipment is available. See 90.7, 110.2, and 110.3 for more information.

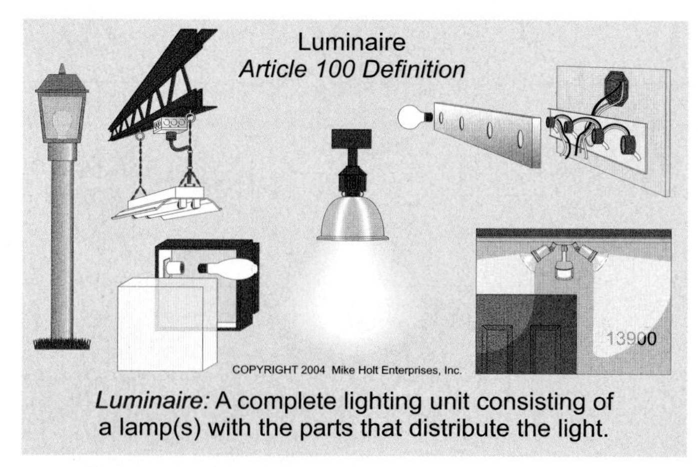

Luminaire: A complete lighting unit consisting of a lamp(s) with the parts that distribute the light.

Figure 100–39

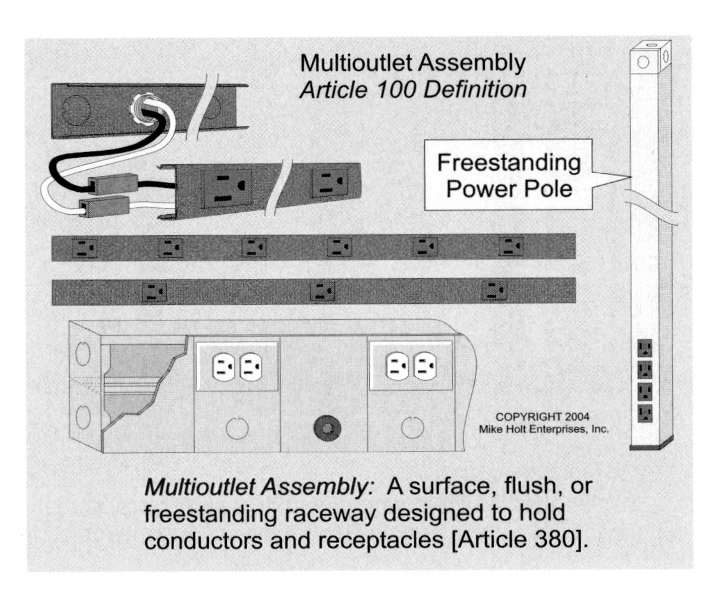
Multioutlet Assembly
Article 100 Definition
Freestanding
Power Pole
COPYRIGHT 2004
Mike Holt Enterprises, Inc.
Multioutlet Assembly: A surface, flush, or freestanding raceway designed to hold conductors and receptacles [Article 380].

Figure 100–40

Location, Damp. Locations protected from weather and not subject to saturation with water or other liquids. This includes locations partially protected under canopies, marquees, roofed open porches, and interior locations subject to moderate degrees of moisture, such as some basements, barns, and cold-storage warehouses.

Location, Dry. An area not normally subjected to dampness or wetness, but which may temporarily be subject to dampness or wetness, such as a building under construction.

Location, Wet. Underground installations, installations in concrete in direct contact with the earth, locations subject to saturation with water, and unprotected locations exposed to weather. **Figure 100–38**

Luminaire. A complete lighting unit that consists of a lamp or lamps together with the parts designed to distribute the light. **Figure 100–39**

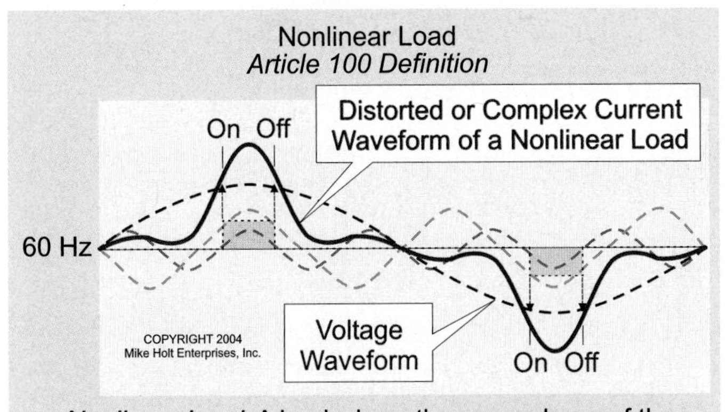
Nonlinear Load
Article 100 Definition
Distorted or Complex Current
Waveform of a Nonlinear Load
On Off
60 Hz
Voltage
Waveform
On Off
COPYRIGHT 2004
Mike Holt Enterprises, Inc.
Nonlinear Load: A load where the wave shape of the current does not follow the wave shape of the voltage.

Figure 100–41

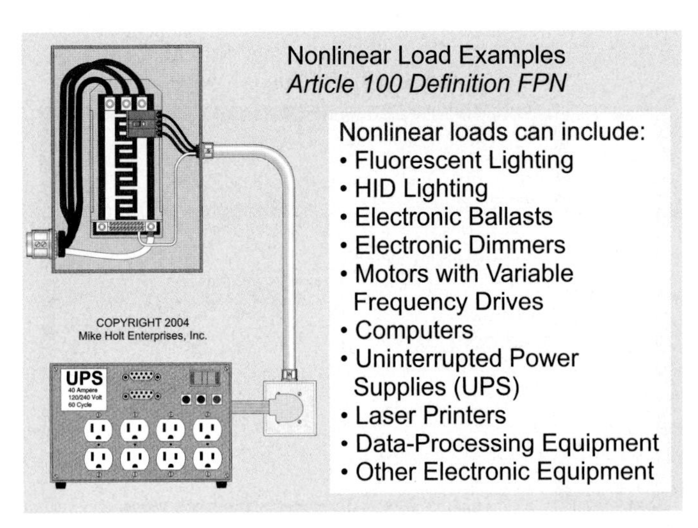
Nonlinear Load Examples
Article 100 Definition FPN
Nonlinear loads can include:
• Fluorescent Lighting
• HID Lighting
• Electronic Ballasts
• Electronic Dimmers
• Motors with Variable Frequency Drives
• Computers
• Uninterrupted Power Supplies (UPS)
• Laser Printers
• Data-Processing Equipment
• Other Electronic Equipment
COPYRIGHT 2004
Mike Holt Enterprises, Inc.
UPS

Figure 100–42

Multioutlet Assembly. A surface, flush, or freestanding raceway designed to hold conductors and receptacles. See Article 380 for additional details. **Figure 100–40**

Nonlinear Load. A load where the current waveform doesn't follow the applied sinusoidal voltage waveform. See 210.4(A) FPN, 220.61(C)(2) FPN 2, 310.15(B)(4)(c), and 450.3 FPN 2. **Figure 100–41**

> **FPN:** Single-phase nonlinear loads include electronic equipment, such as copy machines, laser printers, and electric-discharge lighting. Three-phase nonlinear loads include uninterruptible power supplies (UPSs), induction motors, and electronic switching devices, such as variable-frequency and variable-speed drives (VFD-VSD). **Figure 100–42**

> **Author's Comment:** The subject of nonlinear loads is outside the scope of this textbook. For more information on this topic, visit http://www.MikeHolt.com/news/archive/html/master/necharmonics.htm

Outlet. A point in the wiring system where electric current is taken to supply a load, such as receptacle(s), luminaire(s), and equipment. **Figure 100–43**

Outline Lighting. An arrangement of incandescent lamps, electric-discharge lighting, or other electrically powered light sources to outline or call attention to certain features such as the shape of a building or the decoration of a window.

Overcurrent. Current in amperes that is greater than the rated current of the equipment or conductors, resulting from an overload, short circuit, or ground fault. **Figure 100–44**

> **Author's Comment:** See the definitions of "Ground Fault" in 250.2 and "Overload" in Article 100.

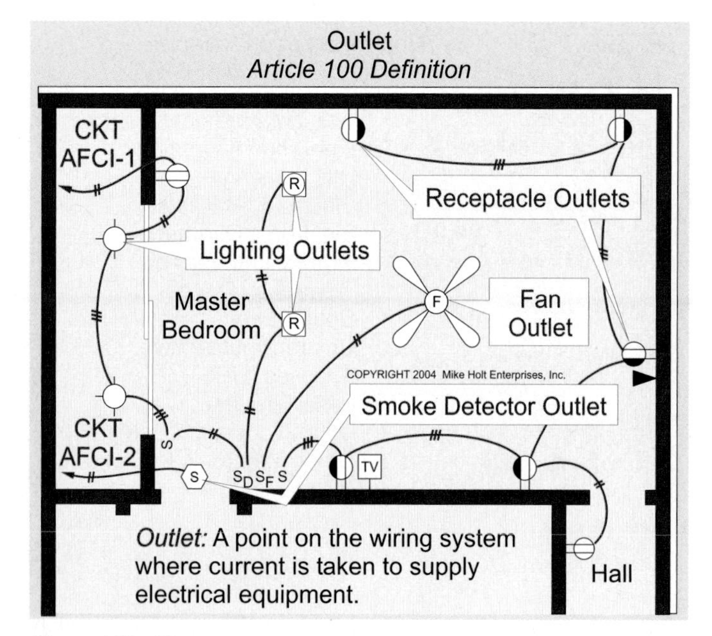

Outlet: A point on the wiring system where current is taken to supply electrical equipment.

Figure 100–43

Overload. The operation of equipment above its ampere current rating or current in excess of conductor ampacity. When an overload condition persists for a sufficient length of time, it could result in equipment failure or a fire from damaging or dangerous overheating. A fault, such as a short circuit or ground fault, isn't an overload.

Panelboard [Article 408]. A distribution point containing overcurrent protection devices and designed to be installed in a cabinet. **Figure 100–45**

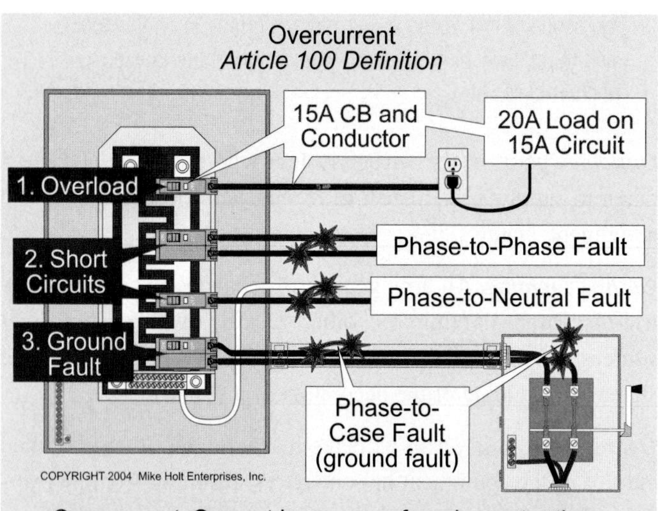

Overcurrent: Current in excess of equipment rating caused from an overload, short circuit, or ground fault.

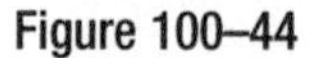
Figure 100–44

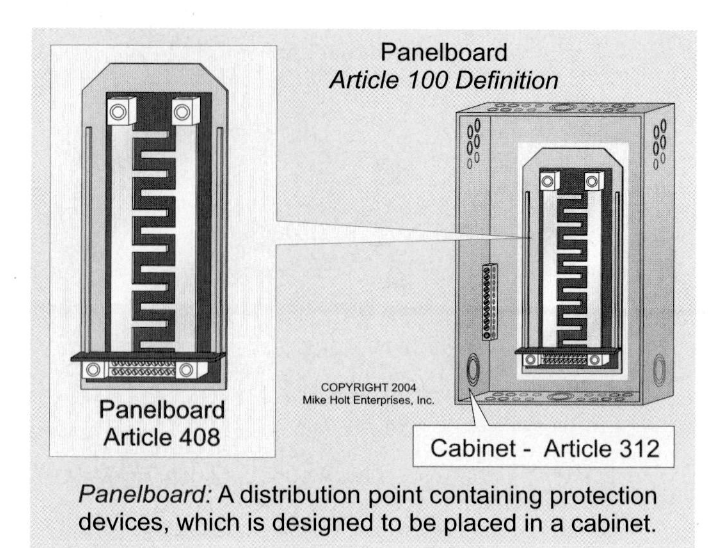

Panelboard: A distribution point containing protection devices, which is designed to be placed in a cabinet.

Figure 100–45

Author's Comments:

- See the definition of "Cabinet" in this article.
- The slang term in the electrical field for a panelboard is "the guts."

Plenum. A compartment or chamber to which one or more ducts are connected and that forms part of the air distribution system.

Premises Wiring. The interior and exterior wiring, including power, lighting, control, and signal circuits, and all associated hardware, fittings, and wiring devices, both permanently and temporarily installed. This doesn't include the internal wiring of electrical equipment and appliances, such as luminaires, dishwashers, water heaters, motors, controllers, motor control centers, A/C equipment, etc. See 90.7 and 300.1(B).

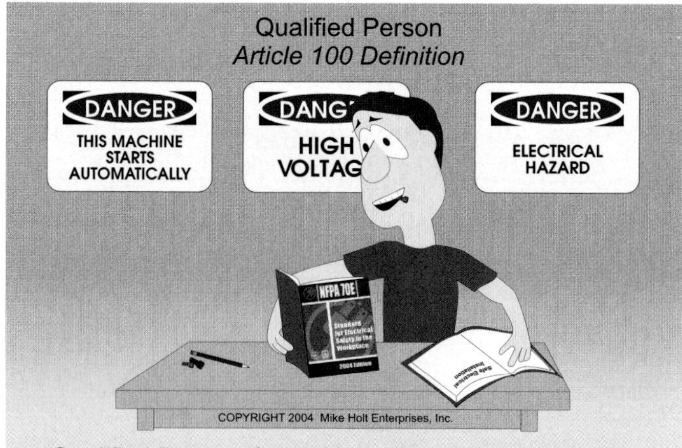

Qualified Person: One who has knowledge and skill related to the construction, operation, and installation of electrical equipment, including safety training on the hazards involved with electrical systems.

Figure 100–46

Qualified Person. A person who has the skill and knowledge related to the construction and operation of the electrical equipment and its installation. This person must have received safety training on the hazards involved with electrical systems. **Figure 100–46**

FPN: Refer to NFPA 70E, *Standard for Electrical Safety in the Workplace*, for electrical safety training requirements.

Author's Comments:

- Examples of this safety training include, but aren't limited to, training in the use of special precautionary techniques, of personal protective equipment, of insulating and shielding materials, and of using insulated tools and test equipment when working on or near exposed conductors or circuit parts that are or can become energized.
- In many parts of the United States, electricians, electrical contractors, electrical inspectors, and electrical engineers must complete from 6 to 24 hours of *NEC* review each year as a requirement to maintain licensing. This in itself doesn't make one qualified to deal with the specific hazards involved.

Raceway. An enclosure designed for the installation of conductors, cables, or busbars. Raceways in the *NEC* include:

Raceway Type	Article
• Busways	368
• Electrical Metallic Tubing	358
• Electrical Nonmetallic Tubing	362
• Flexible Metal Conduit	348
• Intermediate Metal Conduit	342
• Liquidtight Flexible Metal Conduit	350
• Liquidtight Flexible Nonmetallic Conduit	356
• Metal Wireways	376
• Multioutlet Assembly	380
• Nonmetallic Wireways	378
• Rigid Metal Conduit	344
• Rigid Nonmetallic Conduit	352
• Strut-Type Channel Raceways	384
• Surface Metal Raceways	386
• Surface Nonmetallic Raceways	388

Author's Comment: A cable tray system isn't a raceway; it's a support system for cables, raceways, and enclosures. See Article 392.

Receptacle. A contact device installed at an outlet for the connection of an attachment plug. A single receptacle contains one device on a strap (mounting yoke), and a multiple receptacle contains more than one device on a common yoke. **Figure 100–47**

Receptacle Outlet. An opening in an outlet box where one or more receptacles have been installed.

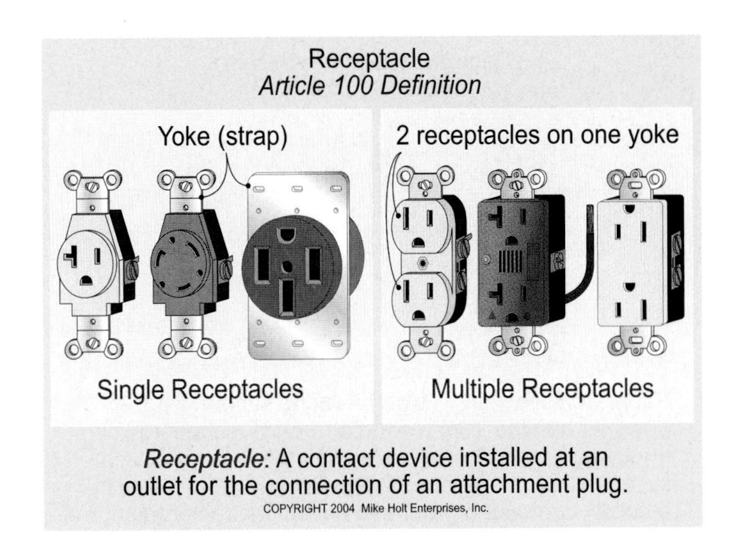
Receptacle
Article 100 Definition
Yoke (strap)
2 receptacles on one yoke
Single Receptacles
Multiple Receptacles
Receptacle: A contact device installed at an outlet for the connection of an attachment plug.
COPYRIGHT 2004 Mike Holt Enterprises, Inc.

Figure 100–47

Remote-Control Circuit [Article 725]. An electric circuit that controls another circuit by a relay or equivalent device installed in accordance with Article 725. **Figure 100–48**

Separately Derived System. A wiring system whose power is derived from a source of electric energy or equipment other than the electric utility service. This includes a battery, a solar photovoltaic system, a transformer, or a converter winding, where there's no direct electrical connection to the supply conductors of another system. **Figure 100–49**

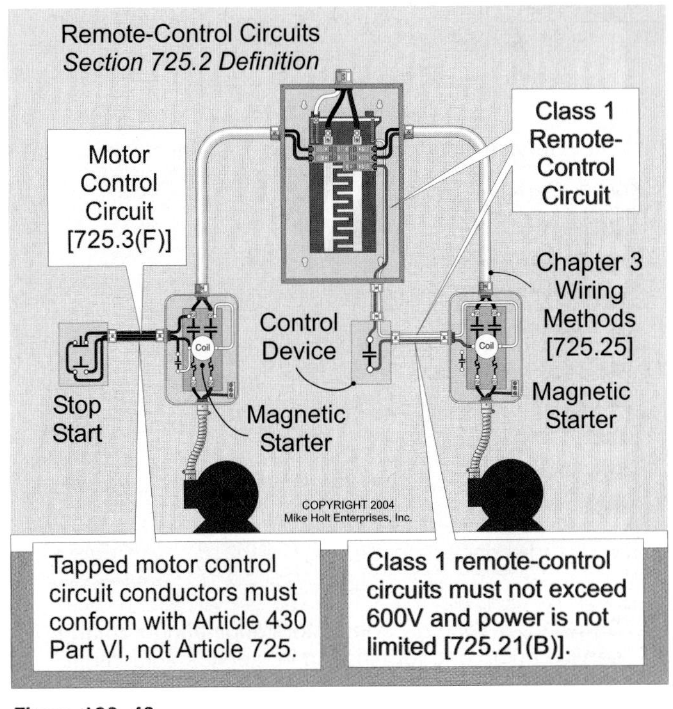
Remote-Control Circuits
Section 725.2 Definition
Motor Control Circuit [725.3(F)]
Class 1 Remote-Control Circuit
Chapter 3 Wiring Methods [725.25]
Control Device
Stop
Start
Magnetic Starter
Magnetic Starter
COPYRIGHT 2004 Mike Holt Enterprises, Inc.
Tapped motor control circuit conductors must conform with Article 430 Part VI, not Article 725.
Class 1 remote-control circuits must not exceed 600V and power is not limited [725.21(B)].

Figure 100–48

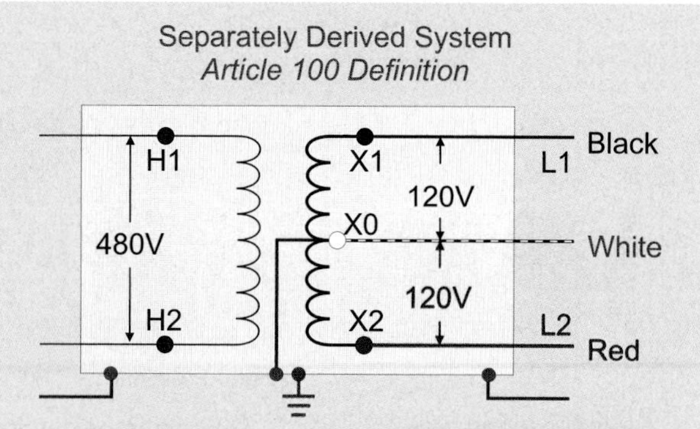

Separately Derived System: A wiring system whose power is derived from a source of electrical energy or equipment other than the electric utility service. Such systems have no direct electrical connection, including a solidly connected grounded circuit conductor to supply conductors originating in another system.

Figure 100–49

Author's Comments:

- The revised definition clarifies that a separately derived system also includes equipment such as transformers, converters, and inverters, which might not be considered a source of energy.
- Separately derived systems are actually a lot more complicated than the above definition suggests, and understanding them requires additional study. For more information, see 250.20(D) and 250.30.

Service. The conductors from the electric utility that deliver electric energy to the premises.

Author's Comment: Conductors from a UPS system, solar photovoltaic system, generator, or transformer are not identified as service conductors. See the definitions for "Feeder" and for "Service Conductors" in this article.

Service Conductors. Conductors originating from the "service point" and terminating in "service equipment." See the definition of Service Point and Service Equipment. **Figure 100–50**

Author's Comment: Conductors from a generator, UPS system, or transformers are feeder conductors, not service conductors. See the definition of Feeder.

Service Equipment. The one to six disconnects connected to the load end of service conductors intended to control and cut off the supply to the building or structure. **Figure 100–51**

Author's Comment: It's important to know where a service begins and where it ends in order to properly apply the *NEC* requirements. Sometimes the service ends before the metering equipment. **Figure 100–52**

Service Point. The point where the electrical utility conductors make contact with premises wiring. **Figure 100–53**

Author's Comments:

- See the definition of "Premises Wiring" in this article.
- The service point can be at the utility transformer, at the service weatherhead, or at the meter enclosure, depending on where the utility conductors terminate. **Figure 100–54**

Signaling Circuit [Article 725]. Any electric circuit that energizes signaling equipment.

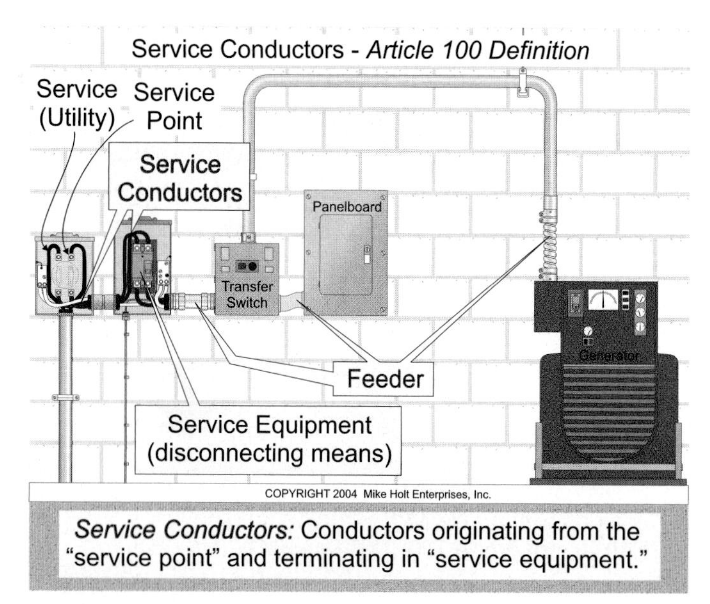

Service Conductors: Conductors originating from the "service point" and terminating in "service equipment."

Figure 100–50

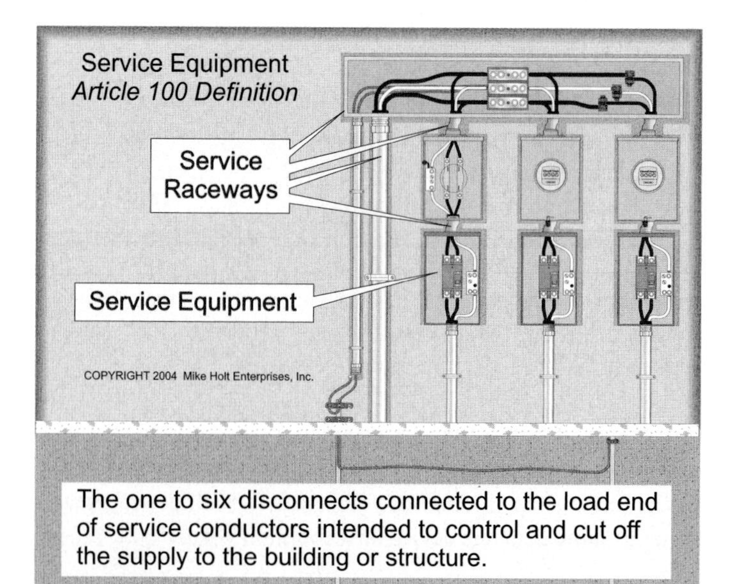

The one to six disconnects connected to the load end of service conductors intended to control and cut off the supply to the building or structure.

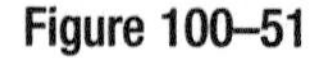

Figure 100–51

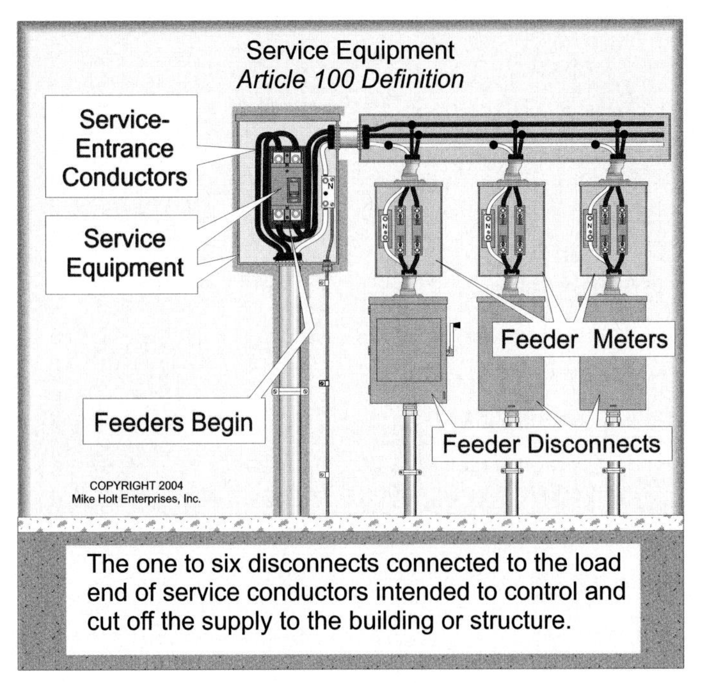

Figure 100–52

Special Permission. Written consent from the authority having jurisdiction.

> **Author's Comment:** See the definition for Authority Having Jurisdiction.

Structure. That which is built or constructed.

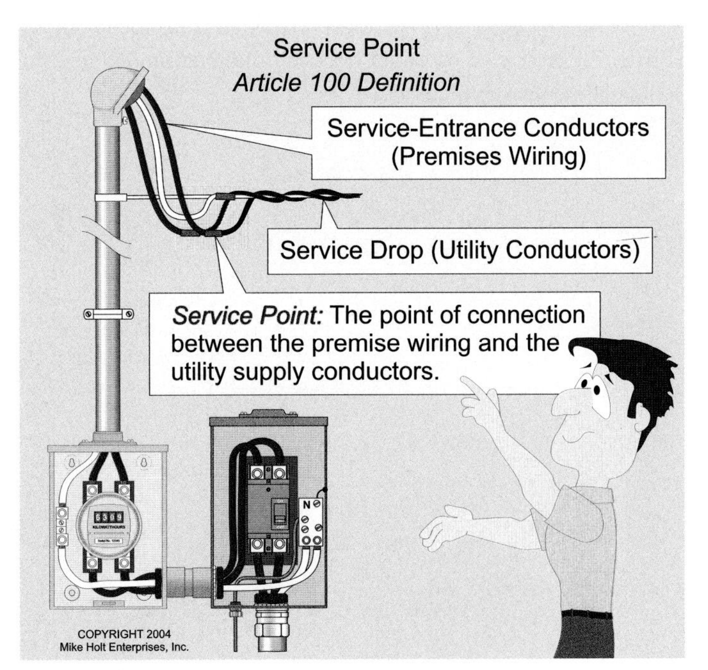

Figure 100–53

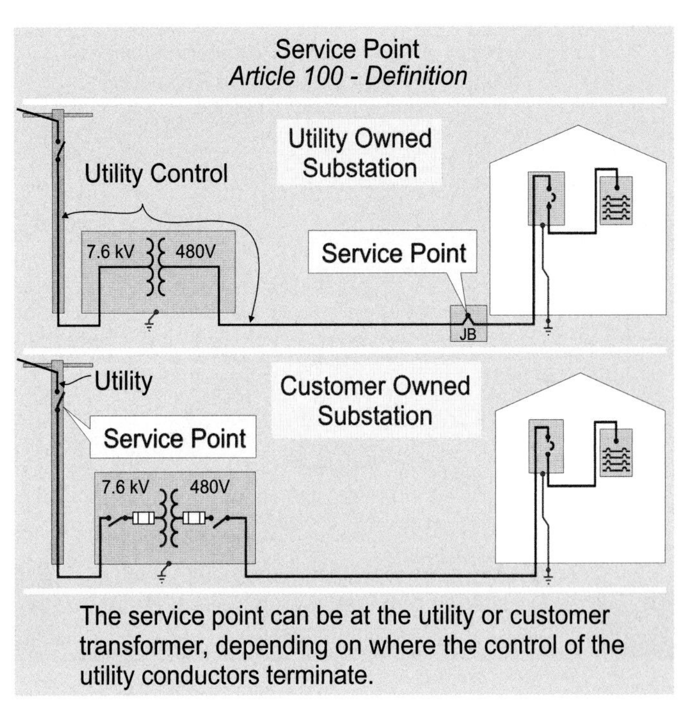

Figure 100–54

Supplementary Overcurrent Protective Device. A device intended to provide limited overcurrent protection for specific applications and utilization equipment, such as luminaires and appliances. This limited protection is in addition to the protection provided in the required branch circuit by the branch-circuit overcurrent protective device. **Figure 100–55**

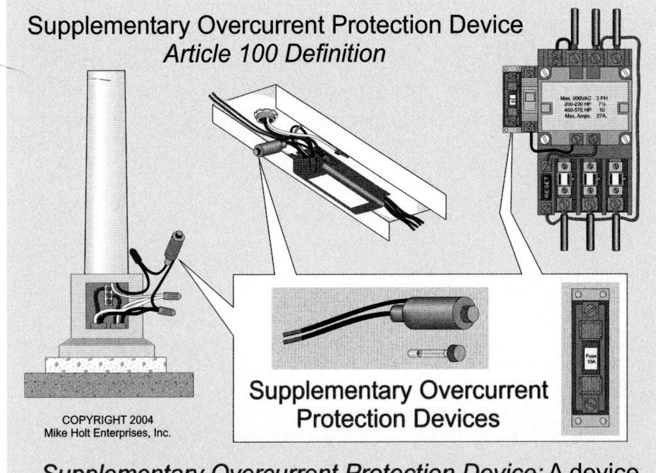

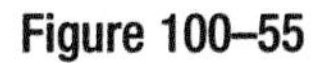
Figure 100–55

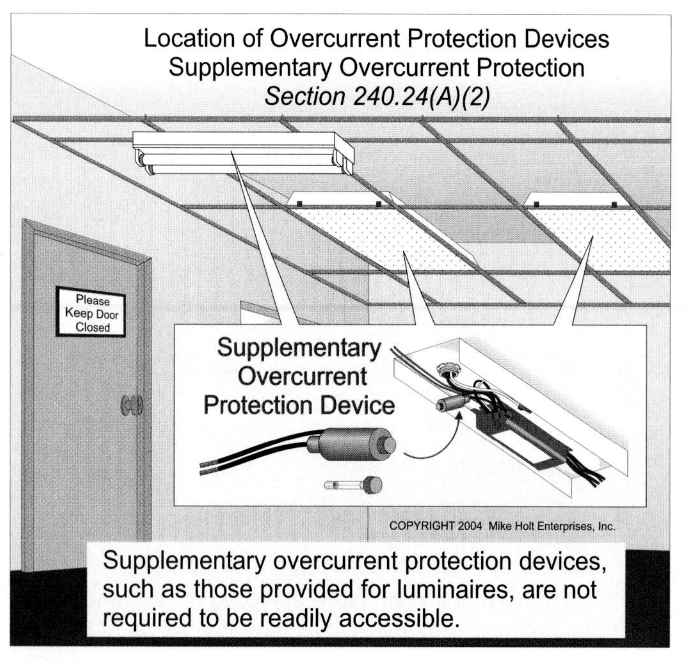
Location of Overcurrent Protection Devices
Supplementary Overcurrent Protection
Section 240.24(A)(2)
Please Keep Door Closed
Supplementary Overcurrent Protection Device
COPYRIGHT 2004 Mike Holt Enterprises, Inc.
Supplementary overcurrent protection devices, such as those provided for luminaires, are not required to be readily accessible.

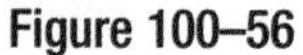
Figure 100–56

Author's Comment: Supplementary overcurrent devices aren't required to be readily accessible [240.10 and 240.24(A)(2)]. **Figure 100–56**

Switch, General-Use Snap. A switch constructed to be installed in a device box or a box cover.

Voltage of a Circuit. The greatest effective root-mean-square (RMS) difference of potential between any two conductors of the circuit.

Author's Comment: Voltages can be reported in many ways: peak, peak-to-peak, average, and root-mean-square (RMS). Electrical voltage measuring equipment that provides readings in either average or RMS are commercially available. RMS meters provide the most accurate reading when used on circuits with nonlinear loads that have a high harmonic content.

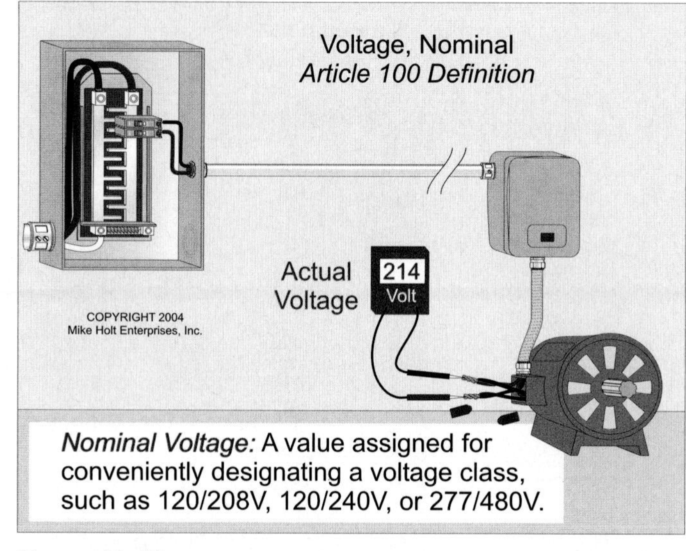
Voltage, Nominal
Article 100 Definition
Actual Voltage
214 Volt
COPYRIGHT 2004 Mike Holt Enterprises, Inc.
Nominal Voltage: A value assigned for conveniently designating a voltage class, such as 120/208V, 120/240V, or 277/480V.

Figure 100–57

Voltage, Nominal. A value assigned for the purpose of conveniently designating voltage class, such as 120/240V, 120/208V, or 277/480V. See 220.5(A). **Figure 100–57**

Author's Comment: The actual voltage at which a circuit operates can vary from the nominal within a range that permits satisfactory operation of equipment. In addition, common voltage ratings of electrical equipment are 115, 200, 208, 230, and 460. The electrical power supplied might be at the nominal voltage (240), but the voltage at the equipment will be less than this voltage value (230). Therefore, electrical equipment is rated at a value less than the nominal system voltage.

Voltage-to-Ground. The greatest effective root-mean-square (RMS) difference of potential between the ungrounded and the grounded neutral conductor.

Article 100 Questions

(• Indicates that 75% or fewer exam takers get the question correct.)

1. Admitting close approach, not guarded by locked doors, elevation, or other effective means, is commonly referred to as _____.

   (a) accessible (equipment)
   (b) accessible (wiring methods)
   (c) accessible, readily
   (d) all of these

2. •A device that, by insertion in a receptacle, establishes a connection between the conductors of the attached flexible cord and the conductors connected permanently to the receptacle is called a(n) _____.

   (a) attachment plug
   (b) plug cap
   (c) plug
   (d) any of these

3. •For a circuit to be considered a multiwire branch circuit, it must have _____.

   (a) two or more ungrounded conductors with a voltage potential between them
   (b) a grounded neutral conductor having equal voltage potential between it and each ungrounded conductor of the circuit
   (c) a grounded neutral conductor connected to the grounded neutral terminal of the system
   (d) all of these

4. NM cable is considered _____ if rendered inaccessible by the structure or finish of the building.

   (a) inaccessible
   (b) concealed
   (c) hidden
   (d) enclosed

5. A component of an electrical system that is intended to carry or control but not utilize electric energy is a(n) _____.

   (a) raceway
   (b) fitting
   (c) device
   (d) enclosure

6. The *NEC* term to define wiring methods that are not concealed is _____.

   (a) open
   (b) uncovered
   (c) exposed
   (d) bare

7. _____ is defined as intentionally connected to earth through a ground connection or connections of sufficiently low impedance and having sufficient current-carrying capacity, to prevent the build up of voltages that may result in undue hazards to connected equipment or to persons.

   (a) Effectively grounded
   (b) A proper wiring system
   (c) A lighting rod
   (d) A grounded neutral conductor

8. The grounding electrode conductor is the conductor used to connect the grounding electrode to the equipment grounding (bonding) conductor and the grounded neutral conductor at _____.

   (a) the service
   (b) each building or structure supplied by feeder(s)
   (c) the source of a separately derived system
   (d) all of these

9. Recognized as suitable for the specific purpose, function, use, environment, and application is the definition of _____.

   (a) labeled
   (b) identified (as applied to equipment)
   (c) listed
   (d) approved

10. A _____ location may be temporarily subject to dampness and wetness.

    (a) dry (b) damp (c) moist (d) wet

11. Outline lighting may not include light sources such as light emitting diodes (LEDs).

    (a) True (b) False

12. NFPA 70E, *Standard for Electrical Safety in the Workplace,* provides information to help determine the electrical safety training requirements expected of a "qualified person."

    (a) True (b) False

13. When one electrical circuit controls another circuit through a relay, the first circuit is called a _____.

    (a) control circuit (b) remote-control circuit (c) signal circuit (d) controller

14. The _____ is the necessary equipment, usually consisting of a circuit breaker(s) or switch(es) and fuse(s) and their accessories, connected to the load end of service conductors to a building or other structure, or an otherwise designated area, and intended to constitute the main control and cutoff of the supply.

    (a) service equipment (b) service
    (c) service disconnect (d) service overcurrent protection device

15. A form of general-use switch constructed so that it can be installed in device boxes or on box covers, or otherwise used in conjunction with wiring systems recognized by the *Code,* is called a _____ switch.

    (a) transfer (b) motor-circuit (c) general-use snap (d) bypass isolation

ARTICLE

Requirements for Electrical Installations

Introduction

Article 110 sets the stage for how you will implement the rest of the *NEC*. This article contains a few of the most important and yet neglected parts of the *Code*. For example:

- What do you do with unused openings in enclosures?
- What's the right working clearance for a given installation?
- How should you terminate conductors?
- What kinds of warnings, markings, and identification does a given installation require?

It's critical that you master Article 110, and that's exactly what this Illustrated Guide is designed to help you do. As you read this article, remember that doing so helps build your foundation for correctly applying much of the *NEC*. In fact, the article itself is a foundation for much of the *Code*. You may need to read something several times to understand it. The time you take to do that will be well spent. The illustrations will also help. But if you find your mind starting to wander, take a break. What matters is how well you master the material and how safe your work is—not how fast you blazed through a book.

PART I. GENERAL REQUIREMENTS

110.1 Scope. Article 110 covers the general requirements for the examination and approval, installation and use, access to and spaces about electrical equipment; as well as general requirements for enclosures intended for personnel entry (manholes, vaults, and tunnels).

110.2 Approval of Equipment. The authority having jurisdiction must approve all electrical conductors and equipment. **Figure 110–1**

Author's Comment: For a better understanding of product approval, review 90.4, 90.7, 110.3 and the definitions for Approved, Identified, Labeled, and Listed in Article 100.

Approval of Equipment
Section 110.2
Unlisted Gutter
Can I use this unlisted gutter?
Okay
Conductors and equipment can be installed only if they are approved.
COPYRIGHT 2004
Mike Holt Enterprises, Inc.
Approved: Acceptable to the authority having jurisdiction.

Figure 110–1

110.3 Examination, Identification, Installation, and Use of Equipment.

(A) Guidelines for Approval. The authority having jurisdiction must approve equipment, and consideration must be given to the following:

(1) Listing or labeling
(2) Mechanical strength and durability
(3) Wire-bending and connection space
(4) Electrical insulation
(5) Heating effects under conditions of use
(6) Arcing effects

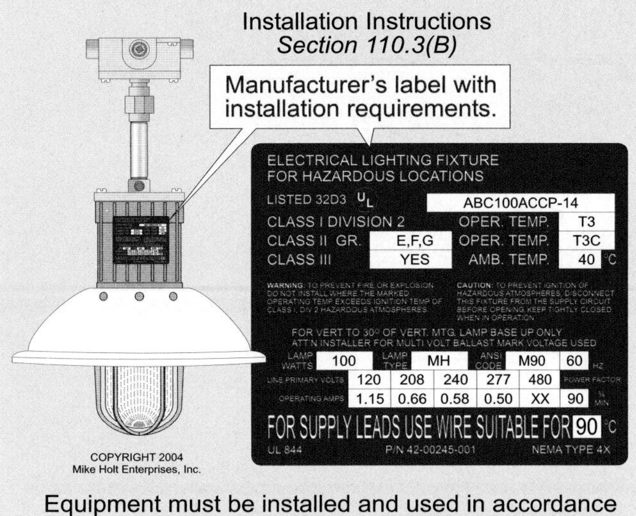

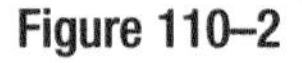
Figure 110–2

(7) Classification by voltage, current capacity, and specific use

(8) Other factors contributing to the practical safeguarding of persons using or in contact with the equipment

(B) Installation and Use. Equipment must be installed and used in accordance with any instructions included in the listing or labeling requirements.

Author's Comments:

- See Article 100 for the definitions of "Labeling" and "Listing."
- Equipment is listed for a specific condition of use, operation, or installation, and it must be installed and used in accordance with those listed instructions. Failure to follow product listing instructions, such as torquing of terminals and sizing of conductors, is a violation of this *Code* rule [110.3(B)]. **Figure 110–2**

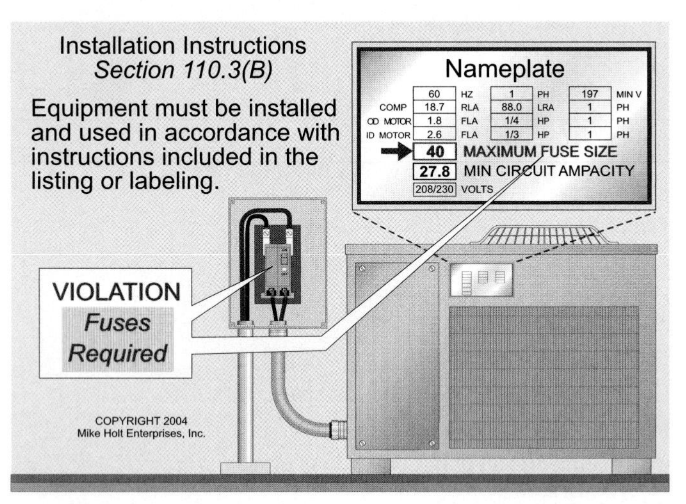

Figure 110–3

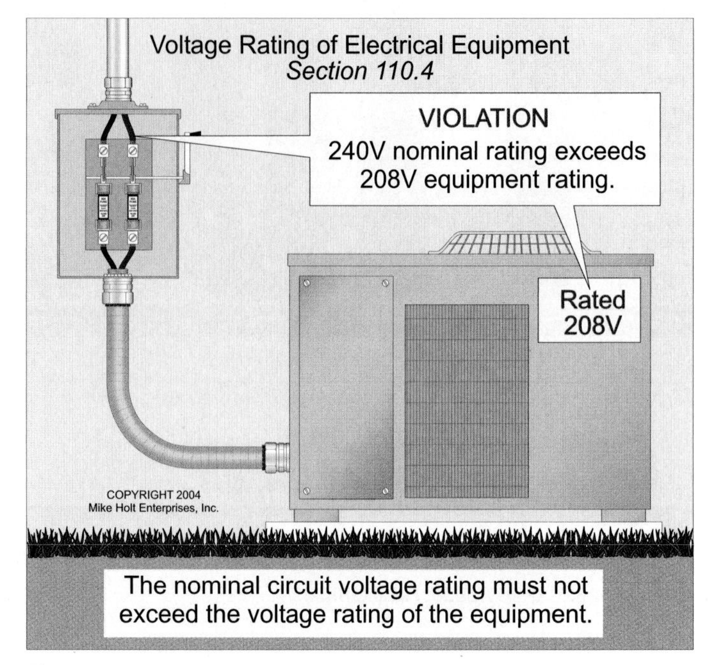

Figure 110–4

- When an air conditioner nameplate specifies "Maximum Fuse Size," one-time or dual-element fuses must be used to protect the equipment. **Figure 110–3**

110.4 Voltages. The voltage rating of electrical equipment must not be less than the nominal voltage of a circuit. Put another way, electrical equipment must be installed on a circuit where the nominal system voltage doesn't exceed the voltage rating of the equipment. This rule is intended to prohibit the installation of 208V rated motors on a 240V nominal voltage rated circuit. **Figure 110–4**

Author's Comments:

- See Article 100 for the definition of "Nominal Voltage."
- According to 110.3(B), equipment must be installed in accordance with any instructions included in the listing or labeling. Therefore, equipment must not be connected to a circuit where the nominal voltage is less than the rated voltage of the electrical equipment. For example, you cannot place a 230V rated motor on a 208V system. **Figure 110–5**

110.5 Copper Conductors. Where conductor material isn't specified in a rule, the material and the sizes given in the *Code* (and this textbook) are based on copper.

110.6 Conductor Sizes. Conductor sizes are expressed in American Wire Gage (AWG), typically from 18 AWG up to 4/0 AWG. Conductor sizes larger than 4/0 AWG are expressed in kcmil (thousand circular mils). **Figure 110–6**

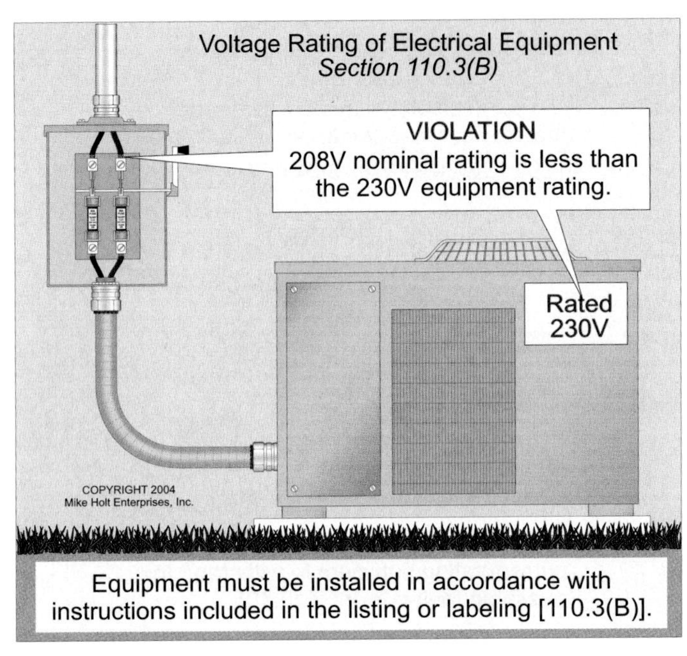
Voltage Rating of Electrical Equipment
Section 110.3(B)
VIOLATION
208V nominal rating is less than the 230V equipment rating.
Rated 230V
COPYRIGHT 2004
Mike Holt Enterprises, Inc.
Equipment must be installed in accordance with instructions included in the listing or labeling [110.3(B)].

Figure 110–5

110.7 Conductor Insulation. All wiring must be installed so as to be free from short circuits and ground faults.

Author's Comments:

- A ground fault is an unintentional electrical connection between an ungrounded or grounded neutral conductor and metallic enclosures, raceways, or equipment [250.2].
- Short circuits and ground faults often arise from insulation failure due to mishandling or improper installation. This happens when, for example, wire is dragged on a sharp edge, when insulation is scraped on boxes and enclosures, when wire is pulled hard, when insulation is nicked while being stripped, or when cable clamps and/or staples are installed too tightly.

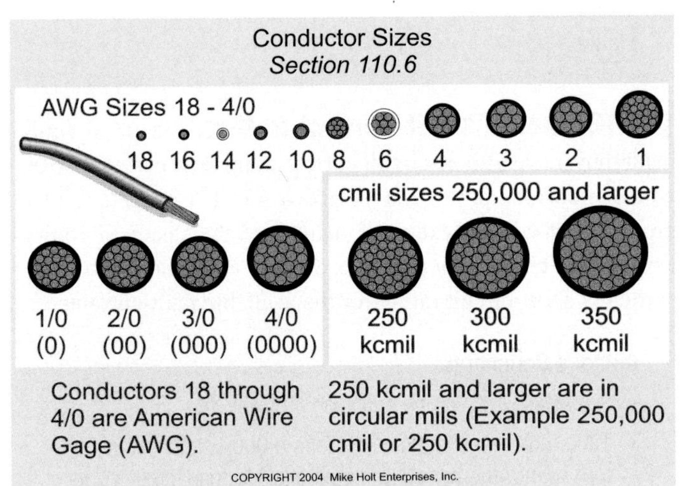
Conductor Sizes
Section 110.6
AWG Sizes 18 - 4/0
18 16 14 12 10 8 6 4 3 2 1
cmil sizes 250,000 and larger
1/0 (0) 2/0 (00) 3/0 (000) 4/0 (0000)
250 kcmil 300 kcmil 350 kcmil
Conductors 18 through 4/0 are American Wire Gage (AWG).
250 kcmil and larger are in circular mils (Example 250,000 cmil or 250 kcmil).
COPYRIGHT 2004 Mike Holt Enterprises, Inc.

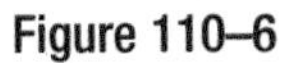
Figure 110–6

- To protect against accidental contact with energized conductors, the ends of abandoned conductors must be covered with an insulating device identified for the purpose, such as a twist-on or push-on wire connector [110.14(B)].
- See Article 100 for the definition of "Device."

110.8 Suitable Wiring Methods. Only wiring methods recognized as suitable are included in the *NEC*, and they must be installed in accordance with the *Code*.

Author's Comment: See Chapter 3 for power and lighting wiring methods, Chapter 7 for signaling circuits, and Chapter 8 for communications circuits. **Figure 110–7**

110.9 Interrupting Protection Rating. Overcurrent protection devices such as circuit breakers and fuses are intended to interrupt the circuit, and they must have an interrupting rating sufficient for the short-circuit current available at the line terminals of the equipment. **Figure 110–8**

Author's Comments:

- See Article 100 for the definition of "Interrupting Rating."
- Unless marked otherwise, the ampere interrupting rating for circuit breakers is 5,000A [240.83(C)], and for fuses is 10,000A [240.60(C)(3)]. **Figure 110–9**

Available Short-Circuit Current. Available short-circuit current is the current in amperes that is available at a given point in the electrical system. This available short-circuit current is first determined at the secondary terminals of the utility transformer.

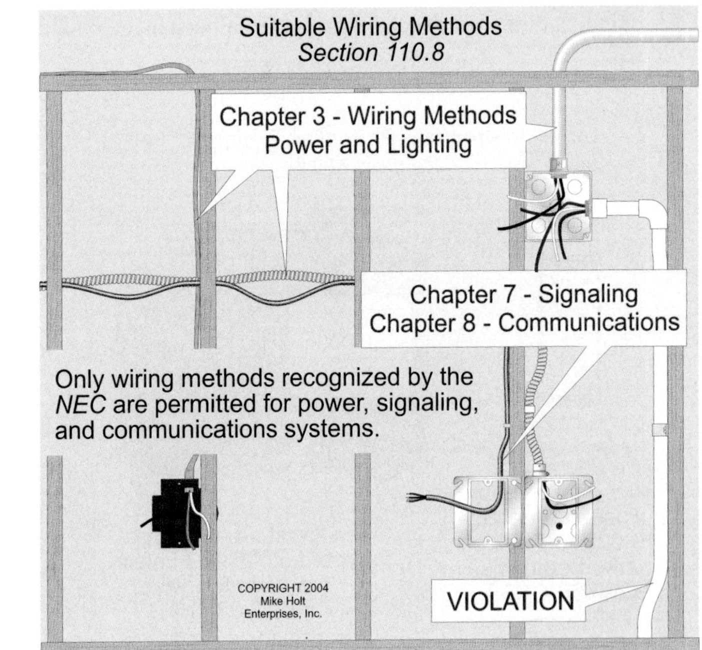
Suitable Wiring Methods
Section 110.8
Chapter 3 - Wiring Methods
Power and Lighting
Chapter 7 - Signaling
Chapter 8 - Communications
Only wiring methods recognized by the NEC are permitted for power, signaling, and communications systems.
COPYRIGHT 2004
Mike Holt
Enterprises, Inc.
VIOLATION

Figure 110–7

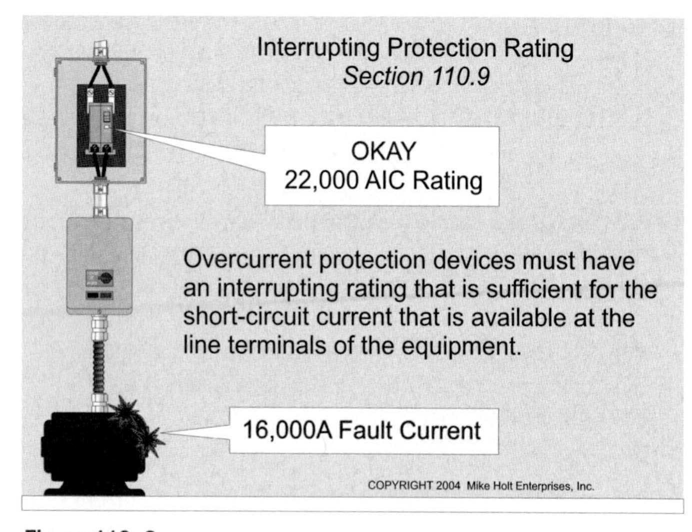
Interrupting Protection Rating
Section 110.9
OKAY
22,000 AIC Rating
Overcurrent protection devices must have an interrupting rating that is sufficient for the short-circuit current that is available at the line terminals of the equipment.
16,000A Fault Current
COPYRIGHT 2004 Mike Holt Enterprises, Inc.

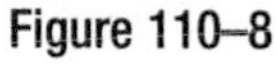
Figure 110–8

Thereafter, the available short-circuit current is calculated at the terminals of service equipment, then at branch-circuit panelboards and other equipment. The available short-circuit current is different at each point of the electrical system. It's highest at the utility transformer and lowest at the branch-circuit load.

The available short-circuit current depends on the impedance of the circuit, which increases moving downstream from the utility transformer. The greater the circuit impedance (utility transformer and the additive impedances of the circuit conductors), the lower the available short-circuit current. **Figure 110–10**

Factors that impact the available short-circuit current at the utility transformer include the system voltage, the transformer kVA rating, and its impedance (expressed in a percentage on the

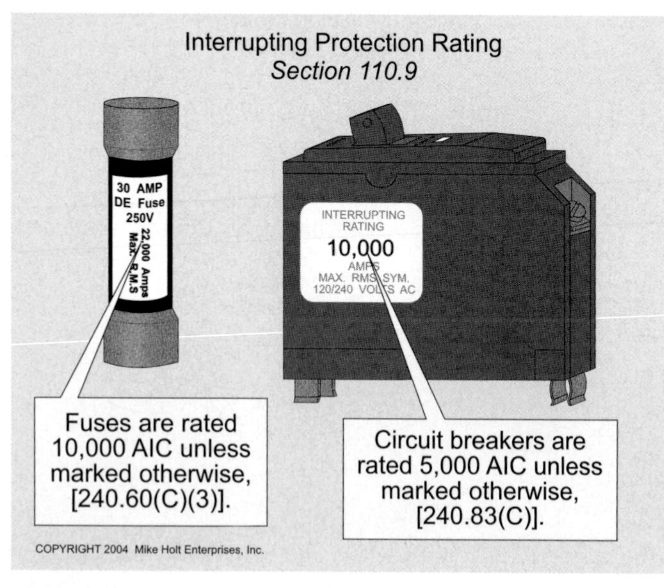
Interrupting Protection Rating
Section 110.9
30 AMP
DE Fuse
250V
22,000 Amps
Max. R.M.S
INTERRUPTING
RATING
10,000
AMPS
MAX. RMS SYM.
120/240 VOLTS AC
Fuses are rated 10,000 AIC unless marked otherwise, [240.60(C)(3)].
Circuit breakers are rated 5,000 AIC unless marked otherwise, [240.83(C)].
COPYRIGHT 2004 Mike Holt Enterprises, Inc.

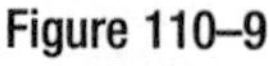
Figure 110–9

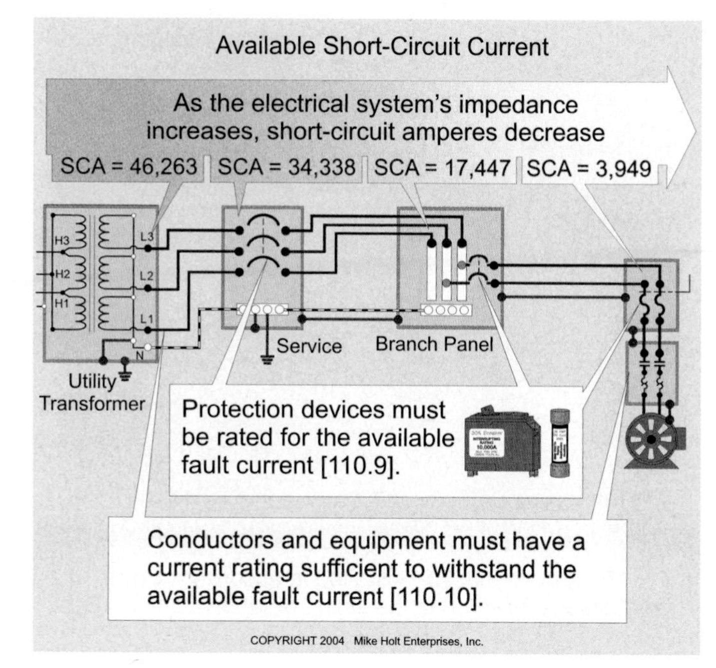
Available Short-Circuit Current
As the electrical system's impedance increases, short-circuit amperes decrease
SCA = 46,263
SCA = 34,338
SCA = 17,447
SCA = 3,949
H3
H2
H1
L3
L2
L1
N
Utility Transformer
Service
Branch Panel
Protection devices must be rated for the available fault current [110.9].
Conductors and equipment must have a current rating sufficient to withstand the available fault current [110.10].
COPYRIGHT 2004 Mike Holt Enterprises, Inc.

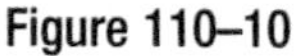
Figure 110–10

equipment nameplate). Properties that impact the impedance of the circuit include the conductor material (copper versus aluminum), conductor size, and conductor length.

> **Author's Comment:** Many in the industry describe Amperes Interrupting Rating (AIR) as "Amperes Interrupting Capacity" (AIC).

> **DANGER:** *Extremely high values of current flow (caused by short circuits or ground faults) produce tremendously destructive thermal and magnetic forces. If the circuit overcurrent protection device isn't rated to interrupt the current at the available fault values at its listed voltage rating, it could explode while attempting to clear a fault. Naturally this can cause serious injury or death, as well as property damage.* **Figure 110–11**

110.10 Short-Circuit Current Rating. Electrical equipment must have a short-circuit current rating that permits the circuit overcurrent protection device to clear a short circuit or ground fault without extensive damage to the electrical components of the circuit. For example, a motor controller must have a sufficient short-circuit rating for the available fault-current.

> **Author's Comments:**
>
> - See Article 100 for the definition of "Controller."
> - If the fault exceeds the controller's 5,000A short-circuit current rating, the controller could explode, endangering persons and property. **Figure 110–12**

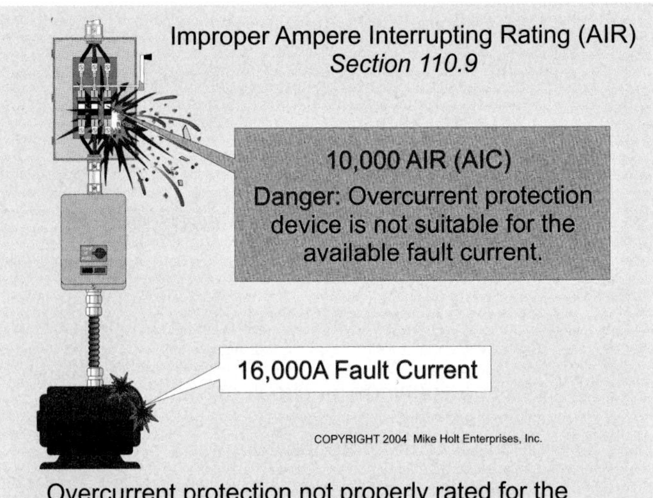

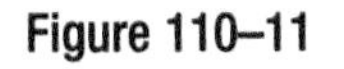
Figure 110–11

To solve this problem, a current-limiting protection device (fast-clearing fuse) can be used to reduce the let-through current to less than 5,000A. **Figure 110–13**

Author's Comment: For more information on the application of current limiting devices, see 240.2 and 240.60(B).

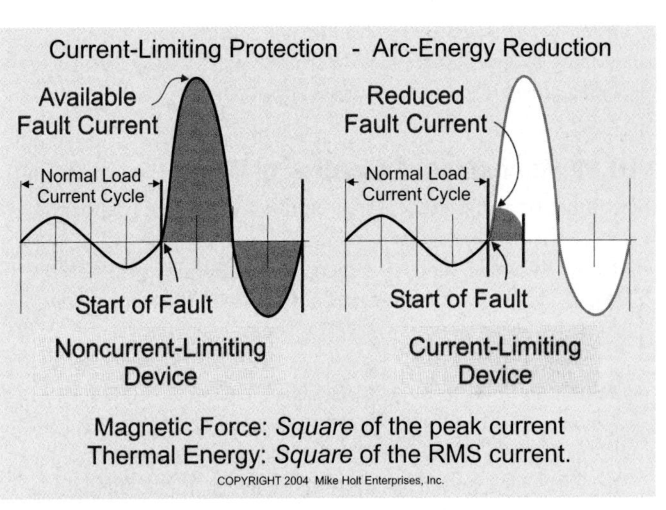

Figure 110–13

110.11 Deteriorating Agents.

Electrical equipment and conductors must be suitable for the environment and conditions of use, and consideration must be given to the presence of corrosive gases, fumes, vapors, liquids, or chemicals that can have a deteriorating effect on the conductors or equipment. **Figure 110–14**

Author's Comment: Conductors must not be exposed to ultraviolet rays from the sun unless identified for the purpose [310.8(D)].

FPN No. 1: Raceways, cable trays, cablebus, auxiliary gutters, cable armor, boxes, cable sheathing, cabinets, elbows, couplings, fittings, supports, and support hardware must be of materials suitable for the environment in which they are to be installed, in accordance with 300.6.

Author's Comment: See Article 100 for the definition of "Raceway."

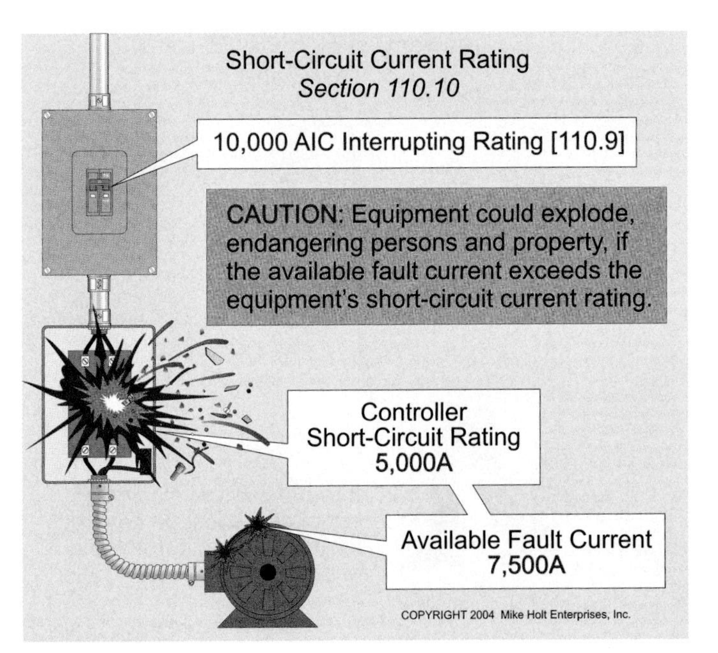

Figure 110–12

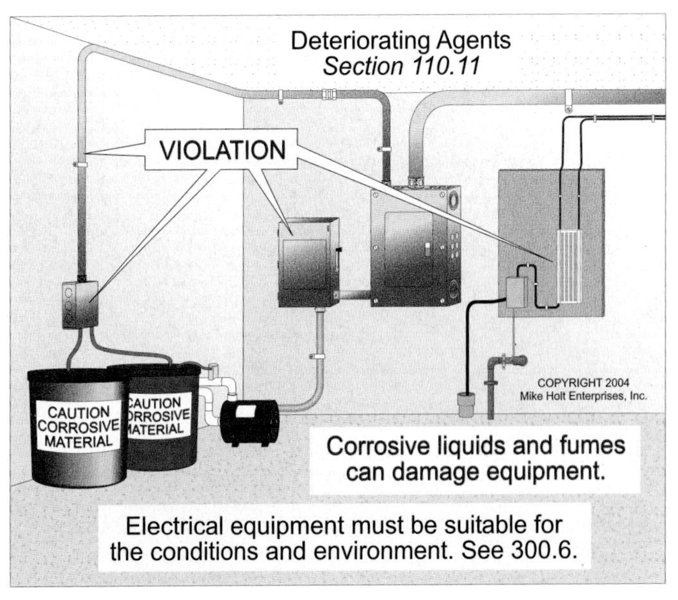

Figure 110–14

FPN No. 2: Some spray cleaning and lubricating compounds contain chemicals that can deteriorate plastic used for insulating and structural applications in equipment.

110.12 Mechanical Execution of Work.

Electrical equipment must be installed in a neat and workmanlike manner.

FPN: Accepted industry practices are described in ANSI/NECA 1-2000, *Standard Practices for Good Workmanship in Electrical Contracting.*

Author's Comment: The National Electrical Contractors Association (NECA) has created a series of National Electrical Installation Standards (NEIS)™ that establish the industry's first quality guidelines for electrical installations. These standards define a benchmark or baseline of quality and workmanship for installing electrical products and systems. They explain what installing electrical products and systems in a "neat and workmanlike manner" means. For more information about these standards, visit http://www.neca-neis.org/.

(A) Unused Openings. Unused cable or raceway openings in electrical equipment must be effectively closed by fittings that provide protection substantially equivalent to the wall of the equipment. **Figure 110–15**

Author's Comments:

- See Article 100 for the definition of "Fitting."
- Unused openings for circuit breakers must be closed using identified closures, or other means approved by the authority having jurisdiction that provide protection substantially equivalent to the wall of the enclosure [408.7]. **Figure 110–16**

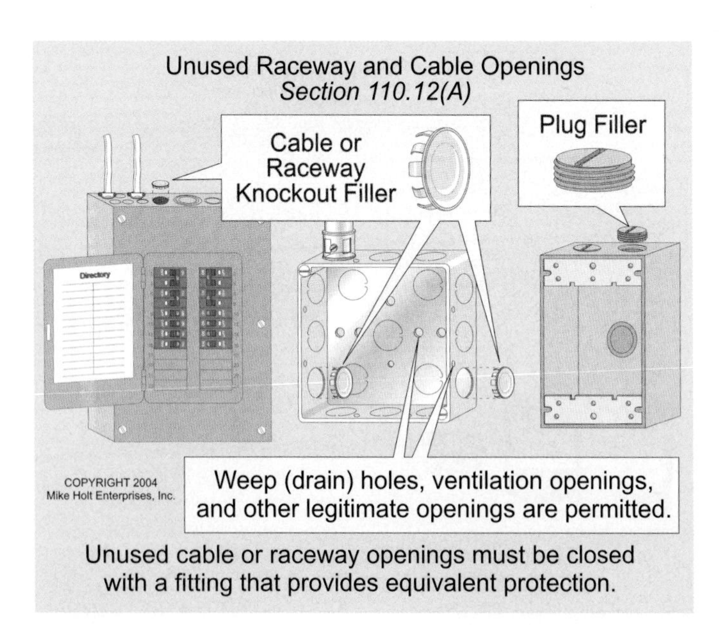

Figure 110–15

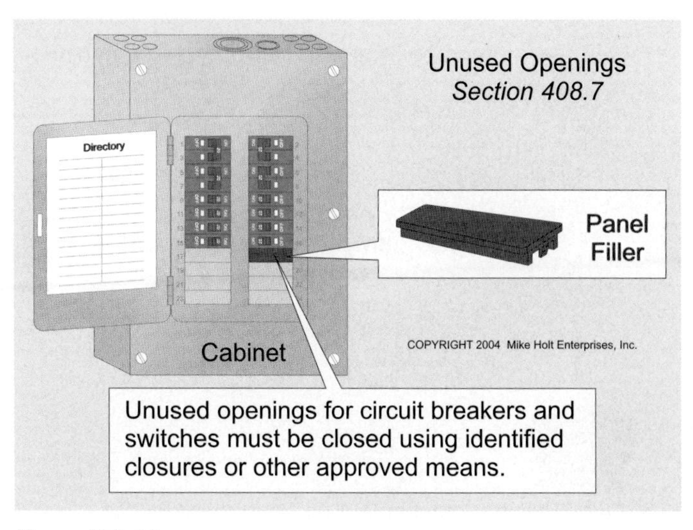

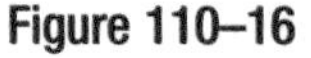
Figure 110–16

(C) Integrity of Electrical Equipment. Internal parts of electrical equipment must not be damaged or contaminated by foreign material, such as paint, plaster, cleaners, etc.

Author's Comment: Precautions must be taken to provide protection from contaminating the internal parts of panelboards and receptacles during the building construction. **Figure 110–17**

Electrical equipment that contains damaged parts that may adversely affect safe operation or mechanical strength of the equipment must not be installed. This would include parts that are broken, bent, cut, or deteriorated by corrosion, chemical action, or overheating.

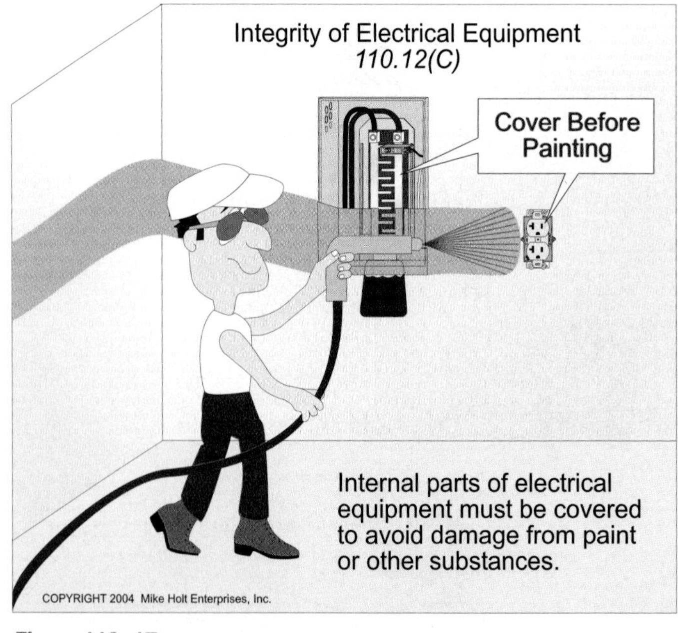

Figure 110–17

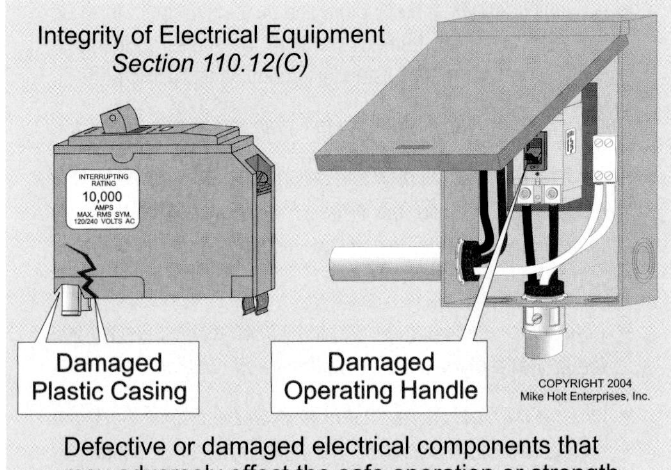
Integrity of Electrical Equipment
Section 110.12(C)
INTERRUPTING RATING
10,000
AMPS
MAX. RMS SYM.
120/240 VOLTS AC
Damaged Plastic Casing
Damaged Operating Handle
COPYRIGHT 2004
Mike Holt Enterprises, Inc.
Defective or damaged electrical components that may adversely effect the safe operation or strength of the equipment must be replaced.

Figure 110–18

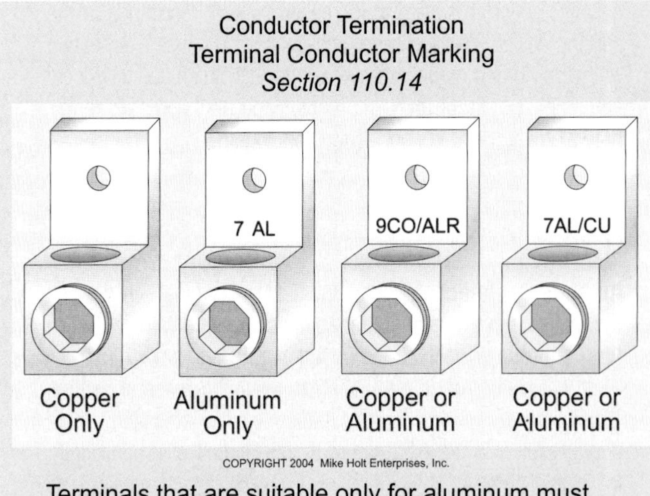
Conductor Termination
Terminal Conductor Marking
Section 110.14
7 AL
9CO/ALR
7AL/CU
Copper Only
Aluminum Only
Copper or Aluminum
Copper or Aluminum
COPYRIGHT 2004 Mike Holt Enterprises, Inc.
Terminals that are suitable only for aluminum must be marked AL. Terminals suitable for both copper and aluminum must be marked CO/ALR or AL/CU.

Figure 110–19

Author's Comment: Damaged parts include anything broken, bent, cut, or deteriorated by corrosion or chemical action, or by overheating, that may adversely affect the safe operation or the mechanical strength of the equipment. This includes cracked insulators, arc shields not in place, overheated fuse clips, and damaged or missing switch or circuit breaker handles. **Figure 110–18**

110.13 Mounting and Cooling of Equipment.

(A) Mounting. Electrical equipment must be firmly secured to the surface on which it's mounted. See 314.23(A)

(B) Cooling. Electrical equipment that depends on natural air circulation must be installed so that walls or equipment do not prevent airflow over the surfaces. The clearances between top surfaces and side surfaces must be maintained to dissipate rising warm air for equipment designed for floor mounting.

Electrical equipment that is constructed with ventilating openings must be installed so that free air circulation isn't inhibited.

Author's Comment: Transformers with ventilating openings must be installed so that the ventilating openings aren't blocked, and the required wall clearances are clearly marked on the transformer case [450.9].

110.14 Conductor Termination. Terminal Conductor Material. Conductor terminal and splicing devices must be identified for the conductor material and they must be properly installed and used. Devices that are suitable only for aluminum must be marked AL, and devices that are suitable for both copper and aluminum must be marked CO/ALR [404.14(C) and 406.2(C)]. **Figure 110–19**

Author's Comments:

- See Article 100 for the definition of "Identified."
- Existing inventories of equipment or devices might be marked AL/CU to indicate a terminal suitable for both copper and aluminum conductors.
- Conductor terminations must comply with manufacturer's instructions as required by 110.3(B). For example, if the instructions for the device state "Suitable for 18-2 AWG Stranded," then only stranded conductors can be used with the terminating device. If the instructions state "Suitable for 18-2 AWG Solid," then only solid conductors are permitted, and if the instructions state "Suitable for 18-2 AWG," then either solid or stranded conductors can be used with the terminating device.

Aluminum: To reduce the contact resistance between the aluminum conductor and the terminal, terminals listed for aluminum conductors are often filled with an antioxidant gel.

Copper: Some terminal manufacturers sell a compound intended to reduce corrosion and heat at copper conductor terminations that is especially helpful at high-amperage terminals. This compound is messy, but apparently it's effective.

Copper and Aluminum Mixed: Copper and aluminum conductors must not make contact with each other in a device unless the device is listed and identified for this purpose.

Author's Comment: Few terminations are listed for the mixing of aluminum wire and copper, but if they are, they will be marked on the product package or terminal device. The reason copper and aluminum should not be in contact with each other is because corrosion will develop between the two different metals due to galvanic action, resulting in increased contact resistance at the splicing device. This increased resistance can cause overheating of the splice and cause a fire. See http://tis-hq.eh.doe.gov/docs/sn/nsh9001.html for more information on how to properly terminate aluminum and copper conductors together.

FPN: Many terminations and equipment are marked with a tightening torque.

Author's Comment: All conductors must terminate in devices that have been properly tightened in accordance with manufacturer's torque specifications included with equipment instructions. Failure to torque terminals can result in excessive heating of terminals or splicing devices (due to loose connection), which could result in a fire because of a short circuit or ground fault. In addition, this is a violation of 110.3(B), which requires all equipment to be installed in accordance with listed or labeling instructions. **Figure 110–20**

Question: *What do you do if the torque value isn't provided with the device?*

Answer: *Call the manufacturer, visit the manufacturer's website, or have the supplier make a copy of the installation instructions.*

Author's Comment: Terminating conductors without a torque tool can result in an improper and unsafe installation. If a torque screwdriver is not used, there's a good chance the conductors are not properly terminated.

(A) Terminations. Conductor terminals must ensure a good connection without damaging the conductors and must be made by pressure connectors (including set-screw type) or splices to flexible leads.

Author's Comments:

- See Article 100 for the definition of "Pressure Connector."
- Grounding (earthing) conductors and bonding jumpers must be connected by exothermic welding, pressure connectors, clamps, or other means listed for grounding (earthing) [250.8].

Question: *What if the wire is larger than the terminal device?*

Answer: *This condition needs to be anticipated in advance, and the equipment should be ordered with terminals that will accommodate the larger wire. However, if you're in the field, you should:*

- *Contact the manufacturer and have them express deliver you the proper terminals, bolts, washers and nuts, or*
- *Order a terminal device that crimps on the end of the larger conductor and reduces the termination size, or splice the conductors to a smaller wire.*

One Wire Per Terminal: Terminals for more than one wire must be identified for this purpose, either within the equipment instructions or on the terminal itself. **Figure 110–21**

Author's Comment: Split-bolt connectors are commonly listed for only two conductors although some are listed for three conductors. However, it's a common industry practice to terminate as many conductors as possible within a split-bolt connector, even though this violates the *NEC*. **Figure 110–22**

Split-bolt connectors for aluminum-to-aluminum or aluminum-to-copper conductors must be identified as suitable for the application.

(B) Conductor Splices. Conductors must be spliced by a splicing device identified for the purpose or by exothermic welding.

Author's Comment: Conductors are not required to be twisted together prior to the installation of a twist-on wire connector. **Figure 110–23**

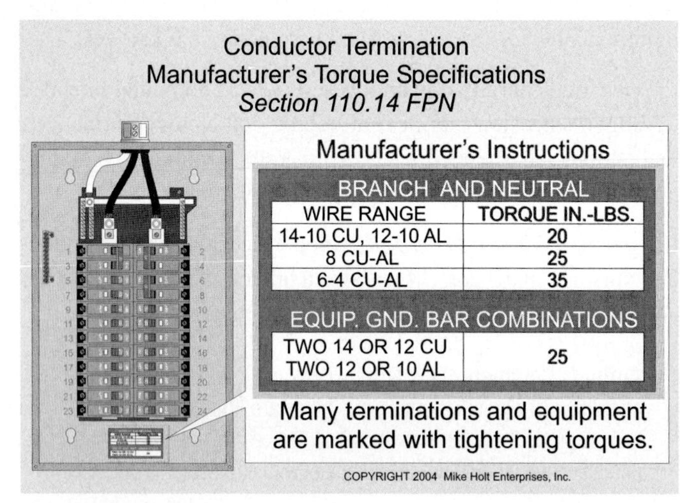

Figure 110–20

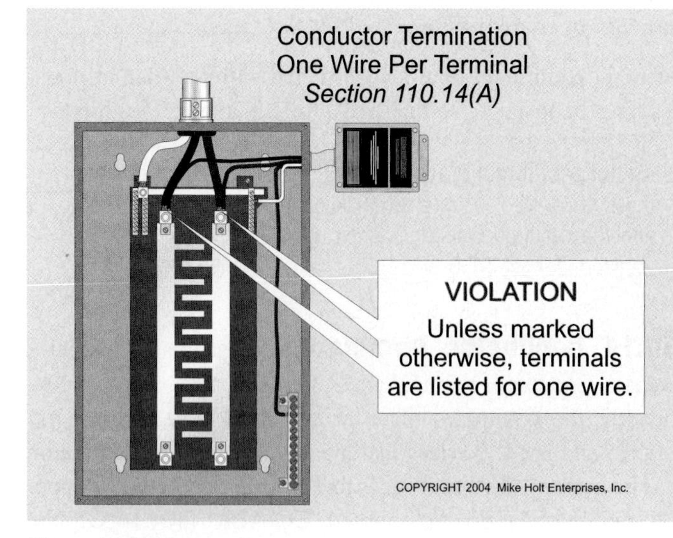

Figure 110–21

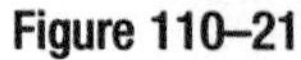

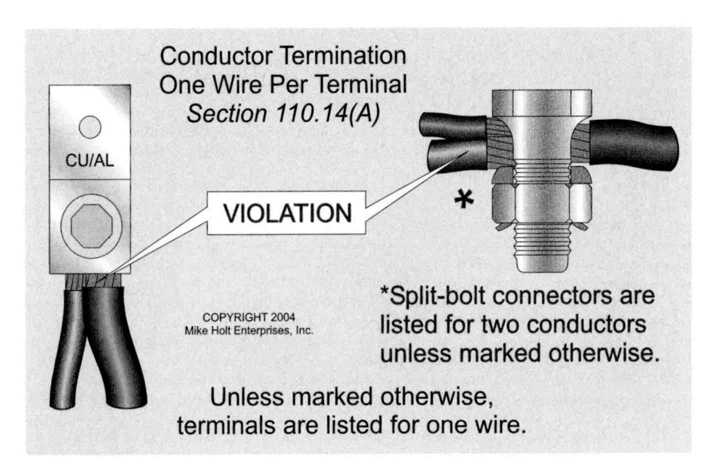
Conductor Termination
One Wire Per Terminal
Section 110.14(A)
CU/AL
VIOLATION
*
*Split-bolt connectors are listed for two conductors unless marked otherwise.
COPYRIGHT 2004
Mike Holt Enterprises, Inc.
Unless marked otherwise, terminals are listed for one wire.

Figure 110–22

Splices, joints, and the free ends of all conductors must be covered with insulation that is equal to that of the conductor.

Author's Comments:

- Circuit conductors not being used are not required to be removed. However, to prevent an electrical hazard, the free ends of the conductors must be insulated to prevent the exposed end of the conductor from touching energized parts. This requirement can be met by the use of an insulated twist-on or push-on wire connector. **Figure 110–24**
- See Article 100 for the definition of "Energized."

Underground Splices: Single Conductors: Single direct burial conductors of Type UF or USE can be spliced underground without a junction box, but the conductors must be spliced with a device that is listed for direct burial. See 300.5(E) and 300.15(G). **Figure 110–25**

Multiconductor Cable: Multiconductor Type UF or Type USE cable can have the individual conductors spliced underground with a listed splice kit that encapsulates the conductors and the cable jacket.

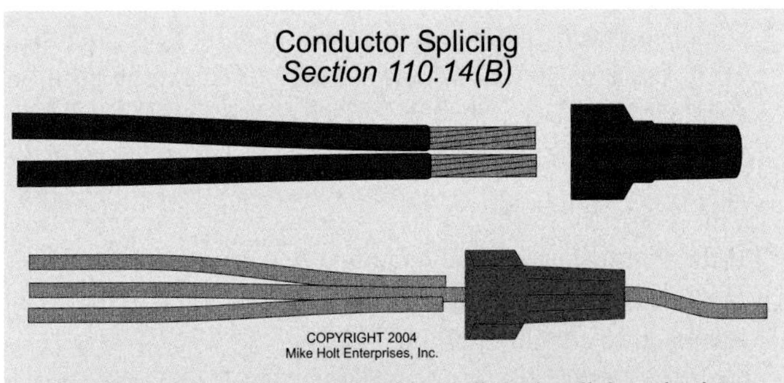
Conductor Splicing
Section 110.14(B)
COPYRIGHT 2004
Mike Holt Enterprises, Inc.
Conductors must be spliced by a listed splicing device and they are not required to be twisted together prior to the installation of a twist-on wire connector.

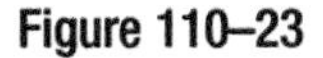

Figure 110–23

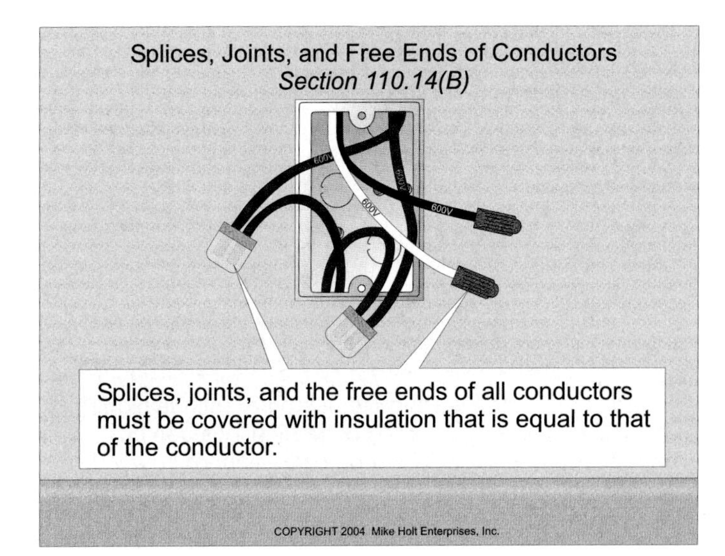
Splices, Joints, and Free Ends of Conductors
Section 110.14(B)
600V
Splices, joints, and the free ends of all conductors must be covered with insulation that is equal to that of the conductor.
COPYRIGHT 2004 Mike Holt Enterprises, Inc.

Figure 110–24

(C) Temperature Limitations (Conductor Size). Conductors are to be sized to the lowest temperature rating of any terminal, device, or conductor of the circuit in accordance with (1) for terminals of equipment, and (2) for independent pressure connectors on a bus.

Conductor Ampacity. Conductors with insulation temperature ratings higher than the termination's temperature rating can be used for conductor ampacity adjustment, correction, or both.

Author's Comments:

- See Article 100 for the definition of "Ampacity."
- This means that conductor ampacity must be based on the conductor's insulation temperature ratings listed in Table 310.16, as adjusted for ambient temperature correction factors, conductor bundling adjustment factors, or both. This

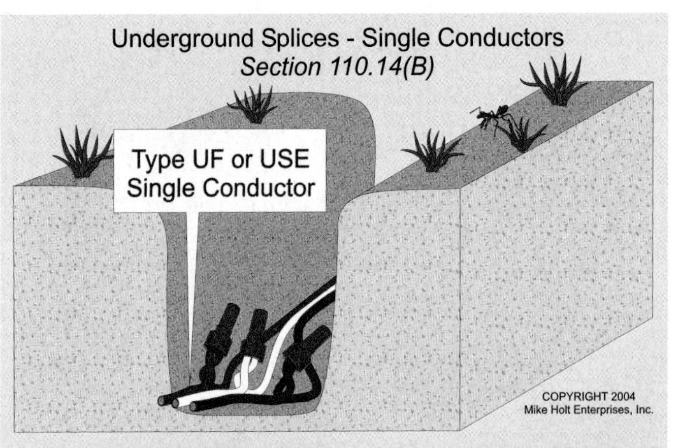
Underground Splices - Single Conductors
Section 110.14(B)
Type UF or USE
Single Conductor
COPYRIGHT 2004
Mike Holt Enterprises, Inc.
Single Type UF or USE conductors can be spliced underground with a device that is listed for direct burial.

Figure 110–25

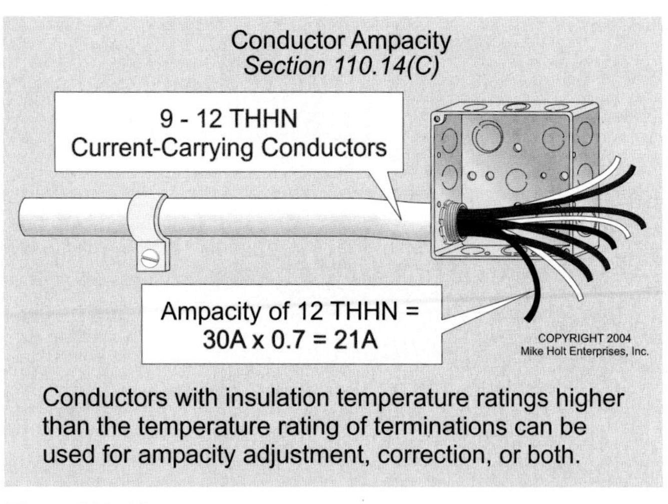

Conductor Ampacity
Section 110.14(C)
9 - 12 THHN
Current-Carrying Conductors
Ampacity of 12 THHN =
30A x 0.7 = 21A
COPYRIGHT 2004
Mike Holt Enterprises, Inc.
Conductors with insulation temperature ratings higher than the temperature rating of terminations can be used for ampacity adjustment, correction, or both.

Figure 110–26

means that conductor ampacity, when required to be adjusted, is based on the conductor insulation temperature rating in accordance with Table 310.16. For example, the ampacity of each 12 THHN conductor is 30A, based on the values listed in the 90°C column of Table 310.16.

If we bundle nine current-carrying 12 THHN conductors in the same raceway or cable, the ampacity for each conductor (30A at 90°C, Table 310.16) needs to be adjusted by a 70 percent adjustment factor [Table 310.15(B)(2)(a)]. **Figure 110–26**

Adjusted Conductor Ampacity = 30A x 0.70
Adjusted Conductor Ampacity = 21A

See *necdigest* magazine, winter 2003 issue, page 32, and the *NEC Handbook*, 310.15(B)(2)(a) Ex. 5, for examples of 90°C ampacity for conductor ampacity adjustment.

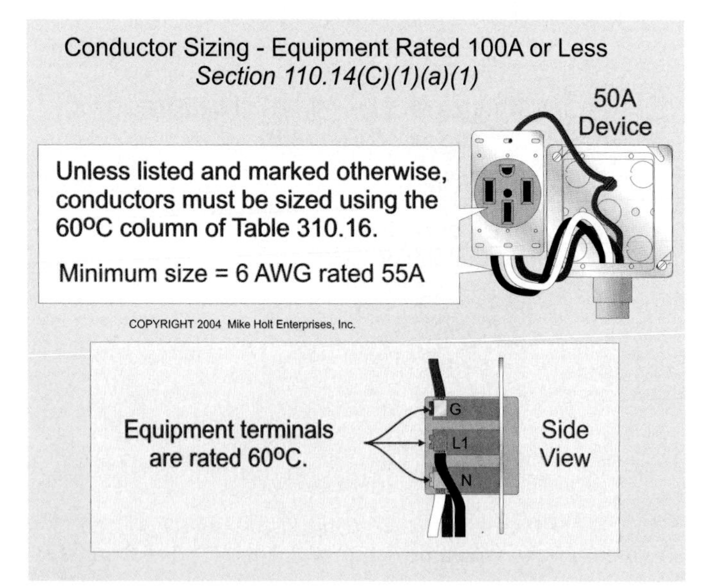

Figure 110–27

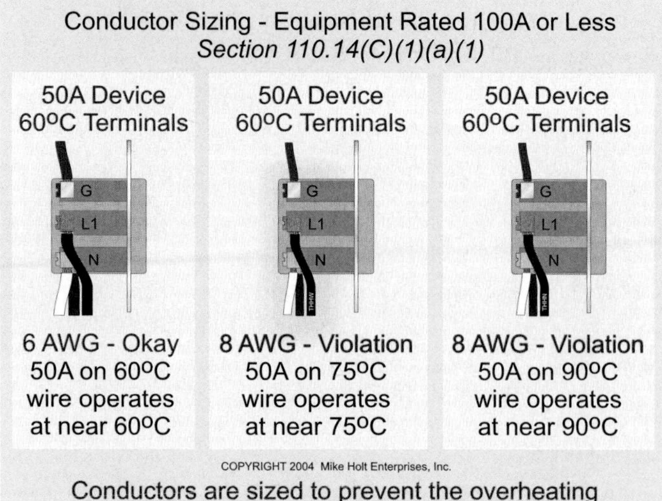

Figure 110–28

(1) Equipment Provisions. Unless the equipment is listed and marked otherwise, conductor sizing for equipment termination must be based on Table 310.16 in accordance with (a) or (b):

(a) Equipment Rated 100A and Less.

(1) Conductor sizing for equipment rated 100A or less must be sized using the 60°C temperature column of Table 310.16. **Figure 110–27**

Author's Comment: Conductors are sized to prevent the overheating of terminals, in accordance with listing standards. For example, a 50A circuit with 60°C terminals requires the circuit conductors to be sized not smaller than 6 AWG, in accordance with the 60°C ampacity listed in Table 310.16. However, an 8 THHN insulated conductor has a 90°C ampacity of 50A, but 8 AWG cannot be used for this circuit because the conductor's operating temperature at full-load ampacity (50A) will be near 90°C, which is well in excess of the 60°C terminal rating. **Figure 110–28**

(2) Conductors with an insulation temperature rating greater than 60°C, such as THHN, which is rated 90°C, can be used on terminals that are rated 60°C, but the conductor must be sized based on the 60°C temperature column of Table 310.16. **Figure 110–29A**

(3) If the terminals are listed and identified as suitable for 75°C, then conductors rated at least 75°C can be sized to the 75°C temperature column of Table 310.16. **Figure 110–29B**

(4) For motors marked with design letters B, C, or D, conductors having an insulation rating of 75°C or higher can be used provided the ampacity of such conductors doesn't exceed the 75°C ampacity.

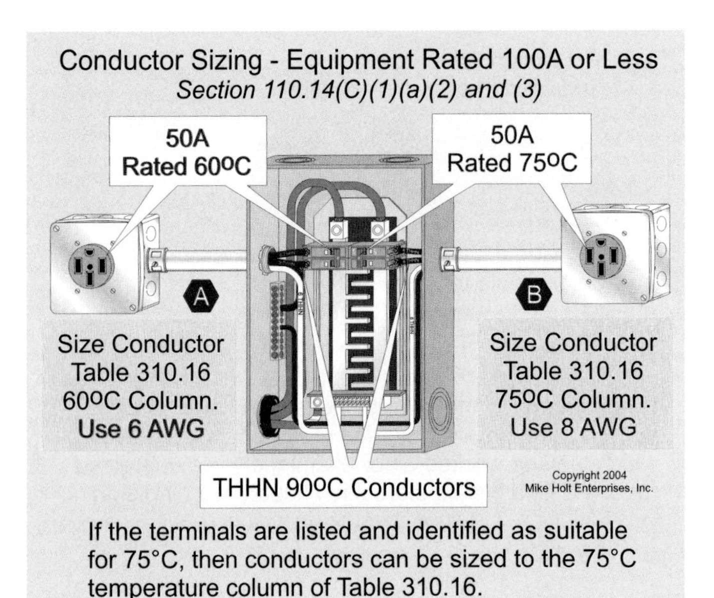

Figure 110–29

(b) Equipment Rated Over 100A.

(1) Conductors for equipment rated over 100A must be sized based on the 75°C temperature column of Table 310.16. **Figure 110–30**

(2) Conductors with an insulation temperature rating greater than 75°C can be used on terminals that are rated 75°C, but the conductor must be sized based on the 75°C temperature column of Table 310.16. **Figure 110–31**

(2) Separate Connector Provisions. Conductors can be sized to the 90°C ampacity rating of THHN, if the conductor terminates to a bus connector that is rated 90°C. **Figure 110–32**

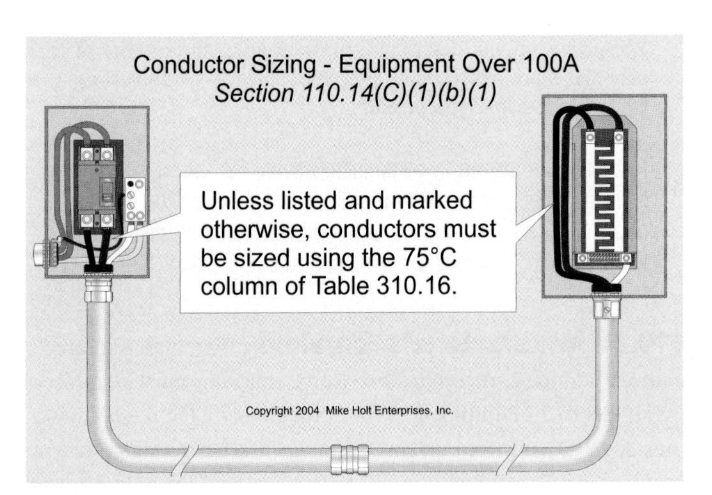

Figure 110–30

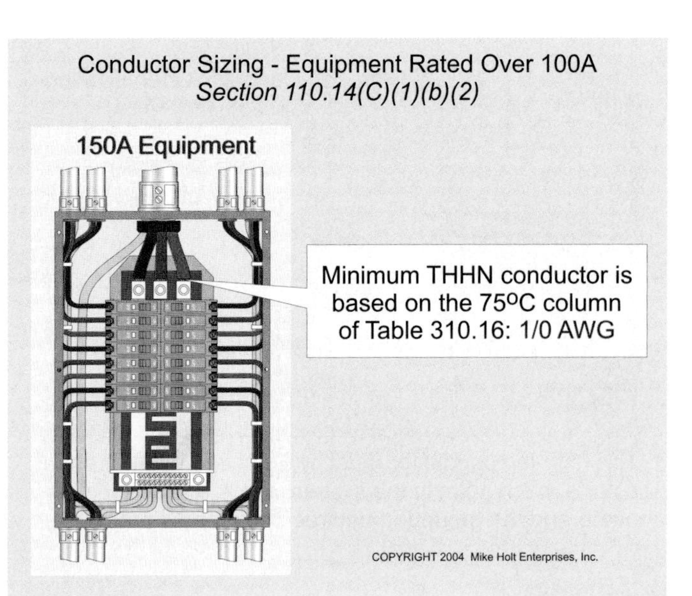

Conductor Sizing - Equipment Rated Over 100A
Section 110.14(C)(1)(b)(2)
150A Equipment
Minimum THHN conductor is based on the 75ºC column of Table 310.16: 1/0 AWG
COPYRIGHT 2004 Mike Holt Enterprises, Inc.

Figure 110–31

110.15 High-Leg Conductor Identification.

Identification. On a 4-wire three-phase delta-connected system, where the midpoint of one phase winding is grounded, the conductor with the higher phase voltage-to-ground (208V) must be durably and permanently marked by an outer finish that is orange in color, or other effective means. Such identification must be placed at each point on the system where a connection is made if the grounded neutral conductor is present [110.15, 215.8, and 230.56]. **Figure 110–33**

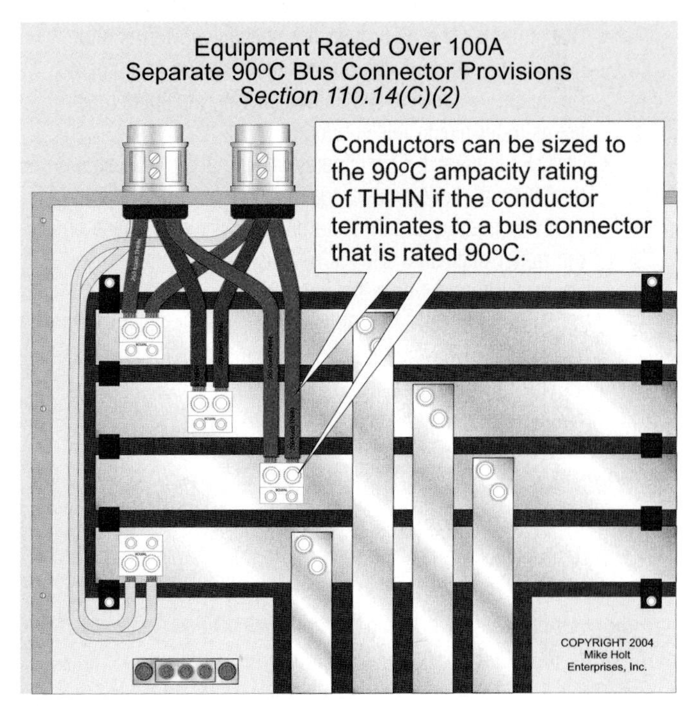

Figure 110–32

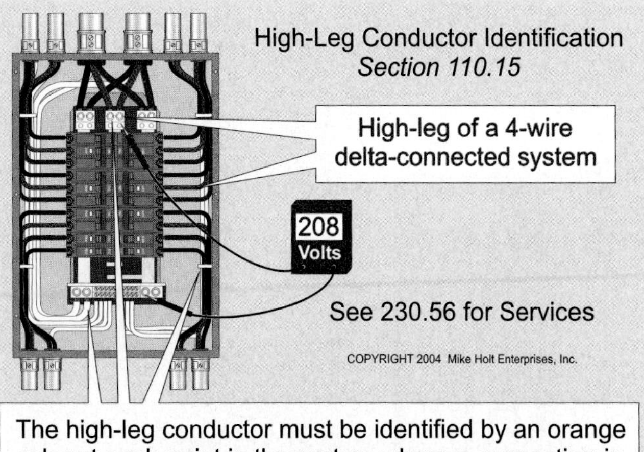

Figure 110–33

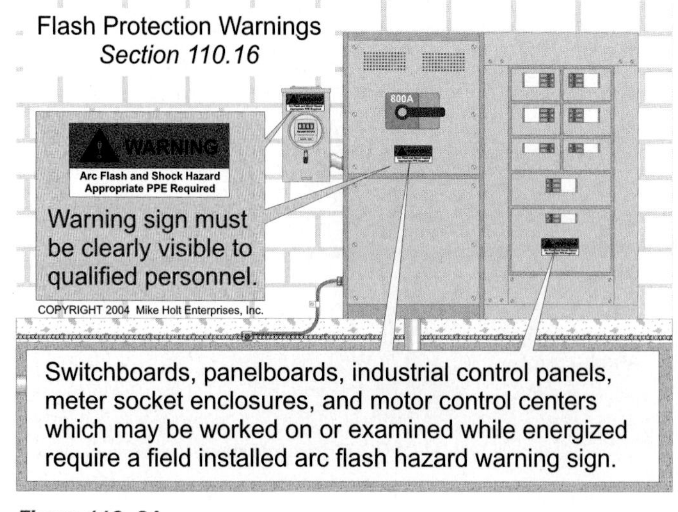

Flash Protection Warnings
Section 110.16
WARNING
Arc Flash and Shock Hazard
Appropriate PPE Required
800A
Warning sign must be clearly visible to qualified personnel.
COPYRIGHT 2004 Mike Holt Enterprises, Inc.
Switchboards, panelboards, industrial control panels, meter socket enclosures, and motor control centers which may be worked on or examined while energized require a field installed arc flash hazard warning sign.

Figure 110–34

Author's Comments:

- The high-leg conductor is also called the "wild-leg," "stinger-leg," or "bastard-leg."
- Other important *NEC* rules relating to the high-leg are as follows:
  - Panelboards. Since 1975, panelboards supplied by a 4-wire three-phase delta-connected system must have the high-leg conductor (208V) terminate to the "B" (center) phase of a panelboard [408.3(E)].

    An exception to 408.3(E) permits the high-leg conductor to terminate to the "C" phase when the meter is located in the same section of a switchboard or panelboard.
  - Disconnects. The *NEC* does not specify the termination location for the high-leg conductor in switch equipment (Switches—Article 404), but the generally accepted practice is to terminate this conductor to the "B" phase.
  - Utility Equipment: It's my understanding that the ANSI standard for meter equipment requires the high-leg conductor (208V-to-neutral) to terminate on the "C" (right) phase of the meter enclosure. This is because the demand meter needs 120V and it gets this from the "B" phase.

    Also hope the utility lineman is not color blind and doesn't inadvertently cross the "orange" high-leg conductor (208V) with the red (120V) service conductor at the weatherhead. It's happened before…

WARNING: ***When replacing equipment in existing facilities that contain a high-leg conductor, care must be taken to ensure that the high-leg conductor is replaced in the original location. Prior to 1975, the high-leg conductor was required to terminate on the "C" phase of panelboards and switchboards. Failure to re-terminate the high-leg in accordance with the existing installation can result in 120V circuits inadvertently connected to the 208V high-leg, with disastrous results.***

110.16 Flash Protection Warning. Switchboards, panelboards, industrial control panels, meter socket enclosures, and motor control centers in commercial and industrial occupancies that are likely to require examination, adjustment, servicing, or maintenance while energized must be field marked to warn qualified persons of the danger associated with an arc flash from line-to-line or ground faults. The field marking must be clearly visible to qualified persons before they examine, adjust, service, or perform maintenance on the equipment. **Figure 110–34**

Author's Comments:

- See Article 100 for the definitions of "Panelboard" and "Qualified Persons."
- This rule is meant to warn qualified persons who work on energized electrical systems that an arc flash hazard exists so they will select proper personal protective equipment (PPE) in accordance with industry accepted safe work practice standards.

FPN No. 1: NFPA 70E, *Standard for Electrical Safety in the Workplace*, provides assistance in determining the severity of potential exposure, planning safe work practices, and selecting personal protective equipment.

Author's Comment: In some installations, the use of current-limiting protection devices may significantly reduce the degree of arc flash hazards. For more information about flash protection, visit http://bussmann.com/safetybasics. **Figure 110–35**

110.21 Manufacturer's Markings. The manufacturer's name, trademark, or other descriptive marking must be placed on all electrical equipment. Where required by the *Code*, markings such as voltage, current, wattage, or other ratings must be provided with sufficient durability to withstand the environment involved.

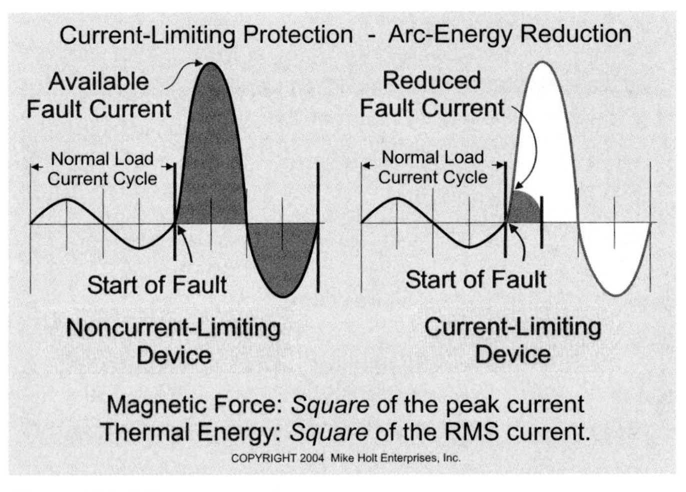

Figure 110–35

110.22 Identification of Disconnecting Means. All installed disconnecting means must be legibly marked to indicate their purpose unless they are located and arranged so their purpose is evident. In addition, the marking must be of sufficient durability to withstand the environment involved. **Figure 110–36**

Author's Comment: See Article 100 for the definition of "Disconnecting Means."

PART II. 600V, NOMINAL OR LESS

110.26 Spaces About Electrical Equipment. For the purpose of safe operation and maintenance of equipment, sufficient access and working space must be provided. Enclosures housing electrical apparatus that are controlled by locks are considered accessible to qualified persons who require access. **Figure 110–37**

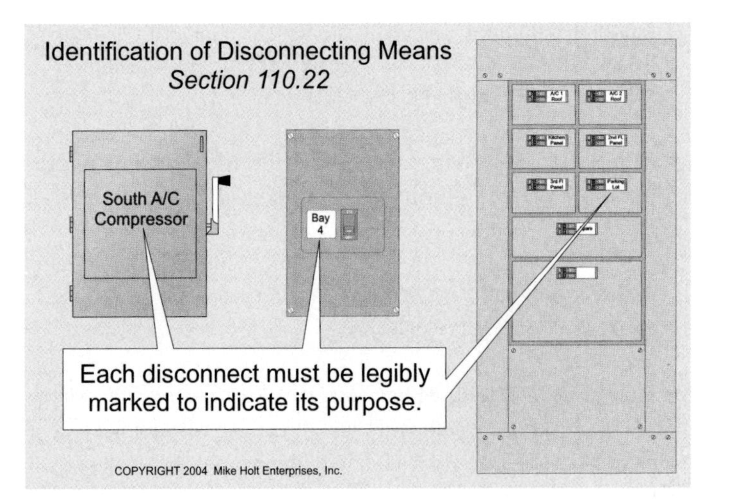

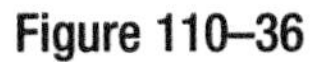
Figure 110–36

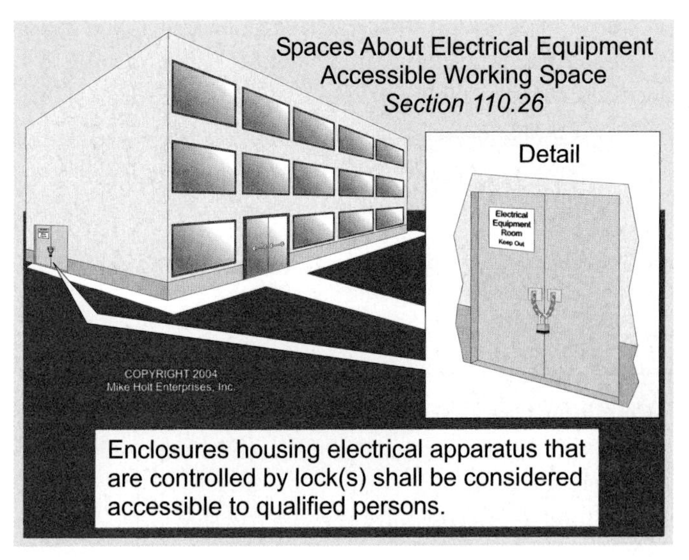

Figure 110–37

Author's Comments:

- See Article 100 for the definition of "Accessible" as it applies to equipment.
- It might be unwise to use an electrically operated lock, if it locks in the de-energized condition!

(A) Working Space. Working space for equipment that may need examination, adjustment, servicing, or maintenance while energized must have sufficient working space in accordance with (1), (2), and (3):

Author's Comment: The phrase "while energized" is the root of many debates. Since electric power to almost all equipment can be turned off, one could argue that working space is never required!

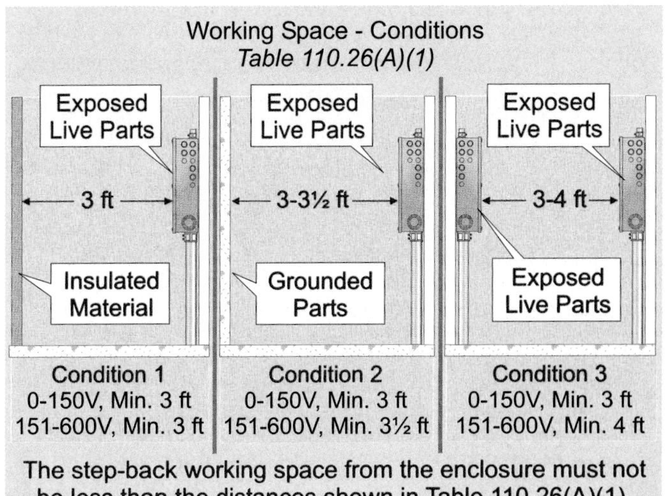

Figure 110–38

(1) Depth of Working Space. The step-back working space, measured from the enclosure front must not be less than the distances contained in Table 110.26(A)(1). **Figure 110–38**

Table 110.26(A)(1) Step-Back Working Space

Voltage-to-Ground	Condition 1	Condition 2	Condition 3
0–150V	3 ft	3 ft	3 ft
151–600V	3 ft	3½ ft	4 ft

- **Condition 1**—Exposed live parts on one side of the working space and no live or grounded parts on the other side of the working space.
- **Condition 2**—Exposed live parts on one side of the working space and grounded parts on the other side of the working space. For this table, concrete, brick, or tile walls are considered grounded.
- **Condition 3**—Exposed live parts on both sides of the working space.

(a) Rear and Sides. Step-back working space isn't required for the back or sides of assemblies where all connections are accessible from the front. **Figure 110–39**

(b) Low Voltage. Where special permission is granted in accordance with 90.4, working space for equipment that operates at not more than 30V ac or 60V dc can be smaller than the distance in Table 110.26(A)(1). **Figure 110–40**

Author's Comment: See Article 100 for the definition of "Special Permission."

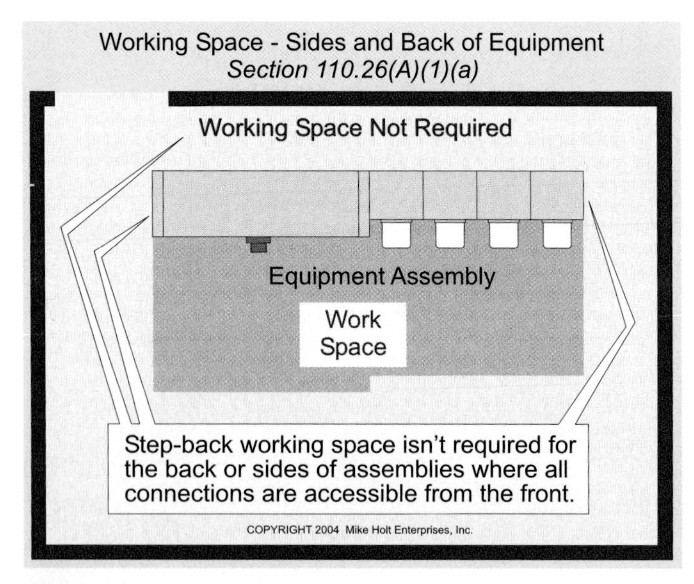

Figure 110–39

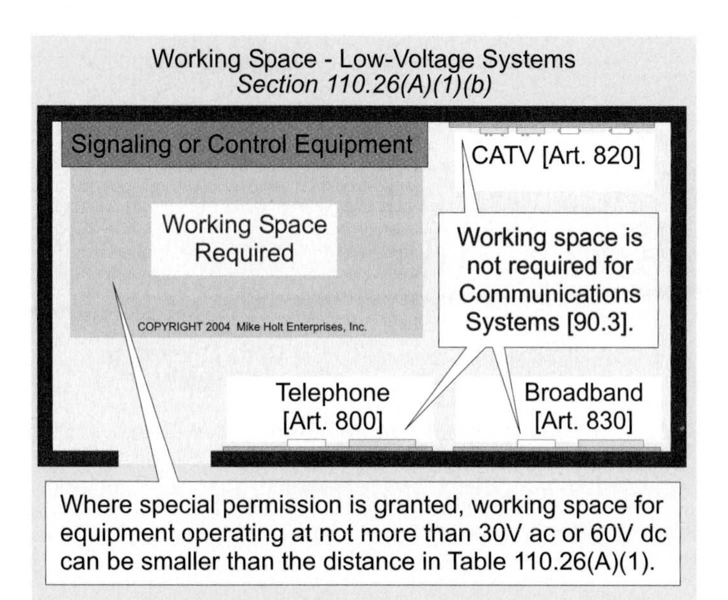

Figure 110–40

(c) Existing Buildings. Where electrical equipment is being replaced, Condition 2 working clearance is permitted between dead-front switchboards, panelboards, or motor control centers located across the aisle from each other where conditions of maintenance and supervision ensure that written procedures have been adopted to prohibit equipment on both sides of the aisle to be open at the same time and only authorized, qualified persons will service the installation.

Author's Comment: The step-back working space requirements of 110.26 do not apply to equipment included in Chapter 8 Communications Circuits [90.3]. **See Figure 110–40.**

(2) Width of Working Space. The width of the working space must be a minimum of 30 in., but in no case less than the width of the equipment. **Figure 110–41**

Author's Comment: The width of the working space can be measured from left-to-right, from the right-to-left, or simply centered on the equipment. **Figure 110–42**

In all cases, the working space must be of sufficient width, depth, and height to permit all equipment doors to open 90°. **Figure 110–43**

Author's Comment: Working space can overlap the working space for other electrical equipment.

(3) Height of Working Space (Headroom). For service equipment, switchboards, panelboards, and motor control equipment, the height of working space in front of equipment must not be less than 6½ ft, measured from the grade, floor, or platform [110.26(E)].

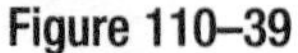

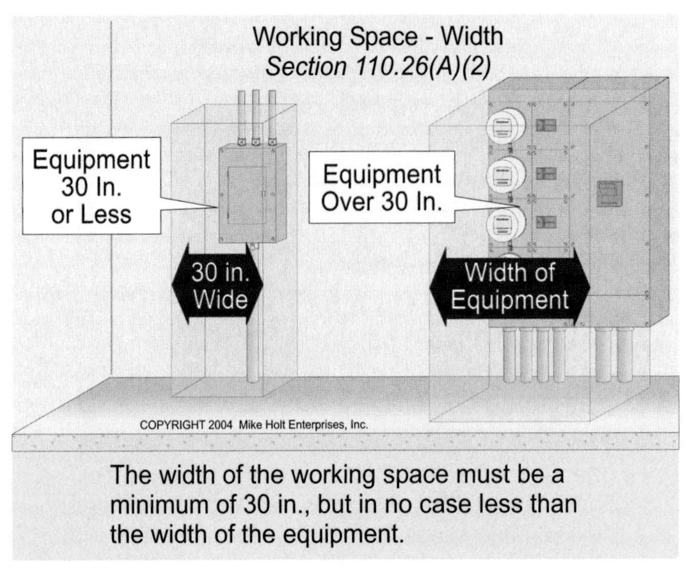

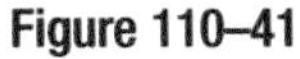
Figure 110–41

Equipment such as raceways, cables, wireways, cabinets, panels, etc., can be located above or below electrical equipment, but it must not extend more than 6 in. into the equipment's working space. **Figure 110–44**

(B) Clear Working Space. The working space required by this section must be clear at all times. Therefore, this space is not permitted for storage.

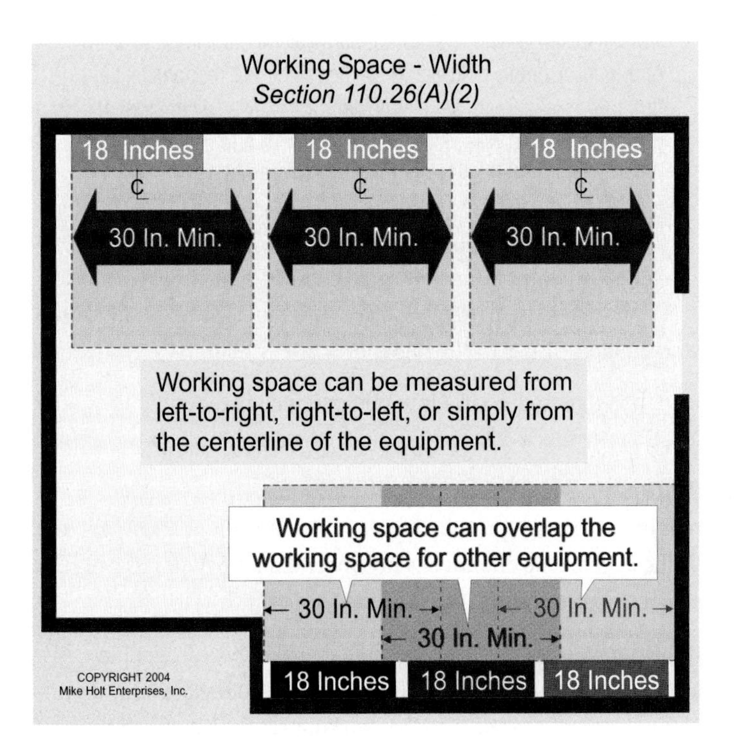

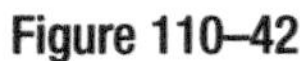
Figure 110–42

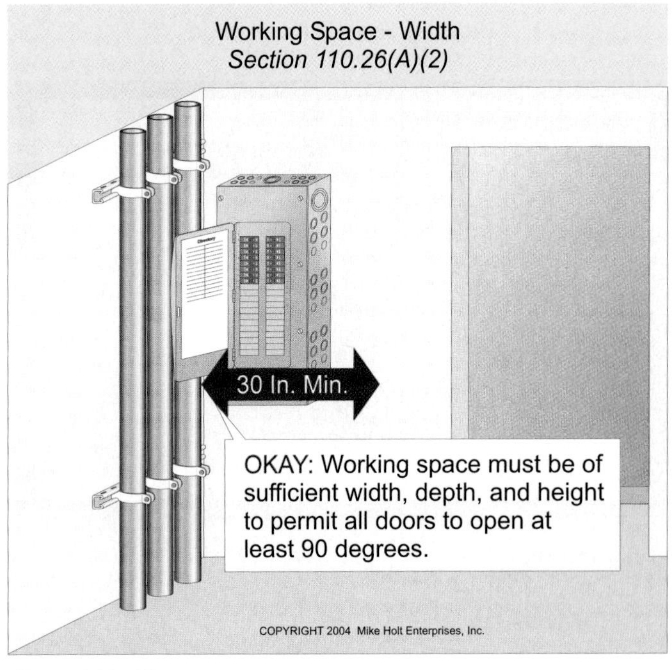

Figure 110–43

CAUTION: *It's very dangerous to service energized parts in the first place, and it's unacceptable to be subjected to additional dangers by working about, around, over, or under bicycles, boxes, crates, appliances, and other impediments.* **Figure 110–45**

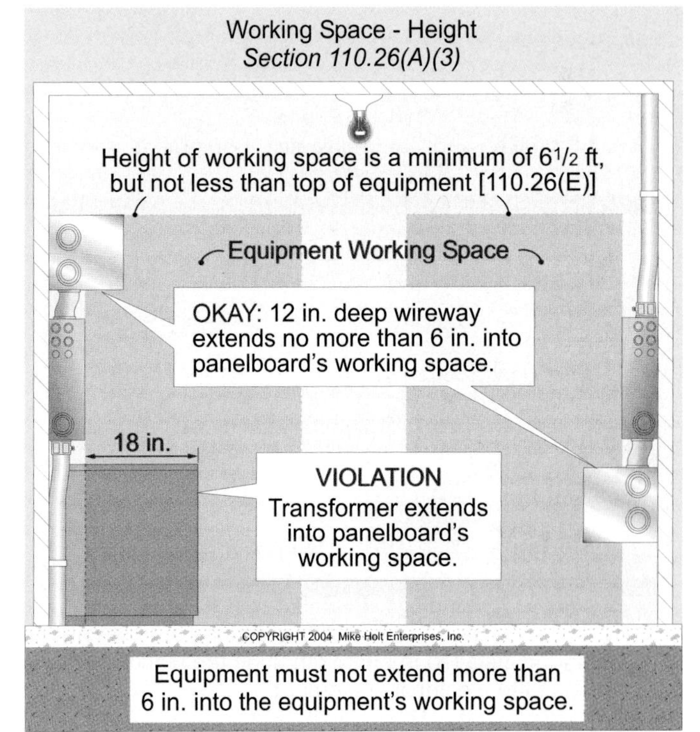

Figure 110–44

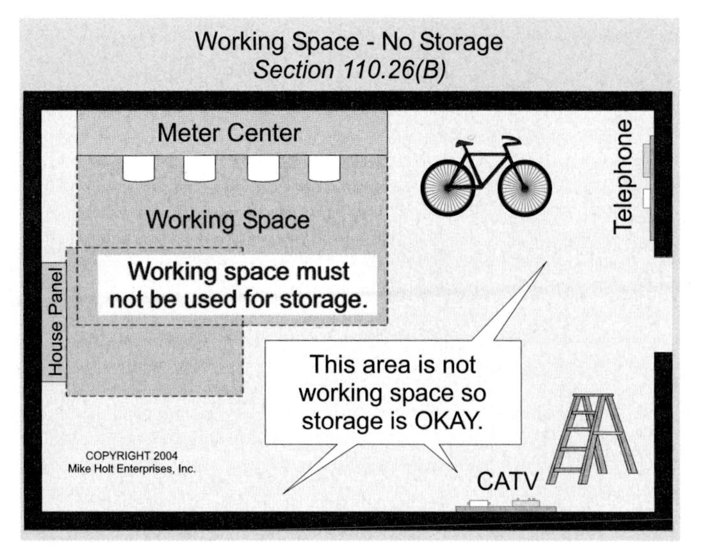
Working Space - No Storage
Section 110.26(B)
Meter Center
Working Space
Telephone
House Panel
Working space must not be used for storage.
This area is not working space so storage is OKAY.
COPYRIGHT 2004 Mike Holt Enterprises, Inc.
CATV

Figure 110–45

Author's Comment: Signaling and communications equipment must not be installed to encroach on the working space of the electrical equipment. **Figure 110–46**

(C) Entrance to Working Space.

(1) Minimum Required. At least one entrance of sufficient area must provide access to the working space.

Author's Comment: Check to see what the authority having jurisdiction considers "sufficient area."

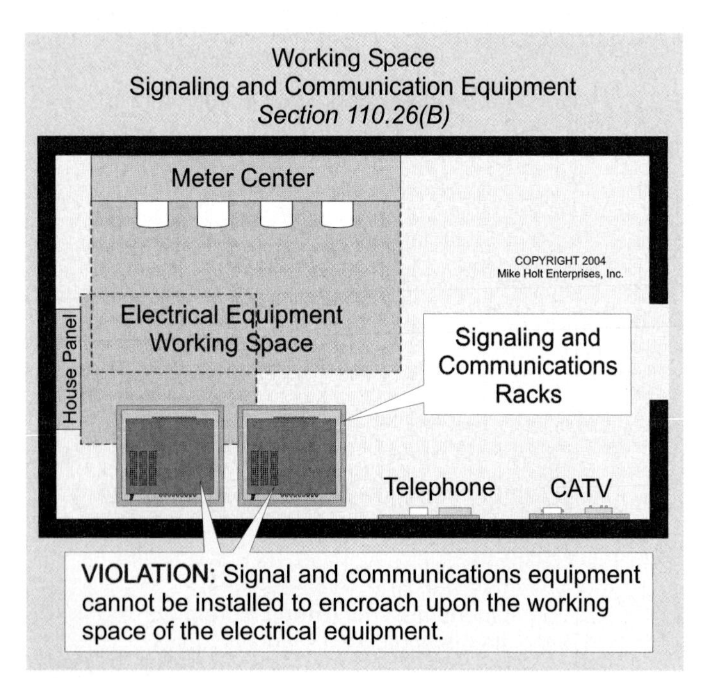
Working Space
Signaling and Communication Equipment
Section 110.26(B)
Meter Center
COPYRIGHT 2004 Mike Holt Enterprises, Inc.
House Panel
Electrical Equipment Working Space
Signaling and Communications Racks
Telephone
CATV
VIOLATION: Signal and communications equipment cannot be installed to encroach upon the working space of the electrical equipment.

Figure 110–46

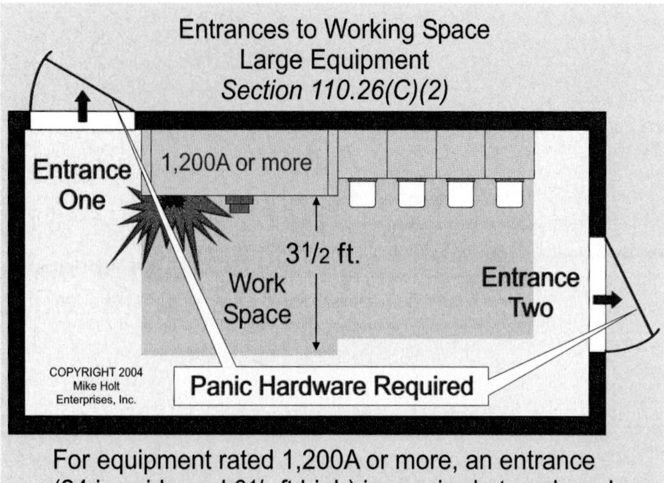
Entrances to Working Space
Large Equipment
Section 110.26(C)(2)
Entrance One
1,200A or more
3 1/2 ft.
Work Space
Entrance Two
COPYRIGHT 2004 Mike Holt Enterprises, Inc.
Panic Hardware Required
For equipment rated 1,200A or more, an entrance (24 in. wide and 6 1/2 ft high) is required at each end of the working space. Doors (if installed) must swing out and be equipped with panic hardware.

Figure 110–47

(2) Large Equipment. For equipment rated 1,200A or more, an entrance measuring not less than 24 in. wide and 6½ ft high is required at each end of the working space. Where the entrance to the working space has a door, the door must open out and be equipped with panic hardware or other devices that open under simple pressure. **Figure 110–47**

Author's Comment: Since this requirement is in the *NEC*, the electrical contractor is responsible for ensuring that panic hardware is installed where required. Some electrical contractors are offended at being held liable for nonelectrical responsibilities, but this rule should be a little less offensive, given that it's designed to save electricians' lives. For this and other reasons, many construction professionals routinely hold "pre-construction" or "pre-con" meetings to review potential opportunities for miscommunication—before the work begins.

A single entrance to the required working space is permitted where either of the following conditions is met.

(a) Unobstructed Exit. Only one entrance is required where the location permits a continuous and unobstructed way of exit travel.

(b) Double Workspace. Only one entrance is required where the required working space is doubled, and the equipment is located so the edge of the entrance is no closer than the required working space distance. **Figure 110–48**

(D) Illumination. Service equipment, switchboards, panelboards, as well as motor control centers located indoors must have illumination located in or next to the working space. Illumination must not be controlled by automatic means only. **Figure 110–49**

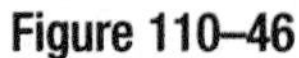

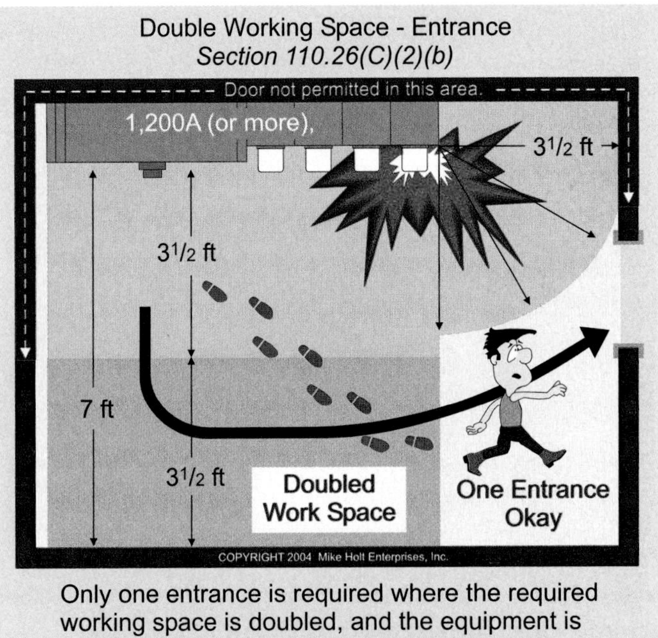

Only one entrance is required where the required working space is doubled, and the equipment is located so the edge of the entrance is no closer than the required working space distance.

Figure 110–48

(E) Headroom. For service equipment, panelboards, switchboards, or motor control centers, the minimum working space headroom must not be less than 6½ ft. When the height of the equipment exceeds 6½ ft, the minimum headroom must not be less than the height of the equipment.

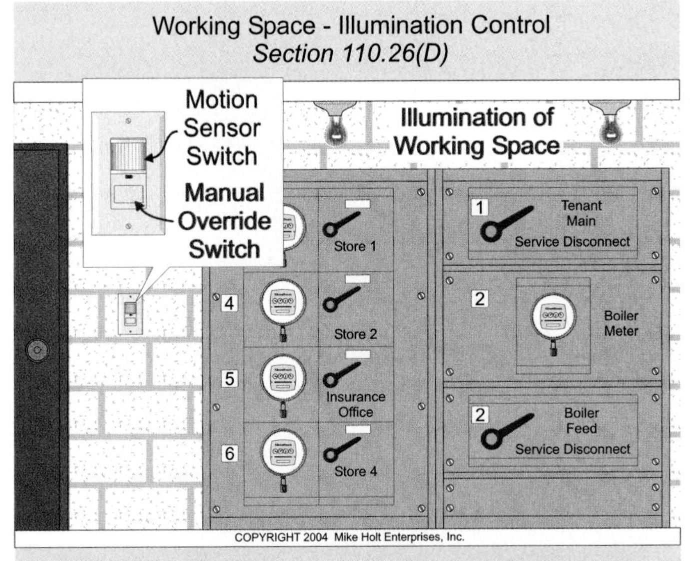

Service equipment, switchboards, panelboards, and motor control centers located indoors must not have illumination controlled by automatic means only.

Figure 110–49

Exception: The minimum headroom requirement doesn't apply to service equipment or panelboards rated 200A or less located in an existing dwelling unit.

Author's Comment: See Article 100 for the definition of "Dwelling Unit."

(F) Dedicated Equipment Space. Switchboards, panelboards, distribution boards, and motor control centers must comply with the following:

(1) Indoors.

(a) Dedicated Electrical Space. The footprint space (width and depth of the equipment) extending from the floor to a height of 6 ft above the equipment or to the structural ceiling, whichever is lower, must be dedicated for the electrical installation. No piping, duct, or other equipment foreign to the electrical installation can be installed in this dedicated footprint space. **Figure 110–50**

Exception: Suspended ceilings with removable panels can be within the dedicated footprint space.

Author's Comment: Electrical raceways and cables not associated with the dedicated space can be within the dedicated space. It isn't considered "equipment foreign to the electrical installation." See *necdigest* magazine, winter 2003, page 30. **Figure 110–51**

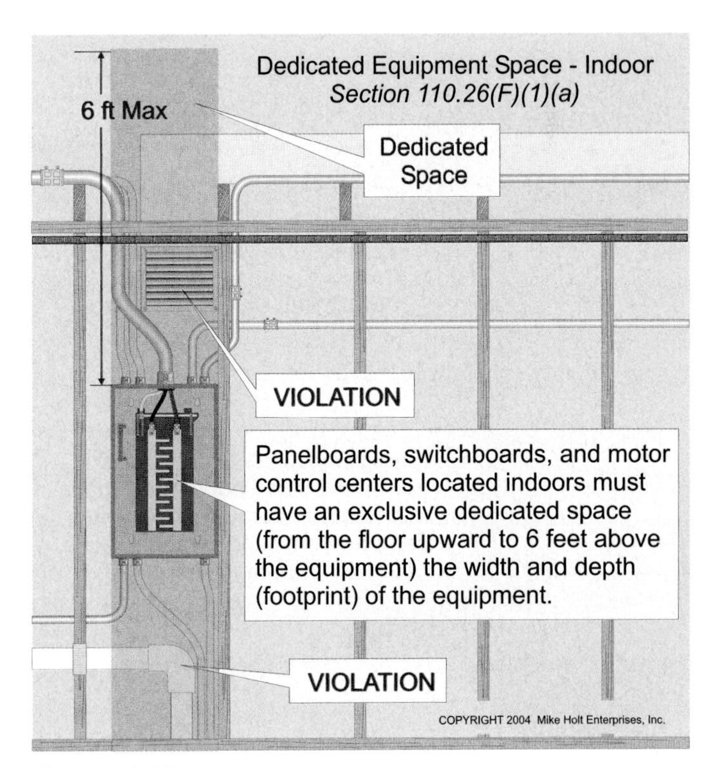

Panelboards, switchboards, and motor control centers located indoors must have an exclusive dedicated space (from the floor upward to 6 feet above the equipment) the width and depth (footprint) of the equipment.

Figure 110–50

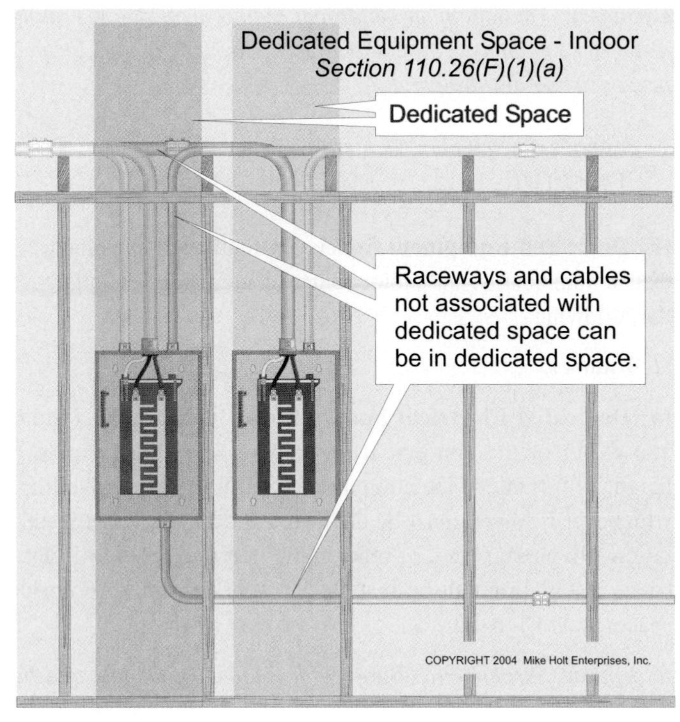
Dedicated Equipment Space - Indoor
Section 110.26(F)(1)(a)
Dedicated Space
Raceways and cables not associated with dedicated space can be in dedicated space.
COPYRIGHT 2004 Mike Holt Enterprises, Inc.

Figure 110–51

(b) Foreign Systems. Foreign systems can be located above the dedicated space if protection is installed to prevent damage to the electrical equipment from condensation, leaks, or breaks in the foreign systems. **Figure 110–52**

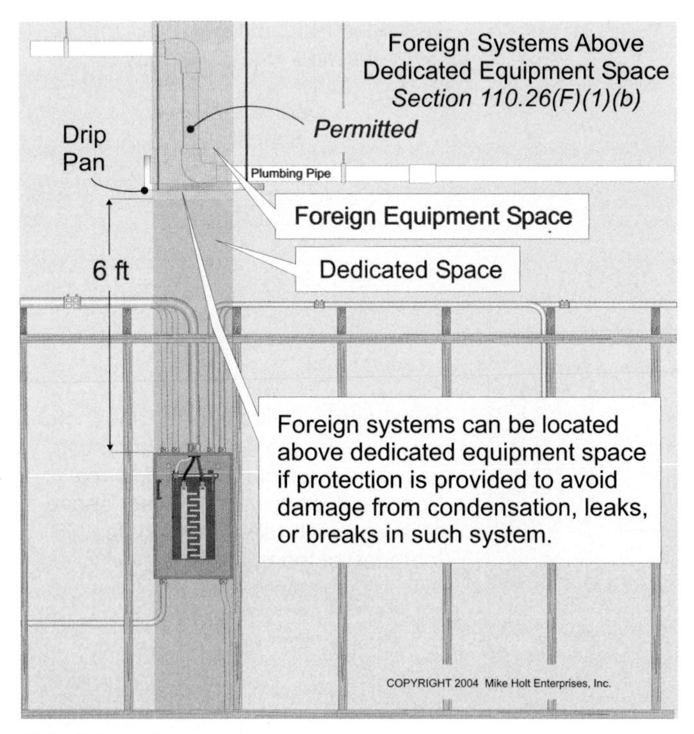
Foreign Systems Above Dedicated Equipment Space
Section 110.26(F)(1)(b)
Permitted
Drip Pan
Plumbing Pipe
Foreign Equipment Space
Dedicated Space
6 ft
Foreign systems can be located above dedicated equipment space if protection is provided to avoid damage from condensation, leaks, or breaks in such system.
COPYRIGHT 2004 Mike Holt Enterprises, Inc.

Figure 110–52

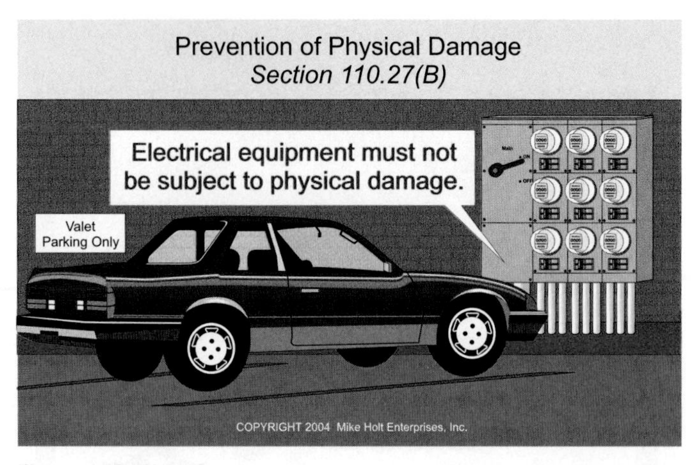
Prevention of Physical Damage
Section 110.27(B)
Electrical equipment must not be subject to physical damage.
Valet Parking Only
COPYRIGHT 2004 Mike Holt Enterprises, Inc.

Figure 110–53

(c) Sprinkler Protection. Sprinkler protection piping isn't permitted in the dedicated space, but the *NEC* doesn't prohibit sprinklers from spraying water on electrical equipment.

(d) Suspended Ceilings. A dropped, suspended, or similar ceiling isn't considered a structural ceiling.

110.27 Guarding.

(A) Guarding Live Parts. Live parts of electrical equipment operating at 50V or more must be guarded against accidental contact.

(B) Prevent Physical Damage. Electrical equipment must not be installed where it could be subject to physical damage. **Figure 110–53**

Article 110 Questions

(• Indicates that 75% or fewer exam takers get the question correct.)

1. In determining equipment to be installed, considerations such as the following should be evaluated:

   (a) Mechanical strength (b) Cost (c) Arcing effects (d) a and c

2. A wiring method included in the *Code* is recognized as being a(n) _____ wiring method.

   (a) expensive (b) efficient (c) suitable (d) cost-effective

3. •Equipment approved for use in dry locations only must be protected against permanent damage from the weather during _____.

   (a) design (b) building construction (c) inspection (d) none of these

4. The *Code* prohibits damage to the internal parts of electrical equipment by foreign material such as paint, plaster, cleaners, etc. Precautions must be taken to provide protection from the detrimental effects of paint, plaster, cleaners, etc. on internal parts such as _____.

   (a) busbars (b) wiring terminals (c) insulators (d) all of these

5. Soldered splices must first be spliced or joined so as to be mechanically and electrically secure without solder and then be soldered.

   (a) True (b) False

6. Conductors must have their ampacity determined using the _____ column of Table 310.16 for circuits rated over 100A or marked for conductors larger than 1 AWG, unless the equipment terminals are listed for use with higher temperature rated conductors.

   (a) 60°C (b) 75°C (c) 30°C (d) 90°C

7. Identification of the high leg of a 3Ø, 4-wire delta connected system is required _____.

   (a) at the service disconnect only
   (b) at each point on the system where a connection is made if the equipment grounding (bonding) conductor is also present
   (c) at each point on the system where a connection is made if the grounding electrode conductor is also present
   (d) at each point on the system where a connection is made if the grounded neutral conductor is also present

8. Sufficient access and _____ must be provided and maintained about all electrical equipment to permit ready and safe operation and maintenance of such equipment.

   (a) ventilation (b) cleanliness (c) circulation (d) working space

9. •The required working clearance for access to live parts operating at 300V to ground, where there are exposed live parts on one side and grounded parts on the other side, is _____ according to Table 110.26(A).

   (a) 3 ft (b) 3½ ft (c) 4 ft (d) 4½ ft

10. When normally-enclosed live parts are exposed for inspection or servicing, the working space, if in a passageway or general open space, must be suitably _____.

    (a) accessible (b) guarded (c) open (d) enclosed

11. The minimum headroom of working spaces about motor control centers must be _____.

    (a) 3 ft (b) 5 ft (c) 6 ft (d) 6½ ft

CHAPTER 5

Special Occupancies

Introduction

Chapter 5, which covers special occupancies, is the first of four *NEC* chapters that deal with special topics.

Chapters 6, 7, and 8 cover special equipment, special conditions, and communications systems, respectively. Remember, the first four Chapters of the *NEC* are sequential and form a foundation for each of the subsequent four Chapters.

What exactly is a "special occupancy?" It's a location where the physical facility or use of the physical facility creates specific conditions that require additional measures to ensure the "safeguarding of people and property" mission of the *NEC*, as put forth in Article 90.

The *NEC* groups these logically, as you might expect. Here are the general groupings:

- Environments that pose additional hazards. Articles 500—510. Examples include Hazardous (Classified) Locations.
- Specific types of facilities that pose additional hazards. Articles 511—516. Examples include motor fuel dispensing facilities, aircraft hangars, and bulk storage plants.
- Facilities that pose evacuation difficulties. Articles 517—525. Examples include hospitals, theaters, and carnivals.
- Motion picture related. Articles 530 and 540.
- Specific types of buildings. Articles 545—553. Examples include park trailers and floating buildings.
- Marinas and boatyards. Article 555.
- Temporary installations. Article 590.

Many people struggle to understand the requirements for Special Occupancies, mostly because of the narrowness of application. However, if you study the illustrations and explanations here, you will clearly understand them.

Article 500. Hazardous (Classified) Locations. A hazardous (classified) location is an area where the possibility of fire or explosion exists because of the presence of flammable gases or vapors, combustible dusts, or ignitible fibers or flyings. The three components necessary to create a fire or explosion are fuel, oxygen, and a source of ignition.

Article 501. Class I Hazardous (Classified) Locations. A Class I hazardous (classified) location is an area where flammable gases or vapors may be present in quantities sufficient to produce an explosive or ignitible mixture.

Article 502. Class II Hazardous (Classified) Locations. A Class II hazardous (classified) location is an area where combustible dust may be suspended in the air in quantities sufficient to ignite or explode.

Article 503. Class III Hazardous (Classified) Locations. Class III locations are hazardous due to the presence of easily ignitible fibers or flyings, but these materials aren't likely to be suspended in the air in quantities sufficient to produce ignitible mixtures. This includes materials such as cotton and rayon, which are found in textile mills and clothing manufacturing plants. It can also include establishments and industries such as woodworking plants. There are no "Group" classifications for Class III locations as there are for Class I and Class II locations.

Article 504. Intrinsically Safe Systems. This article covers the installation of intrinsically safe apparatus, wiring, and systems for Class I, II, and III locations. An intrinsically safe circuit doesn't develop sufficient electrical energy to cause ignition of a specified gas or vapor under normal or abnormal operating conditions. An intrinsically safe system reduces the risk of ignition by electrical equipment or circuits and offers an optional wiring method in hazardous (classified) locations.

Article 511. Commercial Garages, Repair, and Storage. These occupancies include locations used for service and repair operations in connection with self-propelled vehicles (including, but not limited to, passenger automobiles, buses, trucks, and tractors) in which petroleum-based chemicals (volatile organic compounds) are used for fuel or power.

Article 513. Aircraft Hangars. This article applies to buildings or structures in any part of which aircraft are housed or stored containing Class I (flammable) liquids or Class II (combustible) liquids whose temperatures are above their flash points, and in which aircraft might undergo service, repairs, or alterations. It isn't necessary to classify areas where only Class II combustible liquids are used or stored below the flash point. Article 513 doesn't apply to areas used exclusively for aircraft that have never contained fuel, or to unfueled aircraft.

Article 514. Motor Fuel Dispensing Facilities. This article applies to gasoline dispensing and service stations where gasoline or other volatile flammable liquids or liquefied flammable gases are transferred to fuel tanks of self-propelled vehicles. Wiring and equipment in the area of service and repair rooms of service stations must comply with the installation requirements in Article 511.

Article 517. Health Care Facilities. This article applies to electrical wiring in health care facilities such as hospitals, nursing homes, limited-care facilities, clinics, medical and dental offices, and ambulatory care, whether permanent or movable. This article isn't intended to apply to animal veterinary facilities.

Article 518. Assembly Occupancies. This article covers all buildings or portions of buildings or structures specifically designed or intended for the assembly of 100 or more persons.

Article 525. Carnivals, Circuses, Fairs, and Similar Events. This article covers the installation of portable wiring and equipment for carnivals, circuses, exhibitions, fairs, traveling attractions, and similar functions, including wiring in or on all structures.

Article 547. Agricultural Buildings. The provisions of this article apply to agricultural buildings or those parts of buildings or adjacent areas where excessive dust or dust with water may accumulate, or where a corrosive atmosphere exists.

Article 550. Mobile Homes, Manufactured Homes, and Mobile Home Parks. The provisions of this article cover the electrical conductors and equipment installed within or on mobile or manufactured homes, the conductors that connect mobile or manufactured homes to a supply of electricity, and the installation of electrical wiring, fixtures, and equipment.

Article 551. Recreational Vehicles and Recreational Vehicle Parks. Requirements for recreational vehicle park wiring and the wiring in recreational vehicles (other than the automotive low voltage wiring) are covered by this article.

Article 555. Marinas and Boatyards. This article covers the installation of wiring and equipment in the areas that comprise fixed or floating piers, wharves, docks, and other areas in marinas, boatyards, boat basins, boathouses, and similar occupancies that are used, or intended to be used, for the purpose of repair, berthing, launching, storing, or fueling of small craft and the mooring of floating buildings. This article doesn't apply to docks or boathouses for single-family dwelling units.

Article 590. Temporary Installations. This article applies to temporary power and lighting for construction, remodeling, maintenance, repair, demolitions, and decorative lighting. This article also applies when temporary installations are necessary.

ARTICLE

500 Hazardous (Classified) Locations

Introduction

A hazardous (classified) location is an area where the possibility of fire or explosion can be created by the presence of flammable gases or vapors, combustible dusts, or ignitible fibers or flyings. Sparks and/or heated surfaces can serve as a source of ignition in such environments.

Article 500 provides a foundation for applying Article 501 (Class I Locations), Article 502 (Class II Locations), Article 503 (Class III Locations), and Article 504 (Intrinsically Safe Systems)—all of which immediately follow Article 500. It also provides a foundation for applying Articles 510 through 516.

Before you apply any of the articles just mentioned, you must understand and apply Article 500. It's a fairly long and detailed article. But don't worry; we'll help you master the concepts.

A Fire Triangle (fuel, oxygen, and ignition) often illustrates this concept. **Figure 500-1**

- Fuel—Flammable gases or vapors, combustible dusts, and ignitible fibers or flyings.
- Oxygen—Air and oxidizing atmospheres.
- Ignition—Electric arcs or sparks, heat-producing equipment such as luminaires and motors, conductor insulation, failure of transformers, coils, or solenoids, as well as sparks caused by metal tools dropping on a metal surface.

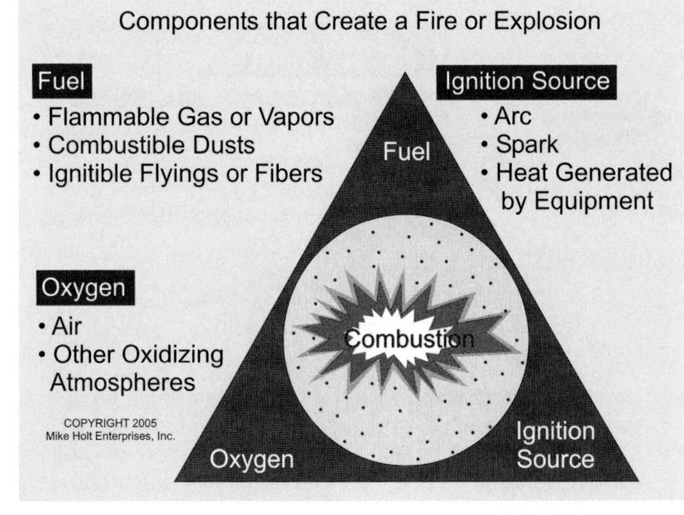

Figure 500–1

500.1 Scope—Articles 500 Through 504. To prevent injury, death, or extensive damage to building structures from fires or explosions, the *NEC* contains stringent requirements for equipment and its installation in a hazardous (classified) location. The specific requirements for electrical installations in hazardous (classified) locations are contained in:

- Article 501. Class I—Flammable Gases or Vapors
- Article 502. Class II—Combustible Dust
- Article 503. Class III—Ignitible Fibers, Particles, or Combustible Flyings
- Article 504. Intrinsically Safe Systems

Author's Comments:

- See Article 100 for the definition of "Structure."
- Locating electrical wiring and equipment outside the classified location provides the safest electrical installation and is often more cost-effective [500.5(A) FPN].

- Many of the graphics contained in Chapter 5 use two shades of red to identify a Class I Division location (medium red for Division 1 and lighter red to identify Division 2). In some cases, these color schemes are used as a background color to help you tell if it applies to Division 1, Division 2, or both (split color background).
- The *NEC* doesn't classify specific hazardous (classified) locations, except as identified in Articles 511 through 517. Determining the classification of a specific hazardous area is the responsibility of those who understand the dangers of flammable, combustible, or ignitible products, such as the fire marshal, plant facility engineer, or insurance adjuster. It isn't the responsibility of the electrical designer, electrical contractor, or electrical inspector. Prior to performing any wiring in or near a hazardous (classified) location, contact the plant facility and design engineer to ensure that proper installation and materials are used. Be sure to review 500.4(B) for additional standards that might need to be consulted.

Other articles in Chapter 5 containing specific installation requirements include:

- Article 505. Class I, Zone 0, 1, and 2 Locations
- Article 511. Commercial Garages, Repair, and Storage
- Article 513. Aircraft Hangars
- Article 514. Motor Fuel Dispensing Facilities
- Article 515. Bulk Storage Plants
- Article 516. Spray Application, Dipping, and Coating Processes
- Article 517. Health Care Facilities

500.2 Definitions. The definitions contained in 500.2 apply to Articles 500 through 504 and Articles 510 through 516.

Dust-Ignitionproof. Equipment enclosed in a manner that excludes dust and doesn't permit arcs, sparks, or heat inside the enclosure to ignite accumulations or atmospheric suspensions of a specified dust on or in the vicinity outside of the enclosure.

Dusttight. Enclosures constructed so that dust will not enter. Examples of dusttight enclosures include FS boxes and Bell boxes. **Figure 500-2**

Electrical and Electronic Equipment. Materials, fittings, devices, and appliances that are part of, or in connection with, the electrical installation.

Author's Comment: See Article 100 for the definitions of "Appliances," "Devices," and "Fittings."

FPN: Portable or transportable equipment with a self-contained power supply, such as battery-operated equipment, could potentially become an ignition source in hazardous (classified) locations. **Figure 500-3**

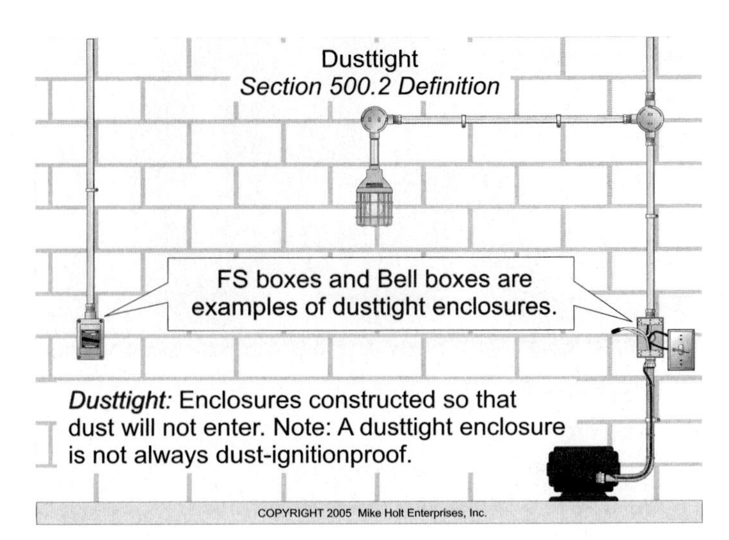

Figure 500–2

Author's Comment: The *NEC* is an "installation standard" and battery-operated equipment does not fall within the scope of this *Code.* Workplace safety enforcement agencies, such as OSHA, can and do regulate the use of battery-operated equipment in hazardous (classified) locations.

Explosionproof Apparatus. Apparatus enclosed in a case that is capable of withstanding an internal explosion of a specified gas or vapor and of preventing the ignition of a specified gas or vapor surrounding the enclosure. It operates so that the external temperature will not ignite the surrounding flammable atmosphere. **Figure 500-04**

Hermetically Sealed. Equipment sealed against the entrance of an external atmosphere.

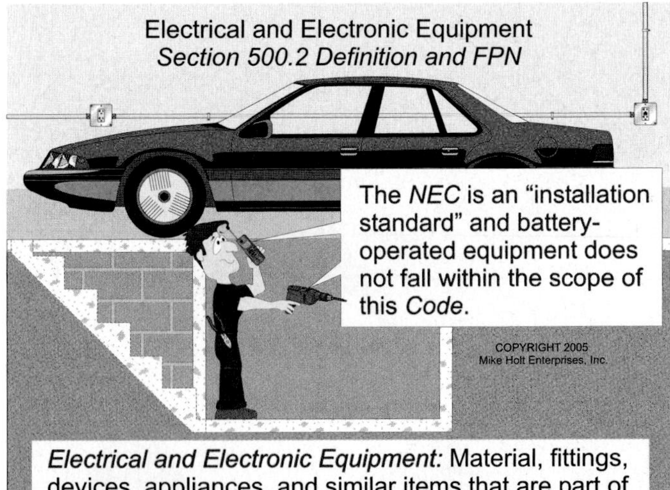

Figure 500–3

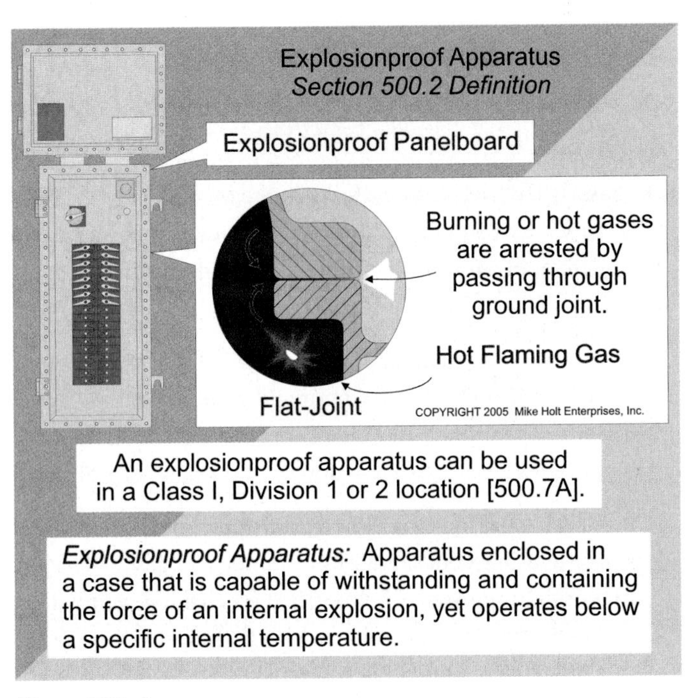

Explosionproof Apparatus
Section 500.2 Definition
Explosionproof Panelboard
Burning or hot gases are arrested by passing through ground joint.
Hot Flaming Gas
Flat-Joint
COPYRIGHT 2005 Mike Holt Enterprises, Inc.
An explosionproof apparatus can be used in a Class I, Division 1 or 2 location [500.7A].
Explosionproof Apparatus: Apparatus enclosed in a case that is capable of withstanding and containing the force of an internal explosion, yet operates below a specific internal temperature.

Figure 500–4

Oil Immersion. Electrical equipment immersed in a protective liquid.

Purged and Pressurized.

(1) Purging. Supplying an enclosure with a protective gas at a sufficient flow and positive pressure to reduce the concentration of any flammable gas or vapor.

(2) Pressurization. Supplying an enclosure with a protective gas with or without continuous flow at sufficient pressure to prevent the entrance of a flammable gas or vapor, a combustible dust, or an ignitible fiber.

500.3 Other Articles. Except as modified in Articles 500 through 504, all installation requirements contained in Chapters 1 through 4 of the *NEC* apply to electrical equipment and wiring installed in hazardous (classified) locations.

500.4 General.

(A) Classification Documentation. The *NEC* requires all hazardous (classified) locations to be properly documented. The documentation must be available to those who are authorized to design, install, inspect, maintain, or operate the electrical equipment.

Author's Comments:

- Proper documentation of hazardous areas assists the designer, installer, and authority having jurisdiction in ensuring compliance with the stringent requirements contained in Chapter 5 of the *NEC.*
- To ensure compliance with the above requirements, some authorities having jurisdiction require "area classification drawings."

(B) Other Standards. To ensure a proper and safe installation, the authority having jurisdiction must be familiar with the industry and standards of the National Fire Protection Association (NFPA), the American Petroleum Institute (API), and the Instrumentation, Systems, and Automation Society (ISA).

FPN: The following standards should assist in area classification, the determination of adequate ventilation, and the protection requirements necessary to protect against static electricity or lightning.

- Classification of Hazardous (Classified) Locations, NFPA 497
- Dipping and Coatings—Processes Using Flammable or Combustible Liquids, NFPA 34
- Dust—Area Classification in Hazardous (Classified) Dust Locations, ISA 12.10
- Lightning Protection Code, NFPA 780
- Liquids—Flammable and Combustible Liquids Code, NFPA 30
- Petroleum—Classification of Electrical Installations at Petroleum Facilities, ANSI/API RP 500
- Spray Applications Using Flammable and Combustible Materials, NFPA 33
- Static Electricity, NFPA 77
- Static Lightning and Stray Currents, API RP 2003
- Liquefied Petroleum Gases Code, NFPA 58
- Wastewater Treatment and Collection Facilities, NFPA 820

500.5 Classifications of Locations.

(A) Classifications of Locations. Locations are classified according to the properties of the volatile flammable gases or vapors, or combustible dusts or fibers that may be present, and the likelihood that a flammable or combustible concentration will be present.

Each room, section, or area is considered individually in determining its classification. The same building or structure might contain both Class I, Division 1 and 2 locations, or Class II, Division 1 and 2 locations, or Class III, Division 1 and 2 locations. **Figure 500-5**

Author's Comment: See Article 100 for the definition of "Building."

FPN: To reduce expensive equipment and expensive wiring methods, locate as much electrical equipment as possible in an unclassified location.

(B) Identification of a Class I Location. A Class I location is an area where flammable gases or vapors may be present in the air and in quantities sufficient to produce explosive or ignitible mixtures.

(1) Class I, Division 1 Location. An area where ignitible concentrations of flammable gases or vapors may exist in the course of normal operations: **Figure 500-6**

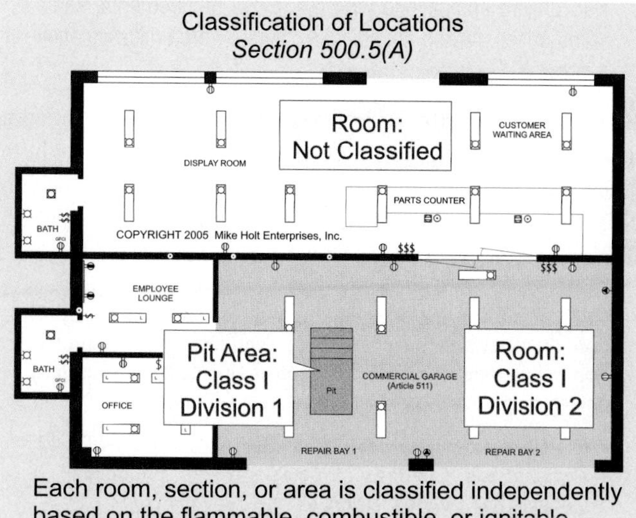

Figure 500–5

(1) Continuously or periodically under normal operating conditions.

(2) Because of repair or maintenance operations or leakage.

(3) Where faulty equipment releases ignitible concentrations of flammable gases or vapors and the equipment becomes a source of ignition.

FPN: Class I, Division 1 locations include:

- Areas where volatile flammable liquids or gases are transferred from one container to another, such as at gasoline storage and dispensing areas.
- Interiors of spray booths and in the vicinity of spraying and painting operations where volatile flammable solvents are used to coat products with paint or plastics.

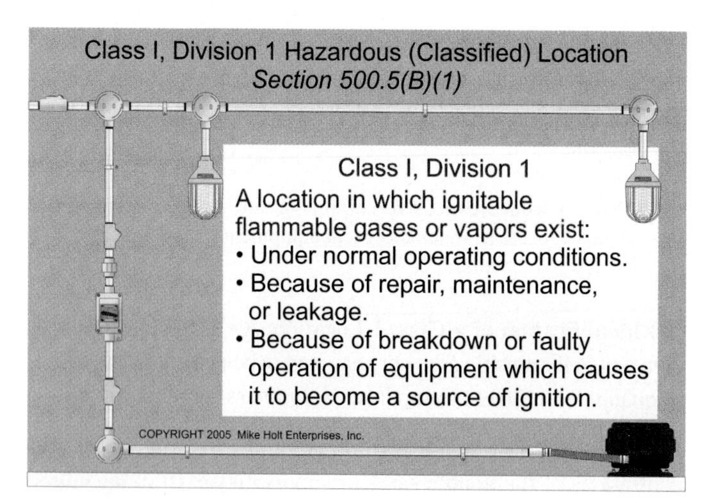

Class I, Division 1 Hazardous (Classified) Location
Section 500.5(B)(1)
Class I, Division 1
A location in which ignitable flammable gases or vapors exist:
• Under normal operating conditions.
• Because of repair, maintenance, or leakage.
• Because of breakdown or faulty operation of equipment which causes it to become a source of ignition.
COPYRIGHT 2005 Mike Holt Enterprises, Inc.

Figure 500–6

- Locations containing open tanks or vats of volatile flammable liquids, or dip tanks for parts cleaning or other operations.
- See 500.5(B)(3) FPN No. 1 and 2 in the *NEC* for additional details.

(2) Class I, Division 2 Location. An area where volatile flammable gases or vapors would become hazardous only in case of an accident or of some unusual operating condition, or under any of the following conditions: **Figure 500-7**

(1) Where volatile flammable liquids or gases are handled, processed, or used, but are normally confined within closed containers and the gases would only escape in the case of accidental rupture or breakdown, or in case of abnormal operation of equipment.

(2) Where ignitible concentrations of flammable gases or vapors are normally prevented by positive mechanical ventilation, but might become hazardous through failure or abnormal operation of the ventilating equipment.

(3) Areas adjacent to a Class I, Division 1 location to where flammable gases or vapors might occasionally be communicated unless prevented by adequate positive-pressure ventilation with effective safeguards against ventilation failure.

FPN No. 1: The quantity of volatile flammable liquids or gases that might escape in case of accident, the adequacy of ventilating equipment, the total area involved, and the record of the industry with respect to explosions or fires are all factors that must be taken into consideration.

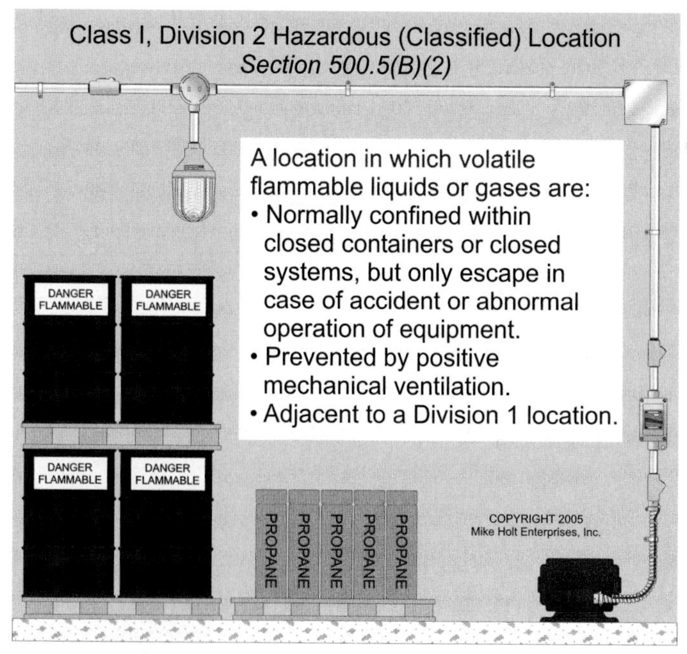

Figure 500–7

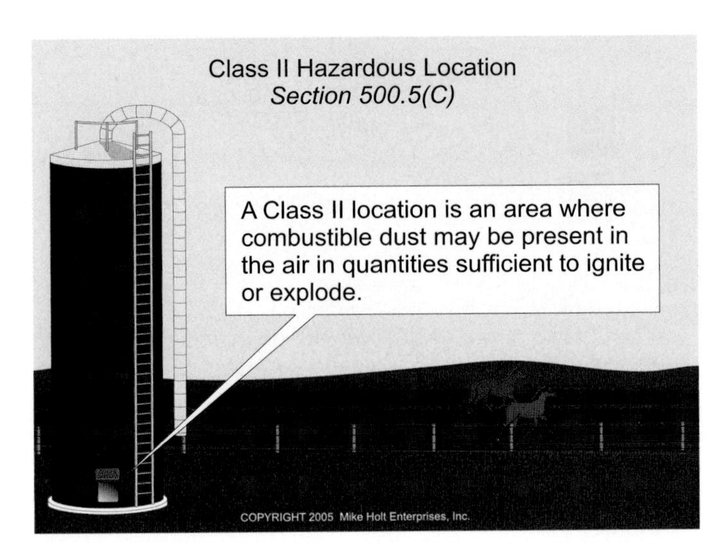
Class II Hazardous Location
Section 500.5(C)
A Class II location is an area where combustible dust may be present in the air in quantities sufficient to ignite or explode.
COPYRIGHT 2005 Mike Holt Enterprises, Inc.

Figure 500–8

(C) Identification of a Class II Location. An area where the presence of combustible dust may be suspended in the air with quantities sufficient to ignite or explode. **Figure 500-8**

(1) Class II, Division 1 Location. A Class II, Division 1 location is an area where combustible dust may exist under any of the following conditions:

(1) Nonconductive combustible dust is continuously or periodically suspended in the air in sufficient quantities to produce mixtures that will ignite or explode.

(2) Where faulty equipment releases ignitible mixtures of dust and the equipment becomes a source of ignition.

(3) Where Group E electrically conductive combustible dust may be present in sufficient quantities to ignite or explode.

> **FPN:** Dusts containing magnesium or aluminum are particularly hazardous, and the use of extreme caution is necessary to avoid ignition and explosion.

(2) Class II, Division 2 Location. An area where combustible dust would become hazardous under any of the following conditions:

(1) Where combustible dust, due to abnormal operations, may be present in the air in quantities sufficient to produce explosive or ignitible mixtures, or

(2) Where combustible dust accumulation is normally insufficient to interfere with the normal operation of electrical equipment or apparatus, but where malfunctioning of equipment may result in combustible dust being suspended in the air, or

(3) Where combustible dust accumulations on, in, or near electrical equipment could be sufficient to interfere with the safe dissipation of heat from electrical equipment, or could be ignitible by abnormal operation or failure of electrical equipment.

> **FPN No. 1:** The quantity of combustible dust that may be present and the adequacy of dust removal systems should be considered when determining the area classification.

(D) Identification of a Class III Location. A Class III location is an area where easily ignitible fibers or flyings aren't likely to be suspended in the air in quantities sufficient to produce ignitible mixtures.

(1) Class III, Division 1 Location. An area where ignitible fibers or flyings are manufactured, handled, or used, such as within textile mills or clothing manufacturing plants as well as in facilities that create sawdust and flyings by pulverizing or cutting wood.

(2) Class III, Division 2 Location. An area where ignitible fibers or flyings are stored or handled other than in the manufacturing process.

500.6 Material Groups.

The explosion characteristics of air mixtures of gases, vapors, or dusts vary with the specific material involved. Equipment located in a hazardous (classified) location must have a "group" rating to indicate that it has been tested and identified for the specific explosion characteristics of the gas, vapor, or dust that may be present.

> **Author's Comment:** See Article 100 for the definition of "Identified."

(A) Class I Group Ratings.

> **FPN No. 2:** Equipment installed in a Class I location must be designed and constructed to meet the maximum explosion pressure and clearance between parts to ensure the minimum ignition temperature of the atmosphere isn't exceeded. It is necessary that equipment be identified for the class and the specific group of the gas or vapor that will be present. Group ratings for Class I locations for flammable gases and vapors include:
>
> - Group A: Atmospheres containing acetylene.
> - Group B: Atmospheres containing manufactured gas, hydrogen, butadiene, ethylene oxide, and propylene oxide.
> - Group C: Atmospheres containing ethyl ether, ethylene, and acetaldehyde.
> - Group D: Atmospheres containing cyclopropane, gasoline, propane, natural gas, methane, benzene, butane, and ethane.

(B) Class II Group Ratings. Equipment installed in a Class II location must be designed and constructed to prevent the entrance of dust within the enclosure. Group ratings for Class II locations include:

- Group E: Atmospheres containing combustible metal dusts such as magnesium or aluminum powders.

- Group F: Atmospheres containing carbon black, charcoal, coal, or coke dusts.
- Group G: Atmospheres containing combustible dusts such as flour, grain, wood, or plastic. Locations include grain elevators, flour and feed mills, and pharmaceutical plants.

500.7 Protection Techniques. Electrical equipment and wiring within hazardous (classified) locations must be protected by any of the following techniques.

(A) Explosionproof Enclosures (Class I Locations). Explosionproof equipment is designed to be capable of withstanding and containing the force of an internal explosion and the hot gases within the enclosure cool as they escape [500.2]. **Figure 500-9**

(B) Dust-Ignitionproof Enclosures (Class II Locations). Dust-ignitionproof enclosures are designed to exclude dusts and will not permit arcs, sparks, or heat within the enclosure to cause ignition of exterior dust [500.2]. **Figure 500-10**

(C) Dusttight Enclosures (Class II, Division 2 and Class III Locations). Dusttight enclosures prevent the entrance of dust or flyings and have no openings to allow electrical sparks or burning material to escape [500.2 and 502.115(B)]. **Figure 500-11**

(D) Purged and Pressurized Systems.

- **Purging System (Class I Locations).** Purging systems permit general-purpose enclosures in an area that contains flammable gases or vapors [500.2].
- **Pressurized System (Class II Locations).** Pressurizing is the process of supplying positive pressure to general-purpose enclosures in an area that contains combustible dust [500.20].

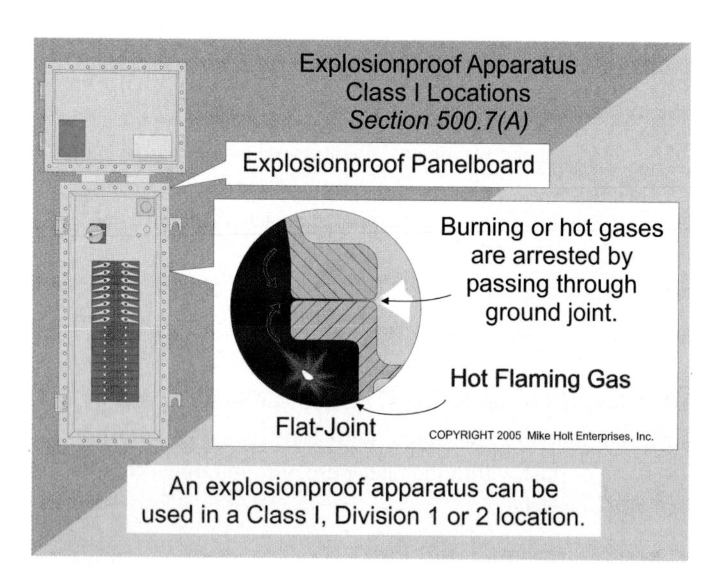

Figure 500–9

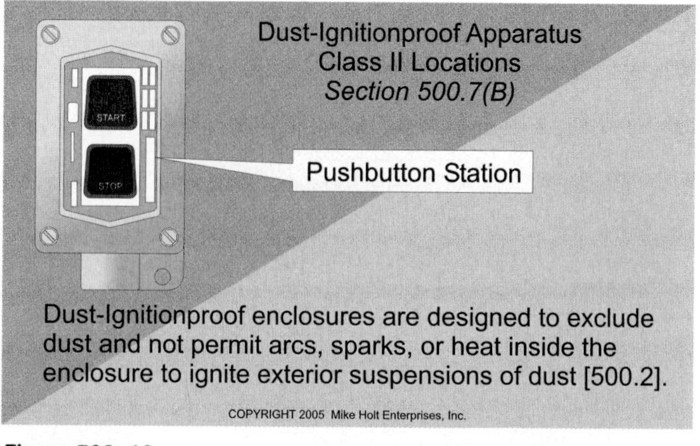

Dust-Ignitionproof Apparatus
Class II Locations
Section 500.7(B)
START
STOP
Pushbutton Station
Dust-Ignitionproof enclosures are designed to exclude dust and not permit arcs, sparks, or heat inside the enclosure to ignite exterior suspensions of dust [500.2].
COPYRIGHT 2005 Mike Holt Enterprises, Inc.

Figure 500–10

(E) Intrinsically Safe Systems (All Locations). These systems are incapable of releasing sufficient electrical or thermal energy to cause ignition of flammable gases or vapors. This system of protection is permitted in any hazardous (classified) location [500.2].

None of the requirements contained in Articles 501 through 503, or 510 through 516 apply to intrinsically safe system installations, except as required by Article 504.

(F) Nonincendive Circuits (Class I, Division 2; Class II, Division 2; or Class III, Division 1 and 2 Locations). Nonincendive circuits and equipment are incapable of releasing sufficient electrical or thermal energy to cause ignition of flammable gases, vapors, or dust [500.2].

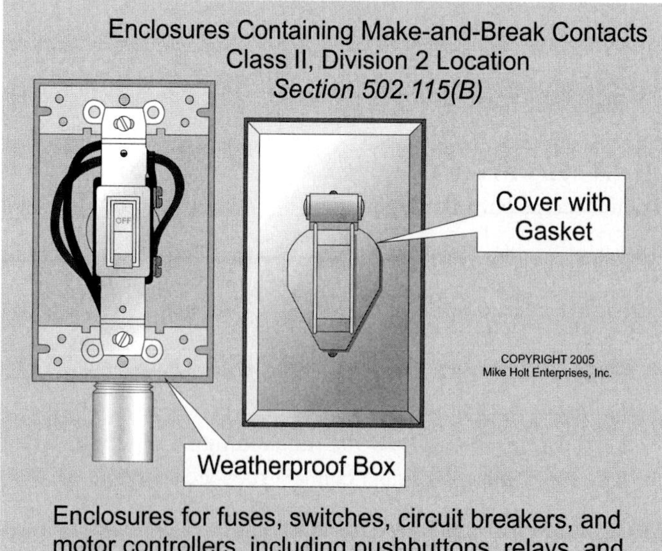

Figure 500–11

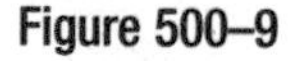

(I) Oil-Immersed Contacts (Class I, Division 2). Oil-immersed make-and-break contacts can be installed in a general-purpose enclosure in an area that doesn't contain explosive or ignitible mixtures under normal conditions [500.2].

(J) Hermetically Sealed Contacts (Class I, Division 2; Class II, Division 2; or Class III, Division 1 and 2 Locations). Hermetically sealed devices can be installed in a general-purpose enclosure in an area that doesn't contain explosive or ignitible mixtures under normal conditions [500.2].

(L) Other Protection Techniques. Other protection techniques used in equipment identified for use in hazardous (classified) locations are permitted.

> **Author's Comment:** The *NEC* is silent on the meaning of this section. The key is the use of the term "identified," which is defined in Article 100 as, "Recognizable as suitable for the specific purpose, function, use, environment, application, and so forth, where described in a particular *Code* requirement."

500.8 Equipment.

(A) Approved for Class and Properties.

(1) Identified for Use. Equipment installed in any hazardous (classified) location must be identified for the "Class" and explosive, combustible, or ignitible properties of the specific gas, vapor, dust, fiber, or flyings that will be present (Group) [500.6].

The suitability of equipment is determined by any of the following:

(1) Equipment listing or labeling. **Figure 500-12**

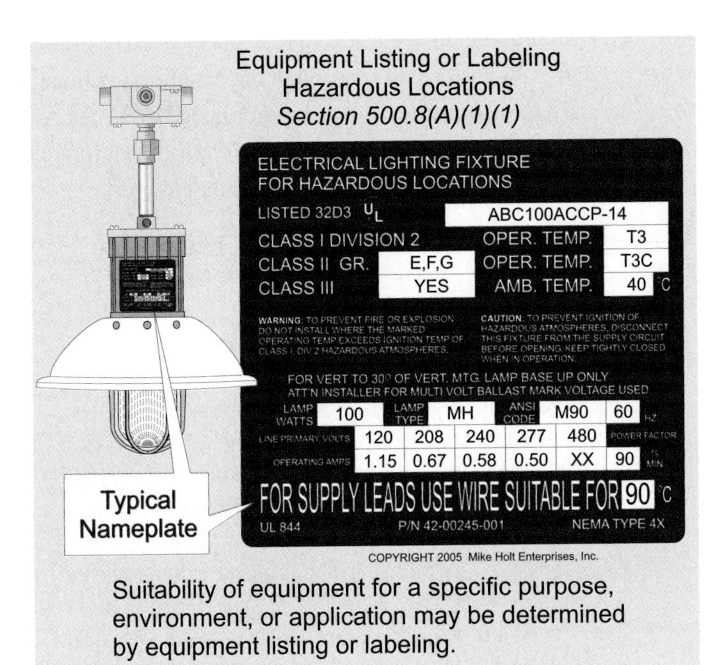

Suitability of equipment for a specific purpose, environment, or application may be determined by equipment listing or labeling.

Figure 500–12

(2) Evidence of evaluation by a qualified testing laboratory or inspection agency concerned with product evaluation.

(3) Evidence acceptable to the authority having jurisdiction, such as manufacturer's self-evaluation or owner's engineering judgment.

(2) Divisions. Equipment listed for Class I, Division 1 locations can be installed in a Class I, Division 2 location of the same class and group.

> **Author's Comment:** 500.5(B) through (D) explain the differences between Class I, Class II, and Class III locations, as well as the differences between Division 1 and Division 2.

(3) General-Purpose Enclosures. General-purpose enclosures not containing make-and-break contacts can be installed in a Class I location that doesn't contain explosive or ignitible mixtures under normal conditions (Class I, Division 2). **Figure 500-13**

> **Author's Comment:** General-purpose enclosures are permitted for signaling, alarm, remote control, communications systems and motors, instruments, and relays in Class II locations where the quantities of combustible dust aren't sufficient to produce a fire or explosion under normal conditions. See 502.150(A)(2) Ex, 502.150(A)(3) Ex, 502.150(B)(1) Ex, and 502.150(B)(3) Ex.

General-purpose enclosures are permitted for intrinsically safe systems in any hazardous (classified) location. See 500.7(E), 504.10(B), and 504.20.

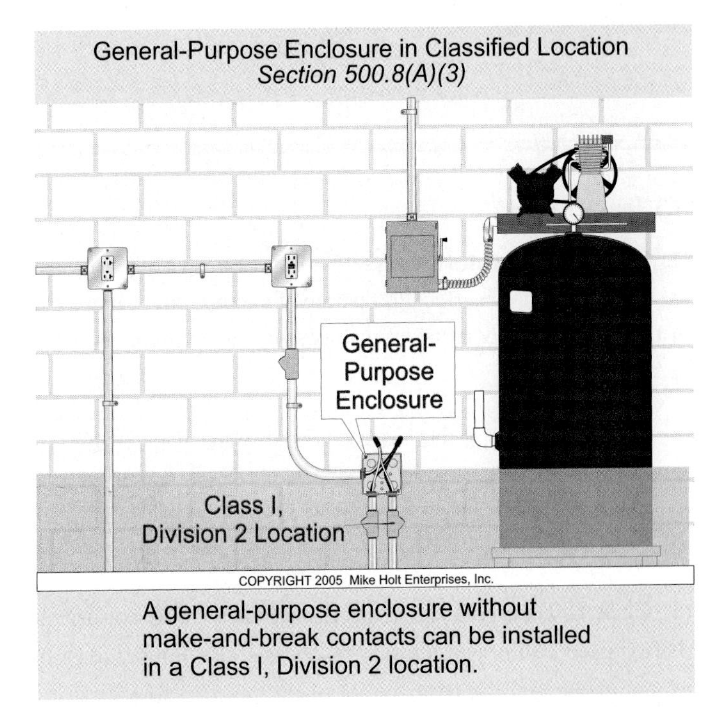

A general-purpose enclosure without make-and-break contacts can be installed in a Class I, Division 2 location.

Figure 500–13

(B) Equipment Markings. In addition to the class, division, and group rating, all heat-producing equipment, such as luminaires, motors, etc., installed in a hazardous (classified) location must be marked to indicate its operation temperature or temperature range (T-Rating).

> **Author's Comment:** The T-Ratings of explosionproof luminaires (Class I, Division 1) are based on the hottest exposed surface area of the luminaire, and the T-Ratings of luminaires installed in Class I, Division 2 locations are based on the hottest surface temperature of the lamp. **Figure 500-14**

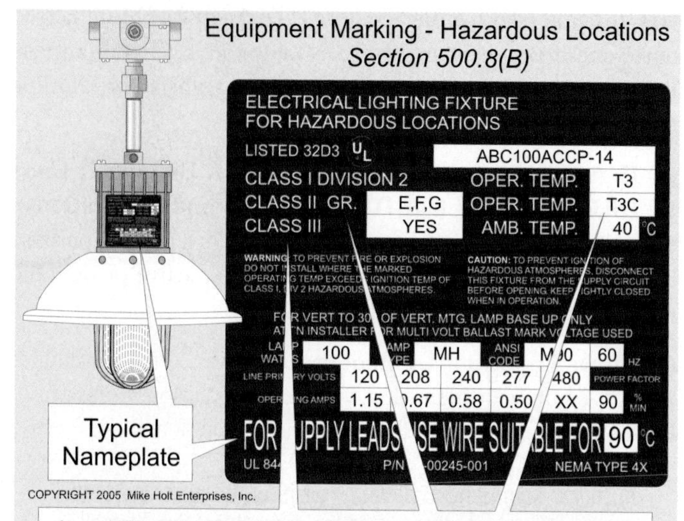

Figure 500–14

Table 500.8(B) Classification of Maximum Surface Temperature

Maximum °C	Temperature °F	Identification Number
450	842	T1
300	572	T2
280	536	T2A
260	500	T2B
230	446	T2C
215	419	T2D
200	392	T3
180	356	T3A
165	329	T3B
160	320	T3C
135	275	T4
120	248	T4A
100	212	T5
85	185	T6

(C) Temperature.

(1) Class I Temperature. Class I equipment must not permit the exposed equipment surface to operate at a temperature in excess of the ignition temperature of the specific gas or vapor, as listed in 500.8(B).

> **Author's Comment:** This is accomplished by ensuring that the temperature marking on the equipment (especially luminaires and motors) doesn't exceed the ignition temperature of the specific gas or vapor to be encountered.

(2) Class II Temperature. Class II equipment must not permit the exposed equipment surface to operate at a temperature in excess of the ignition temperature of the specific dust, as listed in 500.8(B).

> **Author's Comment:** This is accomplished by ensuring that the temperature marking on the equipment (especially luminaires) doesn't exceed the ignition temperature of the specific dust to be encountered.

(D) Threaded Conduit. Where threaded conduit is required in Articles 501 through 517, it must be threaded with a National Pipe Thread (NPT) taper of ¾ in. per foot. In addition, all threaded conduits must be made wrenchtight to prevent arcing when ground-fault current flows through the conduit system and to ensure the explosion or dust-ignitionproof integrity of the conduit system. Threaded entries into explosionproof equipment must be made up with at least five threads fully engaged.

Exception: For listed explosionproof equipment, factory threaded entries must be made up with at least 4½ threads fully engaged.

> **Author's Comments:**
>
> - This requirement ensures that if an explosion occurs within a conduit or enclosure, the expanding gas will sufficiently cool as it dissipates through the threads. This prevents hot flaming gases from igniting the surrounding atmosphere of a hazardous (classified) location.
> - Keep in mind - it is assumed that the flammable atmosphere outside the conduit will seep into the conduit system over time. The goal of the *Code* is to contain any explosion that occurs inside the conduit so the event will not ignite the flammable mixture outside the conduit system.

(E) Optical Fiber Cable. Optical fiber cable without current-carrying conductors can be installed in accordance with Article 770. They need not be installed in accordance with the requirements of Articles 500, 501, 502, or 503. **Figure 500-15**

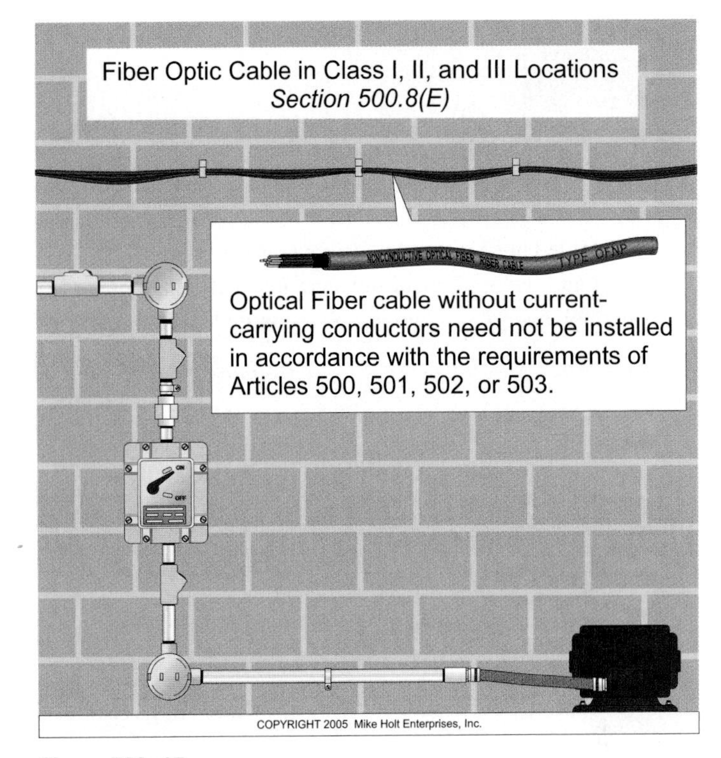

Figure 500–15

500.9 Specific Occupancies. In addition, the following articles also apply:

- Article 511—Commercial Garages, Repair, and Storage
- Article 513—Aircraft Hangars
- Article 514—Motor Fuel Dispensing Facilities
- Article 515—Bulk Storage Plants
- Article 516—Spray Application, Dipping, and Coating Processes
- Article 517—Health Care Facilities

Article 500 Questions

1. Portable battery-operated devices such as cordless drills, cell phones, etc., must be listed for use in a hazardous (classified) location and the equipment must comply with the requirements of Chapter 5. This is because battery-operated devices can produce enough energy in a hazardous (classified) location to ignite a fire or cause an explosion.

   (a) True (b) False

2. Class I, Division 2 usually includes locations where volatile flammable liquids or flammable gases or vapors are used but that, in the judgment of the authority having jurisdiction, would become hazardous only in case of an accident or of some unusual operating condition.

   (a) True (b) False

3. Locations in which easily ignitible combustible fibers are stored or handled other than in the process of manufacturing are designated as _____.

   (a) Class II, Division 2 (b) Class III, Division 1 (c) Class III, Division 2 (d) nonhazardous

4. Equipment is required to be identified not only for the class of location but also for the explosive, combustible, or ignitible properties of the specific _____ that will be present.

   (a) gas or vapor (b) dust (c) fiber or flyings (d) all of these

5. A fiber optic cable assembly that contains current-carrying conductors must be installed according to the applicable requirements of Articles 500, 501, 502, and 503.

   (a) True (b) False

ARTICLE

501 Class I Hazardous (Classified) Locations

Introduction

If enough flammable gases or vapors are present to produce an explosive or ignitible mixture, you have a Class I location. Examples of such locations include fuel storage areas, certain solvent storage areas, grain processing (where hexane is used), plastic extrusion where oil removal is part of the process, refineries, and paint storage areas.

Article 500 contains a general background on hazardous (classified) locations as well as describing the differences between Class I, II, and III locations and the difference between Division 1 and Division 2 in each of the three classifications.

Article 501 contains the actual Class I, Division 1 and Division 2 installation requirements, including wiring methods, seals, and specific equipment requirements.

A Class I hazardous (classified) location is an area where flammable gases or vapors may be present in quantities sufficient to produce an explosive or ignitible mixture.

PART I. GENERAL

501.1 Scope. Article 501 covers the requirements for electrical and electronic equipment and wiring for all voltages in Class I, Division 1 and 2 locations where fire or explosion hazards may exist due to flammable gases, vapors, or flammable liquids.

501.5 General. Electric wiring and equipment installed in a Class I hazardous (classified) location must be installed in accordance with the installation requirements contained in Chapters 1 through 4, as well as the requirements contained in this article.

> **Author's Comment:** 501.140 contains specific installation requirements on the use of flexible cords. In addition to those requirements, the use of cords in Class I locations must comply with the general installation requirements contained in Article 400.

PART II. WIRING

501.10 Wiring Methods.

(A) Class I, Division 1.

(1) Wiring Methods. Only the following wiring methods are permitted within a Class I, Division 1 location. **Figure 501-1**

(a) Fixed Wiring. Threaded rigid metal conduit or intermediate metal conduit with explosionproof fittings.

(b) Type MI cable with termination fittings listed for the location.

(c) Type MC-HL cable with fittings listed for a Class I, Division 1 location is permitted in industrial establishments with restricted public access where only qualified persons will service the installation. **Figure 501-2**

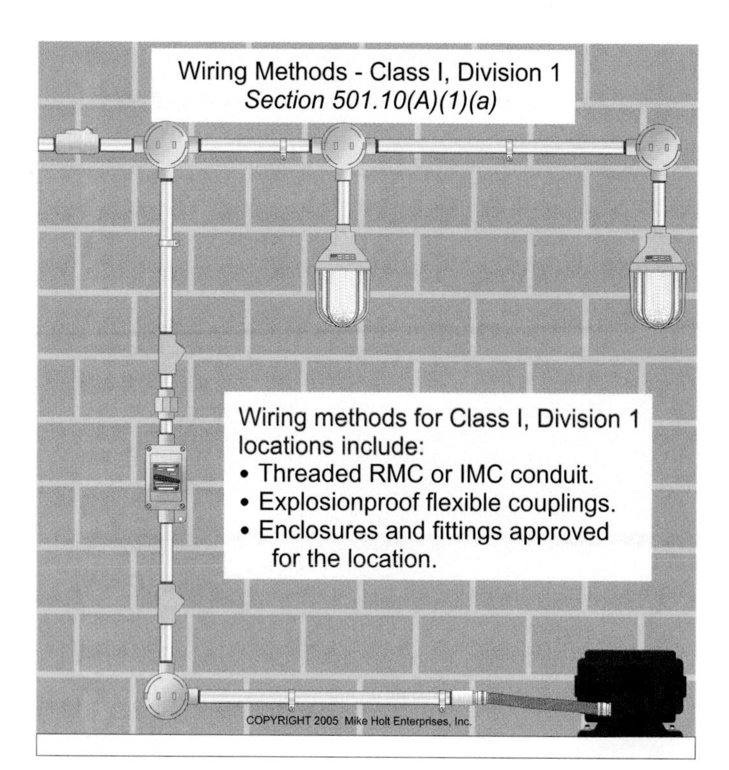
Wiring Methods - Class I, Division 1
Section 501.10(A)(1)(a)
Wiring methods for Class I, Division 1 locations include:
• Threaded RMC or IMC conduit.
• Explosionproof flexible couplings.
• Enclosures and fittings approved for the location.
COPYRIGHT 2005 Mike Holt Enterprises, Inc.

Figure 501–1

Author's Comment: Article 100 defines a "Qualified Person" as one who has skills and knowledge related to the construction and operation of the equipment and has received safety training on the hazards involved.

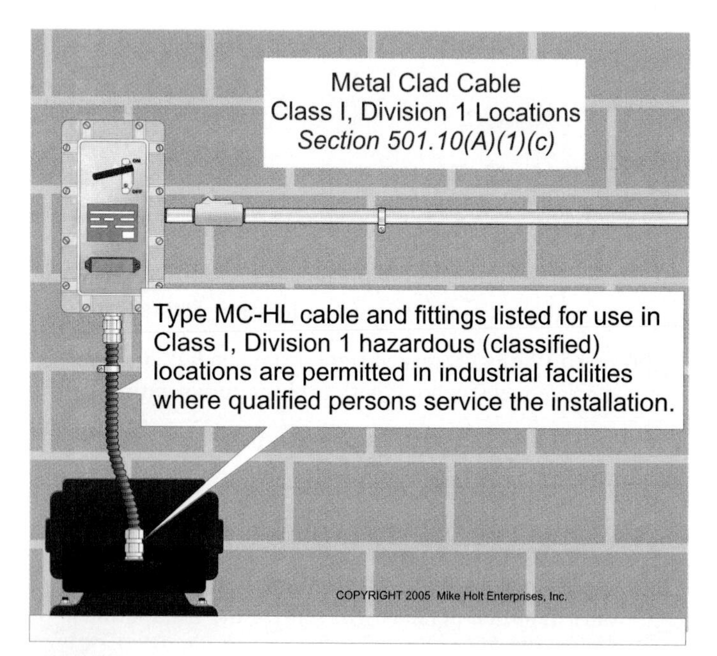
Metal Clad Cable
Class I, Division 1 Locations
Section 501.10(A)(1)(c)
Type MC-HL cable and fittings listed for use in Class I, Division 1 hazardous (classified) locations are permitted in industrial facilities where qualified persons service the installation.
COPYRIGHT 2005 Mike Holt Enterprises, Inc.

Figure 501–2

(d) Type ITC-HL cable with fittings listed for a Class I, Division 1 location is permitted in industrial establishments where only qualified persons will service the installation.

(2) Flexible Wiring. Explosionproof flexible metal couplings are permitted when necessary for vibration, movement, or for difficult bends. Flexible cords can be installed in a Class I, Division 1 location if installed in accordance with 501.140.

(3) Boxes and Fittings. All boxes and fittings must be approved by the authority having jurisdiction for Class I, Division 1 locations.

Author's Comment: See Article 100 for the definition of "Approved."

(B) Class I, Division 2.

(1) General. Only the following wiring methods are permitted within a Class I, Division 2 location. **Figure 501-3**

(1) Wiring methods permitted in Class I, Division 1 areas [501.10(A)].

(2) Threaded rigid metal or intermediate metal conduit.

(3) Enclosed gasketed busways and enclosed gasketed wireways.

(4) Type PLTC cable in accordance with the provisions of Article 725 installed in a manner to avoid tensile stress at the termination fittings.

(5) Type ITC cable as permitted in 727.4.

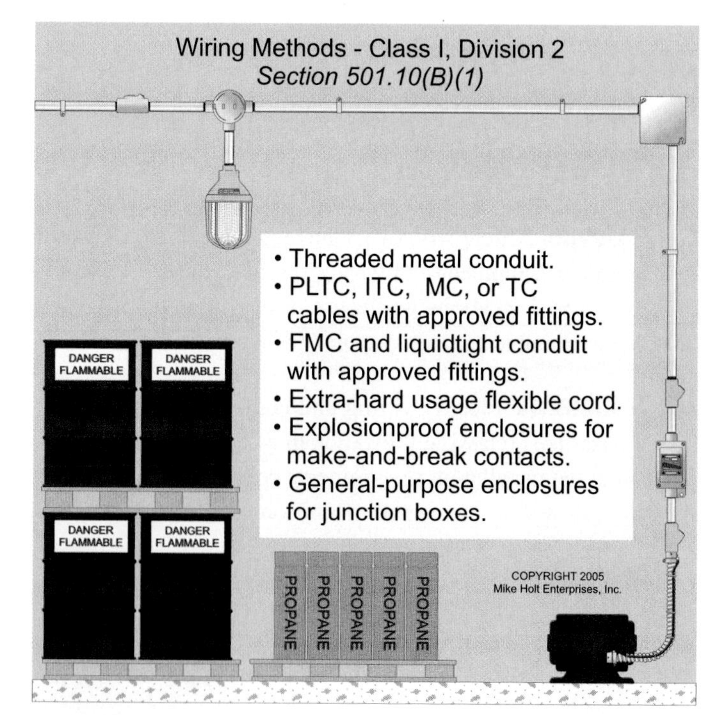
Wiring Methods - Class I, Division 2
Section 501.10(B)(1)
• Threaded metal conduit.
• PLTC, ITC, MC, or TC cables with approved fittings.
• FMC and liquidtight conduit with approved fittings.
• Extra-hard usage flexible cord.
• Explosionproof enclosures for make-and-break contacts.
• General-purpose enclosures for junction boxes.
DANGER FLAMMABLE
DANGER FLAMMABLE
DANGER FLAMMABLE
DANGER FLAMMABLE
PROPANE
PROPANE
PROPANE
PROPANE
PROPANE
COPYRIGHT 2005 Mike Holt Enterprises, Inc.

Figure 501–3

(6) Cables—MI, MC, MV, or TC cable with termination fittings.

(2) Flexible Wiring. Where limited flexibility is necessary, one or more of the following are also permitted:

(1) Flexible metal fittings

(2) Flexible metal conduit with listed fittings

(3) Liquidtight flexible metal conduit with listed fittings

(4) Liquidtight flexible nonmetallic conduit with listed fittings

(5) Flexible cords listed for extra-hard usage, containing an equipment grounding (bonding) conductor, and terminated with a listed bushed fitting

Author's Comments:

- See Article 100 for the definition of "Grounding Conductor, Equipment."
- Where flexible cords are used, they must comply with 501.140.

FPN: See 501.30(B) for grounding (bonding) requirements where flexible conduit is used.

(4) Boxes and Fittings. General-purpose enclosures and fittings are permitted unless the enclosure contains make-and-break contacts for meters, instruments, and relays [501.105(B)(1)], switches, circuit breakers, or motor controllers [501.115(B)(1)], signaling, alarm, remote-control, and communications systems [501.150(B)].

Author's Comment: See Article 100 for the definition of "Controller."

501.15 Conduit and Cable Seals.

FPN: Conduit and cable seals must be installed to:

- Minimize the passage of gases and vapors from one portion of electrical equipment to another through the conduit or cable.
- Minimize the passage of flames from one portion of electrical equipment to another through the conduit or cable.
- Limit internal explosions to within the explosionproof enclosure.

(A) Conduit Seal—Class I, Division 1. In Class I, Division 1 locations, conduit seals must be located as follows:

(1) Entering Enclosures. A conduit seal is required in each conduit that enters an explosionproof enclosure where either:

(1) Enclosure with Make-and-Break Contacts. A conduit seal fitting must be installed in each conduit that enters an explosionproof enclosure that contains make-and-break contacts.

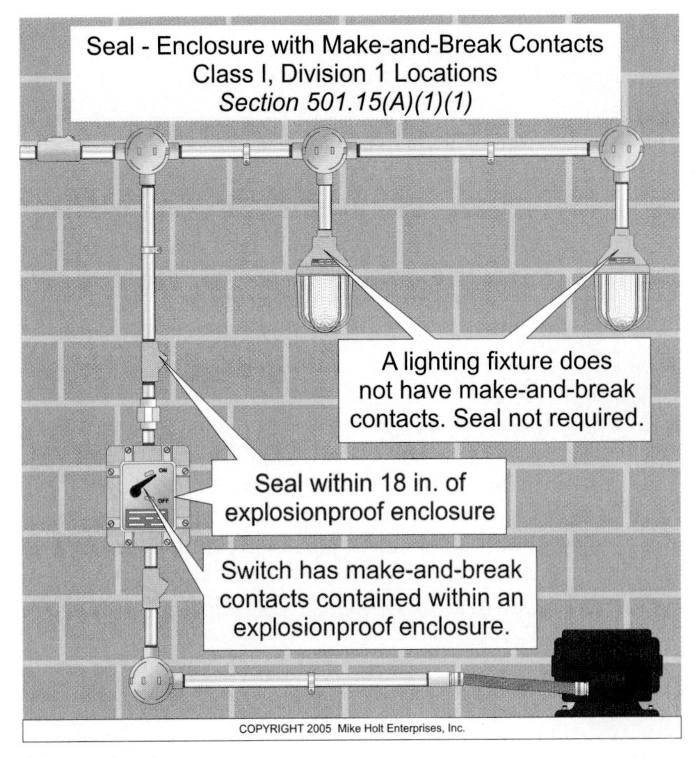

Figure 501–4

The seal fitting must be installed within 18 in. of the explosionproof enclosure. **Figure 501-4**

(2) Enclosures without Make-and-Break Contacts. A conduit seal fitting isn't required for trades size ½, ¾, 1, 1¼, or 1½ conduits that enter an explosionproof enclosure that doesn't contain any make-and-break contacts (junction and splice box). An example would be an enclosure that only contains terminals, splices, or taps. However, trade size 2 or larger conduit that enters any explosionproof enclosure must have a conduit seal fitting installed within 18 in. of the explosionproof enclosure [501.15(A)(1)(1)].

Author's Comment: An explosionproof enclosure that contains trade size 2 and larger, as well as smaller diameter conduits only requires a conduit seal for the conduits that are trade size 2 or larger. **Figure 501-5**

Fittings Between Conduit Seal and Enclosure. Only explosionproof unions, couplings, reducers, elbows, capped elbows, and conduit bodies are permitted between the conduit seal and the explosionproof enclosure. **Figure 501-6**

(2) Purge Systems. A conduit seal fitting must be installed in each conduit that isn't pressurized where the conduit enters a pressurized enclosure. The conduit seal fitting must be installed within 18 in. for each conduit that terminates to the pressurized enclosure.

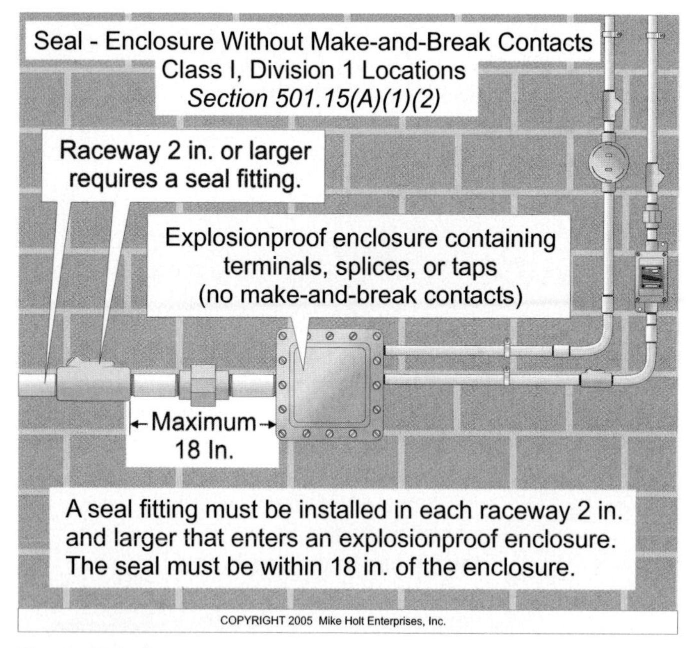

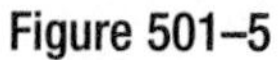
Figure 501–5

(3) Between Explosionproof Enclosures. A single conduit seal is permitted between two explosionproof enclosures containing make-and-break contacts if the conduit seal fitting is located not more than 18 in. from either explosionproof enclosure. **Figure 501-7**

(4) Boundary. A conduit seal fitting must be installed in each conduit that leaves a Class I, Division 1 location within 10 ft of the Class I, Division 1 location on either side of the boundary. There must be no fitting, except a listed explosionproof reducer, between the seal fitting and the point at which the conduit leaves the Class I, Division 1 location. **Figure 501-8**

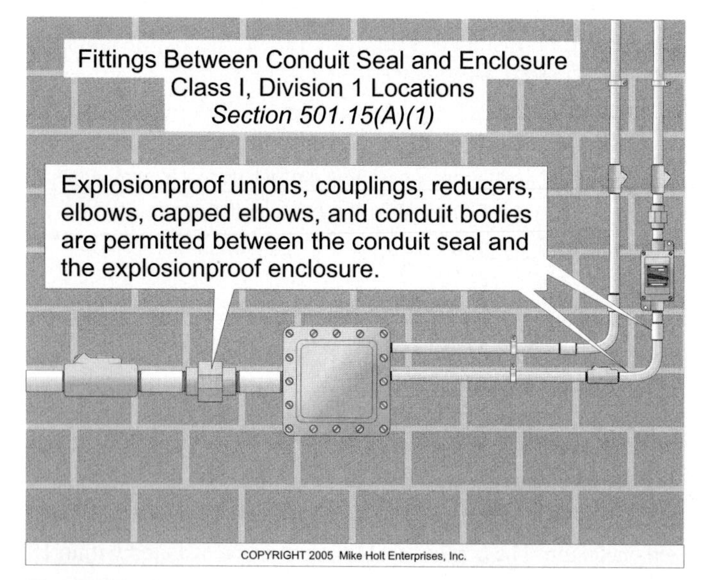

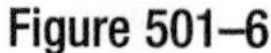
Figure 501–6

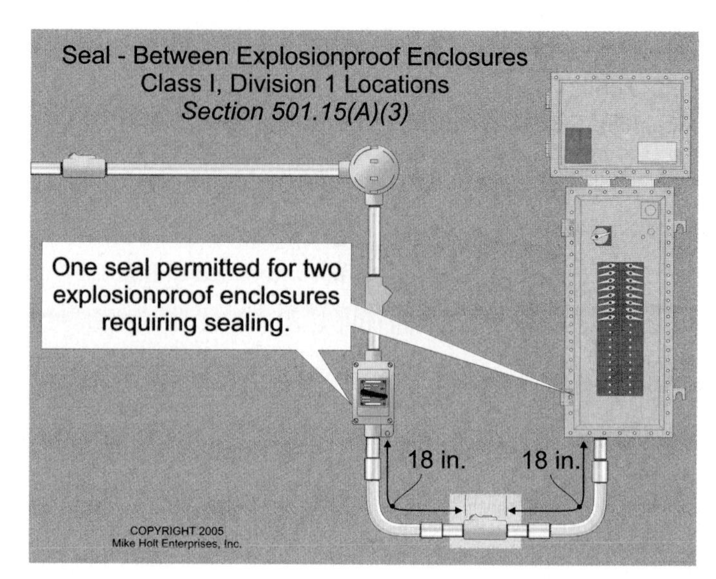

Figure 501–7

Exception 1: A conduit boundary seal fitting isn't required for a conduit that passes unbroken through the Class I, Division 1 area with no fittings less than 1 ft beyond the boundary. **Figure 501-9**

Exception 2: The conduit seal can be located aboveground, after the conduit leaves the ground, where underground conduit is installed in accordance with 300.5 and the boundary is underground. **Figure 501-10**

(B) Conduit Seal—Class I, Division 2. In Class I, Division 2 locations, conduit seals must be located as follows:

(1) Enclosures with Make-and-Break Contacts. A conduit seal fitting must be installed in each conduit that enters an enclosure that must be explosionproof and that contains make-and-break contacts. The seal fitting must be installed within 18 in. of the explosionproof enclosure.

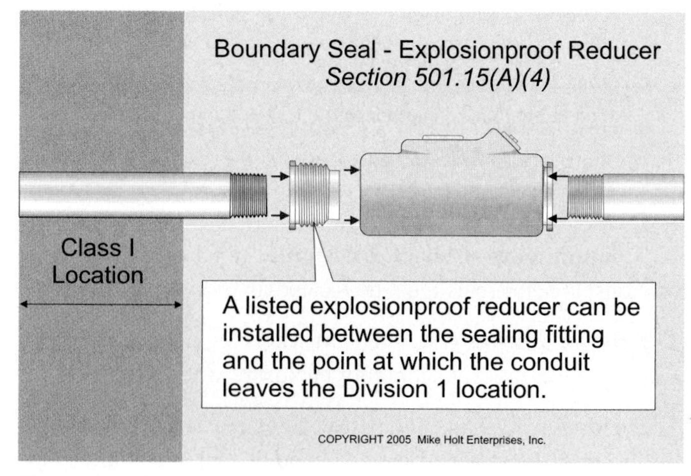

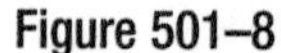
Figure 501–8

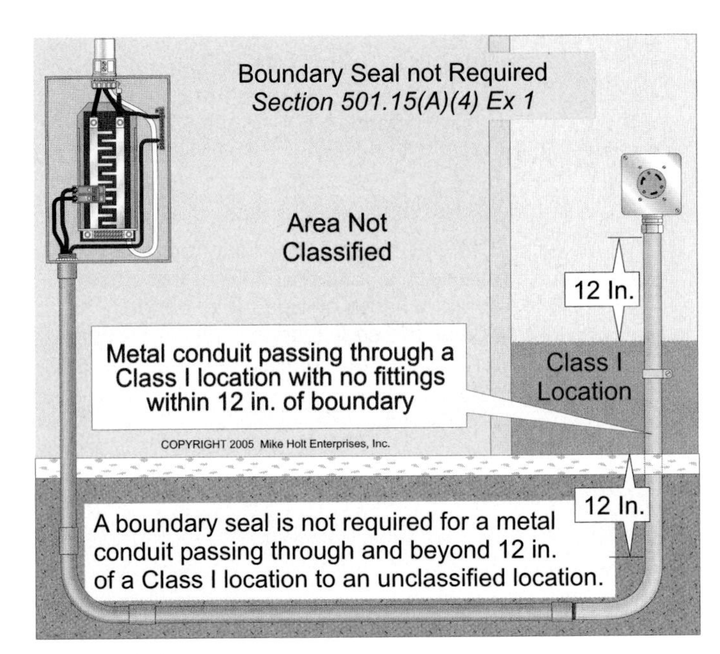
Boundary Seal not Required
Section 501.15(A)(4) Ex 1
Area Not
Classified
12 In.
Metal conduit passing through a
Class I location with no fittings
within 12 in. of boundary
Class I
Location
COPYRIGHT 2005 Mike Holt Enterprises, Inc.
12 In.
A boundary seal is not required for a metal
conduit passing through and beyond 12 in.
of a Class I location to an unclassified location.

Figure 501–9

Exception: A conduit seal isn't required if the make-and-break contacts are contained within a hermetically sealed chamber or immersed in oil in accordance with 501.115(B)(1)(2), or enclosed within a factory-sealed explosionproof chamber.

A single conduit seal is permitted between two explosionproof enclosures containing make-and-break contacts, if the conduit seal fitting is located not more than 18 in. from either explosionproof enclosure.

(2) Conduit Boundary Seal at Unclassified Location. A conduit seal fitting must be installed in each conduit that passes from a Class I, Division 2 location into an unclassified location within 10 ft of the Class I, Division 2 area on either side of the boundary. Rigid metal conduit or threaded steel intermediate metal conduit must be used between the sealing fitting and the point at which the conduit leaves the Division 2 location, and a threaded connection must be used at the sealing fitting. Except for listed reducers at the conduit seal, there must be no union, coupling, box, or fitting between the conduit seal and the point at which the conduit leaves the Division 2 location.

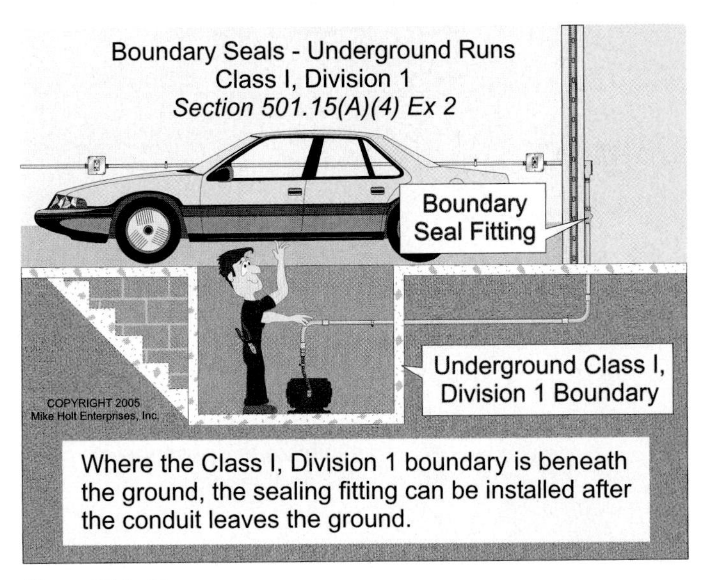
Boundary Seals - Underground Runs
Class I, Division 1
Section 501.15(A)(4) Ex 2
Boundary
Seal Fitting
Underground Class I,
Division 1 Boundary
COPYRIGHT 2005
Mike Holt Enterprises, Inc.
Where the Class I, Division 1 boundary is beneath
the ground, the sealing fitting can be installed after
the conduit leaves the ground.

Figure 501–10

Conduit boundary seals aren't required to be explosionproof, but must be identified for the purpose of minimizing the passage of gases under normal operating conditions and must be accessible. **Figure 501-11**

Author's Comments:

- See Article 100 for the definition of "Accessible" as it relates to wiring methods.
- The conduit boundary seal at unclassified locations is to minimize the passage of gases or vapors, not to contain explosions in the conduit system.

Exception 1: A conduit boundary seal fitting isn't required for a conduit that passes through the Class I, Division 2 area unbroken with no fittings less than 1 ft beyond the boundary to an unclassified location. See **Figure 501-9**.

Exception 2: A conduit boundary seal fitting isn't required for conduits that terminate at an outdoor unclassified location for cable trays, cablebus, ventilated busway, Type MI cable, or open wiring.

Exception 3: A boundary seal fitting isn't required for conduit that passes from an enclosure or room that is unclassified as a result of pressurization into a Class I, Division 2 location.

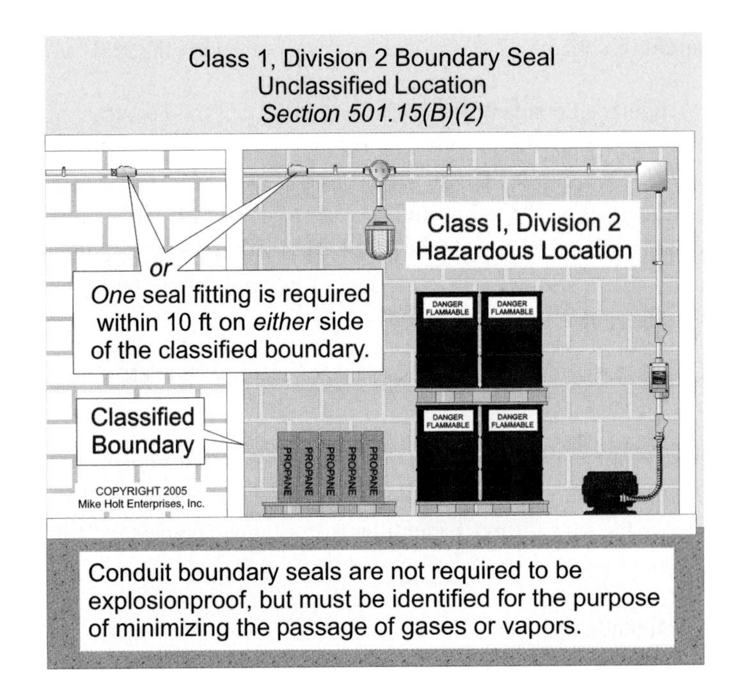
Class 1, Division 2 Boundary Seal
Unclassified Location
Section 501.15(B)(2)
Class I, Division 2
Hazardous Location
or
One seal fitting is required
within 10 ft on either side
of the classified boundary.
Classified
Boundary
COPYRIGHT 2005
Mike Holt Enterprises, Inc.
PROPANE
Conduit boundary seals are not required to be
explosionproof, but must be identified for the purpose
of minimizing the passage of gases or vapors.

Figure 501–11

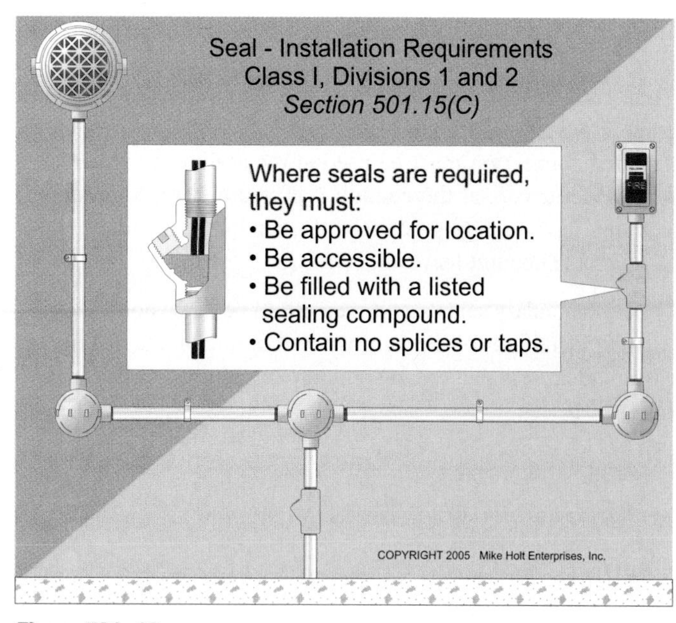
Seal - Installation Requirements
Class I, Divisions 1 and 2
Section 501.15(C)
Where seals are required, they must:
• Be approved for location.
• Be accessible.
• Be filled with a listed sealing compound.
• Contain no splices or taps.
COPYRIGHT 2005 Mike Holt Enterprises, Inc.

Figure 501–12

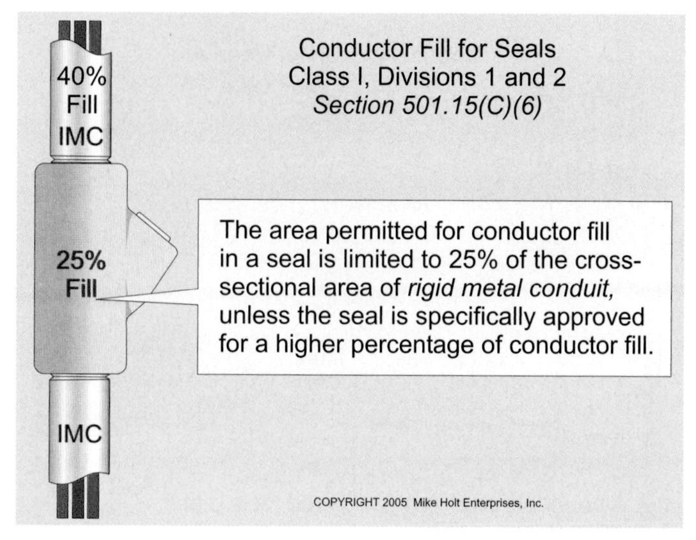
40% Fill IMC
25% Fill
IMC
Conductor Fill for Seals
Class I, Divisions 1 and 2
Section 501.15(C)(6)
The area permitted for conductor fill in a seal is limited to 25% of the cross-sectional area of rigid metal conduit, unless the seal is specifically approved for a higher percentage of conductor fill.
COPYRIGHT 2005 Mike Holt Enterprises, Inc.

Figure 501–13

(C) Conduit Seal—Installation Requirements. Where explosionproof sealing fittings are required in Class I, Division 1 and 2 locations, they must comply with (1) through (6): **Figure 501-12**

Exception: Seals are not required to be explosionproof in accordance with 501.15(B)(2).

(1) Fittings. Conduit seal fittings must be listed for the specific sealing compounds and Class I location.

(2) Compound. The conduit seal compound must be mixed and installed in accordance with manufacturer's instructions so it minimizes the passage of gases and vapors through the seal.

> **Author's Comment:** The sealing compound must be from the same manufacturer as the conduit seal.

(3) Thickness of Compounds. Except for listed cable sealing fittings, the minimum thickness of the conduit seal compound must not be less than the trade size of the seal fitting, but in no case less than ⅝ in.

(4) Splices and Taps. Splices and taps must not be made within a conduit seal fitting.

(6) Number of Conductors. The conductor area must not exceed 25 percent of the cross-sectional area of rigid metal conduit, unless the seal fitting is identified for a higher percentage fill. **Figure 501-13**

> **Author's Comment:** The cross-sectional area of intermediate metal conduit is approximately 7 percent greater than that of rigid metal conduit because the wall thickness of intermediate metal conduit is less than rigid metal conduit. If the cross-sectional area of intermediate metal conduit were used for conductor fill calculations, the 25 percent of "rigid metal conduit conductor fill" could be exceeded.

(D) Cable Seal—Class I, Division 1. In Class I, Division 1 locations, cable seals must be located as follows:

(1) Terminations. Type MC-HL cable is inherently gas/vaportight by the construction of the cable, but the termination fittings must permit the sealing compound to surround each individual insulated conductor to exclude moisture and other fluids. **Figure 501-14**

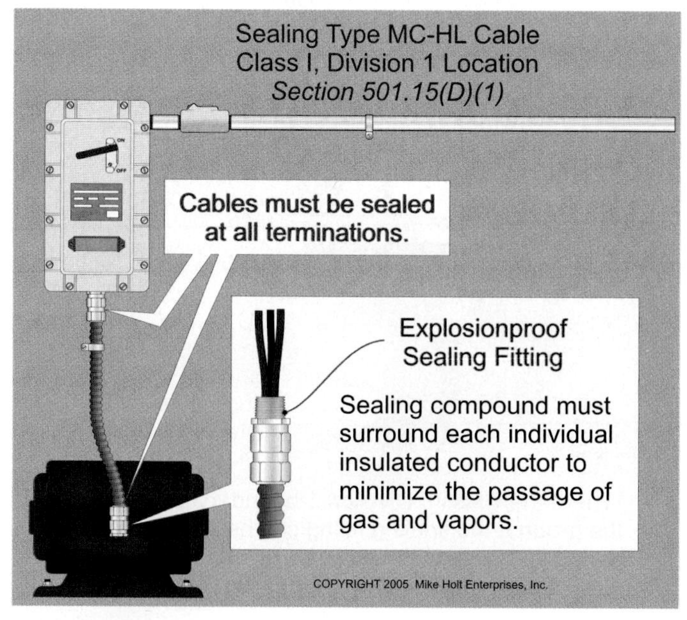
Sealing Type MC-HL Cable
Class I, Division 1 Location
Section 501.15(D)(1)
Cables must be sealed at all terminations.
Explosionproof Sealing Fitting
Sealing compound must surround each individual insulated conductor to minimize the passage of gas and vapors.
COPYRIGHT 2005 Mike Holt Enterprises, Inc.

Figure 501–14

Exception: The removal of shielding material or the separation of the twisted pairs isn't required within the cable seal fitting.

(2) Multiconductor Cable Within a Conduit. Conduits containing cables must be sealed after removing the jacket and any other coverings so that the sealing compound surrounds each individual insulated conductor in a manner so as to minimize the passage of gases and vapors.

Exception: The removal of shielding material or the separation of the twisted pairs isn't required within the conduit seal fitting. **Figure 501-15**

(3) Multiconductor Cables Within a Conductor. Each multiconductor cable in conduit is considered to be a single conductor if the cable is incapable of transmitting gases or vapors through the cable core.

(E) Cable Seal—Class I, Division 2. In Class I, Division 2 locations, cable seals must be located as follows:

(1) Multiconductor Cable. Multiconductor cables that enter an explosionproof enclosure must be sealed after removing the jacket and any other coverings so that the sealing compound will surround each individual insulated conductor in a manner so as to minimize the passage of gases and vapors.

Multiconductor cables installed in a conduit must be sealed in accordance with 501.15(D)(2) or (3).

Exception: The removal of shielding material or the separation of the twisted pairs isn't required within the cable seal fitting.

(4) Cable Seal—Boundary. Cables without gas/vaportight continuous sheath must be sealed at the boundary of the Class I, Division 2 location in a manner so as to minimize the passage of gases or vapors into an unclassified location.

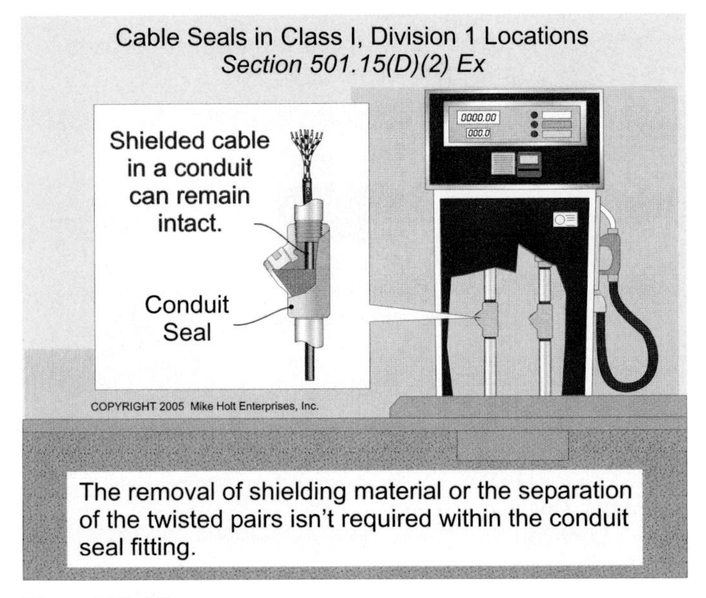
Cable Seals in Class I, Division 1 Locations
Section 501.15(D)(2) Ex
Shielded cable in a conduit can remain intact.
Conduit Seal
COPYRIGHT 2005 Mike Holt Enterprises, Inc.
The removal of shielding material or the separation of the twisted pairs isn't required within the conduit seal fitting.

Figure 501–15

501.20 Conductor Insulation. Conductors installed in Class I, Division 1 and 2 locations must be identified for use under such conditions.

> **Author's Comment:** According to the UL *Electrical Construction Material Directory*, nylon-jacketed THWN conductors are suitable for use where exposed to oil or gasoline if marked "Type THWN Gasoline and Oil Resistant I."

501.30 Bonding. Wiring and equipment in a Class I, Division 1 or 2 location must be grounded (bonded) to an effective ground-fault current path in accordance with Article 250 [250.100] and bonded in accordance with (A) and (B) below.

(A) Bonding—Metal Raceway. All metal raceways, enclosures, and fittings between a Class I, Division 1 or 2 hazardous (classified) location and the service or separately derived system must be bonded to an effective ground-fault current path. This is accomplished by bonding the metal raceway to the enclosure or fitting by one of the following methods:

- Threaded conduit entry
- Bonding bushing with a bonding jumper sized to Table 250.122
- A bonding-type locknut

> **Author's Comment:** See Article 100 for the definitions of "Bond," "Bonding Jumper," "Raceway," "Separately Derived System," and "Service."

Standard locknuts alone are not suitable for this purpose. **Figure 501-16**

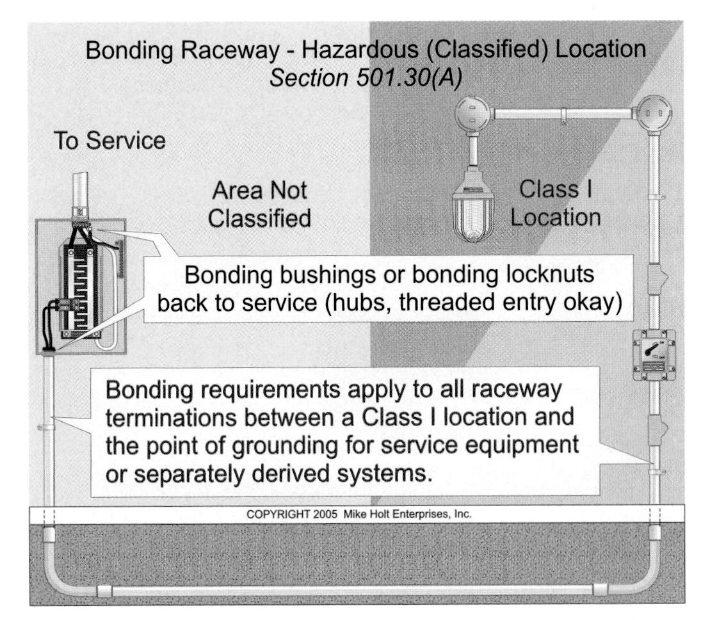
Bonding Raceway - Hazardous (Classified) Location
Section 501.30(A)
To Service
Area Not Classified
Class I Location
Bonding bushings or bonding locknuts back to service (hubs, threaded entry okay)
Bonding requirements apply to all raceway terminations between a Class I location and the point of grounding for service equipment or separately derived systems.
COPYRIGHT 2005 Mike Holt Enterprises, Inc.

Figure 501–16

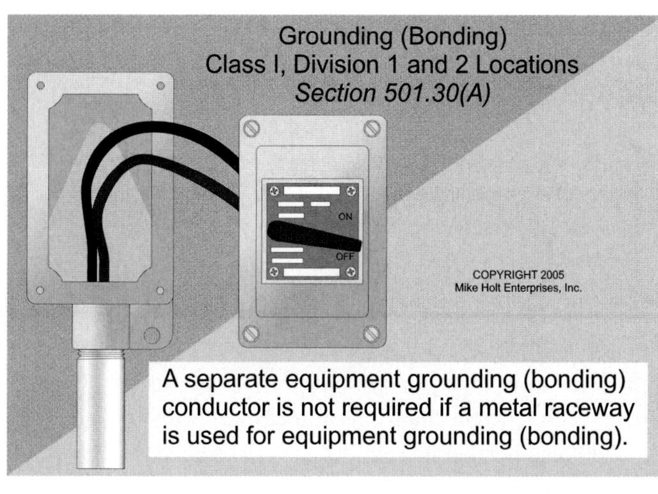
Grounding (Bonding)
Class I, Division 1 and 2 Locations
Section 501.30(A)
ON
OFF
COPYRIGHT 2005
Mike Holt Enterprises, Inc.
A separate equipment grounding (bonding) conductor is not required if a metal raceway is used for equipment grounding (bonding).

Figure 501–17

> **Author's Comment:** A separate equipment grounding (bonding) conductor is not required in a metal raceway that is used as the effective ground-fault current path in accordance with 250.118. **Figure 501-17**

(B) Bonding—Flexible Raceway. Where flexible metallic conduit or liquidtight flexible metal conduit is used in a Class I, Division 2 location as permitted in 501.10(B), it must be installed with an internal or external equipment bonding jumper. If external, the bonding jumper must not be longer than 6 ft, and it must be routed with the flexible conduit. The equipment bonding jumper is sized based on the circuit overcurrent protection device rating in accordance with 250.122 [250.102(E)]. **Figure 501-18**

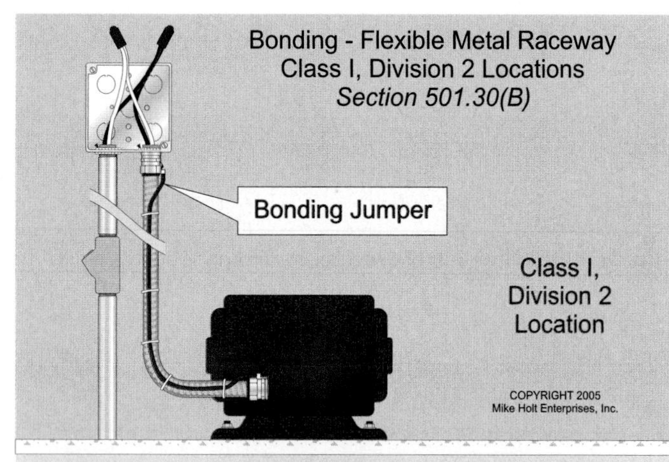
Bonding - Flexible Metal Raceway
Class I, Division 2 Locations
Section 501.30(B)
Bonding Jumper
Class I, Division 2 Location
COPYRIGHT 2005
Mike Holt Enterprises, Inc.
Flexible metal or liquidtight metal conduit can be installed in a Class I, Division 2 location [501.10(B)], but it must have an internal or external bonding jumper routed with the raceway.

Figure 501–18

> **Author's Comment:** The *Code* doesn't limit the length of an equipment grounding (bonding) conductor if it's installed inside a flexible raceway.

501.35 Surge Protection.

(A) Class I, Division 1. Surge arresters, transient voltage surge suppressors (TVSSs), and capacitors must be installed in enclosures identified for Class I, Division 1 locations.

(B) Class I, Division 2. Surge arresters and transient voltage surge suppressors (TVSSs) must be of a type designed for the specific duty, and they can be installed in general-purpose enclosures.

501.40 Multiwire Branch Circuits. Multiwire branch circuits cannot be used in Class I, Division 1 locations.

Exception: Multiwire branch circuits are permitted if all ungrounded conductors are opened simultaneously, such as with 2- or 3-pole circuit breakers. **Figure 501-19**

> **Author's Comment:** See Article 100 for the definition of "Branch Circuit, Multiwire."

PART III. EQUIPMENT

501.100 Transformers and Capacitors.

(A) Class I, Division 1.

(1) Containing Liquid That Will Burn. Transformers and capacitors located in a Class I, Division 1 location and containing a liquid that will burn must be installed in a vault in accordance with Article 450.

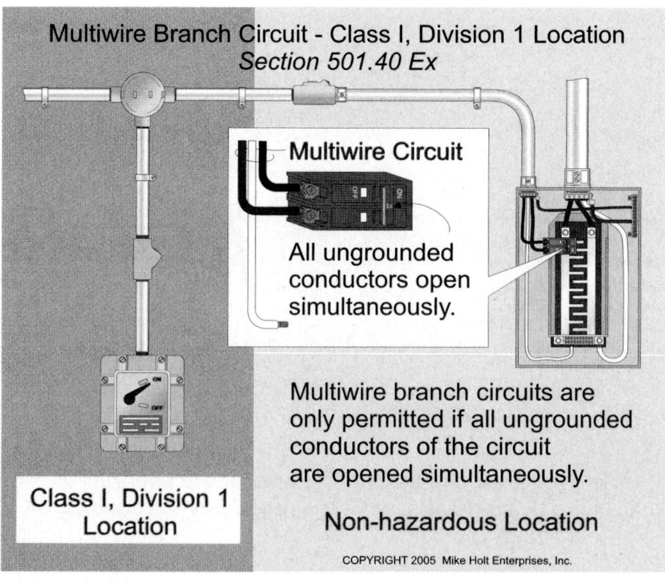
Multiwire Branch Circuit - Class I, Division 1 Location
Section 501.40 Ex
Multiwire Circuit
All ungrounded conductors open simultaneously.
Multiwire branch circuits are only permitted if all ungrounded conductors of the circuit are opened simultaneously.
Class I, Division 1 Location
Non-hazardous Location
COPYRIGHT 2005 Mike Holt Enterprises, Inc.

Figure 501–19

(2) Not Containing Liquid That Will Burn. Transformers and capacitors that don't contain a liquid that will burn must be installed in a vault, or the authority having jurisdiction must approve the equipment for a Class I, Division 1 location.

(B) Class I, Division 2. Transformers and capacitors installed in a Class I, Division 2 location must be installed in accordance with 450.21 through 450.27.

501.105 Meters, Instruments, and Relays.

(A) Class I, Division 1. Meters, instruments, and relays installed in a Class I, Division 1 location must be installed within an enclosure identified for a Class I, Division 1 location.

(B) Class I, Division 2.

(1) With Make-and-Break Contacts. Meters, instruments, and relays installed in a Class I, Division 2 location containing make-and-break contacts must be installed within an enclosure identified for a Class I, Division 1 location.

Exception: General-purpose enclosures are permitted if the make-and-break contacts are immersed in oil, enclosed within a chamber that is hermetically sealed against the entrance of gases or vapors, part of a nonincendive circuit, or listed for Class I, Division 2 locations.

501.115 Enclosures.

(A) Class I, Division 1. Enclosures for switches, circuit breakers, motor controllers, and fuses, including pushbuttons, relays, and similar devices, must be identified for use in a Class I, Division 1 location. **Figure 501-20**

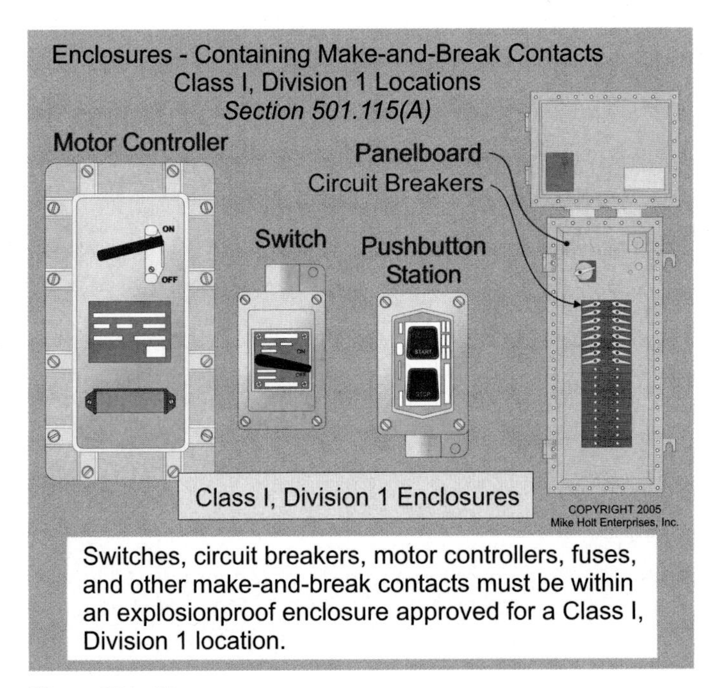

Figure 501–20

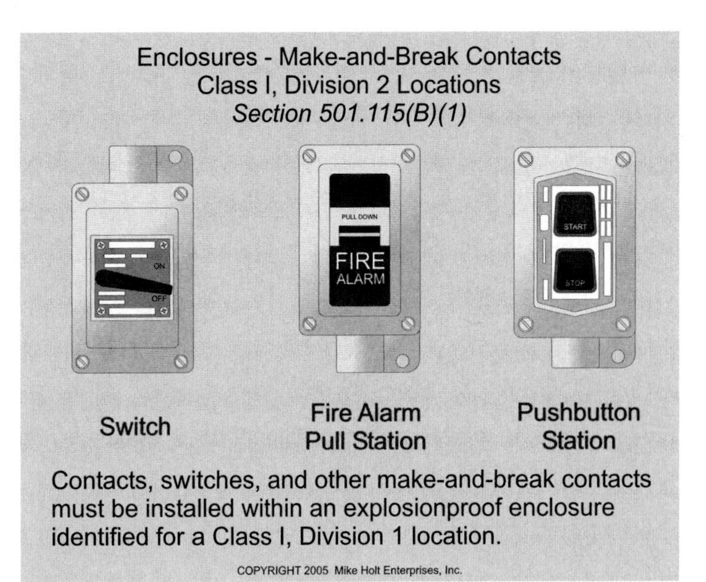

Figure 501–21

(B) Class I, Division 2.

(1) Type Required. Enclosures containing make-and-break contacts for switches, circuit breakers, motor controllers, fuses, pushbuttons, relays, and other make-and-break contact devices must be installed in an explosionproof enclosure in accordance with 501.105(A). **Figure 501-21**

General-purpose enclosures can be installed in a Class 1, Division 2 location if:

(1) The interruption of current occurs within a hermetically sealed chamber.

(2) The make-and-break contacts are oil-immersed.

(3) The interruption of current occurs within a factory-sealed explosionproof chamber.

501.120 Control Transformers and Relays.

(A) Class I, Division 1. Control transformers, coils, and relays must be installed within an explosionproof enclosure that is identified for a Class I, Division 1 location.

(B) Class I, Division 2. General-purpose enclosures are permitted for control transformers, solenoids, or coils. However, make-and-break contacts must be installed within an enclosure identified for a Class I, Division 1 location.

501.125 Motors and Generators.

(A) Class I, Division 1. Motors and generators installed in a Class I, Division 1 location must be:

(1) Identified for Class I, Division 1 locations; or be

(2) Of the totally enclosed type supplied with positive-pressure ventilation and arranged to automatically de-energize if the air supply fails; or be

(3) Of the totally enclosed inert gas-filled type and arranged to automatically de-energize if the gas supply fails; or be

(4) Submerged in a liquid that is flammable only when vaporized and mixed with air, and arranged to automatically de-energize if the liquid is reduced to atmospheric pressure (vaporized).

Totally enclosed motors of the types specified in 501.125(A)(2) or (A)(3) must be designed so no external surface has an operating temperature in excess of 80 percent of the ignition temperature of the gas or vapor involved.

(B) Class I, Division 2. Motors and generators installed in a Class I, Division 2 location may be of the open type (squirrel-cage induction motors without arcing devices) if the motor or generator doesn't contain any brushes, switching mechanisms, or similar arc-producing devices. However, motors and generators containing switching mechanisms must be of the type identified for a Class I, Division 1 location [501.125(A)]. **Figure 501-22**

501.130 Luminaires.

(A) Class I, Division 1. Luminaires installed in a Class I, Division 1 location must comply with the following:

(1) Luminaires. Luminaires must be identified for use in a Class I, Division 1 location. The luminaire must be completely enclosed and capable of both withstanding an explosion and preventing the ignition of the gas or vapor surrounding the luminaire. These luminaires must be marked to indicate their maximum lamp wattage.

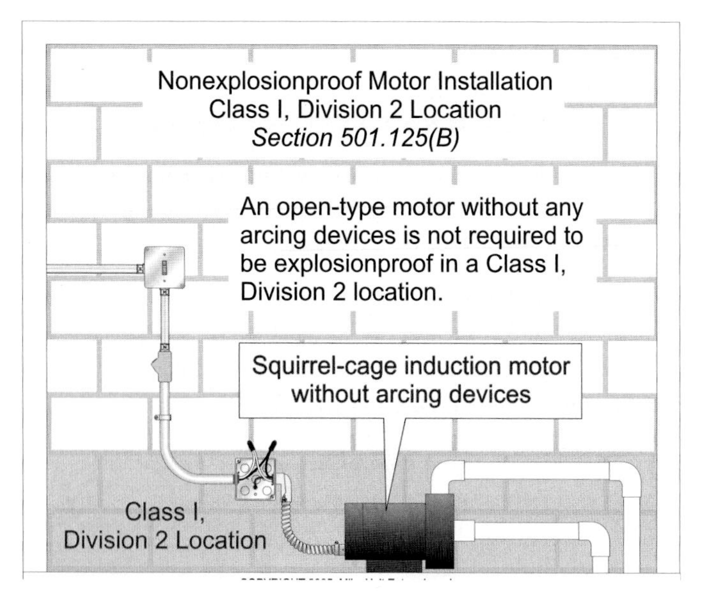

Nonexplosionproof Motor Installation
Class I, Division 2 Location
Section 501.125(B)
An open-type motor without any arcing devices is not required to be explosionproof in a Class I, Division 2 location.
Squirrel-cage induction motor without arcing devices
Class I, Division 2 Location

Figure 501–22

Author's Comment: Conduit seals aren't required for listed Class I, Division 1 explosionproof luminaires because the lamp compartment is separated or sealed from the wiring compartment in accordance with the listing requirements.

(2) Physical Damage. Fixed luminaires must be protected against physical damage by a suitable guard or by location.

(3) Pendant Luminaires. Pendant luminaires must be suspended by and supplied through threaded conduit stems, and threaded joints must be provided with set screws or other means to prevent loosening. Stems longer than 1 ft must be provided with permanent and effective lateral bracing or with a flexible fitting or connector identified for the Class I, Division 1 location. **Figure 501-23**

(4) Boxes and Fittings. Boxes or fittings used to support luminaires must be identified for a Class I, Division 1 location.

(B) Class I, Division 2. Luminaires installed in a Class I, Division 2 location must comply with the following:

(2) Physical Damage. Fixed luminaires must be protected from physical damage by suitable guards or by location. Where the lamp temperature exceeds 80 percent of the ignition temperature of the gas or vapor, luminaires must be identified for a Class I, Division 1 location.

(3) Pendant Luminaires. Pendant luminaires must be suspended by and supplied through threaded conduit stems, and threaded joints must be provided with set screws or other means to prevent loosening. Stems longer than 1 ft must be provided with permanent and effective lateral bracing, or an identified, flexible fitting or connector must be provided.

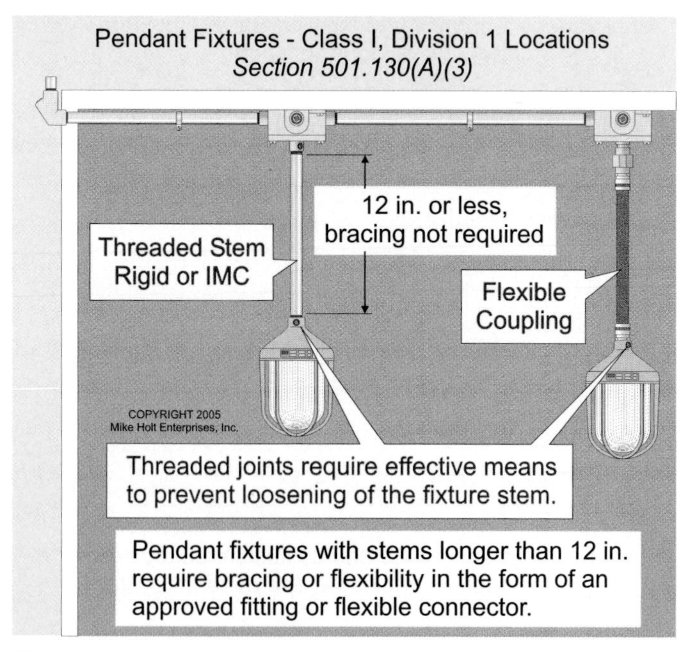

Figure 501–23

(4) Portable Lighting. Portable lighting equipment must be listed for use in a Class I, Division 1 location, unless the luminaire is mounted on movable stands and connected by a flexible cord as provided in 501.140.

(5) Switches. Switches that are a part of an assembled luminaire or lampholder must be installed within an enclosure identified for a Class I, Division 1 location.

501.140 Flexible Cords.

(A) Permitted Uses. Flexible cord is permitted for:

(1) Portable lighting or portable utilization equipment.

(2) In an industrial establishment where conditions of maintenance and engineering supervision ensure that only qualified persons install and service the installation where the fixed wiring methods of 501.10(A) can't provide the flexibility necessary or where necessary for mobile electrical equipment, and the flexible cord is protected by location or suitable guard.

(3) The extension within a suitable raceway between the submersible pump in a wet-pit and the power source.

(4) Electric mixers intended for travel into and out of open-type mixing tanks or vats.

(B) Installation. Where flexible cords are used, the cords must:

(1) Be listed for extra-hard usage.

(2) Contain an equipment grounding (bonding) conductor that complies with 400.23.

(3) Terminate in a manner approved by the authority having jurisdiction.

(4) Be supported so that no tension will be transmitted to the terminal connections.

(5) Be provided with suitable seals where required at explosionproof enclosures.

(6) Be of continuous length.

Author's Comment: 400.8 specifies that flexible cords are not to be used for any of the following:

- As a substitute for fixed wiring.
- Run through holes in walls, structural, suspended or dropped ceilings, or floors.
- Run through doorways, windows, or similar openings.
- Attached to the building surface except as allowed by 368.8 for branches from busways.
- Concealed behind building walls, floors, ceilings, or located above suspended or dropped ceilings.
- Within a raceway.

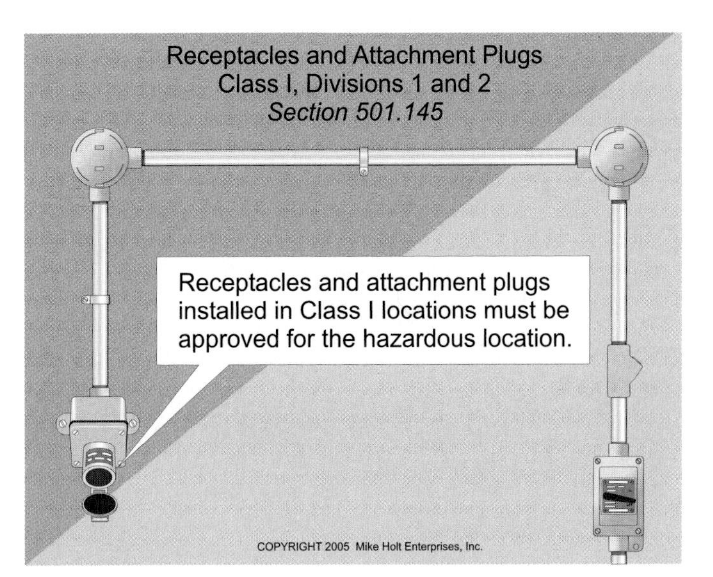
Receptacles and Attachment Plugs
Class I, Divisions 1 and 2
Section 501.145
Receptacles and attachment plugs installed in Class I locations must be approved for the hazardous location.
COPYRIGHT 2005 Mike Holt Enterprises, Inc.

Figure 501–24

501.145 Receptacles and Attachment Plugs. Receptacles and attachment plugs installed in a Class I area must be identified for the location. **Figure 501-24**

Author's Comments:

- See Article 100 for the definitions of "Attachment Plug" and "Receptacle."
- Receptacles listed for Class I locations can be any of the following types:

Interlocked Switch Receptacle—This receptacle contains a built-in rotary switch that is interlocked with the attachment plug. The switch must be off before the attachment plug may be inserted or removed.

Manual Interlocked Receptacle—The attachment plug is inserted into the receptacle, and then it's rotated to operate the receptacle switching contacts.

Delayed Action Receptacle—This receptacle requires an attachment plug and receptacle constructed so that an electrical arc will be confined within the explosionproof chamber of the receptacle.

501.150 Low-Voltage, Limited-Energy, and Communications Systems.

(A) Class I, Division 1. Low-voltage, limited-energy, and communications systems must be installed in accordance with 501.10(A) as follows:

Wiring Methods. Only the following wiring methods are permitted within a Class I, Division 1 location. **Figure 501-25**

- Threaded rigid metal conduit or intermediate metal conduit with explosionproof fittings.

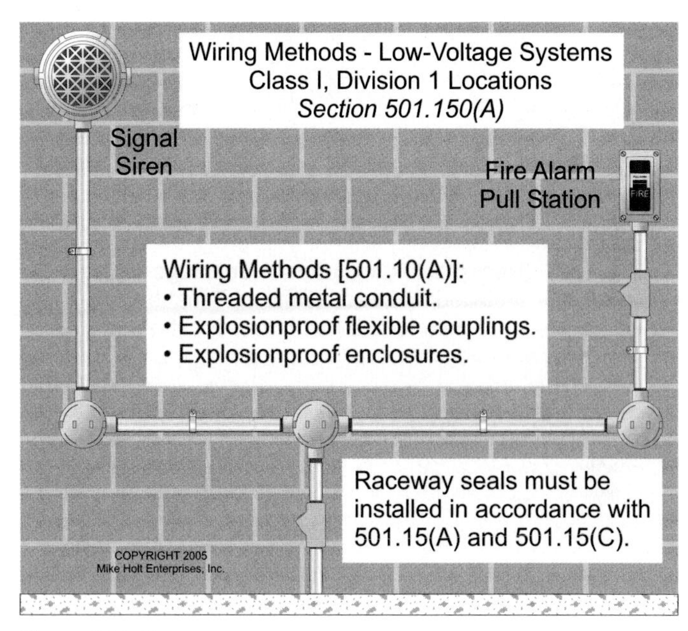
Wiring Methods - Low-Voltage Systems
Class I, Division 1 Locations
Section 501.150(A)
Signal Siren
Fire Alarm Pull Station
Wiring Methods [501.10(A)]:
• Threaded metal conduit.
• Explosionproof flexible couplings.
• Explosionproof enclosures.
Raceway seals must be installed in accordance with 501.15(A) and 501.15(C).
COPYRIGHT 2005
Mike Holt Enterprises, Inc.

Figure 501–25

- Type MI cable with termination fittings listed for the location.
- Type MC-HL cable with fittings listed for a Class I, Division 1 location is permitted in industrial establishments with restricted public access where only qualified persons will service the installation.
- Type ITC-HL cable with fittings listed for a Class I, Division 1 location is permitted in industrial establishments where only qualified persons will service the installation.

Flexible Wiring. Explosionproof flexible metal couplings are permitted when necessary for vibration, movement, or for difficult bends. Flexible cords can be installed in a Class I, Division 1 location if installed in accordance with 501.140.

Boxes and Fittings. All boxes and fittings must be approved by the authority having jurisdiction for Class I, Division 1 locations.

> **Author's Comment:** See Article 100 for the definition of "Approved."

(B) Class I, Division 2.

(1) Enclosures. General-purpose enclosures are permitted unless the enclosure contains make-and-break contacts. An enclosure identified for a Class I, Division 1 location must be used where the enclosure contains make-and-break contacts [501.10(A)].

Exception: General-purpose enclosures are permitted if the contacts are immersed in oil, enclosed within a chamber hermetically sealed against the entrance of gases or vapors, or if the circuits don't release sufficient energy to ignite a specific ignitible atmospheric mixture, such as an intrinsically safe or nonincendive system.

(4) Wiring Methods and Seals. All wiring within a Class I, Division 2 location must be made with threaded rigid metal, intermediate metal conduit, flexible metal conduit, or liquidtight flexible conduit with listed fittings. In addition, Types MI, MC, TC, ITC, and PLTC cable with fittings are also permitted [501.10(A)].

Conduit and cable seals must be installed in accordance with 501.15(B) and 501.15(C).

Article 501 Questions

1. Conduits 1½ in. or smaller entering an explosionproof enclosure that houses switches intended to interrupt current in the normal performance of the function are not required to be sealed, if the current-interrupting contacts are within a chamber hermetically sealed against the entrance of gases and vapors.

   (a) True (b) False

2. A conduit seal fitting must be installed in each conduit that passes from a Class I, Division 2 location into an unclassified location. Conduit boundary seals are not required to be _____, but must be identified for the purpose of minimizing the passage of gases under normal operating conditions.

   (a) listed (b) installed (c) explosionproof (d) accessible

3. When shielded cables and twisted-pair cables are installed in Class I, Division 2 locations, the removal of the shielding material or separation of the twisted pairs is not required, provided the termination is by an approved means to minimize the entrance of _____ and prevent propagation of flame into the cable core.

   (a) gases (b) vapors (c) dust (d) a or b

4. Transformers and capacitors installed in Class I, Division 1 locations that do not contain flammable liquids are not required to be installed in vaults if they are approved for Class I locations.

   (a) True (b) False

5. Motors, generators, or other rotating electric machinery that are identified for Class I, Division 2 locations are allowed to be used in a Class 1, Division 1 location.

   (a) True (b) False

6. Boxes, box assemblies, or fittings used to support luminaires in Class I, Division 1 locations must be identified for Class 1 locations.

   (a) True (b) False

7. In Class I, Division 1 and 2 locations, receptacles and attachment plugs must be of the type providing for _____ a flexible cord and must be identified for the location.

   (a) sealing compound around (b) quick connection to
   (c) connection to the grounding conductor of (d) none of these

8. In Class I, Division 1 locations, all apparatus and equipment of signaling, alarm, remote-control, and communications systems, _____, must be identified for Class I, Division 1 locations.

   (a) above 50V (b) above 100V-to-ground (c) regardless of voltage (d) except under 24V

The Most Informative Electrical Resource on the Web Today!

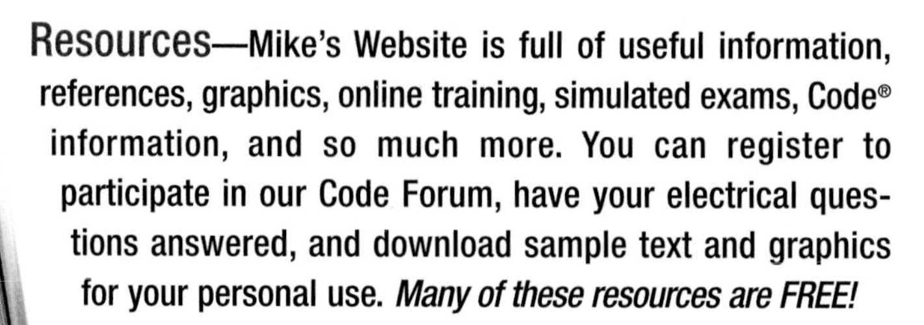

Resources—Mike's Website is full of useful information, references, graphics, online training, simulated exams, Code® information, and so much more. You can register to participate in our Code Forum, have your electrical questions answered, and download sample text and graphics for your personal use. *Many of these resources are FREE!*

Products—You can check out Mike's products by viewing sample pages and the table of contents for his most popular books before purchasing using the Website's online secure shopping cart.

FREE online tools, designed to help you succeed!

- Arc Blast Calculator
- Electrical Formulas (Chart)
- Electrical Instructor's Training Materials
- Fault-Current Calculator
- Graphics and Video Downloads
- Journeyman Simulated Online Exam
- Low-Voltage Book
- Master/Contractor Simulated Online Exam
- NEC® Quiz
- Newsletter
- Online Training Course
- Product Catalog
- Technical Calculators
- Touch Voltage Calculator
- TVSS Manual
- Video Clips
- Wiring and Raceway Chart

And Much More!

For an entire listing of the FREE resources available online, just click on the FREE STUFF link at www.NECcode.com

Interactive Online Training

Interactive Online Training is available. Many states are now accepting online testing for Continuing Education credits. Go online today and see if your state is on the list and *view selected chapters for FREE.* Don't get frustrated trying to find a local class or miss valuable time at work if your state accepts this convenient solution. Try an online quiz and see just how quick and easy it is to meet your Continuing Education requirements. FREE Online Simulated Journeyman and Master/Contractor exams are also available. Discover your strengths and weaknesses before you take your exam.

ARTICLE 502 Class II Hazardous (Classified) Locations

Introduction

If an area has enough combustible dust *suspended in the air* to ignite or explode, it's classified as a Class II location. **Figure 502-1**

Examples of such locations include grain silos, coal bins, wood pulp storage areas, and munitions plants.

Article 502 follows a logical arrangement similar to that of Article 501. Article 502 pertains strictly to locations containing combustible dust *in the air*, rather than simply the presence of such material.

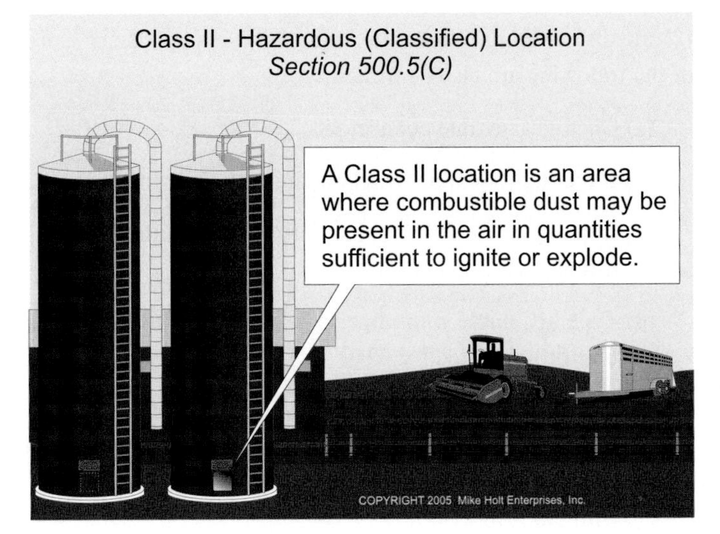

Class II - Hazardous (Classified) Location
Section 500.5(C)
A Class II location is an area where combustible dust may be present in the air in quantities sufficient to ignite or explode.
COPYRIGHT 2005 Mike Holt Enterprises, Inc.

Figure 502–1

PART I. GENERAL

502.1 Scope. Article 502 covers the requirements for electrical and electronic equipment and wiring for all voltages in Class II, Division 1 and 2 locations where fire or explosion hazards may exist due to the presence of combustible dust, combustible metal dusts, coal, carbon black, charcoal, coke, flour, grain, wood, plastic, and chemicals in the air in quantities sufficient to produce explosive or ignitible mixtures [500.5(C) and 500.8]. **Figure 502-2**

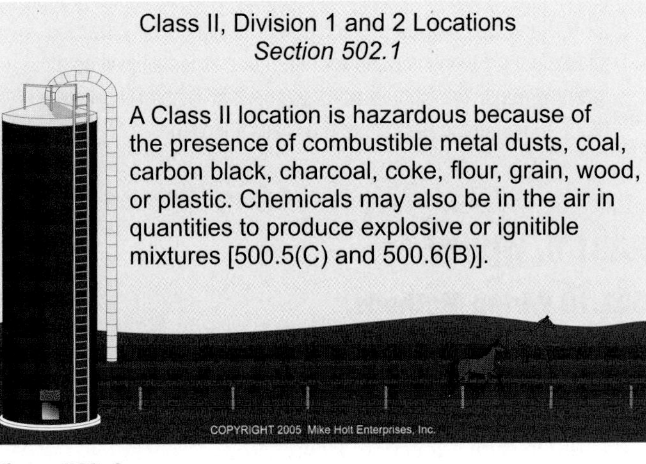

Class II, Division 1 and 2 Locations
Section 502.1
A Class II location is hazardous because of the presence of combustible metal dusts, coal, carbon black, charcoal, coke, flour, grain, wood, or plastic. Chemicals may also be in the air in quantities to produce explosive or ignitible mixtures [500.5(C) and 500.6(B)].
COPYRIGHT 2005 Mike Holt Enterprises, Inc.

Figure 502–2

502.5 General. Electric wiring and equipment installed in a Class II hazardous (classified) location must be installed in accordance with the installation requirements contained in Chapters 1 through 4, as well as with the requirements contained in this article.

Exception: Except as modified by this article.

Explosionproof Class I equipment as defined in Article 100 isn't required or permitted to be installed in a Class II location unless the equipment is specifically identified for the Class II location. **Figure 502-3**

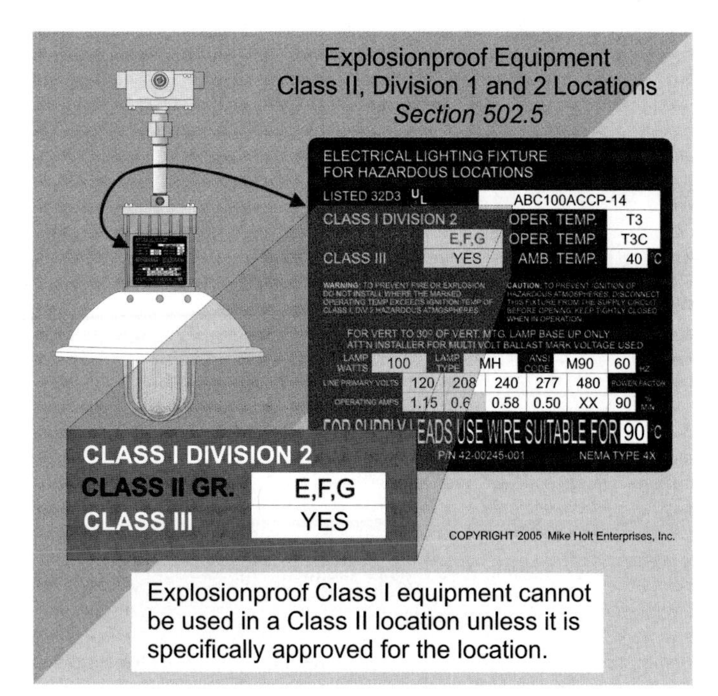

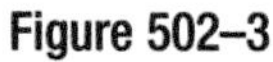
Figure 502–3

Author's Comment: All equipment used in a Class II hazardous (classified) location must be identified for both class and group [500.8 and 502.10]. Some equipment designed for hazardous (classified) locations is rated for different classes as well as different groups. You should not assume that Class I explosionproof equipment is permitted in a Class II location.

PART II. WIRING

502.10 Wiring Methods.

(A) Class II, Division 1.

(1) General. The following wiring methods can be installed in a Class II, Division 1 location: **Figure 502-4**

(1) Threaded rigid metal conduit or threaded intermediate metal conduit.

(2) Type MI cable with termination fittings.

(3) Type MC cable with impervious jacket and termination fittings listed for Class II, Division 1 locations in industrial installations where the conditions of maintenance and supervision ensure that only qualified persons service the installation.

(4) Boxes and Fittings. Boxes and fittings in which taps, joints, or terminal connections are made, or that are used in locations where dusts are of a combustible, electrically conductive nature, must be dust-ignitionproof (identified for Class II locations).

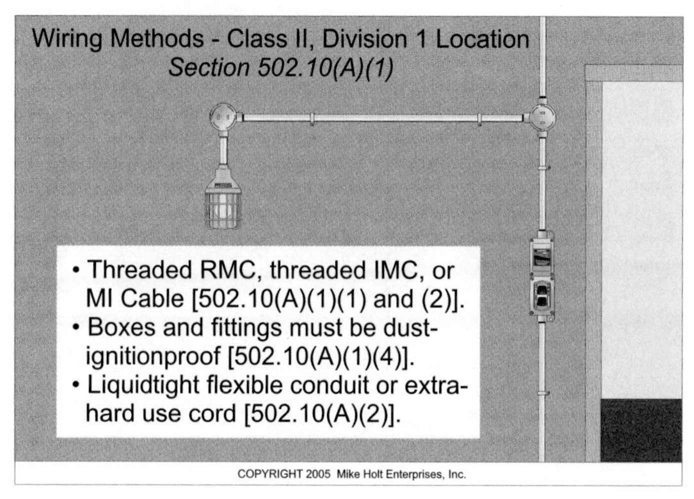

Figure 502–4

Boxes and fittings that don't contain taps, joints, or terminal connections must be dusttight and must have threaded entries for conduit or cable terminations.

Author's Comment: Rigid nonmetallic conduit isn't permitted within a Class II, Division 1 or 2 location.

(2) Flexible Wiring. Where flexibility is necessary, one or more of the following are also permitted:

(a) Dusttight flexible connectors

(b) Liquidtight flexible metal conduit with listed fittings

(c) Liquidtight flexible nonmetallic conduit with listed fittings

(d) Type MC cable with impervious jacket and termination fittings listed for Class II, Division 1 locations.

(e) Flexible cords listed for extra-hard usage containing an equipment grounding (bonding) conductor and terminating in a listed, bushed fitting in accordance with 502.140.

FPN: See 502.30(B) for grounding (bonding) requirements where flexible conduit is used.

(B) Class II, Division 2.

(1) General. In Class II, Division 2 locations, the following wiring methods are permitted: **Figure 502-5**

(1) Any of the wiring methods in 502.10(A).

(2) Rigid metal conduit, intermediate metal conduit (either threaded or with compression fittings), electrical metallic tubing with compression fittings, and dusttight wireways.

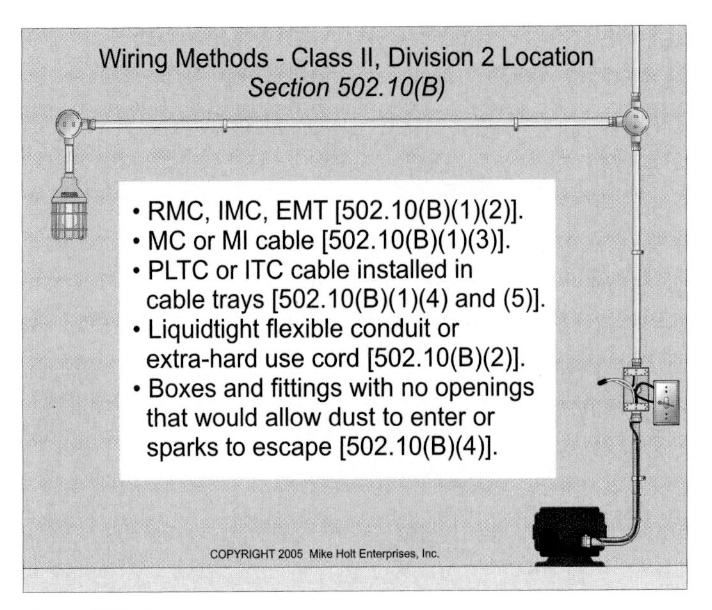

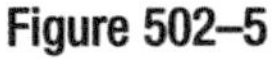

Figure 502–5

(3) Type MC or MI cable with listed termination fittings.

(4) Type PLTC in cable trays.

(5) Type ITC in cable trays.

Author's Comment: Rigid nonmetallic conduit isn't permitted within a Class II, Division 1 or 2 location.

(2) Flexible Wiring. Where flexibility is required, the following wiring methods are permitted [502.10(A)(2)]:

(a) Dusttight flexible connectors

(b) Liquidtight flexible metal conduit with listed fittings

(c) Liquidtight flexible nonmetallic conduit with listed fittings

(d) Interlocked armor Type MC cable having an overall jacket of suitable polymeric material and provided with termination fittings listed for Class II, Division 1 locations.

(e) Flexible cord listed for extra-hard usage, containing an equipment grounding (bonding) conductor, terminated with a listed bushed fitting, and installed in accordance with 502.140.

(4) Boxes and Fittings. Boxes and fittings in Class II, Division 2 areas must be dusttight.

Author's Comment: Standard weatherproof boxes with a gasketed seal can be used to meet this requirement.

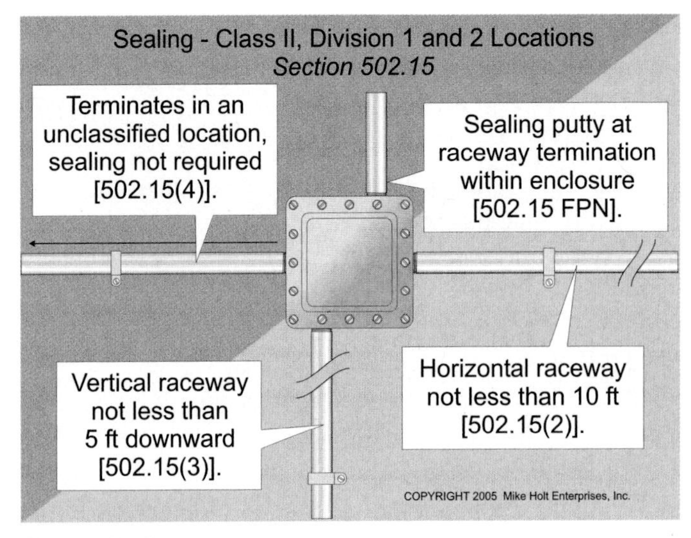

Figure 502–6

502.15 Seals. In Class II, Division 1 locations, dust must be prevented from entering the required dust-ignitionproof enclosure from a raceway by one of the following methods: **Figure 502-6**

(1) A permanent effective seal, such as sealing putty [502.15 FPN]

(2) A horizontal raceway not less than 10 ft long

(3) A vertical raceway that extends downward for not less than 5 ft from the dust-ignitionproof enclosure

(4) A raceway installed in a manner equivalent to (2) or (3) that extends only horizontally and downward from the dust-ignitionproof enclosure

A raceway seal isn't required between a dust-ignitionproof enclosure and an enclosure located in an unclassified location.

Sealing fittings for Class II locations must be accessible, but they aren't required to be explosionproof.

502.30 Bonding. Wiring and equipment in a Class II, Division 1 or 2 location must be grounded (bonded) to an effective ground-fault current path in accordance with Article 250 [250.100] and bonded in accordance with (A) and (B) below.

(A) Bonding—Metal Raceway. All metal enclosures and fittings between a Class II, Division 1 or 2 hazardous (classified) location and the service or separately derived system must be bonded to an effective ground-fault current path. This is accomplished by bonding the metal raceway to the enclosure or fitting by one of the following methods:

- Threaded conduit entry
- Bonding bushing with a bonding jumper sized to Table 250.122
- A bonding-type locknut

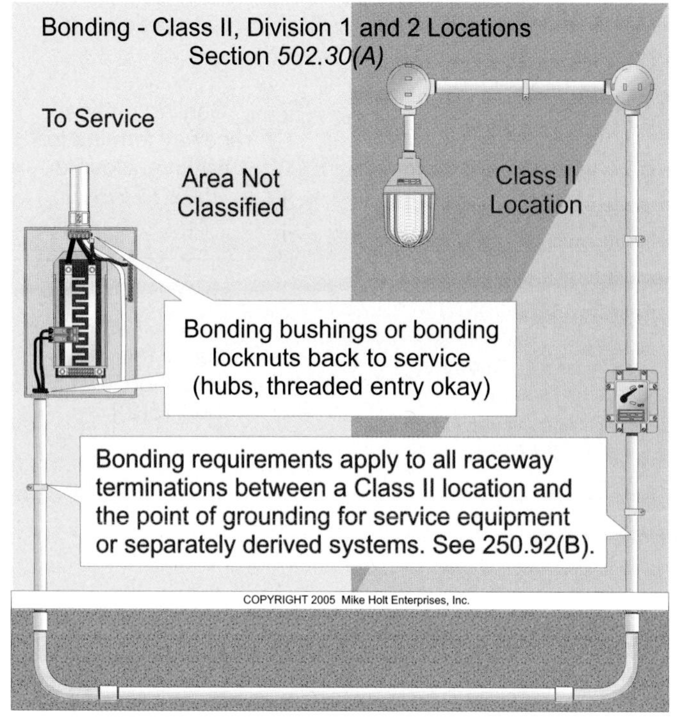
Bonding - Class II, Division 1 and 2 Locations
Section 502.30(A)
To Service
Area Not Classified
Class II Location
Bonding bushings or bonding locknuts back to service (hubs, threaded entry okay)
Bonding requirements apply to all raceway terminations between a Class II location and the point of grounding for service equipment or separately derived systems. See 250.92(B).
COPYRIGHT 2005 Mike Holt Enterprises, Inc.

Figure 502–7

Standard locknuts alone are not suitable for this purpose. **Figure 502-7**

Author's Comment: A separate equipment grounding (bonding) conductor is not required in a metal raceway used as the effective ground-fault current path in accordance with 250.118. **Figure 502-8**

(B) Bonding—Flexible Raceway. Where liquidtight flexible conduit is used in a Class II, Division 1 or 2 location [502.10], it must be installed with an internal or external equipment bonding jumper. If external, the bonding jumper must not be longer than

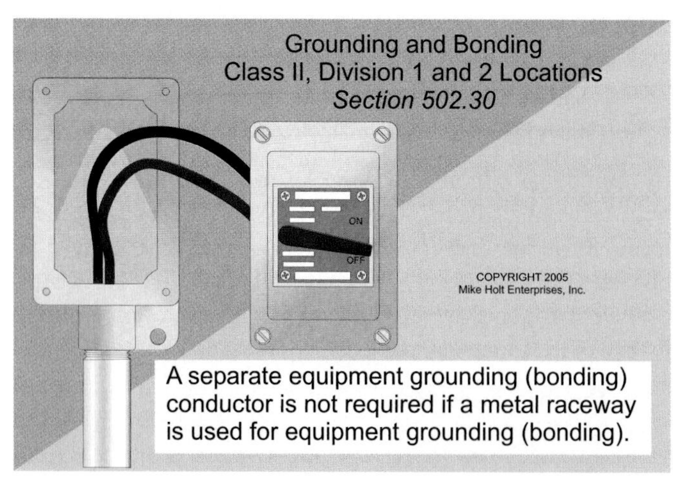
Grounding and Bonding
Class II, Division 1 and 2 Locations
Section 502.30
ON
OFF
COPYRIGHT 2005
Mike Holt Enterprises, Inc.
A separate equipment grounding (bonding) conductor is not required if a metal raceway is used for equipment grounding (bonding).

Figure 502–8

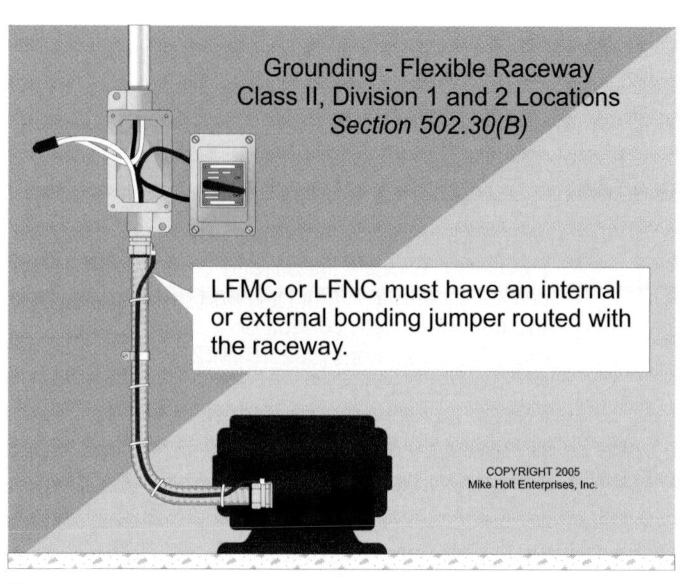
Grounding - Flexible Raceway
Class II, Division 1 and 2 Locations
Section 502.30(B)
LFMC or LFNC must have an internal or external bonding jumper routed with the raceway.
COPYRIGHT 2005
Mike Holt Enterprises, Inc.

Figure 502–9

6 ft, it must be routed with the raceway, and its size must be based on the circuit overcurrent protection device rating in accordance with 250.122 [250.102(E)]. **Figure 502-9**

Author's Comment: The *Code* doesn't limit the length of an equipment grounding (bonding) conductor if it's installed inside a flexible raceway.

502.35 Surge Protection. Surge arresters and transient voltage surge suppressors (TVSSs), must comply with Article 280 or 285. When installed in a Class II, Division 1 location, they must be in suitable enclosures and must be of a type designed for the specific duty.

502.40 Multiwire Branch Circuits. Multiwire branch circuits cannot be used in Class II, Division 1 locations.

Exception: Multiwire branch circuits can be installed in Class II, Division 1 locations where all of the ungrounded conductors of the circuit are opened simultaneously, such as with 2- or 3-pole circuit breakers. **Figure 502-10**

PART III. EQUIPMENT

502.115 Switches, Circuit Breakers, Motor Controllers, and Fuses.

(A) Class II, Division 1. Enclosures for switches, circuit breakers, motor controllers, and fuses must comply with the following:

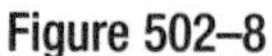

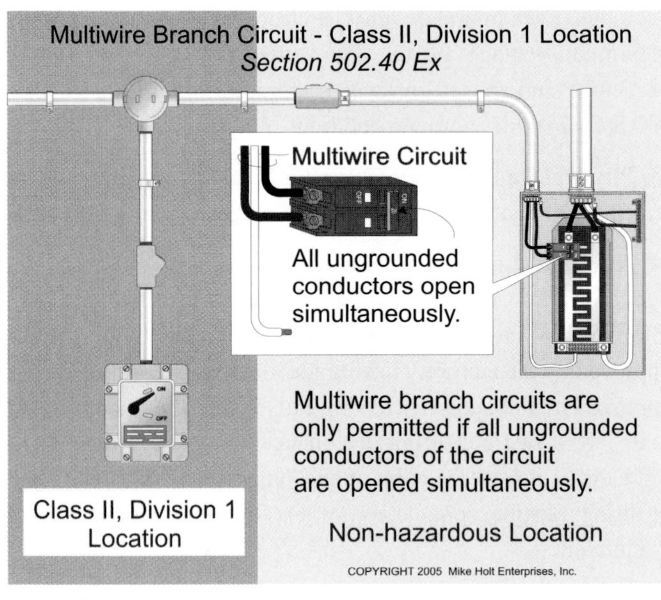
Multiwire Branch Circuit - Class II, Division 1 Location
Section 502.40 Ex
Multiwire Circuit
All ungrounded conductors open simultaneously.
Multiwire branch circuits are only permitted if all ungrounded conductors of the circuit are opened simultaneously.
Class II, Division 1 Location
Non-hazardous Location
COPYRIGHT 2005 Mike Holt Enterprises, Inc.

Figure 502–10

(1) Type Required. Switches, circuit breakers, motor controllers, and fuses, including pushbuttons, relays, and similar devices, must be provided with identified dust-ignitionproof enclosures. **Figure 502-11**

(2) Isolating Switches. Disconnecting and isolating switches that don't contain fuses and aren't intended to interrupt current, must be provided with tight metal enclosures designed to minimize the entrance of dust and, in addition, they must:

(1) Be equipped with close-fitting covers or with other effective means to prevent the escape of sparks or burning material, and

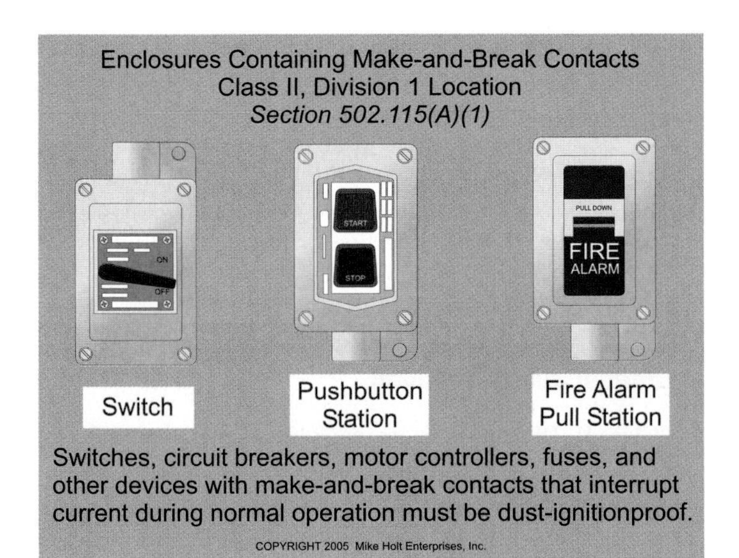
Enclosures Containing Make-and-Break Contacts
Class II, Division 1 Location
Section 502.115(A)(1)
Switch
Pushbutton Station
Fire Alarm Pull Station
FIRE ALARM
Switches, circuit breakers, motor controllers, fuses, and other devices with make-and-break contacts that interrupt current during normal operation must be dust-ignitionproof.
COPYRIGHT 2005 Mike Holt Enterprises, Inc.

Figure 502–11

(2) Have no openings (such as holes for attachment screws) through which, after installation, sparks or burning material might escape or through which exterior accumulations of dust or adjacent combustible material might be ignited.

(B) Class II, Division 2. Enclosures for fuses, switches, circuit breakers, and motor controllers, including pushbuttons, relays, and similar devices in a Class II, Division 2 location, must be dusttight.

Author's Comment: A standard weatherproof box with a cover and gasket meets this requirement. **Figure 502-12**

502.120 Control Transformers.

(A) Class II, Division 1. Control transformers must be installed within dust-ignitionproof enclosures that are identified for Class II locations.

(B) Class II, Division 2. Control transformers must be installed within tight metal housings without ventilating openings.

Author's Comment: A standard weatherproof box with a cover and gasket meets this requirement.

502.125 Motors and Generators.

(A) Class II, Division 1. Motors, generators, and other rotating electrical machinery must be identified for Class II, Division 1 locations, or be totally enclosed or pipe-vented.

(B) Class II, Division 2. Motors, generators, and other rotating electrical equipment must be totally enclosed nonventilated, pipe-ventilated, water-air-cooled, fan-cooled, or be dust-ignitionproof.

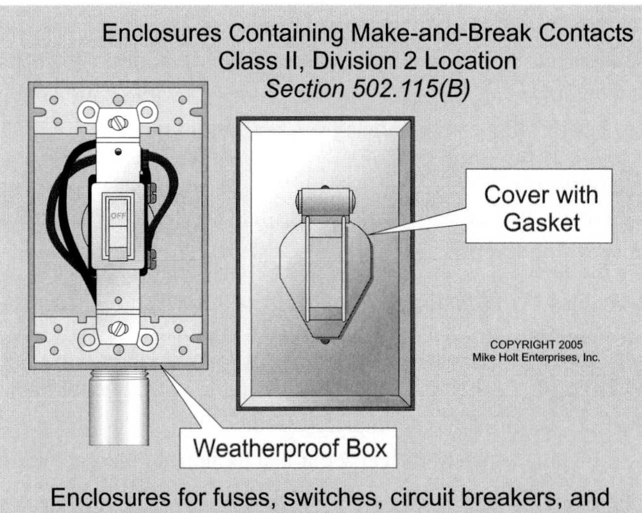
Enclosures Containing Make-and-Break Contacts
Class II, Division 2 Location
Section 502.115(B)
Cover with Gasket
COPYRIGHT 2005 Mike Holt Enterprises, Inc.
Weatherproof Box
Enclosures for fuses, switches, circuit breakers, and motor controllers, including pushbuttons, relays, and similar devices must be dusttight.

Figure 502–12

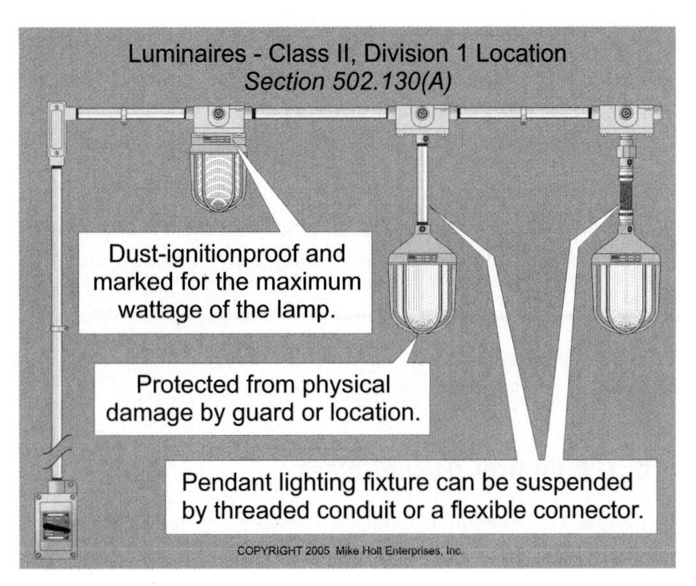
Luminaires - Class II, Division 1 Location
Section 502.130(A)
Dust-ignitionproof and marked for the maximum wattage of the lamp.
Protected from physical damage by guard or location.
Pendant lighting fixture can be suspended by threaded conduit or a flexible connector.
COPYRIGHT 2005 Mike Holt Enterprises, Inc.

Figure 502–13

502.130 Luminaires.

(A) Class II, Division 1. Luminaires must comply with the following: **Figure 502-13**

(1) Fixtures. Fixed luminaires must be identified for the location and marked to indicate the maximum wattage of the lamp for which the luminaire is designed.

(2) Physical Damage. Luminaires for fixed lighting must be protected from physical damage by suitable guards or by location.

(3) Pendant Luminaires. Pendant luminaires must be suspended by and supplied through threaded conduit stems, and threaded joints must be provided with set screws, or other effective means, to prevent loosening. For stems longer than 1 ft, a flexible fitting or connector listed for the Class II, Division 1 location must be provided not more than 1 ft from the point of attachment to the supporting box or fitting. Flexible cord cannot serve as the supporting means for a pendant luminaire.

(4) Supports. Boxes, box assemblies, or fittings used for the support of luminaires must be identified for Class II locations (dust-ignitionproof).

(B) Class II, Division 2. Luminaires must comply with the following:

(1) Portable Lighting Equipment. Portable lighting equipment must be identified as a complete assembly for use in a Class II location, and it must be clearly marked to indicate the maximum wattage of the lamps for which it is designed.

(2) Fixed Lighting. Fixed luminaires not identified for Class II locations must be designed to minimize the deposit of dust on lamps and to prevent the escape of sparks, burning material, or hot metal. Each luminaire must be clearly marked to indicate the maximum wattage of the lamp that is permitted without exceeding an exposed surface temperature in accordance with 500.8(C)(2) under normal conditions of use.

(3) Physical Damage. Luminaires for fixed lighting must be protected from physical damage by suitable guards or by location.

(4) Pendant Luminaires. Pendant luminaires must be suspended by threaded rigid metal conduit stems, threaded steel intermediate metal conduit stems, by chains with fittings approved by the authority having jurisdiction, or by other means approved by the authority having jurisdiction. For stems longer than 1 ft, a flexible fitting or connector must be provided not more than 1 ft from the point of attachment to the supporting box or fitting. Flexible cord cannot serve as the supporting means for a luminaire.

502.140 Flexible Cords. Flexible cords used in Class II, Division 1 and 2 locations must:

(1) Be of a type listed for extra-hard usage.

(2) Contain an equipment grounding (bonding) conductor identified according to 400.23.

(3) Be connected to terminals in a manner approved by the authority having jurisdiction.

(4) Be supported in a manner that there will be no tension on the terminal connections, and

(5) Be provided with suitable seals to prevent the entrance of dust into dust-ignitionproof enclosures.

502.145 Receptacles and Attachment Plugs.

(A) Class II, Division 1. Receptacles and attachment plugs must provide for connection of the equipment grounding (bonding) conductor and be identified for Class II locations.

(B) Class II, Division 2. Receptacles and attachment plugs must provide for connection of the grounding (bonding) conductor and be designed so that connection to the supply circuit cannot be made or broken while live parts are exposed. **Figure 502-14**

502.150 Low-Voltage, Limited-Energy, and Communications Systems.

(A) Class II, Division 1. Signaling, alarm, remote-control, and communications systems must comply with the following: **Figure 502-15**

(1) Wiring Methods. The wiring method must comply with 502.10(A).

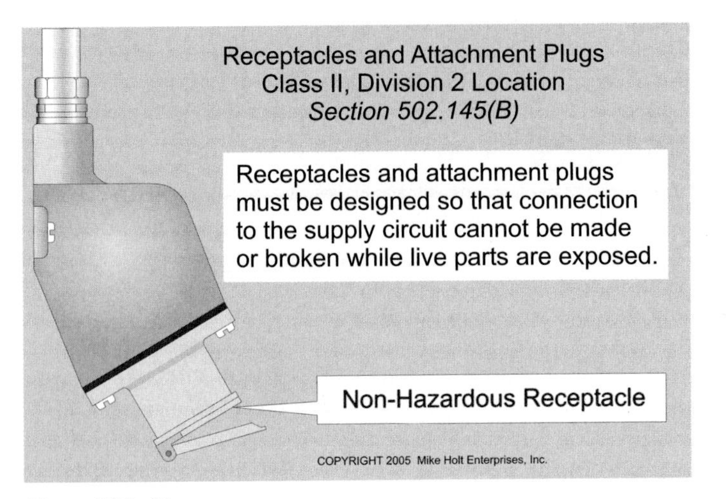
Receptacles and Attachment Plugs
Class II, Division 2 Location
Section 502.145(B)
Receptacles and attachment plugs must be designed so that connection to the supply circuit cannot be made or broken while live parts are exposed.
Non-Hazardous Receptacle
COPYRIGHT 2005 Mike Holt Enterprises, Inc.

Figure 502–14

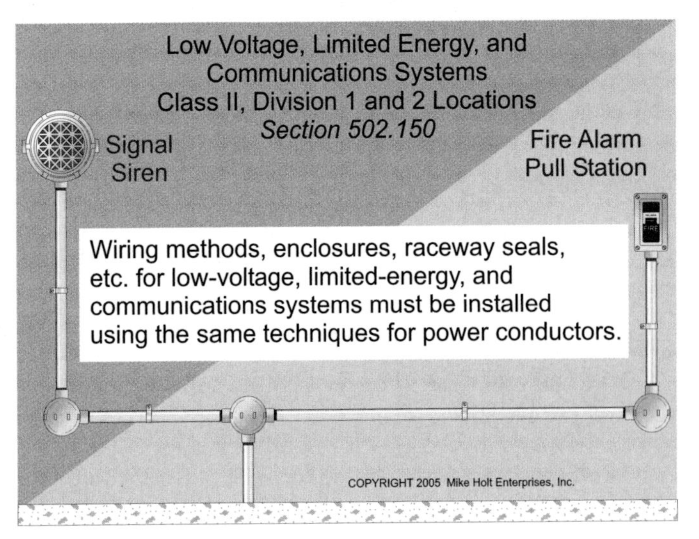
Low Voltage, Limited Energy, and
Communications Systems
Class II, Division 1 and 2 Locations
Section 502.150
Signal Siren
Fire Alarm Pull Station
Wiring methods, enclosures, raceway seals, etc. for low-voltage, limited-energy, and communications systems must be installed using the same techniques for power conductors.
COPYRIGHT 2005 Mike Holt Enterprises, Inc.

Figure 502–15

(2) Contacts. Switches, circuit breakers, relays, contactors, fuses, and current-breaking contacts for bells, horns, howlers, sirens, and other devices in which sparks or arcs may be produced must be within a dust-ignitionproof enclosure identified for a Class II location.

Exception: Enclosures of the general-purpose type are permitted where the contacts are immersed in oil or within a chamber sealed against the entrance of dust.

(B) Class II, Division 2. Signaling, alarm, remote-control, and communications systems must comply with the following:

(1) Contacts. Enclosures for switches, circuit breakers, relays, contactors, fuses, and current-breaking contacts for bells, horns, howlers, sirens, and other devices in which sparks or arcs may be produced must have tight metal enclosures designed to minimize the entrance of dust and must have tight fitting covers and no openings through which, after installation, sparks or burning material might escape.

Author's Comment: Standard weatherproof boxes with a gasketed seal can be used to meet this requirement.

(2) Transformers. The windings and terminal connections of transformers must be provided with tight metal enclosures without ventilating openings.

Author's Comment: Standard weatherproof boxes with a gasketed seal can be used to meet this requirement.

(5) Wiring Methods. Wiring methods must comply with 502.10(B).

Article 502 Questions

1. Raceways permitted as a wiring method in a Class II, Division 1 hazardous (classified) location include _____.

   (a) threaded rigid metal conduit and intermediate metal conduit
   (b) rigid nonmetallic conduit
   (c) electrical metallic tubing
   (d) any of these

2. Seals in a Class II hazardous (classified) location are required to be explosionproof.

   (a) True (b) False

3. In a Class II, Division 2 location, enclosures for fuses, switches, circuit breakers, and motor controllers, including pushbuttons, relays, and similar devices, must be _____.

   (a) dusttight (b) raintight
   (c) rated as Class I, Division 1 explosionproof (d) general duty

4. Luminaires installed in Class II, Division 1 locations must be protected from physical damage by a suitable _____.

   (a) warning label (b) pendant (c) guard or by location (d) all of these

5. In Class II, Division 1 locations, receptacles and attachment plugs must be of the type providing for connection to the grounding conductor of the flexible cord and must be identified _____.

   (a) as explosionproof (b) for Class II locations (c) with laminated tags (d) for general duty

ARTICLE 503 Class III Hazardous (Classified) Locations

Introduction

The Class III location definition is cumbersome, and many people have a hard time grasping what it means. If you have easily ignitible fibers or flyings present, you may have a Class III location. Put another way, if you don't have enough fibers or flyings in the air to produce an ignitible mixture but they are present, then you have a Class III area. Examples of such locations include sawmills, textile mills, and fiber processing plants.

Article 503 follows a logical arrangement similar to that of Article 502.

PART I. GENERAL

503.1 Scope. Article 503 covers the requirements for electrical and electronic equipment and wiring for all voltages in Class III, Division 1 and 2 locations where fire or explosion hazards may exist due to ignitible fibers or flyings.

503.5 General. The general requirements contained in Chapters 1 through 4, as well as the requirements of Article 503, apply to the electric wiring and equipment in Class III hazardous (classified) locations [500.1].

Exception: As modified by this article.

Author's Comment: Equipment installed in Class III locations must be able to function without developing surface temperatures high enough to cause spontaneous ignition.

PART II. WIRING

503.10 Wiring Methods. The wiring methods permitted in Class III locations include:

(A) Class III, Division 1 Location. Rigid metal conduit, steel intermediate metal conduit, rigid nonmetallic conduit, electrical metallic tubing, dusttight wireways, MI, and Type MC cable with listed fittings.

(1) Boxes and Fittings. Boxes and fittings that are dusttight.

Author's Comment: Standard weatherproof boxes with a gasketed seal can be used to meet this requirement.

(2) Flexible Connections. Liquidtight flexible conduit with listed fittings or flexible cord in accordance with 503.140.

(B) Class III, Division 2 Location. Rigid metal conduit, steel intermediate metal conduit, rigid nonmetallic conduit, electrical metallic tubing, dusttight wireways, MI, and Type MC cable with listed fittings.

(1) Boxes and Fittings. Boxes and fittings that are dusttight.

Author's Comment: Standard weatherproof boxes with a gasketed seal could be used to meet this requirement.

(2) Flexible Connections. Liquidtight flexible conduit with listed fittings or flexible cord in accordance with 503.140.

503.30 Bonding. Wiring and equipment in a Class III, Division 1 or 2 location must be grounded (bonded) to an effective ground-fault current path in accordance with Article 250 [250.100] and bonded in accordance with (A) and (B) below.

(A) Bonding—Metal Raceway. All metal enclosures and fittings between a Class III, Division 1 or 2 hazardous (classified) location and the service or separately derived system must be bonded to an effective ground-fault current path. This is accomplished by bonding the metal raceway to the enclosure or fitting by one of the following methods:

- Threaded conduit entry,
- Bonding bushing with a bonding jumper sized to Table 250.122, or
- A bonding-type locknut.

Standard locknuts alone are not suitable for this purpose.

Author's Comments:

- A separate equipment grounding (bonding) conductor is not required in a metal raceway that is used as the effective ground-fault current path in accordance with 250.118.
- The wiring methods permitted in Article 503 don't require threaded conduits. Therefore, threadless connectors are permitted, but bonding bushings with jumpers or bonding locknuts are required for rigid metal conduit, intermediate metal conduit, or electrical metallic tubing fittings where a standard locknut is used.

(B) Bonding—Flexible Raceway. Liquidtight flexible conduit used in a Class III location [503.10] must be installed with an internal or external equipment bonding jumper. If external, the bonding jumper must not be longer than 6 ft, it must be routed with the raceway, and it's sized based on the circuit overcurrent protection device rating in accordance with 250.122 [250.102(E)]. **Figure 503-1**

PART III. EQUIPMENT

503.115 Switches, Circuit Breakers, Motor Controllers, and Fuses. Enclosures for switches, circuit breakers, motor controllers, and fuses, including pushbuttons, relays, and similar devices intended to interrupt current during normal operation in Class III locations, must be dusttight.

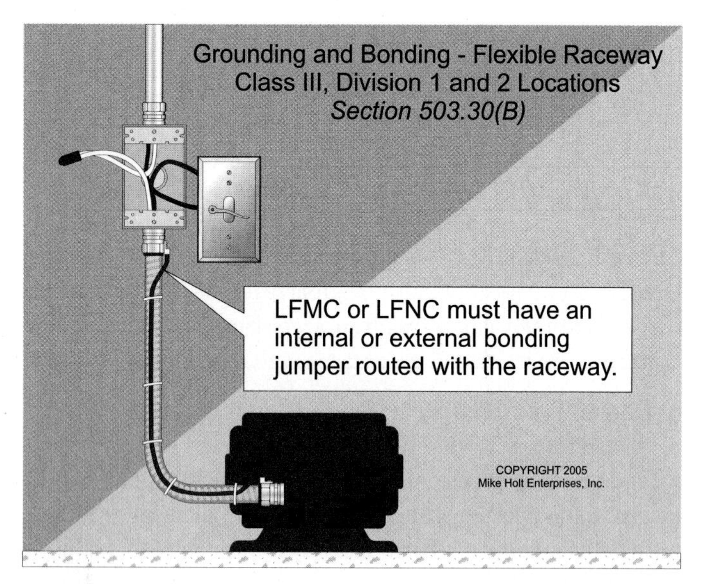

Grounding and Bonding - Flexible Raceway
Class III, Division 1 and 2 Locations
Section 503.30(B)
LFMC or LFNC must have an internal or external bonding jumper routed with the raceway.
COPYRIGHT 2005
Mike Holt Enterprises, Inc.

Figure 503–1

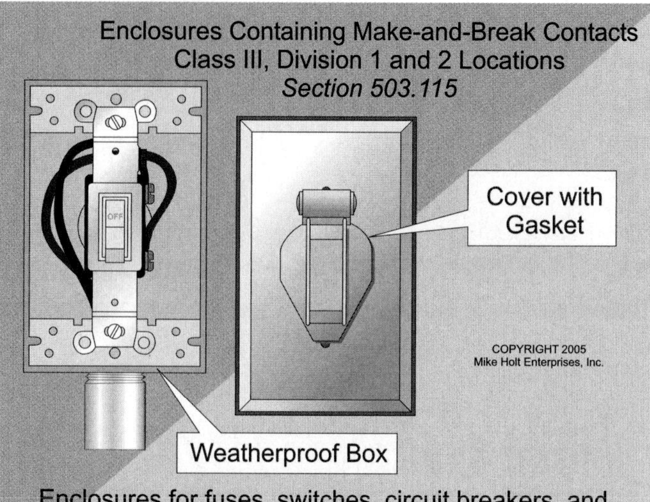

Figure 503–2

Author's Comment: According to Article 100, dusttight means that it has been constructed so dust will not enter the enclosing case under specified test conditions. Standard weatherproof boxes with a gasketed seal can be used to meet this requirement. **Figure 503-2**

503.120 Control Transformers. Control transformers installed in Class III locations must be installed within dusttight enclosures that comply with the temperature limitation in 503.5.

503.125 Motors and Generators. Motors, generators, and other rotating electrical equipment installed in Class III locations must be totally enclosed nonventilated, pipe-ventilated, or fan-cooled.

503.130 Luminaires. Luminaires installed in a Class III location must comply with the following: **Figure 503-3**

(A) Fixed Luminaires. Fixed luminaires must be designed to minimize the entrance of fibers and flyings and to prevent the escape of sparks, burning material, or hot metal. Luminaires must be clearly marked to show the maximum wattage of the lamps.

(B) Physical Damage. A suitable guard must protect all luminaires exposed to physical damage.

(C) Pendant Luminaires. Pendant luminaires must be suspended by stems of threaded rigid metal conduit, threaded intermediate metal conduit, threaded metal tubing of equivalent thickness, or by chains with fittings approved by the authority having jurisdiction. For stems longer than 1 ft, a flexible fitting

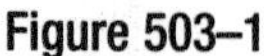

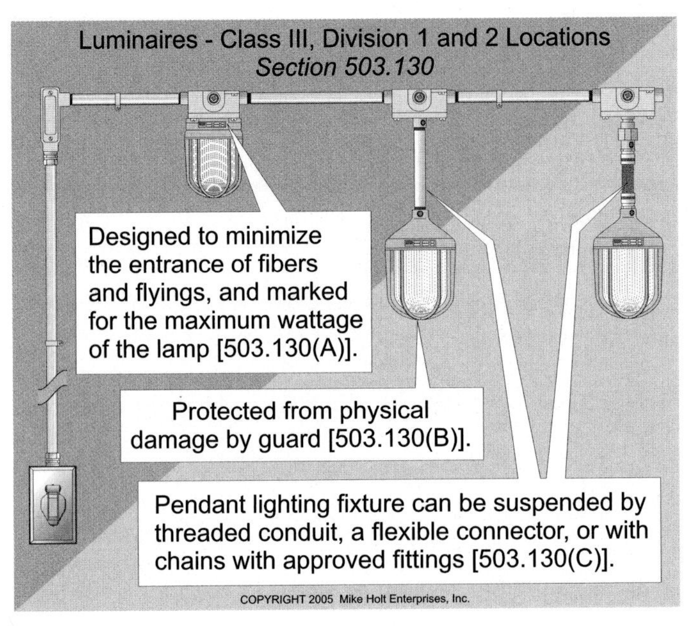

Luminaires - Class III, Division 1 and 2 Locations
Section 503.130
Designed to minimize the entrance of fibers and flyings, and marked for the maximum wattage of the lamp [503.130(A)].
Protected from physical damage by guard [503.130(B)].
Pendant lighting fixture can be suspended by threaded conduit, a flexible connector, or with chains with approved fittings [503.130(C)].
COPYRIGHT 2005 Mike Holt Enterprises, Inc.

Figure 503–3

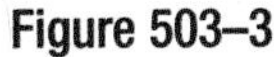

or connector approved for the location must be provided not more than 1 ft from the point of attachment to the supporting box or fitting.

(D) Portable Luminaire Equipment. Portable luminaire equipment must be equipped with handles and protected with substantial guards. Lampholders must be of the unswitched type with no provision for receiving an attachment plug.

503.140 Flexible Cords. Flexible cords must:

(1) Be of the extra-hard usage type.

(2) Contain an equipment grounding (bonding) conductor that complies with 400.23.

(3) Terminate in a manner approved by the authority having jurisdiction.

(4) Be supported so there's no tension on the terminal connections.

(5) Be provided with suitable means to prevent the entrance of fibers or flyings at terminations.

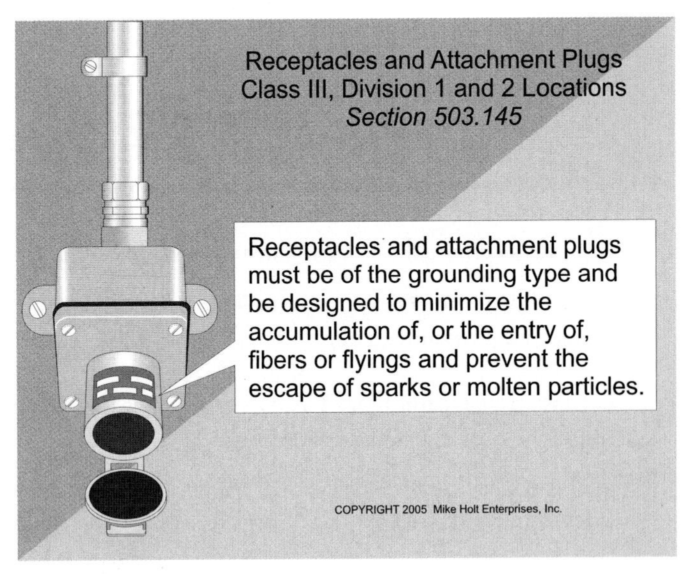

Figure 503–4

503.145 Receptacles and Attachment Plugs. Receptacles and attachment plugs installed in a Class III location must be designed to minimize the accumulation or the entry of fibers or flyings, and prevent the escape of sparks or molten particles. Figure 503-4

Exception: General-purpose grounding-type receptacles that are readily accessible for routine cleaning and mounted to minimize the entry of fibers or flyings can be used.

503.150 Signaling, Alarm, Remote-Control, and Loudspeaker Systems. Signaling, alarm, remote-control, and local loudspeaker systems must comply with Article 503 regarding wiring methods, switches, transformers, resistors, motors, luminaires, and related components.

> **Author's Comment:** Article 725 contains the requirements for Signaling Circuits, Article 760 contains the requirements for Fire Alarm Circuits, and Article 640 contains the requirements for Loudspeakers.

Article 503 Questions

1. Raceways permitted as a wiring method in a Class III hazardous (classified) location include _____.

   (a) rigid metal conduit and intermediate metal conduit
   (b) rigid nonmetallic conduit
   (c) electrical metallic tubing
   (d) any of these

2. In Class III, Division 1 and 2 locations, switches, circuit breakers, motor controllers, and fuses, including pushbuttons, relays, and similar devices, must be provided with _____.

   (a) Class I enclosures
   (b) general duty enclosures
   (c) dusttight enclosures
   (d) seal offs at each enclosure

3. Luminaires in a Class III location that may be exposed to physical damage must be protected by a(n) _____ guard.

   (a) plastic
   (b) metal
   (c) suitable
   (d) explosionproof

4. In Class III, Division 1 and 2 locations, receptacles and attachment plugs must be of the grounding type, must be designed so as to minimize the accumulation or the entry of _____, and must prevent the escape of sparks or molten particles.

   (a) gases or vapors
   (b) particles of combustion
   (c) fibers or flyings
   (d) none of these

5. Signaling, alarm, remote-control, and local loudspeaker communications systems are not required to comply with Article 503 when installed in Class III, Division 1 and 2 locations.

   (a) True
   (b) False

ARTICLE 504 Intrinsically Safe Systems

Introduction

An intrinsically safe circuit doesn't develop sufficient electrical energy to cause ignition of a specified gas or vapor under normal or abnormal operating conditions. An intrinsically safe system reduces the risk of ignition by electrical equipment or circuits and offers an optional wiring method in hazardous (classified) locations. If you're really concerned about ignition in a given location, you have to go back to basic fire prevention theory. Fire requires three elements; fuel, oxygen, and an ignition source. Intrinsically safe systems remove the ignition source part of the equation due to their low energy levels. Of course, if you have fuel and oxygen, you can still have a fire because circuitry isn't the only source of ignition. However, the installation of a 504-compliant system greatly reduces the hazards. Intrinsically safe wiring is often used for control instrumentation because of its low energy level (24V or less with fuse protection of 80 mA).

The decision to apply Article 504 may come out of a formal risk assessment or it may come out of standard industry practice for a given chemical or industrial process. Prime candidates for intrinsically safe systems include grain silos, grain-milling operations, coal pulverizing systems, and oil refineries. Before doing any work in a hazardous (classified) location, ask about the Article 504 requirements for that location.

504.1 Scope. This article covers the installation of intrinsically safe apparatus, wiring, and systems for Class I, II, and III locations.

> **Author's Comment:** Intrinsically safe systems are incapable of releasing sufficient electrical or thermal energy to cause ignition of flammable gases or vapors. Therefore, intrinsically safe apparatus and wiring, installed in accordance with Article 504, can be installed in any hazardous (classified) location area that contains flammable gases or vapors, combustible dust, or easily ignitible fibers or flyings. None of the stringent requirements contained in Articles 501 through 503, and 510 through 516 apply to intrinsically safe system installations [500.7(E)].

504.2 Definitions.

Control Drawing: A drawing or document provided by the manufacturer of the intrinsically safe or associated apparatus that details the interconnections between the intrinsically safe and associated apparatus.

Intrinsically Safe Circuit: A circuit in which any spark or thermal effect is incapable of causing ignition of a mixture of flammable or combustible material in air under prescribed test conditions.

Intrinsically Safe System: An assembly of interconnected intrinsically safe apparatus, associated apparatus, and interconnecting cables designed so that those parts of the system that can be installed in hazardous (classified) locations are intrinsically safe circuits.

504.3 Application of Other Articles. Except as modified by this article, the general requirements contained in Chapters 1 through 4 of this *Code* apply.

> **Author's Comment:** None of the stringent requirements contained in Articles 501 through 503, and 510 through 516 apply to intrinsically safe system installations [500.7(E)].

504.4 Equipment Approval. All intrinsically safe apparatus and associated apparatus must be listed.

504.10 Equipment Installation.

(A) Control Drawing. Intrinsically safe apparatus, associated apparatus, and other equipment must be installed in accordance with the control drawing(s).

(B) Location. Intrinsically safe and associated apparatus can be installed in any hazardous (classified) location for which it has been identified.

504.20 Wiring Methods.

Intrinsically safe apparatus and wiring are permitted using any of the suitable wiring methods contained in Chapter 7 and Chapter 8 of the *NEC*. Separation of the systems must be as provided in 504.30, and seals must comply with 504.70.

504.30 Separation of Intrinsically Safe Circuits.

(A) From Nonintrinsically Safe Circuit Conductors.

(1) In Conduits, Cable Trays, and Cables. Intrinsically safe circuits must not be placed in any conduit, cable tray, or cable that contains conductors of a nonintrinsically safe circuit.

Exception 1: Where intrinsically safe circuits are separated and secured not less than 2 in. from nonintrinsically safe circuit(s), or separated by a grounded metal partition or an insulating partition approved by the authority having jurisdiction.

Exception 2: Where the intrinsically safe circuits or the nonintrinsically safe circuits are in a grounded metal-sheathed or metal-clad cable that is capable of carrying fault current.

(2) Within Enclosures.

(a) Intrinsically safe circuits must be separated and secured not less than 2 in. from nonintrinsically safe circuit(s), and

(b) All conductors must be secured so that any conductor that might come loose from a terminal cannot come in contact with another terminal.

FPN No. 1: A separate compartment for the intrinsically safe and nonintrinsically safe terminals is the preferred method.

FPN No. 2: Grounded metal partitions, insulating partitions approved by the authority having jurisdiction, or restricted access wiring ducts separated from other such ducts approved by the authority having jurisdiction can be used to meet the separation requirements.

(3) Not in Raceway or Cable Tray System. Different intrinsically safe circuits must be in separate cables or must be separated from each other by one of the following means:

(1) The conductors of each circuit are within a grounded metal shield.

(2) The conductors of each circuit have insulation with a minimum thickness of 0.01 in.

(3) The clearance between two terminals for connection of field wiring of different intrinsically safe circuits must be not less than ¼ in., unless permitted by the control drawing.

504.50 Grounding (Earthing).

(A) Intrinsically Safe Apparatus, Associated Apparatus, and Conduits. Intrinsically safe apparatus, associated apparatus, cable shields, enclosures, and conduits, if of metal, must be grounded (bonded) to an effective ground-fault current path in accordance with Article 250 [250.100].

FPN: Supplementary bonding to the grounding electrode may be needed, if specified in the control drawing.

Author's Comment: See Article 100 for the definition of "Grounding Electrode."

(B) Connection to Grounding (Earthing) Electrodes. Where connection to a grounding electrode is required by the manufacturer, the grounding electrode must be as specified in 250.52(A)(1), (2), (3), and (4). Where none of these electrodes are available, one or more of the electrodes specified in 250.52(A)(4) through (A)(7) must be installed and used. The grounding (earthing) conductor must comply with the installation requirements of 250.30(A)(4)(a).

504.60 Bonding.

(A) Hazardous (Classified) Locations. In hazardous (classified) locations, intrinsically safe apparatus must be grounded (bonded) to an effective ground-fault current path in accordance with Part III of Article 250 with an equipment grounding (bonding) conductor of a type specified in 250.118 in accordance with 250.100.

(B) Unclassified. Metal conduits used for intrinsically safe system wiring in hazardous (classified) locations must be grounded (bonded) to an effective ground-fault current path in accordance with Article 250 [250.100] and bonded in accordance with 501.30, 502.30, or 503.30 as applicable.

504.70 Sealing.

To minimize the passage of gases, vapors, or dust, a seal must be installed for each conduit or cable run that leaves a Class I or II location in accordance with 501.15, 502.15, and 505.16. The seal isn't required to be explosionproof or flameproof.

504.80 Identification.

Identification labels required by this section must be suitable for the environment.

(A) Terminals. Intrinsically safe circuits must be identified at terminal and junction locations to prevent unintentional interference during testing and servicing.

(B) Wiring. Conduits, cable trays, and wiring used for intrinsically safe systems must be identified with permanently affixed labels with the wording "Intrinsic Safety Wiring." The labels must be visible after installation, and the spacing between labels must not exceed 25 ft.

Exception: Circuits that run underground must be identified when they are accessible.

FPN No. 1: Identification is required because wiring methods permitted in unclassified locations are permitted for intrinsically safe systems in hazardous (classified) locations.

FPN No. 2: In unclassified locations, the identification is necessary to ensure that nonintrinsically safe wire will not be inadvertently added to existing raceways at a later date.

(C) Color Coding. The color light blue can be used to identify cables, conduits, cable trays, and junction boxes containing intrinsically safe wiring.

Article 504 Questions

1. An intrinsically safe circuit is a circuit in which any spark or thermal effect is incapable of causing ignition of a mixture of flammable or combustible material in air under _____.

   (a) water (b) prescribed test conditions (c) supervision (d) duress

2. All applicable articles of the *Code* apply to intrinsically safe systems except where specifically modified by article 504.

   (a) True (b) False

3. Conductors of intrinsically safe circuits must not be placed in any _____ with conductors of any nonintrinsically safe system.

   (a) raceway (b) cable tray (c) cable (d) any of these

4. Conductors and cables of intrinsically safe circuits not in raceways or cable trays must be separated by at least _____ and secured from conductors and cables of any nonintrinsically safe circuits.

   (a) 6 in. (b) 2 in. (c) 18 in. (d) 12 in.

5. Conduits, cable trays, and open wiring used for intrinsically safe systems must be identified with permanently affixed labels with the wording "Intrinsic Safety Wiring." The labels must be visible after installation and the spacing between labels must not exceed _____ ft.

   (a) 3 (b) 10 (c) 25 (d) 50

ARTICLE 511 Commercial Garages, Repair, and Storage

Introduction

To avoid misunderstandings about what a garage is, refer to the definition in Article 100. Essentially, a commercial garage is a place where people store or repair vehicles that burn volatile liquids, such as gasoline, liquid propane, and alcohol. The requirement is a bit more detailed, but this is the general idea. Article 511 also draws a distinction between a parking garage and a garage used for repair or storage.

Article 511 can be confusing because it provides different rules for five different kinds of classified locations that can potentially be in the same room. This sounds more complicated than it really is. Just remember: 18 inches. The area 18 inches above grade and the area 18 inches below the ceiling require special attention. In either case, the potential problem is the accumulation of vapors.

- *A Pit.* Flammable liquids or vapors can accumulate in any depression below floor grade.
- *Adjacent Areas.* If you ensure that an area adjacent to a classified location meets certain ventilation requirements, or you can satisfy the authority having jurisdiction that the area doesn't present an ignition hazard, it's possible for that to be an unclassified location.

Author's Comment: It's possible for an establishment to have a vehicle service/repair area that must comply with Article 511, another area for fuel dispensing that must comply with Article 514, and other areas such as show rooms, offices, or a parts department that must only comply with the general requirements of the *NEC*, Chapters 1 through 4.

511.1 Scope. Article 511 applies to areas used for service and repair operations of self-propelled vehicles including passenger automobiles, buses, trucks, tractors, etc., in which volatile flammable liquids or gases are used for fuel or power.

Author's Comment: Some installations that could fall within the scope of Article 511 include automobile service/repair centers, service/repair garages for commercial vehicles such as trucks and tractors, and service/repair garages for fleet vehicles such as cars, buses, and trucks.

511.3 Classification of Hazardous Areas.

(A) Unclassified Locations.

(1) Parking Garages. Parking or storage garages aren't classified, and the requirements of Article 511 aren't applicable.

Author's Comment: See Article 100 for the definition of "Garage."

(2) Alcohol-Based Windshield Washer Fluid. The storage, handling, or dispensing into motor vehicles of alcohol-based windshield washer fluid in areas used for the service and repair operations of the vehicles does not cause such areas to be classified as hazardous (classified) locations.

Author's Comment: Windshield washer fluid isn't flammable.

(3) Classification of Adjacent Areas. Areas adjacent to classified locations aren't classified, and the requirements of Article 511 aren't applicable, if mechanically ventilated at a rate of four or more air changes per hour, or when walls or partitions effectively cut off the adjacent area. **Figure 511-1**

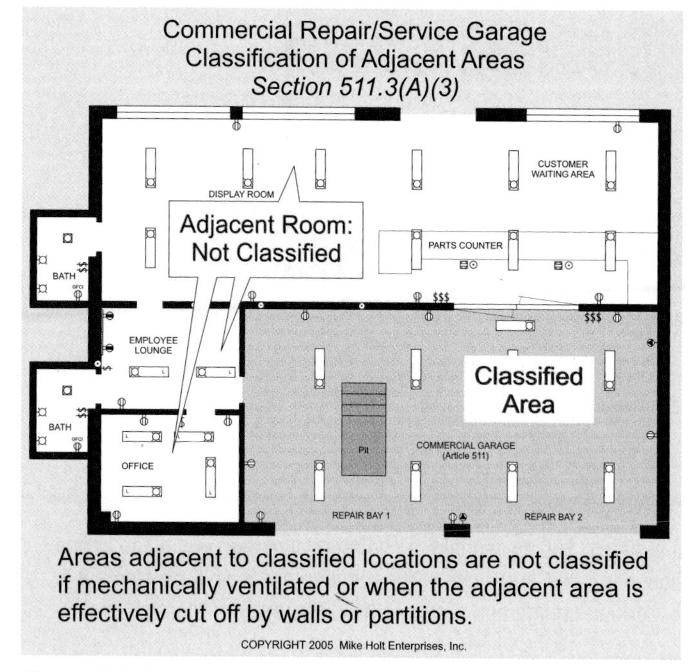

Figure 511–1

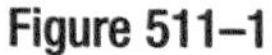

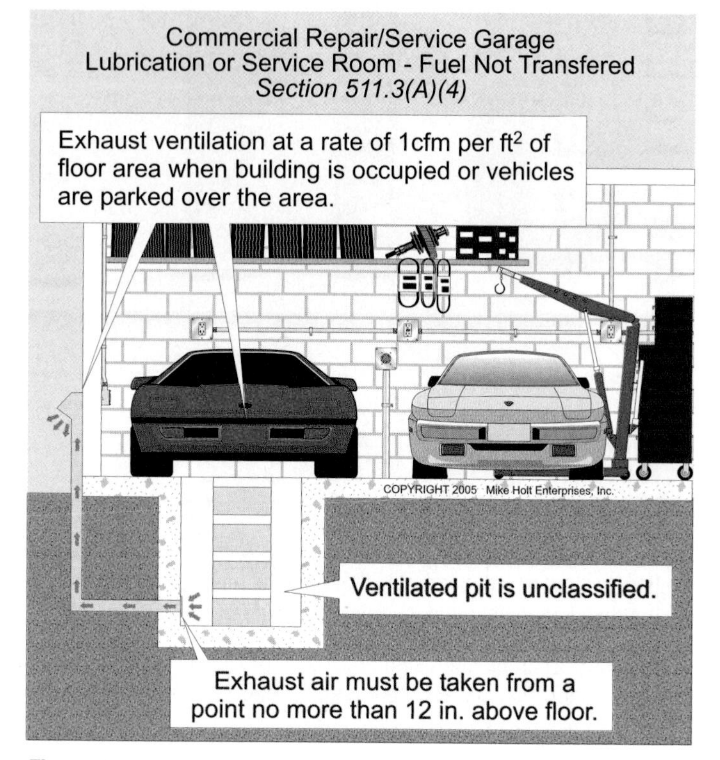

Commercial Repair/Service Garage
Lubrication or Service Room - Fuel Not Transfered
Section 511.3(A)(4)
Exhaust ventilation at a rate of 1cfm per ft2 of floor area when building is occupied or vehicles are parked over the area.
COPYRIGHT 2005 Mike Holt Enterprises, Inc.
Ventilated pit is unclassified.
Exhaust air must be taken from a point no more than 12 in. above floor.

Figure 511–2

(4) Pits in Lubrication or Service Room Where Class 1 Liquids Aren't Transferred. Any pit, belowgrade work area, or subfloor work area that is provided with exhaust ventilation at a rate not less than 1 cfm/ft^2 (cubic foot per minute per square foot) of floor area at times vehicles are parked in or over this area and where exhaust air is taken from a point within 12 in. of the floor of the belowgrade work area, isn't classified and the requirements of Article 511 aren't applicable. **Figure 511-2**

(5) Above the Floor in Lubrication or Service Rooms Where Class 1 Liquids are Transferred. Where mechanical ventilation provides a minimum of four air changes per hour or one cubic foot per minute of exchanged air for each square foot of floor area, the area below 18 in. above the floor is considered unclassified, and the requirements of Article 511 aren't applicable. Ventilation must provide air exchange across the entire floor area, and exhaust air must be taken at a point within 12 in. from the floor. **Figure 511-3**

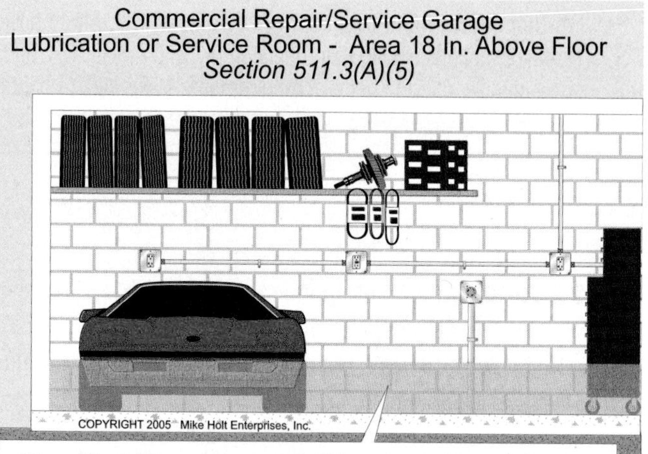

Commercial Repair/Service Garage
Lubrication or Service Room - Area 18 In. Above Floor
Section 511.3(A)(5)
COPYRIGHT 2005 Mike Holt Enterprises, Inc.
Classified: The area up to 18 in. above floor level is a Class I, Division 2 location if fuels are transferred where no ventilation is provided 511.3(B)(3).
Unclassified: The area up to 18 in. above floor level if mechanically ventilated at a rate of at least 1 cfm/ft2 of floor area.
Unclassified: The area up to 18 in. above floor level is not classified in an area where fuels are not transferred [511.3(A)(6)].

Figure 511–3

(6) Flammable Liquids Having Flash Points Below 100°C. Areas where flammable liquids having a flash point below 100°C, such as gasoline or gaseous fuels (such as natural gas or LPG) will not be transferred, are considered unclassified.

(7) Within 18 in. of the Ceiling. In repair garages where lighter-than-air gaseous fuels (such as natural gas) vehicles are repaired or stored, the area within 18 in. of the ceiling is considered unclassified if ventilation of not less than 1 cfm/ft^2 is taken from a point within 18 in. of the highest point in the ceiling.

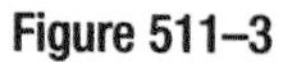

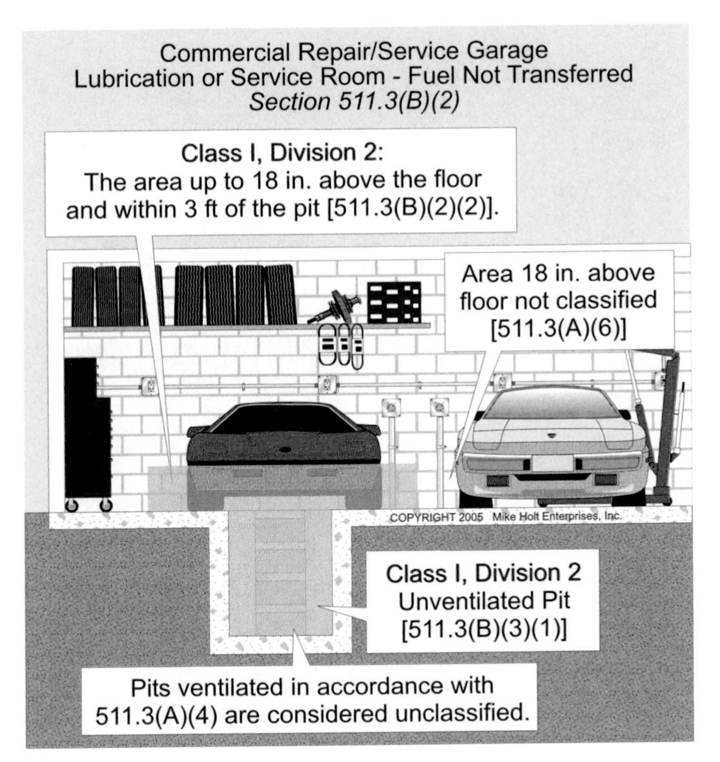
Commercial Repair/Service Garage
Lubrication or Service Room - Fuel Not Transferred
Section 511.3(B)(2)
Class I, Division 2:
The area up to 18 in. above the floor and within 3 ft of the pit [511.3(B)(2)(2)].
Area 18 in. above floor not classified [511.3(A)(6)]
COPYRIGHT 2005 Mike Holt Enterprises, Inc.
Class I, Division 2
Unventilated Pit
[511.3(B)(3)(1)]
Pits ventilated in accordance with 511.3(A)(4) are considered unclassified.

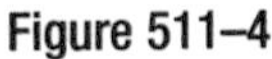
Figure 511–4

(B) Classified Locations.

(1) Flammable Fuel Dispensing Areas. Wiring and equipment in areas where flammable fuel is dispensed into vehicle tanks must be installed in accordance with Article 514.

(2) Lubrication or Service Room (Fuels Not Transferred). Unventilated pits and the work areas directly above them are classified locations. **Figure 511-4**

(1) Unventilated Pit. Any unventilated pit is classified as a Class I, Division 2 location.

(2) Above Unventilated Pit. The space up to 18 in. above the floor, extending 3 ft horizontally from a lubrication pit, is classified as a Class I, Division 2 location.

(3) Lubrication or Service Room (Fuels Transferred). Pits or depressions below the floor area, the work areas above them, and the area around dispensers are classified locations. **Figure 511-5**

(1) Above the Floor. The area up to 18 in. above the floor is classified as a Class I, Division 2 location, unless ventilation is provided in accordance with 511.3(A)(5).

(2) Unventilated Pit. Any unventilated pit is classified as a Class I, Division 1 location.

(3) Ventilated Pit. A ventilated pit in which six air changes per hour are exhausted from a point within 12 in. of the floor level is classified as a Class I, Division 2 location.

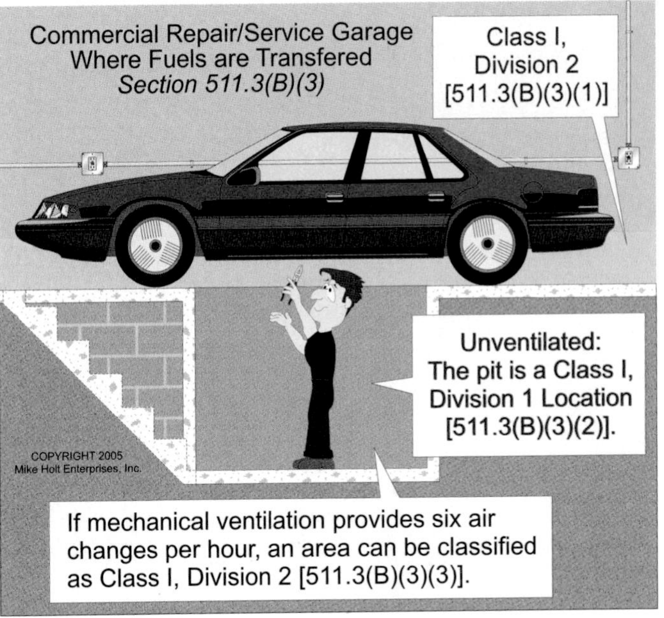
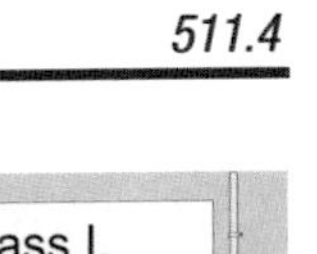
Commercial Repair/Service Garage
Where Fuels are Transfered
Section 511.3(B)(3)
Class I,
Division 2
[511.3(B)(3)(1)]
Unventilated:
The pit is a Class I, Division 1 Location
[511.3(B)(3)(2)].
COPYRIGHT 2005 Mike Holt Enterprises, Inc.
If mechanical ventilation provides six air changes per hour, an area can be classified as Class I, Division 2 [511.3(B)(3)(3)].

Figure 511–5

(4) Within 18 in. of the Ceiling. In repair garages where vehicles fueled with lighter-than-air fuels (such as natural gas) are repaired or stored, ceiling spaces without ventilation are classified as a Class I, Division 2 location. **Figure 511-6**

511.4 Wiring and Equipment in Hazardous (Classified) Locations.

(A) Located in Class I Locations. Wiring and equipment within a Class I location must be installed in accordance with the requirements contained in Article 501. **Figure 511-7**

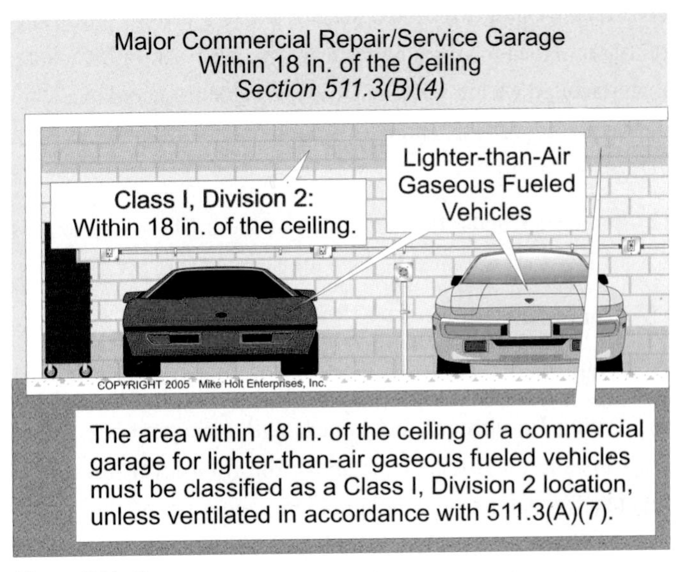
Major Commercial Repair/Service Garage
Within 18 in. of the Ceiling
Section 511.3(B)(4)
Class I, Division 2:
Within 18 in. of the ceiling.
Lighter-than-Air
Gaseous Fueled
Vehicles
COPYRIGHT 2005 Mike Holt Enterprises, Inc.
The area within 18 in. of the ceiling of a commercial garage for lighter-than-air gaseous fueled vehicles must be classified as a Class I, Division 2 location, unless ventilated in accordance with 511.3(A)(7).

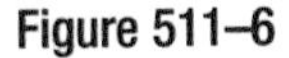
Figure 511–6

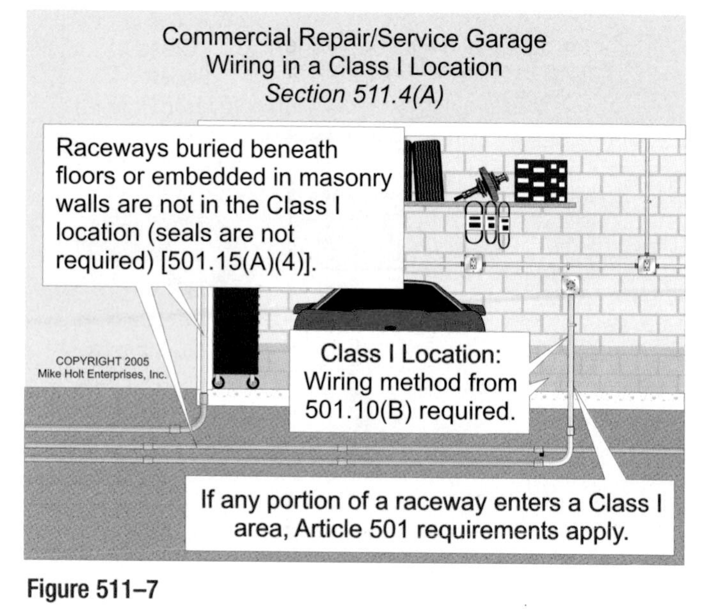

Commercial Repair/Service Garage
Wiring in a Class I Location
Section 511.4(A)
Raceways buried beneath floors or embedded in masonry walls are not in the Class I location (seals are not required) [501.15(A)(4)].
COPYRIGHT 2005 Mike Holt Enterprises, Inc.
Class I Location: Wiring method from 501.10(B) required.
If any portion of a raceway enters a Class I area, Article 501 requirements apply.

Figure 511–7

(B) Equipment Located in Class I Locations.

(1) Fuel-Dispensing Units. The wiring of fuel-dispensing units must comply with Article 514.

(2) Portable Lighting Equipment. The lamp and cord of portable lighting equipment must be supported or arranged in such a manner that it cannot be used in a hazardous (classified) location [511.3(B)], or it must be identified for a Class I, Division 1 location [501.130(B)].

511.7 Wiring and Equipment Above Hazardous (Classified) Locations.

(A) Wiring in Spaces Above Class I Locations.

(1) Fixed Wiring Methods. Wiring above a Class I hazardous (classified) location must be in raceways, Types AC or MC cable, manufactured wiring systems, or PLTC cable in accordance with Article 725, or TC cable in accordance with Article 336. **Figure 511-8**

(2) Pendant Cords. Pendant cords above a Class I hazardous (classified) location must be listed for hard usage. See 400.4

(B) Equipment Above Class I Locations.

(1) Fixed Electrical Equipment.

(a) Arcing Equipment. Equipment with make-and-break contacts installed less than 12 ft above the floor level (excluding receptacles, lamps, and lampholders) must be of the totally enclosed type or be of the type constructed to prevent sparks or hot metal particles from escaping.

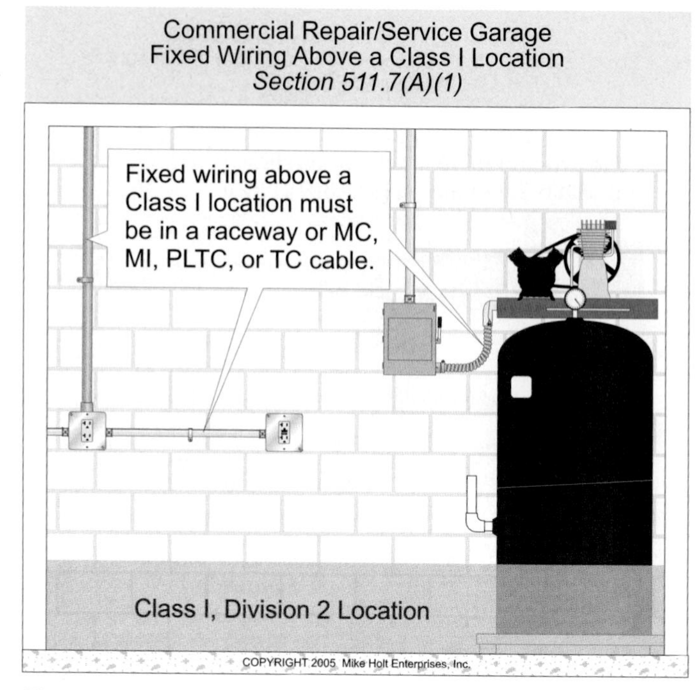

Figure 511–8

(b) Fixed Lighting. Lampholders and lamps for fixed lighting over travel lanes or where exposed to physical damage must be located not less than 12 ft above floor level, unless the luminaires are of the totally enclosed type or constructed to prevent sparks or hot metal particles from escaping. **Figure 511-9**

511.9 Seals. Conduit, cable, and boundary seals must be installed in accordance with the requirements contained in 501.15.

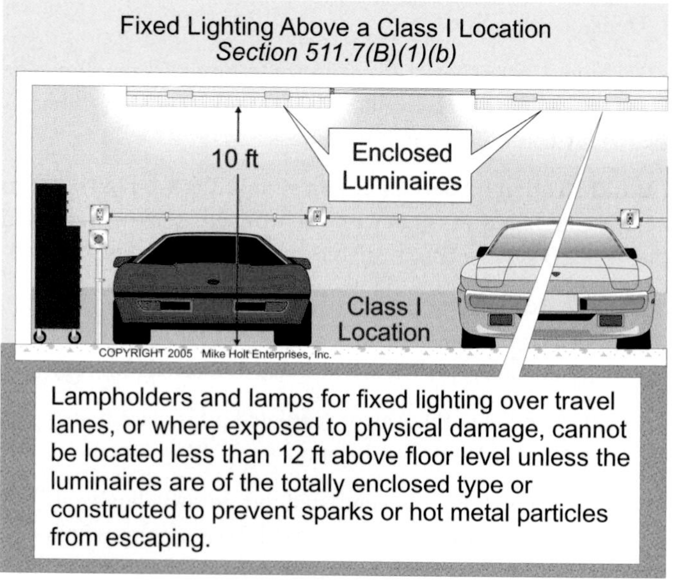

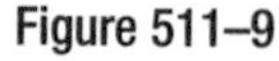
Figure 511–9

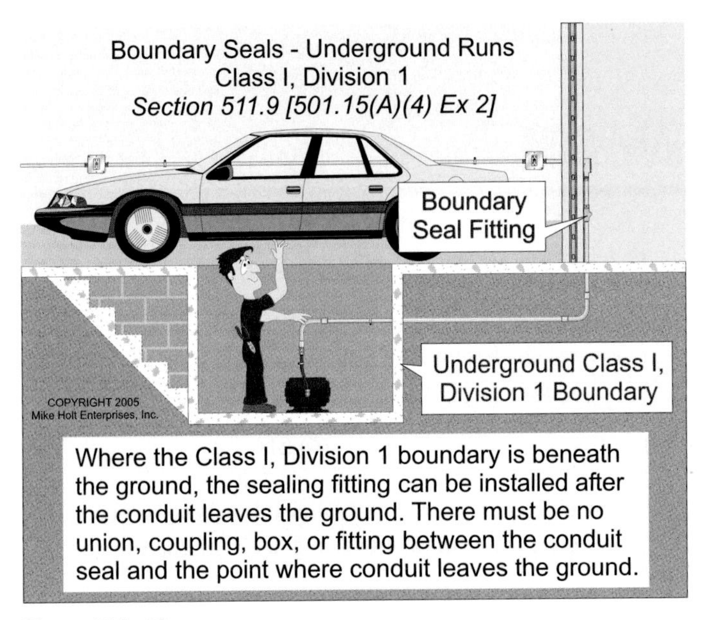

Boundary Seals - Underground Runs
Class I, Division 1
Section 511.9 [501.15(A)(4) Ex 2]
Boundary Seal Fitting
Underground Class I, Division 1 Boundary
COPYRIGHT 2005 Mike Holt Enterprises, Inc.
Where the Class I, Division 1 boundary is beneath the ground, the sealing fitting can be installed after the conduit leaves the ground. There must be no union, coupling, box, or fitting between the conduit seal and the point where conduit leaves the ground.

Figure 511–10

Author's Comment: Where the Class I, Division 1 boundary is beneath the ground, the sealing fitting can be installed after the conduit leaves the ground. However, there must not be any unions, couplings, boxes, or fittings (except explosionproof reducing bushings) between the seal fitting and the point where the conduit leaves the earth [501.15(A)(4) Ex 2]. **Figure 511-10**

511.10 Special Equipment.

(A) Battery Charging Equipment. Battery chargers and batteries being charged must not be located within an area classified in accordance with 511.3(B).

(B) Electric Vehicle Charging.

(1) General. Electrical conductors and equipment that connect an electric automotive-type vehicle must be installed in accordance with Article 625.

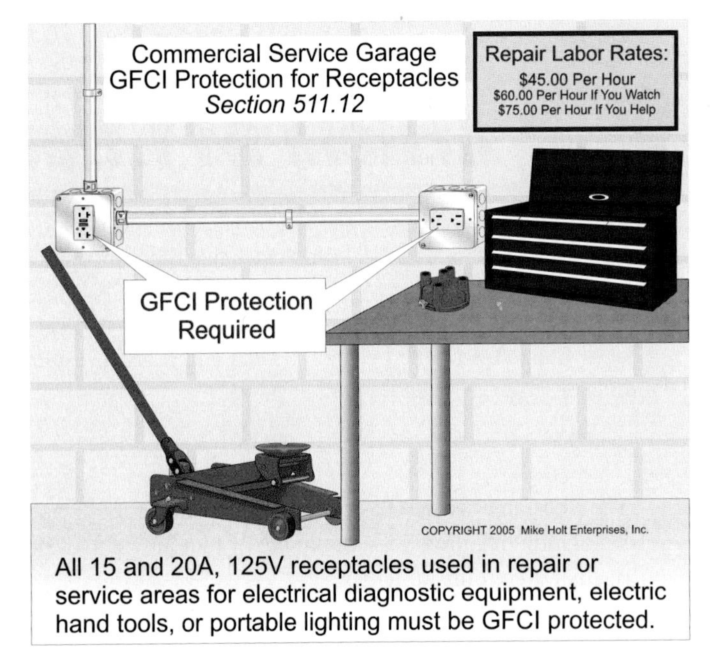

Commercial Service Garage
GFCI Protection for Receptacles
Section 511.12
Repair Labor Rates:
$45.00 Per Hour
$60.00 Per Hour If You Watch
$75.00 Per Hour If You Help
GFCI Protection Required
COPYRIGHT 2005 Mike Holt Enterprises, Inc.
All 15 and 20A, 125V receptacles used in repair or service areas for electrical diagnostic equipment, electric hand tools, or portable lighting must be GFCI protected.

Figure 511–11

511.12 GFCI-Protected Receptacles.

GFCI protection is required for all 15 and 20A, 125V receptacles used for service and repair operations, such as electrical automotive diagnostic equipment, electric hand tools, portable lighting devices, etc. **Figure 511-11**

Author's Comments:

- See Article 100 for the definition of "Ground-Fault Circuit Interrupter."
- Circuits are rated 120V and receptacles are rated 125V.

Article 511 Questions

(• Indicates that 75% or fewer exam takers get the question correct.)

1. Article _____ contains the requirements for the wiring of occupancy locations used for service and repair operations in connection with self-propelled vehicles (including passenger automobiles, buses, trucks, tractors, etc.) in which volatile flammable liquids or gases are used for fuel or power.

   (a) 500 (b) 501 (c) 511 (d) 514

2. The floor area where Class 1 liquids are transferred is not classified if the enforcing agency determines that there is mechanical ventilation that provides a minimum of four air changes per hour or _____ cu ft per minute of exchanged air for each square foot of floor area (cfm/sq. ft).

   (a) 1 (b) 2 (c) 3 (d) 4

3. •Any ventilated pit or depression in a commercial garage lubrication or service room where Class I liquids or gaseous fuels are transferred is classified as a _____ location.

   (a) Class I, Division 2 (b) Class II, Division 2 (c) Class II, Division 1 (d) Class I, Division 1

4. Type NM cable is allowed for wiring above a Class I location in a commercial garage.

   (a) True (b) False

5. For commercial garages, seals conforming to the requirements of 501.5 and 501.5(B)(2) must be provided and apply to _____ boundaries of the defined Class I location.

   (a) vertical (b) horizontal (c) conduit only within the (d) a and b

Aircraft Hangars

Introduction

This article is similar in concept to Article 511. Aircraft burn highly flammable fuel. Therefore, aircraft hangars have their own article.

As with Article 511, Article 513 can be confusing, because it provides different rules for different kinds of classified areas that can potentially be in the same room. In the case of Article 513, there are only four such areas rather than five. You can think of them as these three, because 513.3(D) is just the flip side of 513.3(B):

- *A Pit.* Flammable liquids or vapors can accumulate in any depression below floor grade.
- *Adjacent Areas.* An area adjacent to a hangar also falls under Article 513 unless certain conditions of isolation and separation are met.
- *Next to the Aircraft.* If it's within 5 ft of aircraft or of an aircraft fuel tank, you must classify it as a Class I, Division 2 location or as a Zone 2 location.

513.1 Scope. This article applies to buildings or structures in which aircraft might undergo service, repairs, or alterations.

513.3 Classification of Locations.

(A) Below Floor Level. Any pit or depression below the level of the hangar floor is classified as a Class I, Division 1 location.

(B) Classification. The entire area of the hangar, including any adjacent and communicating areas not suitably cut off from the hangar, is classified as a Class I, Division 2 location up to a level of 18 in. above the floor.

(C) Vicinity of Aircraft. The area within 5 ft horizontally from aircraft power plants or aircraft fuel tanks is classified as a Class I, Division 2 location that extends upward from the floor to a level 5 ft above the upper surface of wings and of engine enclosures.

(D) Adjacent Areas. Areas where flammable liquids or vapors aren't likely to be released, such as stock rooms, electrical control rooms, and other similar locations, aren't classified if adequately ventilated and effectively cut off from the hangar by walls or partitions.

513.4 Wiring and Equipment in Class I Locations.

(A) General. All wiring and equipment within a Class I location as defined in 513.3 must comply with the applicable provisions of Article 501.

Attachment plugs and receptacles in Class I locations must be identified for Class I locations or must be designed so that they cannot be energized while the connections are being made or broken.

513.7 Wiring Not Within Class I Locations.

(A) Fixed Wiring. All wiring in a hangar not within a Class I location as defined in 513.3 must be installed in metal raceways, or be Types MI, TC, or MC cable.

Exception: Wiring in unclassified locations, as defined in 513.3(D), can be any type recognized in Chapter 3.

(B) Pendants. Flexible cords used for pendants must be suitable for the type of service and identified for hard usage with an equipment grounding (bonding) conductor.

(C) Arcing Equipment Above a Classified Location. Equipment that is less than 10 ft above wings and engine enclosures of aircraft and that may produce arcs, sparks, or particles of hot metal must be of the totally enclosed type or constructed so as to prevent the escape of sparks or hot metal particles.

Exception: Equipment in areas described in 513.3(D) can be of the general-purpose type.

(D) Lampholders. Lampholders of metal-shell, fiber-lined types cannot be used for fixed incandescent lighting.

513.8 Wiring in or Under Hangar Floor. All wiring installed in or under the hangar floor must comply with the requirements for Class I, Division 1 locations.

513.9 Sealing. Seals must be provided in accordance with 501.15.

513.12 GFCI-Protected Receptacles. GFCI protection is required for all 15 and 20A, 125V receptacles in aircraft hangars where electrical diagnostic equipment, electric hand tools, and/or portable lighting devices are used. **Figure 513-1**

> **Author's Comment:** Personnel servicing and maintaining aircraft use the same hand tools and equipment that are used in commercial garages, which require GFCI protection [511.12].

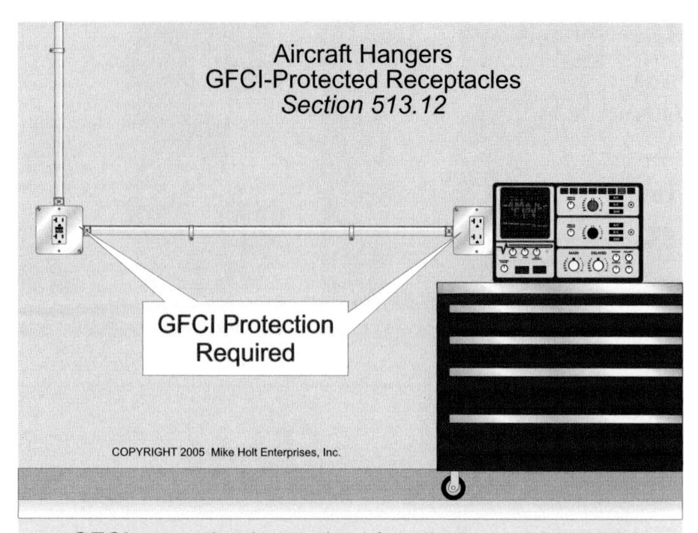

Figure 513–1

513.16 Bonding.

(A) General Requirements. Metal raceways, the metal armor or metallic sheath on cables, and all metal parts of electrical equipment, regardless of voltage, must be grounded (bonded) to an effective ground-fault current path in accordance with 250.100 and bonded in accordance with 501.30.

Article 513 Questions

1. Any pit or depression below the level of the aircraft hangar floor is classified as a _____ location that extends up to said floor level.

   (a) Class I, Division 1 or Zone 1
   (b) Class I, Division 2
   (c) Class II, Division 1
   (d) Class III

2. Stock rooms and similar areas adjacent to classified locations of aircraft hangars, but effectively isolated and adequately ventilated, are designated as _____ locations.

   (a) Class I, Division 2
   (b) Class II, Division 1
   (c) Class II, Division 2
   (d) nonhazardous

3. All fixed wiring in an aircraft hanger not installed in a Class I location must be installed in _____.

   (a) metal raceways
   (b) Type MI, TC, or MC cable
   (c) nonmetallic raceways
   (d) a or b

4. In aircraft hangers, equipment that is less than _____ above wings and engine enclosures of aircraft and that may produce arcs, sparks, or particles of hot metal must be of the totally enclosed type or constructed so as to prevent the escape of sparks or hot metal particles.

   (a) 18 in.
   (b) 5 ft 6 in.
   (c) 10 ft
   (d) 6 ft 6 in.

5. All 125-volt, single-phase, 15 and 20 ampere receptacles installed in aircraft hangers in areas where _____ is (are) used must have ground fault circuit interrupter protection for personnel.

   (a) electrical diagnostic equipment
   (b) electrical hand tools
   (c) portable lighting equipment
   (d) any of these

Notes

ARTICLE

514 Motor Fuel Dispensing Facilities

Introduction

In a facility where fuel is dispensed from storage tanks into the fuel tanks of vehicles, you need to look at Article 514. That facility probably must conform to Article 514 requirements.

What is most striking about Article 514 is the large table that makes up about half of it. This table doesn't provide any electrical requirements, list any electrical specifications, or address any electrical equipment. What it does tell you is how to classify a motor fuel dispensing area based on the equipment contained therein. The rest of Article 514 contains specific provisions, and also refers to other articles that must also be applied.

514.1 Scope. This article applies to motor fuel dispensing facilities, marine fuel dispensing facilities, motor fuel dispensing facilities located inside buildings, and fleet vehicle motor fuel dispensing facilities where gasoline or other volatile flammable liquids or liquefied flammable gases are transferred to self-propelled vehicles or approved containers.

514.2 Definition.

Motor Fuel Dispensing Facility. That portion of a property where gasoline or other volatile flammable liquids or liquefied flammable gases are stored and dispensed into the fuel tanks of motor vehicles or marine craft or into containers approved by the authority having jurisdiction, including equipment used in connection with the dispensing of the fuel. **Figure 514-1**

> **FPN:** Wiring and equipment in the area of commercial service and repair garages of service stations must comply with the installation requirements contained in Article 511.

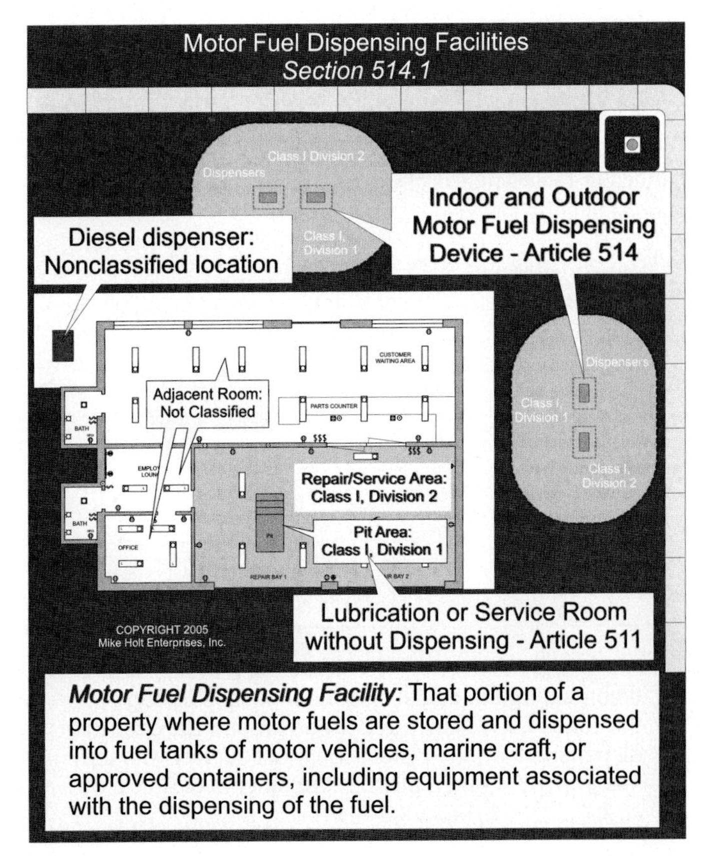

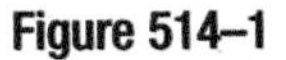
Figure 514–1

514.3 Classification of Locations.

(A) Unclassified Locations. Where the authority having jurisdiction is satisfied that flammable liquids having a flash point below 100°F such as gasoline, will not be handled, such location isn't required to be classified.

Author's Comment: Diesel fuel isn't a flammable liquid having a flash point below 100°F; it's a "combustible" liquid. Therefore, diesel dispensing equipment and associated wiring isn't required to comply with the stringent requirements of this article.

CAUTION: *If the conduit for the diesel dispenser enters a Class I, Division 1 or 2 location, the wiring methods and sealing requirements contained in Article 501 apply.* **Figure 514-2**

(B) Classified Locations.

(1) Class I Locations. The following identifies the most common hazardous (classified) location areas for gasoline dispensing and service stations. See the *NEC* Table 514.3(B)(1) for more details.

A Class I location does not extend beyond an unpierced wall, roof, or other solid barrier.

Author's Comment: The space within 18 in. horizontally in all directions extending to grade from the dispenser enclosure, and the space up to 18 in. above grade level within 20 ft horizontally of any edge of the dispenser is classified as a Class I, Division 2 location. **Figure 514-3**

Sales and storage rooms are not classified if there are no openings from these rooms to a Class I, Division 1 location.

514.4 Wiring and Equipment Within Class I Locations.

Electric equipment and wiring within a Class I location, as defined in Table 514.3(B)(1), must comply with the requirements contained in Article 501.

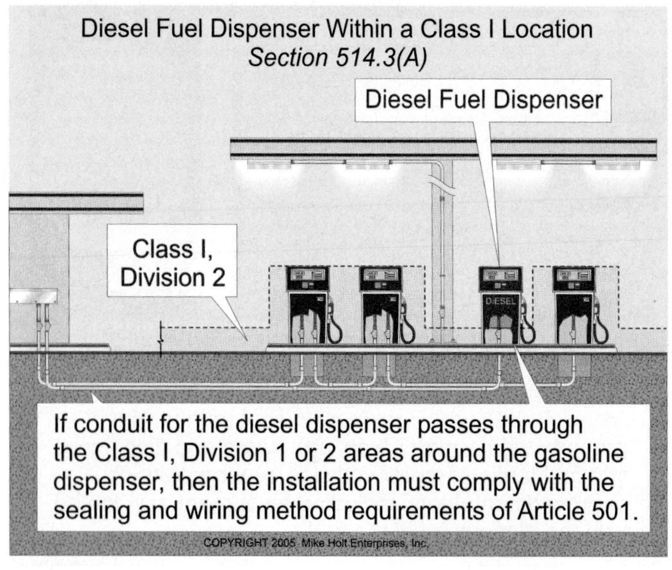

Figure 514–2

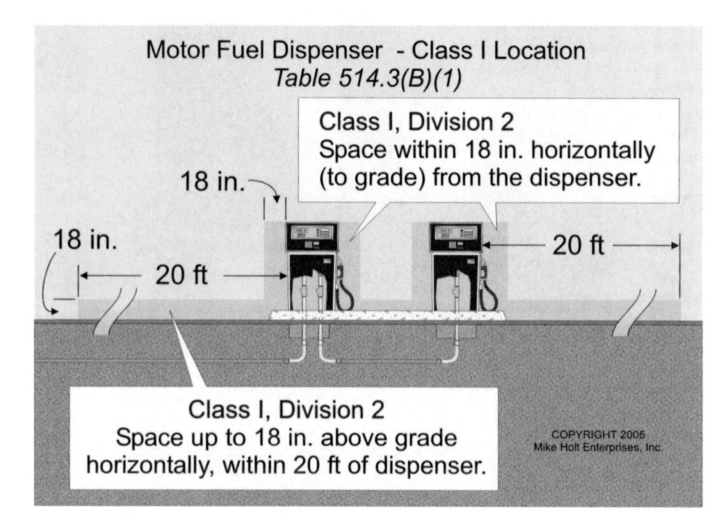

Figure 514–3

Author's Comment: This means that threaded rigid metal conduit and threaded intermediate metal conduit must be used for fixed wiring.

Exception: Rigid nonmetallic conduit is permitted underground in accordance with 514.8.

514.7 Wiring and Equipment Above Class I Locations.

Wiring above a Class I location, as defined in Table 514.3(B)(1), must be installed in a raceway, or in Types AC, MC, MI, PLTC, or TC cable [511.7(A)(1)]. **Figure 514-4**

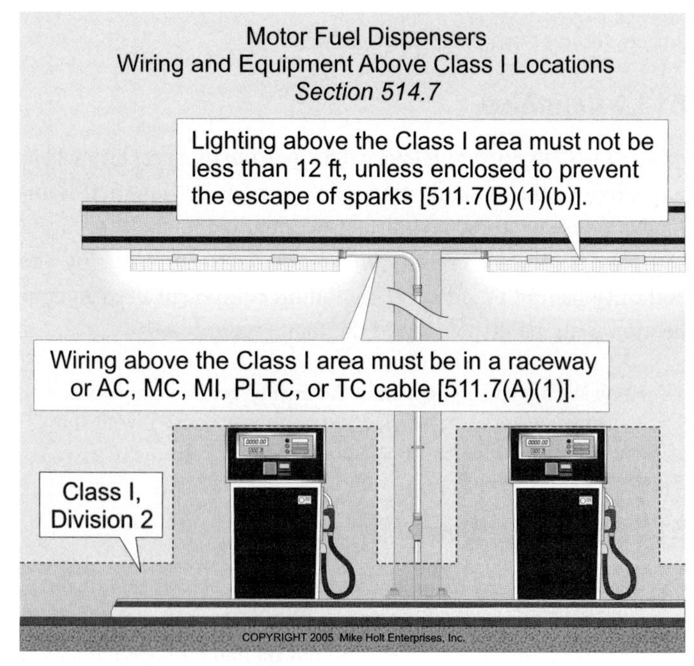

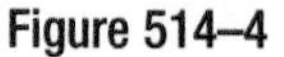
Figure 514–4

Fixed lighting over travel lanes, or where exposed to physical damage, must be located not less than 12 ft above floor level, unless the luminaire is of the totally enclosed type or constructed to prevent sparks or hot metal particles from escaping [511.7(B)].

514.8 Underground Wiring. Underground wiring must be installed in threaded rigid metal conduit or threaded intermediate metal conduit. Electrical conduits located below the surface of a Class I, Division 1 or 2 location as identified in Table 514.3, must be sealed within 10 ft of the point of emergence above grade. **Figure 514-5**

Except for listed explosionproof reducers at the conduit seal, there must be no union, coupling, box, or fitting between the conduit seal and the point of emergence above grade.

Exception 2: Rigid nonmetallic conduit with an equipment grounding (bonding) conductor is permitted underground below a Class I location if buried under not less than 2 ft of earth, concrete, asphalt, etc. Threaded rigid metal conduit or threaded intermediate metal conduit must be used for the last 2 ft of the underground run. **Figure 514-6**

514.9 Raceway Seal.

(A) At Dispenser. A raceway seal must be installed in each conduit run that enters or leaves a dispenser. The raceway seal must be the first fitting after the conduit emerges from the earth or concrete. **Figure 514-7**

(B) At Boundary. A raceway seal that complies with 501.15 must be installed in each conduit run that leaves a Class I location. **Figure 514-8**

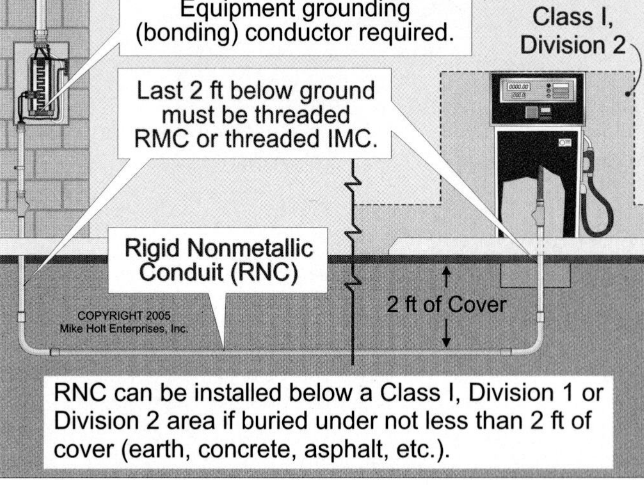
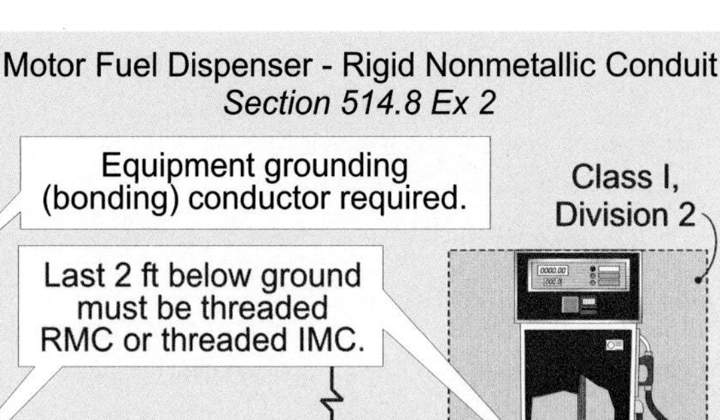

Figure 514–6

Author's Comment: Where the boundary is beneath the ground, the sealing fitting can be installed after the conduit leaves the ground, but there must be no union, coupling, box, or fitting, other than listed explosionproof reducers at the sealing fitting in the conduit between the sealing fitting and the point at which the conduit leaves the earth [501.15(A)(4) Ex 2].

In addition, the raceway seal must be accessible, it must not contain splices, and the total conductor fill area must not exceed 25 percent of the cross-sectional area of rigid metal conduit (even if the raceway is intermediate metal conduit), unless the seal fitting is specifically identified for a higher percentage of fill [501.15(C)(6)]. **Figure 514-9**

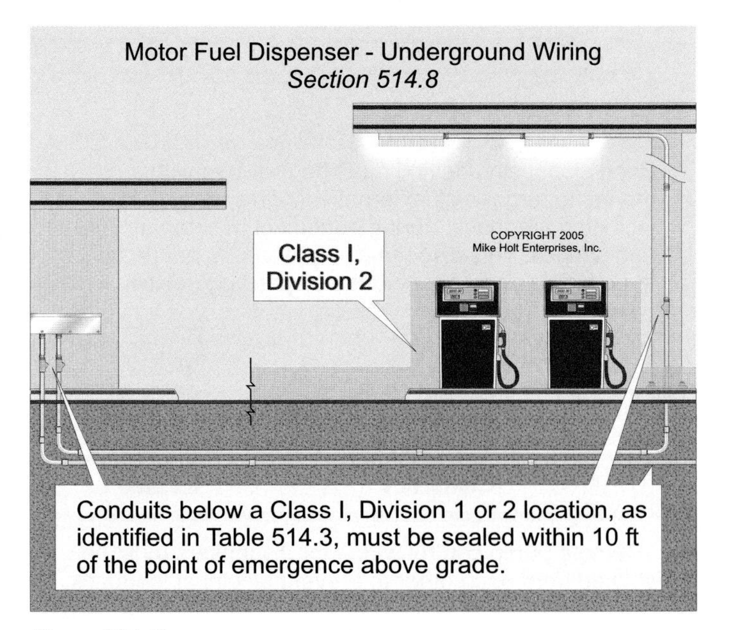

Figure 514–5

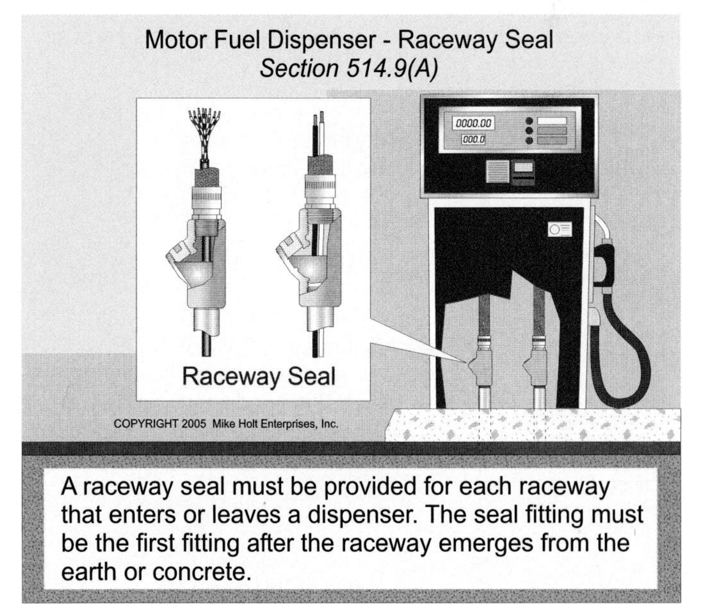

Figure 514–7

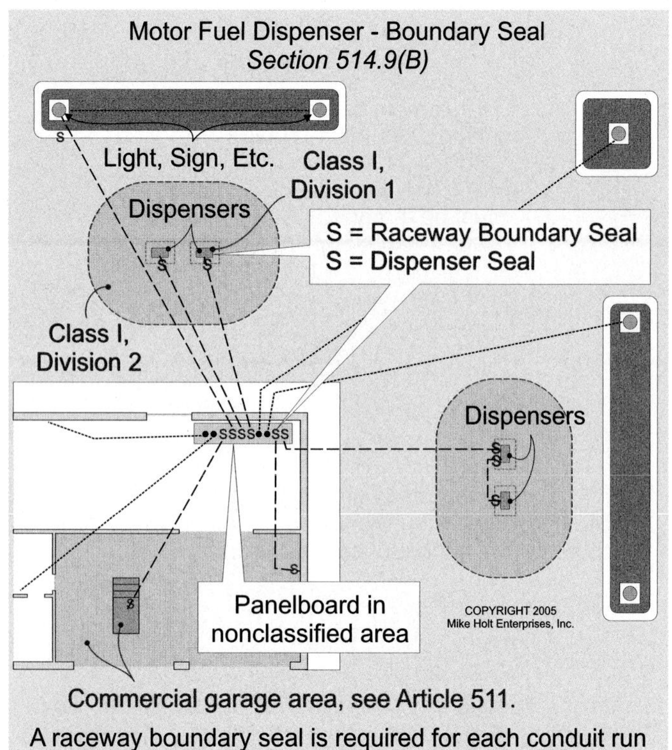

Figure 514–8

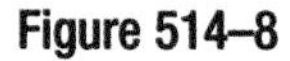

514.11 Emergency Dispenser Disconnects.

(A) General. Each circuit that leads to or through a dispenser (including equipment for remote pumping systems) must have a clearly identified and readily accessible switch located remote from the dispenser to disconnect simultaneously all conductors of the circuit, including the grounded neutral conductor. Single-pole breakers with handle ties cannot be used for this purpose.

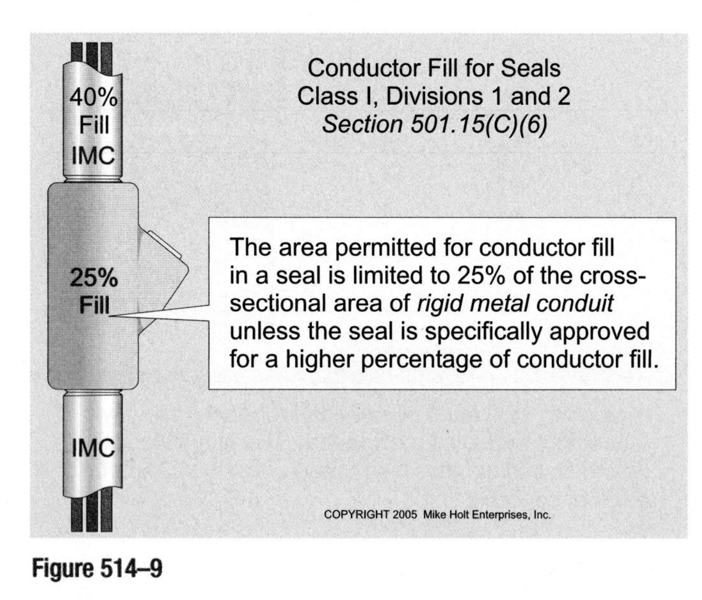

Conductor Fill for Seals
Class I, Divisions 1 and 2
Section 501.15(C)(6)
40% Fill IMC
25% Fill
IMC
The area permitted for conductor fill in a seal is limited to 25% of the cross-sectional area of rigid metal conduit unless the seal is specifically approved for a higher percentage of conductor fill.
COPYRIGHT 2005 Mike Holt Enterprises, Inc.

Figure 514–9

Author's Comment: See Article 100 for the definitions of "Accessible, Readily" and "Grounded Neutral Conductor."

(B) Attended Self-Service Stations. Attended self-service stations must have an emergency dispenser disconnect located no more than 100 ft from the dispenser, at a location that is acceptable to the authority having jurisdiction.

(C) Unattended Self-Service Stations. Unattended self-service stations must have the emergency dispenser disconnect located more than 20 ft, but less than 100 ft from the dispensers, at a location that is acceptable to the authority having jurisdiction.

514.13 Maintenance and Service of Dispensing Equipment.

Each dispensing device must be provided with a means to remove all external voltage sources, including feedback, during periods of maintenance and service of the dispensing equipment. The disconnecting means must be capable of being locked in the open position.

Author's Comment: See Article 100 for the definition of "Disconnecting Means."

Remote pump control wiring for each dispenser must be kept isolated to prevent electrical feedback.

Author's Comment: Dispenser pump control wiring must not supply more than one dispenser because it could cause electrical feedback when another dispenser is in operation. **Figure 514-10**

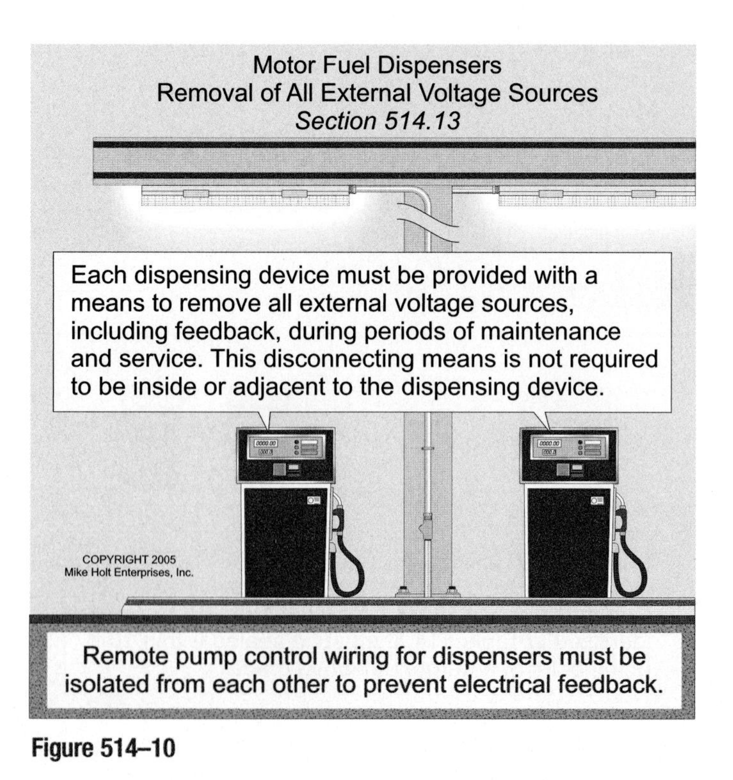

Figure 514–10

514.16 Grounding (Bonding). All metal raceways, enclosures, and fittings between a Class I, Division 1 or 2 hazardous (classified) location, and the service or separately derived system, must be grounded (bonded) to an effective ground-fault current path. This is accomplished by bonding the metal raceway to the enclosure or fitting by one of the following methods:

- Threaded conduit entry,
- Bonding bushing with a bonding jumper sized to Table 250.122, or
- A bonding-type locknut.

Locknuts or double locknuts aren't suitable for this purpose [501.30(A)]. **Figure 514-11**

Author's Comment: Set screw and compression couplings and connectors can be used for electrical metallic tubing, intermediate metal conduit, or rigid metal conduit installed in an unclassified area, provided the circuit doesn't pass through, or isn't part of, any circuit within a hazardous (classified) location.

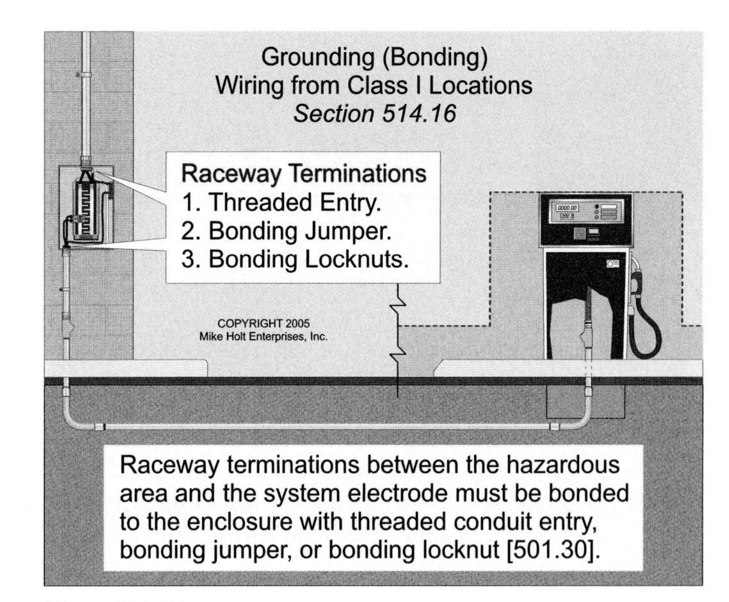

Grounding (Bonding)
Wiring from Class I Locations
Section 514.16
Raceway Terminations
1. Threaded Entry.
2. Bonding Jumper.
3. Bonding Locknuts.
COPYRIGHT 2005
Mike Holt Enterprises, Inc.
Raceway terminations between the hazardous area and the system electrode must be bonded to the enclosure with threaded conduit entry, bonding jumper, or bonding locknut [501.30].

Figure 514–11

Article 514 Questions

1. Article 514 contains requirements for the classification of areas where _____ is stored, handled, or dispensed from motor fuel dispensing facilities.

   (a) compressed natural gas (b) liquefied natural gas (c) liquefied petroleum gas (d) any of these

2. A listed sealing fitting must be _____.

   (a) provided in each conduit run entering a dispenser
   (b) provided in each conduit run leaving a dispenser
   (c) the first fitting after the conduit emerges from the earth or concrete
   (d) all of these

3. Each circuit leading to or through a dispensing pump must be provided with a switch or other acceptable means to disconnect simultaneously from the source of supply all conductors of the circuit, including the _____ conductor, if any.

   (a) grounding (b) grounded (c) bonding (d) all of these

4. Each circuit leading to gasoline dispensing equipment must be provided with a clearly identified and readily accessible switch or other acceptable means to disconnect all conductors of the circuit.

   (a) True (b) False

5. The emergency controls for unattended self-service stations must be located not less than _____ or more than _____ from the gasoline dispensers.

   (a) 10 ft, 25 ft (b) 20 ft, 50 ft (c) 20 ft, 100 ft (d) 50 ft, 100 ft

Introduction

Health care facilities differ from other types of buildings in many important ways. Article 517 is primarily concerned with those parts of health care facilities where patients are examined and treated. Whether those facilities are permanent or movable, they still fall under Article 517. On the other hand, Article 517 wiring and protection requirements don't apply to business offices, patient sleeping areas, or waiting rooms.

Article 517 contains many specialized definitions that only apply to health care facilities. While you do not need to be able to quote these definitions, you should have a clear understanding of what the terms mean.

As you study Parts II and III, keep in mind the special requirements of hospitals and why these requirements exist. The requirements in Parts II and III are highly detailed and not intuitively obvious. But we will show you the key concepts so you can easily understand and apply the requirements. These are three of the main objectives of Article 517, Parts II and III:

- Maximize the physical and electromagnetic protection of wiring by requiring metal raceways.
- Minimize electrical hazards by keeping the voltage potential between patients' bodies and medical equipment low. This involves many specific steps, beginning with 517.11.
- Minimize the negative effects of power interruptions by defining specific requirements for essential electrical systems.

Part IV addresses gas anesthesia stations. The main objective of Part IV is to prevent ignition. Part V addresses X-ray installations and really has two main objectives:

- Provide adequate ampacity and protection for the branch circuits.
- Address the safety issues inherent in high-voltage equipment installations.

Part VI provides requirements for low-voltage communications systems, such as fire alarms and intercoms. The main objective there is to prevent compromising those systems with inductive coupling or other sources of interference.

Part VII provides requirements for isolated power systems. The main objective is to keep them actually isolated.

PART I. GENERAL

517.1 Scope. Article 517 applies to electrical wiring in health care facilities, such as hospitals, nursing homes, limited care and supervisory care facilities, clinics and medical and dental offices, and ambulatory care facilities that provide services to human beings.

> **Author's Comment:** This article doesn't apply to animal veterinary facilities.

517.2 Definitions.

Health Care Facilities. Buildings or portions of buildings in which medical, dental, psychiatric, nursing, obstetrical, or surgical care is provided. Health care facilities include, but aren't limited to, hospitals, nursing homes, limited care facilities, supervisory care facilities, clinics, medical and dental offices, and ambulatory care facilities.

Author's Comment: Business offices, corridors, lounges, day rooms, dining rooms, or similar areas aren't classified as patient care areas. See the FPN for the definition of "Patient Care Area."

Hospital. A building or part thereof used for the medical, psychiatric, obstetrical, or surgical care, on a 24-hour basis, of four or more inpatients.

Limited Care Facility. An area used on a 24-hour basis for the housing of four or more persons who are incapable of self-preservation because of age, physical limitations due to accident or illness, or mental limitations.

Nursing Home. An area used for the lodging, boarding, and nursing care on a 24-hour basis of four or more persons who, because of mental or physical incapacity, may be unable to provide for their own needs and safety without the assistance of another person. This includes nursing and convalescent homes, skilled nursing facilities, intermediate care facilities, and infirmaries of homes for the aged.

Patient Bed Location. The location of an inpatient sleeping bed or the bed or procedure table used in a critical patient care area.

Patient Care Area. The area in a health care facility where patients are intended to be examined or treated. Areas where patient care is administered are classified as general care areas or critical care areas.

General Care Areas. Patient bedrooms, examining rooms, treatment rooms, clinics, and similar areas where patients come in contact with ordinary appliances, or where patients may be connected to electromedical devices.

Critical Care Areas. Special care units where patients are subjected to invasive procedures and connected to electromedical devices. These areas would include intensive care units, coronary care units, delivery rooms, operating rooms, and similar areas.

FPN: Business offices, corridors, lounges, day rooms, dining rooms, or similar areas typically aren't classified as patient care areas.

PART II. WIRING AND PROTECTION

517.10 Applicability.

(B) Not Covered. The requirements contained in Part II of Article 517 don't apply to:

(1) Business offices, corridors, waiting rooms, or similar areas in clinics, medical and dental offices, and outpatient facilities.

(2) Areas of nursing homes and limited care facilities used exclusively for patient sleeping.

Author's Comment: Such as:

- 517.13 – Grounding and Bonding Requirements
- 517.18(B) – Hospital Grade Receptacles
- 517.18(D) – Emergency Wiring Methods

517.12 Wiring Methods. Wiring methods must comply with Chapters 1 through 4, except as modified in this article.

517.13 Grounding (Bonding) Equipment in Patient Care Areas.

Author's Comment: Patient care areas include patient rooms as well as examining rooms, therapy areas, examining and treatment rooms, recreational areas, and some patient corridors. See 517.2.

Wiring in patient care areas must comply with (A) and (B):

(A) Wiring Methods. Branch circuits that serve patient care areas must be installed in a metal raceway or listed cable with a metallic armor or sheath that qualifies as an effective ground-fault current path in accordance with 250.118. **Figure 517-1**

Author's Comments:

- See Article 100 for the definition of "Branch Circuit."

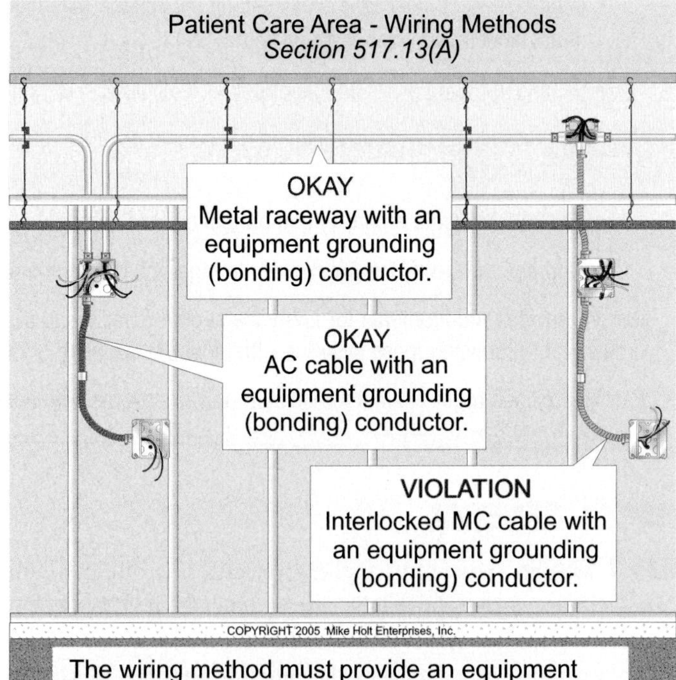

Patient Care Area - Wiring Methods
Section 517.13(A)
OKAY
Metal raceway with an equipment grounding (bonding) conductor.
OKAY
AC cable with an equipment grounding (bonding) conductor.
VIOLATION
Interlocked MC cable with an equipment grounding (bonding) conductor.
COPYRIGHT 2005 Mike Holt Enterprises, Inc.
The wiring method must provide an equipment grounding return path in accordance with 250.118, plus, an equipment grounding conductor is also required in each wiring method [517.13(B)].

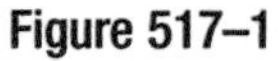

Figure 517–1

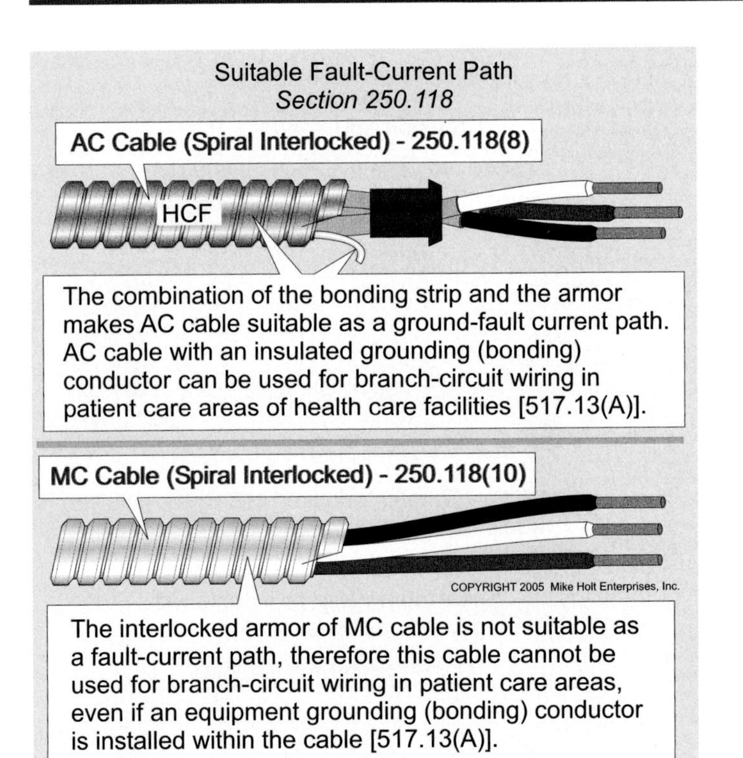

Suitable Fault-Current Path
Section 250.118
AC Cable (Spiral Interlocked) - 250.118(8)
HCF
The combination of the bonding strip and the armor makes AC cable suitable as a ground-fault current path. AC cable with an insulated grounding (bonding) conductor can be used for branch-circuit wiring in patient care areas of health care facilities [517.13(A)].
MC Cable (Spiral Interlocked) - 250.118(10)
COPYRIGHT 2005 Mike Holt Enterprises, Inc.
The interlocked armor of MC cable is not suitable as a fault-current path, therefore this cable cannot be used for branch-circuit wiring in patient care areas, even if an equipment grounding (bonding) conductor is installed within the cable [517.13(A)].

Figure 517–2

- The metal armored sheath of Type AC cable is listed as a suitable ground-fault current path because it contains an internal bonding strip in direct contact with the metal sheath of the cable [250.118(8)]. Typically, the outer metal sheath of interlocked Type MC cable isn't listed as a suitable ground-fault current path [250.118(10]). Therefore, it is not permitted to supply branch circuits in patient care areas of health care facilities. **Figure 517-2**
- Any receptacle in a patient care area that is supplied by an emergency circuit must not use Type AC or MC cable, flexible metallic conduit, or any other flexible wiring methods [517.30(C)(3)].

(B) Insulated Equipment Grounding (Bonding) Conductor. In patient care areas, the grounding terminals of receptacles and conductive surfaces of fixed electrical equipment that operate at over 100V must be grounded (bonded) to an effective ground-fault current path by an insulated copper equipment grounding (bonding) conductor. The equipment grounding (bonding) conductor must be sized in accordance with Table 250.122 and it must be installed in a metal raceway or listed cable with a metallic armor or sheath that qualifies as an effective ground-fault current path in accordance with 250.118 [517.13(A)]. **Figure 517-3**

Author's Comment: 517.13(A) and (B) require two separate types of equipment grounding (bonding) paths. One must be an electrically conductive path that is an integral part of the wiring method [517.13(A)], and the other must be a copper conductor [517.13(B)]. Nonmetallic raceways that contain two equipment grounding (bonding) conductors aren't permitted.

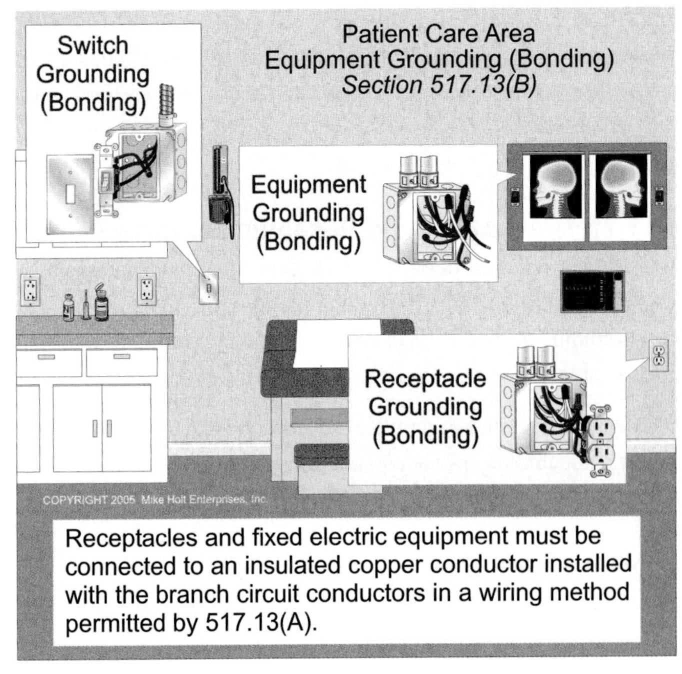

Figure 517–3

Exception 1: Metal faceplates for switches and receptacles are considered grounded (bonded) by the metal mounting screws that secure the faceplate to a grounded (bonded) outlet box or grounded (bonded) wiring device. **Figure 517-4**

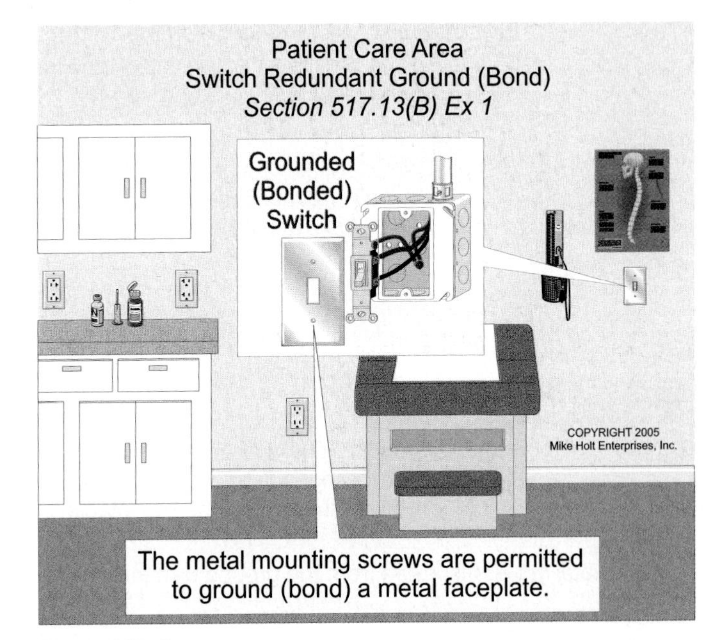

Figure 517–4

Exception 2: Luminaires more than 7½ ft above the floor aren't required to be grounded (bonded) to an insulated equipment grounding (bonding) conductor, but the wiring method must still comply with 517.13(A), which means that MC cable is not permitted.

517.16 Isolated Ground Receptacles.

Isolated ground receptacles installed in patient care areas must have an insulated equipment grounding (bonding) conductor that meets the requirements of 517.13(B) installed in a metal raceway or listed metal cable that meets the requirements of 517.13(A).

The isolated ground receptacle must be identified by an orange triangle located on the face of the receptacle in accordance with 406.2(D). **Figure 517-5**

Author's Comments:

- IGR receptacles are often entirely orange in color, with a triangle molded into the plastic face of the wiring device.
- Nonpatient Care Areas—Type AC cable containing a single insulated equipment grounding (bonding) conductor or Type MC cable with two equipment grounding (bonding) conductors can be used to supply an isolated ground receptacle in a nonpatient care area [250.118(8)]. Figure 517-6
- Patient Care Areas—EMT or Type AC cable with two insulated equipment grounding (bonding) conductors can be used to supply an isolated ground receptacle in a patient care area. One equipment grounding (bonding) conductor serves the isolated ground receptacle and the other grounds (bonds) the box to the effective ground-fault current path in accordance with 517.13(B). **Figure 517-7**
- Because the outer sheath of interlocked Type MC cable is typically not listed as an equipment grounding (bonding) conductor [250.118(10)], it isn't permitted to supply an isolated ground receptacle in a patient care area. See **Figure 517-7**.

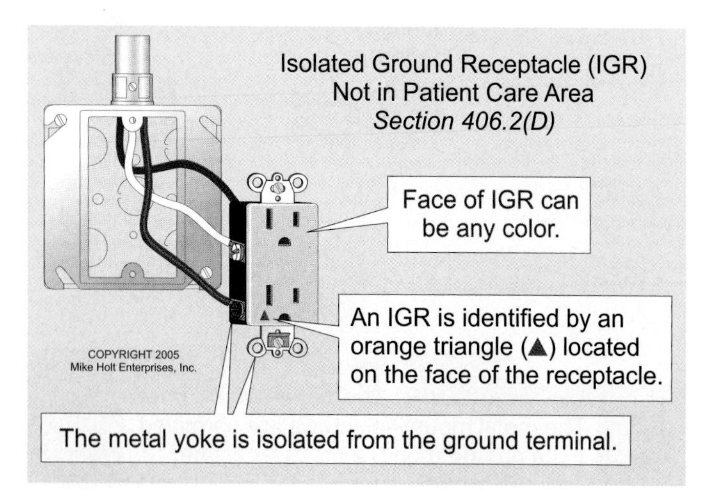
Isolated Ground Receptacle (IGR)
Not in Patient Care Area
Section 406.2(D)
Face of IGR can be any color.
An IGR is identified by an orange triangle (▲) located on the face of the receptacle.
COPYRIGHT 2005 Mike Holt Enterprises, Inc.
The metal yoke is isolated from the ground terminal.

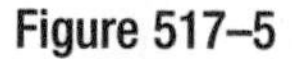
Figure 517–5

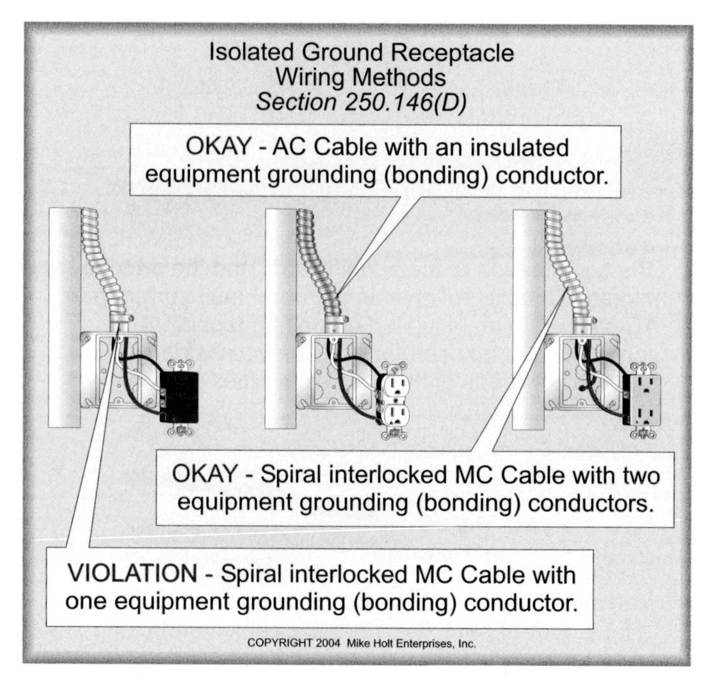
Isolated Ground Receptacle
Wiring Methods
Section 250.146(D)
OKAY - AC Cable with an insulated equipment grounding (bonding) conductor.
OKAY - Spiral interlocked MC Cable with two equipment grounding (bonding) conductors.
VIOLATION - Spiral interlocked MC Cable with one equipment grounding (bonding) conductor.
COPYRIGHT 2004 Mike Holt Enterprises, Inc.

Figure 517–6

517.18 General Care Areas.

(B) Hospital Grade Receptacles. Receptacles for inpatient sleeping beds or procedure table beds used in a critical patient care area (patient bed location) must be listed as "hospital grade." **Figure 517-8**

Author's Comment: "Hospital grade" receptacles aren't required in treatment rooms of clinics, medical and dental offices, or outpatient facilities. This is because these facilities don't have a "patient bed location" as defined in 517.2. **Figure 517-9**

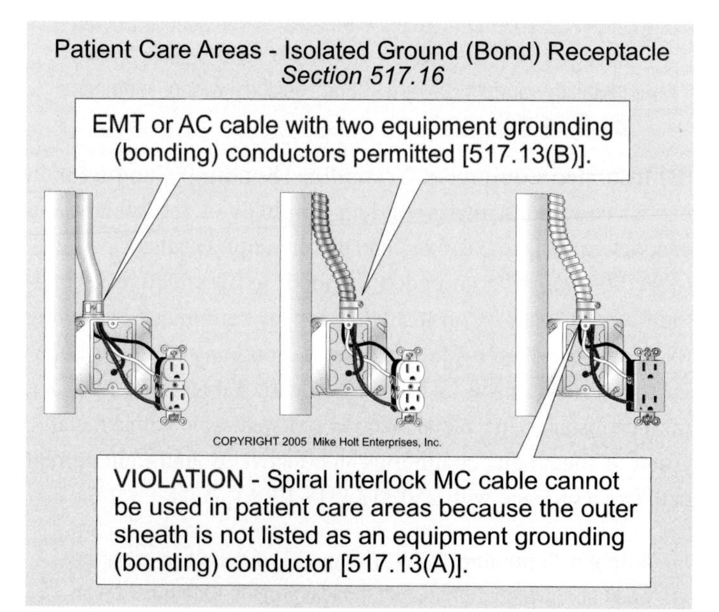
Patient Care Areas - Isolated Ground (Bond) Receptacle
Section 517.16
EMT or AC cable with two equipment grounding (bonding) conductors permitted [517.13(B)].
COPYRIGHT 2005 Mike Holt Enterprises, Inc.
VIOLATION - Spiral interlock MC cable cannot be used in patient care areas because the outer sheath is not listed as an equipment grounding (bonding) conductor [517.13(A)].

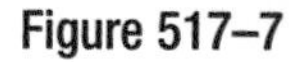
Figure 517–7

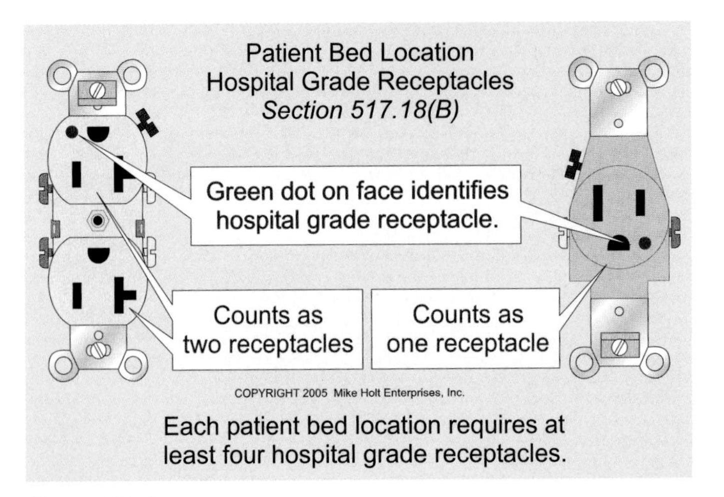
Patient Bed Location
Hospital Grade Receptacles
Section 517.18(B)
Green dot on face identifies hospital grade receptacle.
Counts as two receptacles
Counts as one receptacle
COPYRIGHT 2005 Mike Holt Enterprises, Inc.
Each patient bed location requires at least four hospital grade receptacles.

Figure 517–8

(C) Pediatric Locations. Receptacles in rooms, bathrooms, playrooms, activity rooms, and patient care areas of pediatric wards must be "tamper resistant" or they must have a tamper-resistant cover. **Figure 517-10**

Author's Comment: See Article 100 for the definition of "Bathroom."

517.30 Essential Electrical Systems for Hospitals.

(C) Wiring Requirements.

(3) Mechanical Protection of Emergency Circuits. Emergency circuit conductors must be installed in one of the following wiring methods:

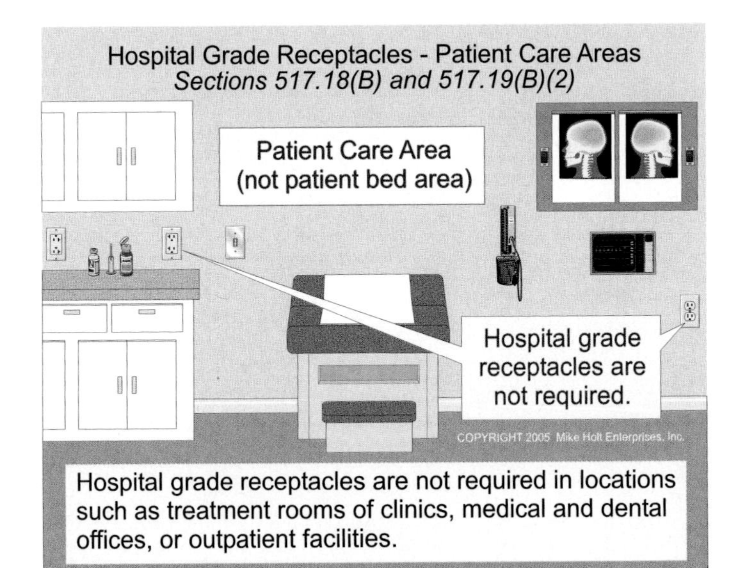
Hospital Grade Receptacles - Patient Care Areas
Sections 517.18(B) and 517.19(B)(2)
Patient Care Area
(not patient bed area)
Hospital grade receptacles are not required.
COPYRIGHT 2005 Mike Holt Enterprises, Inc.
Hospital grade receptacles are not required in locations such as treatment rooms of clinics, medical and dental offices, or outpatient facilities.

Figure 517–9

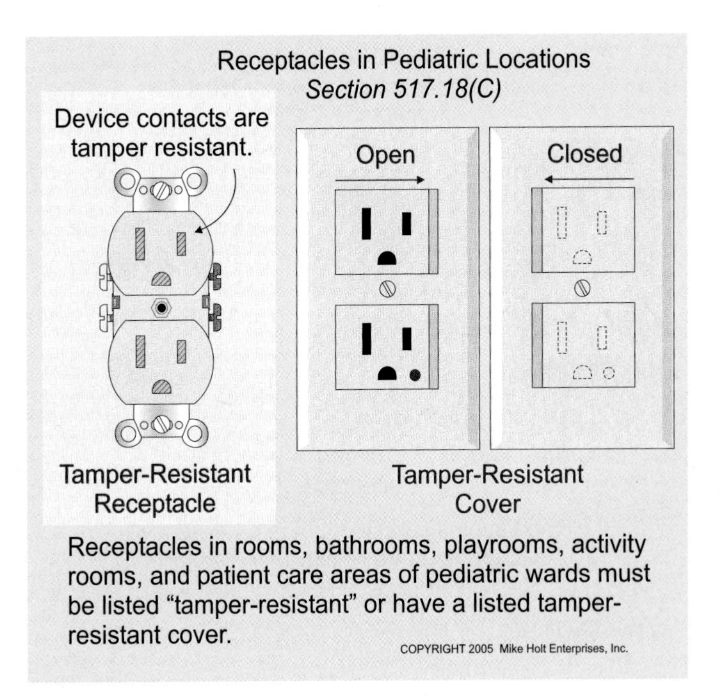
Receptacles in Pediatric Locations
Section 517.18(C)
Device contacts are tamper resistant.
Open
Closed
Tamper-Resistant Receptacle
Tamper-Resistant Cover
Receptacles in rooms, bathrooms, playrooms, activity rooms, and patient care areas of pediatric wards must be listed "tamper-resistant" or have a listed tamper-resistant cover.
COPYRIGHT 2005 Mike Holt Enterprises, Inc.

Figure 517–10

(1) Nonflexible metal raceways, Type MI cable, or Schedule 80 rigid nonmetallic conduit, if not used to supply patient care area branch circuits [517.13(A)].

(2) Schedule 40 rigid nonmetallic conduit or flexible nonmetallic raceways encased in not less than 2 in. of concrete, if not used to supply patient care area branch circuits [517.13(A)].

(3) Listed flexible metal raceways or listed metal-sheathed cables:

(a) When installed in listed prefabricated medical headwalls

(b) When installed in listed office furnishings

(c) Where fished into existing walls or ceilings, and not subject to physical damage

(d) Where necessary for flexible connection to equipment

(4) Flexible power cords of appliances, or other utilization equipment connected to the emergency system aren't required to be enclosed in raceways.

FPN: See 517.13 for the grounding (bonding) requirements for equipment located in patient care areas.

(E) Receptacle Identification. Receptacle faceplates or receptacles supplied from the emergency system must have a distinctive color or marking so as to be readily identifiable. **Figure 517-11**

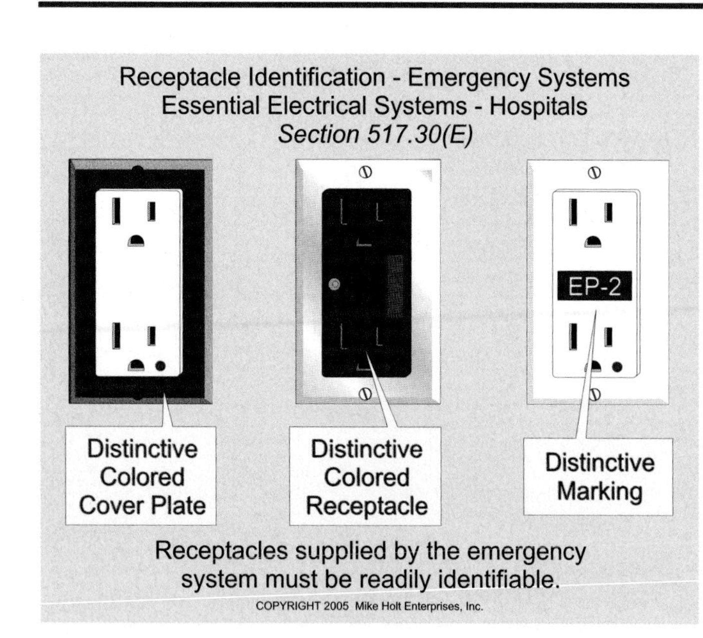

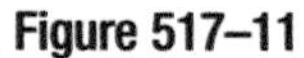
Figure 517–11

Author's Comment: Typically, the color used is red for the plate and/or receptacle. Some manufacturers make a standard receptacle faceplate stamped with the word "Emergency."

Article 517 Questions

1. The requirements of Article 517 (Health Care Facilities) apply to buildings or portions of buildings in which medical, _____, or surgical care is provided.

   (a) psychiatric (b) nursing (c) obstetrical (d) any of these

2. A nursing home is an area used for the lodging, boarding, and nursing care, on a 24-hour basis of _____ or more persons who, because of mental or physical incapacity may be unable to provide for their own needs and safety without assistance.

   (a) 4 (b) 100 (c) 10 (d) 2

3. Patient vicinity is the space with surfaces likely to be contacted by the patient or an attendant who can touch the patient. This encloses a space not less than 6 ft beyond the perimeter of the patient bed in its normal location and extending vertically not less than _____ above the floor.

   (a) 7½ ft (b) 5 ft 6 in. (c) 18 in. (d) 6 ft

4. In areas used for patient care, the grounding terminals of all receptacles and all noncurrent-carrying conductive surfaces of fixed electric equipment _____ must be grounded by an insulated copper equipment grounding conductor.

   (a) operating at over 100V (b) likely to become energized (c) subject to personal contact (d) all of these

5. The cover plates for receptacles, or the receptacles themselves, supplied from the emergency system of essential electrical systems in hospitals must have a distinctive color or marking so as to be readily identifiable.

   (a) True (b) False

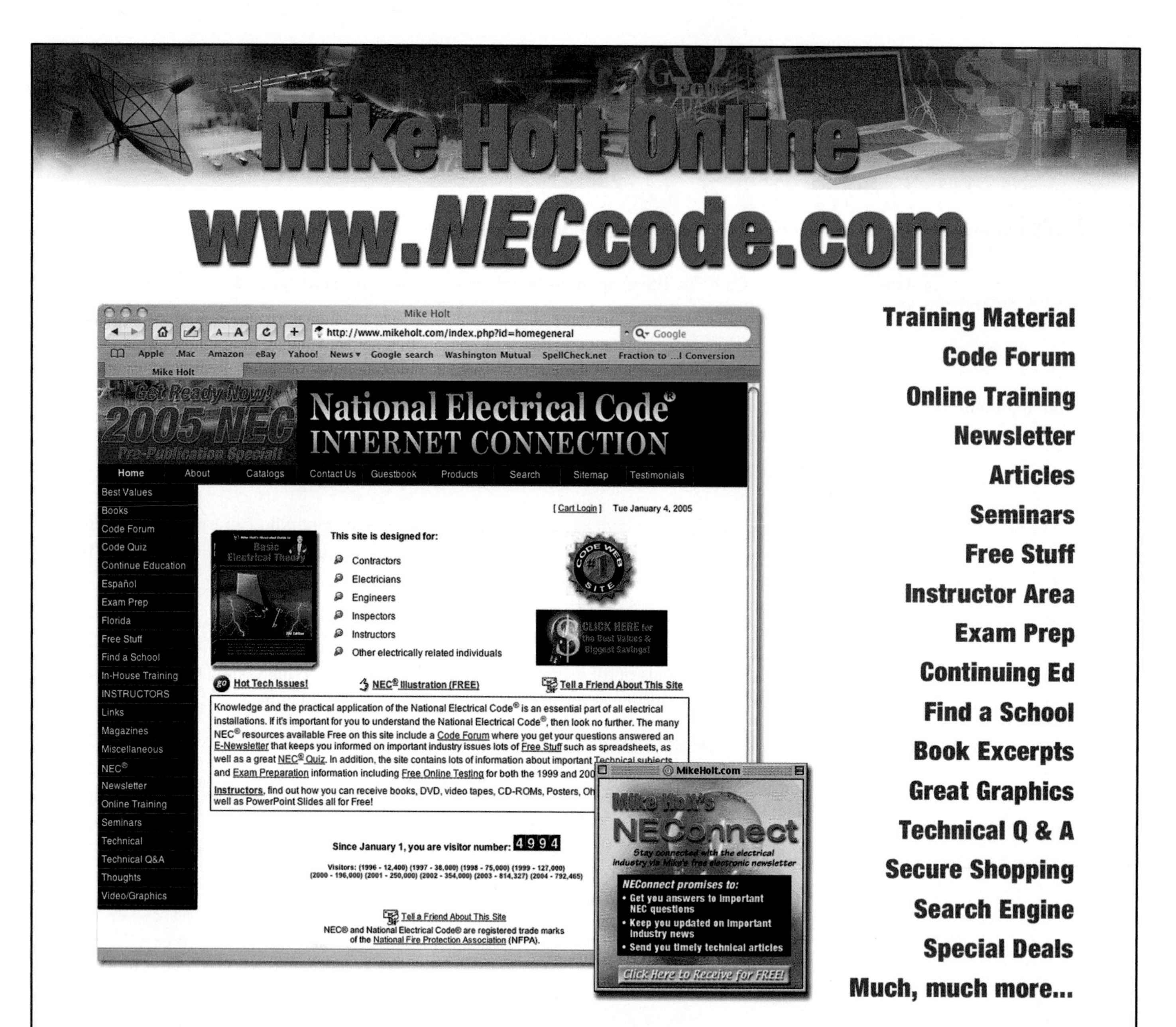

Visit Mike Holt Online Today!

This is the most comprehensive electrical site on the Web today. At www.NECcode.com you'll discover a wealth of information, free tools, a secure shopping cart, free newsletters, a Code® forum, online exams, seminar information, and much, much more.

All available to you at a click of a mouse!

Call us today at 1.888.NEC.Code, or visit us online at www.NECcode.com, for the latest information and pricing.

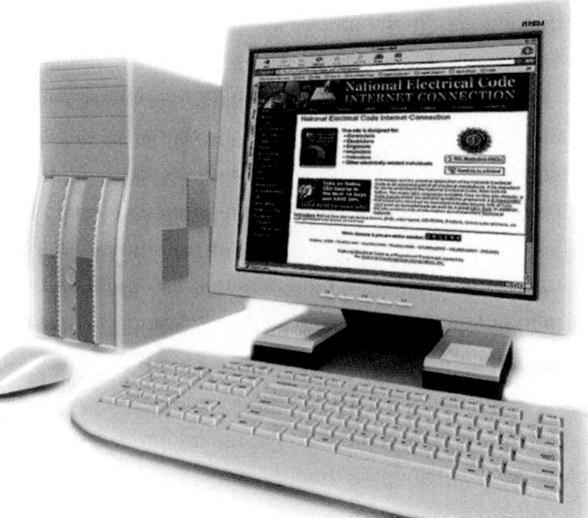

ARTICLE

518 Assembly Occupancies

Introduction

If a building or portion of a building is specifically designed or intended for the assembly of 100 or more people, it falls under Article 518. It goes out of its way to eliminate confusion; see 518.2 for a list of examples of what occupancies Article 518 covers. This article recognizes that it's much harder to evacuate 100 or more people from a burning building than it is to evacuate just a few people. That concept underlies much of the reasoning behind Article 518 requirements.

While Article 518 appears mostly to reference requirements in other articles, it does have several of its own. For example, if you want to install nonmetallic tubing or conduit in spaces with a finish rating, you must satisfy one of two conditions—which we will explain to you.

518.1 Scope. Except for the assembly occupancies explicitly covered by 520.1, this article covers all buildings or portions of buildings or structures designed or intended for the gathering together of 100 or more persons for such purposes as deliberation, worship, entertainment, eating, drinking, amusement, awaiting transportation, or similar purposes.

> **Author's Comment:** Occupancy capacity is determined in accordance with NFPA 101, *Life Safety Code*, or applicable building code.

518.2 General Classifications.

(A) Examples. Assembly occupancies include, but aren't limited to:

- Armories
- Auditoriums
- Club rooms
- Courtrooms
- Dining facilities
- Gymnasiums
- Multipurpose rooms
- Places of awaiting transportation
- Pool rooms
- Skating rinks
- Assembly halls
- Bowling lanes
- Conference rooms
- Dance halls
- Exhibition halls
- Mortuary chapels
- Museums
- Places of religious worship
- Restaurants

(B) Multiple Occupancies. The requirements contained in Article 518 only apply to that portion of the buildings or structures specifically designed or intended for the assembly of 100 or more persons.

518.3 Other Articles.

(B) Temporary Installations. Wiring for display booths in exhibit halls must be installed in accordance with Article 590. However, the GFCI-protection requirements contained in 590.6 do not apply.

Hard or extra-hard usage cords and cables can be laid on floors where protected from the general public.

518.4 Wiring Methods.

(A) Spaces With Fire-Rated Construction. In the fire-rated portions of a building used for assembly occupancy, only metal raceways, nonmetallic conduits encased in not less than 2 in. of concrete, or Types MC and AC cables containing an insulated equipment grounding (bonding) conductor are permitted. **Figure 518-1**

Exception: Wiring methods for the following systems are permitted within fire-rated construction in accordance with:

(a) Sound Systems—Article 640

(b) Communications Systems—Article 800

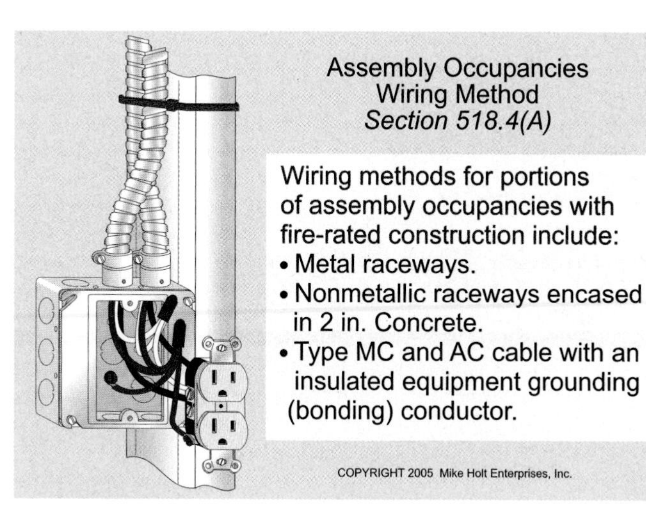

Assembly Occupancies
Wiring Method
Section 518.4(A)
Wiring methods for portions of assembly occupancies with fire-rated construction include:
• Metal raceways.
• Nonmetallic raceways encased in 2 in. Concrete.
• Type MC and AC cable with an insulated equipment grounding (bonding) conductor.
COPYRIGHT 2005 Mike Holt Enterprises, Inc.

Figure 518–1

(c) Class 2 and Class 3 Remote-Control and Signaling Circuits—Article 725

(d) Fire Alarm Circuits—Article 760

(B) Spaces With Nonfire-Rated Construction. In addition to 518.4(A) wiring methods, Type NM cable, Type AC cable, electrical nonmetallic tubing, and rigid nonmetallic conduit can be installed in those portions of an assembly occupancy building that aren't of fire-rated construction.

(C) Spaces With Finish Ratings. Electrical nonmetallic tubing and rigid nonmetallic conduit can be installed in club rooms, conference and meeting rooms in hotels or motels, courtrooms, dining facilities, restaurants, mortuary chapels, museums, libraries, and places of religious worship where:

(1) Electrical nonmetallic tubing or rigid nonmetallic conduit is installed concealed within walls, floors, or ceilings that provide a thermal barrier with not less than a 15-minute finish rating. **Figure 518-2**

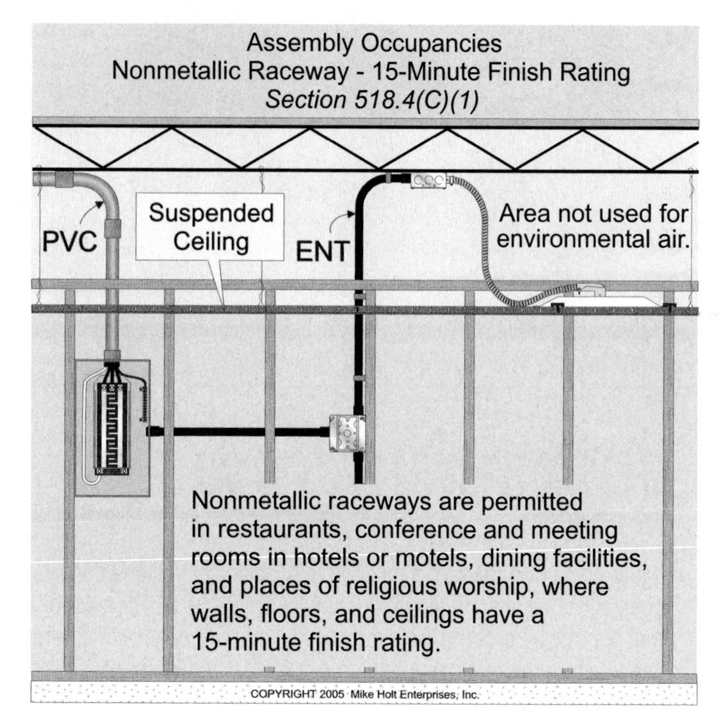

Assembly Occupancies
Nonmetallic Raceway - 15-Minute Finish Rating
Section 518.4(C)(1)
PVC
Suspended Ceiling
ENT
Area not used for environmental air.
Nonmetallic raceways are permitted in restaurants, conference and meeting rooms in hotels or motels, dining facilities, and places of religious worship, where walls, floors, and ceilings have a 15-minute finish rating.
COPYRIGHT 2005 Mike Holt Enterprises, Inc.

Figure 518–2

(2) Electrical nonmetallic tubing or rigid nonmetallic conduit is installed above suspended ceilings where the ceilings provide a thermal barrier not with less than a 15-minute finish rating.

Author's Comment: Electrical nonmetallic tubing and rigid nonmetallic conduit aren't permitted in dropped/suspended ceiling spaces used for environmental air [300.22(C].

Article 518 Questions

1. An assembly occupancy is a building, portion of a building, or structure designed or intended for the assembly of _____ or more persons.

   (a) 50 (b) 100 (c) 150 (d) 200

2. Examples of assembly occupancies include, but are not limited to _____.

   (a) restaurants (b) conference rooms (c) pool rooms (d) all of these

3. For temporary wiring in assembly occupancies, such as exhibition halls used for display booths, the wiring must be installed in accordance with Article 590, except _____.

   (a) the GFCI requirements of 590.6 do not apply
   (b) hard or extra hard usage cords and cables are permitted to be laid on floors where protected from the general public
   (c) no cords are allowed
   (d) a and b

4. Which of the following wiring methods are permitted to be installed in an assembly occupancy?

   (a) Metal raceways.
   (b) Type MC cable.
   (c) Type AC cable containing an insulated equipment grounding conductor.
   (d) all of these

5. In assembly occupancies, nonmetallic raceways encased in not less than _____ of concrete are permitted.

   (a) 1 in. (b) 2 in. (c) 3 in. (d) none of these

Notes

ARTICLE 525 Carinvals, Circuses, Fairs, and Similar Events

Introduction

These locations (carnivals, circuses, fairs, and similar events) are similar to assembly occupancies [Article 518], but there's a big difference: assembly occupancies covered by Article 518 aren't temporary. Another big difference is that the items covered by Article 525 include such things as amusement rides and attractions.

See if you can spot other similarities, as well as differences, between Article 518 and Article 525 as you study. Being aware of these will help you understand both articles better.

PART I. GENERAL REQUIREMENTS

525.1 Scope. This article covers the installation of portable wiring and equipment for carnivals, circuses, exhibitions, fairs, traveling attractions, and similar functions.

525.3 Other Articles.

(A) Portable Wiring. The requirements of Article 525 apply to the installation of portable wiring.

(C) Audio Equipment. The requirements of Article 640 apply to the wiring of audio equipment.

(D) Attractions Utilizing Water. Attractions that utilize water must be installed in accordance with Article 680 *Swimming Pools, Fountains, and Similar Installations*.

> **Author's Comment:** Water attractions include such things as "bumper boats," "dunking tanks," and "duck ponds" where electric motors are used to circulate water in a tank in which guests may come in direct contact. For example, 680.21(A)(1) permits a GFCI-protected single locking-type receptacle for a water-pump motor between 5 ft and 10 ft from the water. **Figure 525-1**

Carnivals, Circuses, and Fairs
Attractions Using Water
Section 525.3(D)
Twist-lock receptacle and GFCI protection are required, [680.22(A)(1)(2)].
COPYRIGHT 2005 Mike Holt Enterprises, Inc.
Wiring and equipment for attractions utilizing water must comply with the applicable requirements of Article 680.

Figure 525–1

525.5 Overhead Conductor Clearances.

(A) Vertical Clearances. Overhead conductors installed outside tents and concession areas must have a vertical clearance not less than [225.18]: **Figure 525-2**

(1) 10 ft above finished grade, sidewalks, platforms, or projections from which they might be accessible to pedestrians for 120V, 120/208V, 120/240V, or 240V circuits.

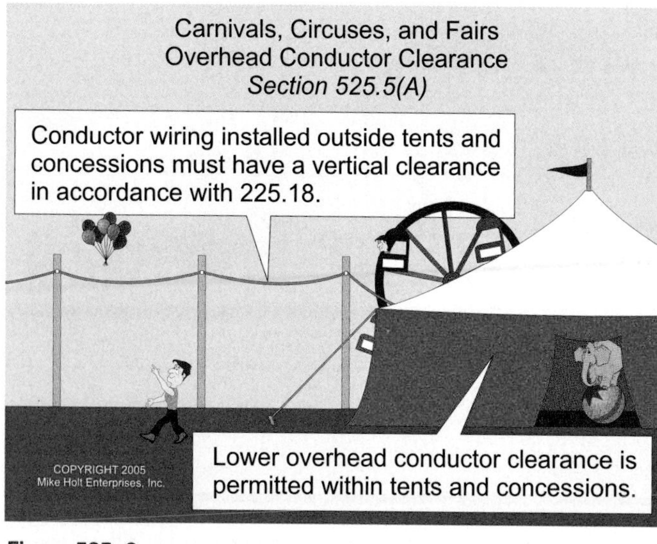

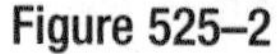
Figure 525–2

(2) 12 ft above residential property and driveways, and those commercial areas not subject to truck traffic for 120V, 120/208V, 120/240V, 240V, 277V, 277/480V, or 480V circuits.

(4) 18 ft over public streets, alleys, roads, parking areas subject to truck traffic, driveways on other than residential property, and other areas traversed by vehicles (such as those used for cultivation, grazing, forestry, and orchards).

(B) Clearance to Rides and Attractions. Overhead conductors, except for the conductors that supply the amusement ride or attraction, must have a clearance of 15 ft from amusement rides and amusement attractions.

525.6 Protection of Electrical Equipment. Electrical equipment and wiring must be provided with mechanical protection where subject to physical damage.

PART II. POWER SOURCES

525.10 Services. Services must comply with (A) and (B):

(A) Guarding. Service equipment must not be installed where accessible to unqualified persons, unless the equipment is lockable.

> **Author's Comment:** See Article 100 for the definition of "Accessible" as it relates to equipment and service equipment.

(B) Mounting and Location. Service equipment must be mounted on solid backing and be installed so as to be protected from the weather, unless of weatherproof construction.

525.11 Multiple Sources of Supply. Where multiple services or separately derived systems or both supply rides, attractions, and other structures, all sources of supply that serve rides, attractions, or other structures separated by less than 12 ft must be bonded to the same grounding electrode (earthing) system.

> **Author's Comment:** The 12-ft separation of structures is simply a nominal value; there is no strong technical reason for this measurement.

525.20 Wiring Methods.

(A) Type. Flexible cords or flexible cables in compliance with Article 400 must be listed for extra-hard usage. When used outdoors they must be listed for wet locations and must be sunlight resistant. **Figure 525-3**

(B) Single Conductor. Single conductor cable is permitted in sizes 2 AWG or larger.

(C) Open Conductors. Open conductors are permitted as part of a listed assembly or part of festoon lighting in accordance with Article 225. **Figure 525-4**

(D) Splices. Flexible cords or flexible cables must be continuous without splice or tap between boxes or fittings. **Figure 525-5**

(E) Cord Connectors. Cord connectors must not be laid on the ground unless listed for wet locations. Connectors placed in audience traffic paths or areas accessible to the public must be guarded. **Figure 525-6**

(F) Support. A ride or structure must not support wiring for an amusement ride, attraction, tent, or similar structure unless the ride or structure is specifically designed for this purpose.

Figure 525–3

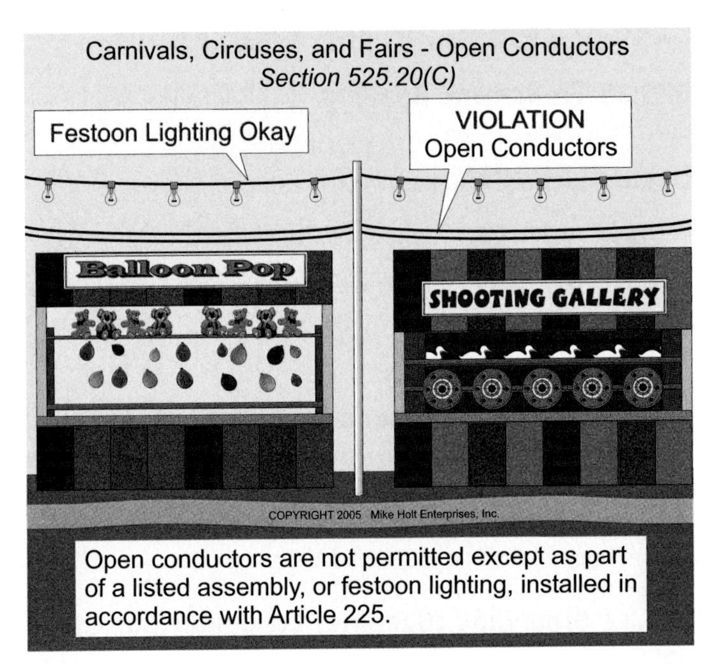

Figure 525–4

525.21 Rides, Tents, and Concessions.

(A) Disconnecting Means. Each ride and concession must have a fused disconnect switch or circuit breaker within sight and within 6 ft of the operator's station.

(B) Inside Tent and Concessions. Electrical wiring for lighting inside tents and concession areas must be securely installed. Where subject to physical damage, the wiring must be provided with mechanical protection and all lamps for general illumination must be protected from accidental breakage by a suitable luminaire or lampholder with a guard.

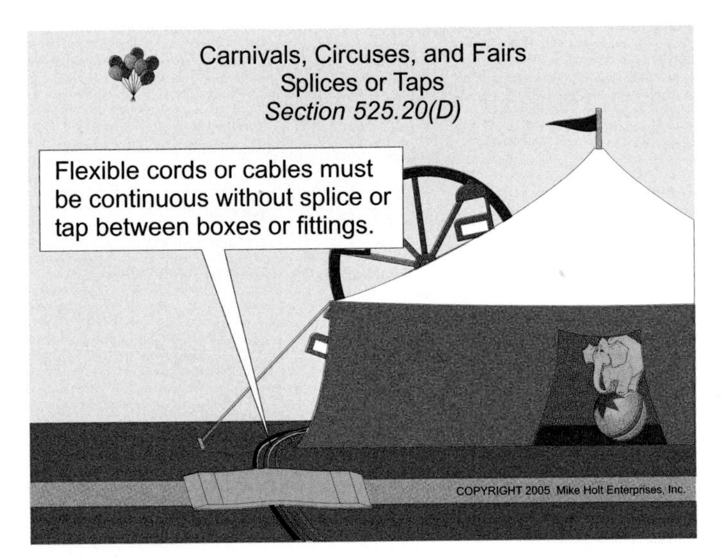

Figure 525–5

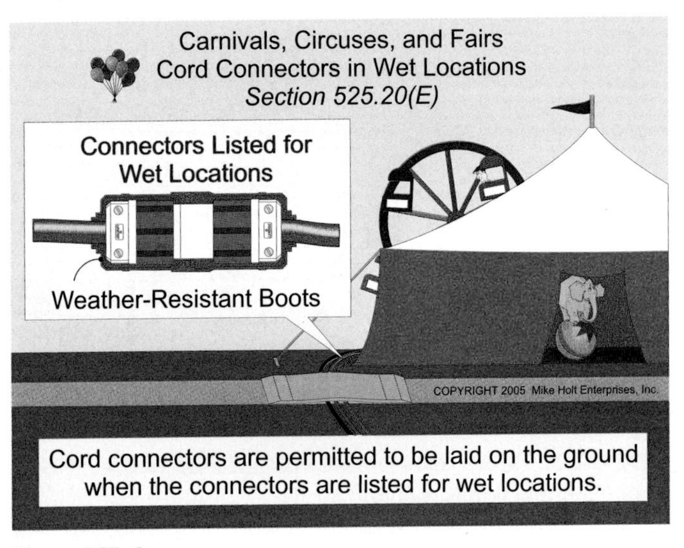

Carnivals, Circuses, and Fairs
Cord Connectors in Wet Locations
Section 525.20(E)
Connectors Listed for Wet Locations
Weather-Resistant Boots
COPYRIGHT 2005 Mike Holt Enterprises, Inc.
Cord connectors are permitted to be laid on the ground when the connectors are listed for wet locations.

Figure 525–6

525.22 Outdoor Portable Distribution or Termination Boxes.

(A) Terminal Boxes. Portable distribution and/or terminal boxes installed outdoors must be weatherproof, and the bottom of the enclosure must be not less than 6 in. above the ground.

525.23 GFCI-Protected Receptacles and Equipment.

(A) Where GFCI Protection is Required. Ground-fault circuit-interrupter protection for personnel can be an integral part of the attachment plug or located in the power-supply cord, within 12 in. of the attachment plug. Listed cord sets incorporating GFCI protection can be used to meet this requirement. **Figure 525-7**

(1) GFCI protection is required for all 15 and 20A, 125V non-locking type receptacles used for disassembly and reassembly of amusement rides and attractions, or readily accessible to the general public.

(2) GFCI protection is required for all equipment that is readily accessible to the general public if it is supplied from a 15 or 20A, 120V branch circuit.

(B) GFCI Protection Not Required. GFCI protection is not required for receptacles of the locking type.

(C) GFCI Protection Not Permitted. GFCI protection is not permitted for egress lighting.

Author's Comment: The purpose of not permitting egress lighting to be GFCI protected is to ensure that exit lighting remains energized and stays illuminated.

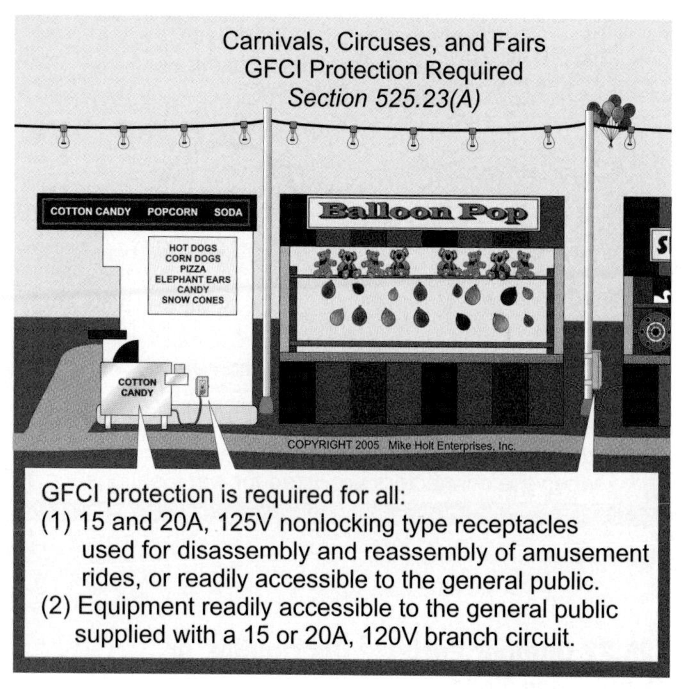

Carnivals, Circuses, and Fairs
GFCI Protection Required
Section 525.23(A)
COTTON CANDY POPCORN SODA
HOT DOGS
CORN DOGS
PIZZA
ELEPHANT EARS
CANDY
SNOW CONES
Balloon Pop
COTTON CANDY
COPYRIGHT 2005 Mike Holt Enterprises, Inc.
GFCI protection is required for all:
(1) 15 and 20A, 125V nonlocking type receptacles used for disassembly and reassembly of amusement rides, or readily accessible to the general public.
(2) Equipment readily accessible to the general public supplied with a 15 or 20A, 120V branch circuit.

Figure 525–7

PART IV. GROUNDING AND BONDING

525.30 Equipment Bonding. The following equipment must be bonded:

(1) Metal raceways and metal-sheath cables

(2) Metal enclosures

(3) Metal frames and metal parts of rides, concessions, tents, trailers, trucks, or other equipment that contain or support electrical equipment

525.31 Equipment Grounding (Bonding). The metal parts of all electrical equipment must be grounded (bonded) to the grounded neutral conductor at the service disconnecting means in accordance with 250.24(C), or the separately derived system in accordance with 250.30(A)(1).

525.32 Grounding (Bonding) Conductor Continuity Assurance. The continuity of the equipment grounding (bonding) conductor must be verified each time the portable electrical equipment is connected.

> **Author's Comment:** Verification of the equipment grounding (bonding) conductor is necessary to ensure electrical safety at carnivals, circuses, or fairs. However, the content of this requirement is vague. It doesn't specify how the grounding (bonding) conductor is to be verified, what circuits are required to be verified, how the verification is to be recorded, and who is required or qualified to perform the verification.

Article 525 Questions

1. Electrical wiring in and around water attractions such as bumper boats for carnivals, circuses, and fairs must comply with the requirements of Article 680—Swimming Pools, Fountains, and Similar Installations.

   (a) True (b) False

2. At carnivals, circuses, and similar events, service equipment must be mounted on a solid backing and be installed so as to be protected from the weather, unless _____.

   (a) the location is a mild climate
   (b) installed for less than 90 days
   (c) of weatherproof construction
   (d) a or b

3. Wiring for an amusement ride, attraction, tent, or similar structure must not be supported by any other ride or structure unless specifically designed for the purpose.

   (a) True (b) False

4. GFCI protection for personnel is required at carnivals, circuses, and fairs for all 15 and 20A, 125V, single-phase receptacle outlets that are readily accessible to the general public.

   (a) True (b) False

5. The continuity of the grounding conductor system used to reduce electrical shock hazards at carnivals, fairs, and similar locations must be verified each time the portable electrical equipment is connected.

   (a) True (b) False

Notes

ARTICLE 547 Agricultural Buildings

Introduction

Two factors have a tremendous influence on the lifespan of agricultural equipment: dust and moisture.

Dust gets into mechanisms and causes premature wear. But with electricity on the scene, dust adds two other dangers: fire and explosion. Dust from hay, grain, and fertilizer is highly flammable. Litter materials, such as straw, are also highly flammable. The excrement from farm animals may cause corrosive vapors that eat at mechanical equipment but can also cause electrical equipment to fail. For these reasons, Article 547 includes requirements for dealing with dust and corrosion.

Another factor to consider in agricultural buildings is moisture, which causes corrosion. Water is present for many reasons, including wash down. Thus, Article 547 has requirements for dealing with wet and damp environments, and also includes other requirements. For example, it requires you to install equipotential (stray voltage) planes in all concrete floor confinement areas of livestock buildings containing metallic equipment accessible to animals and likely to become energized.

Livestock animals have a very low tolerance to small levels of stray electrical current, which can cause loss of milk production and, at times, livestock fatality. As a result, the *NEC* contains specific requirements for an equipotential (stray voltage) plane in buildings that house livestock.

547.1 Scope. Article 547 applies to agricultural buildings or to that part of a building or adjacent areas of similar nature as specified in (A) and (B).

(A) Excessive Dust and Dust with Water. Buildings or areas where excessive dust or dust with water may accumulate, such as all areas of poultry, livestock, and fish confinement systems where litter or feed dust may accumulate.

(B) Corrosive Atmosphere. Buildings or areas where a corrosive atmosphere exists, and where the following conditions exist:

(1) Poultry and animal excrement.

(2) Areas where corrosive particles combine with water.

(3) Areas made damp or wet by periodic washing.

547.2 Definitions.

Equipotential Plane. An equipotential (stray voltage) plane is an area where wire mesh or other conductive elements are embedded in or placed under concrete, and are bonded to the electrical system to prevent a difference in voltage from developing within the plane [547.10(B)]. **Figure 547-1**

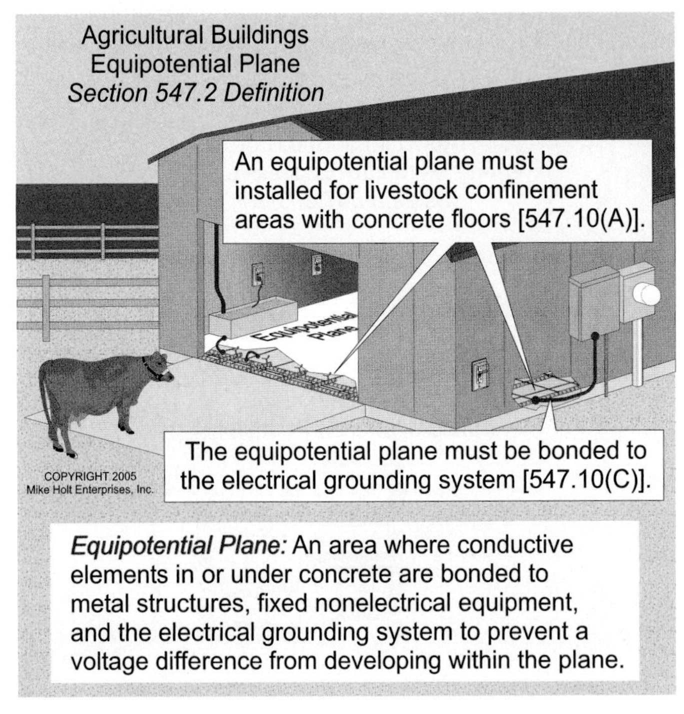

Agricultural Buildings
Equipotential Plane
Section 547.2 Definition
An equipotential plane must be installed for livestock confinement areas with concrete floors [547.10(A)].
Equipotential Plane
The equipotential plane must be bonded to the electrical grounding system [547.10(C)].
COPYRIGHT 2005 Mike Holt Enterprises, Inc.
Equipotential Plane: An area where conductive elements in or under concrete are bonded to metal structures, fixed nonelectrical equipment, and the electrical grounding system to prevent a voltage difference from developing within the plane.

Figure 547–1

547.5 Wiring Methods.

(A) Wiring Systems. Types UF, NMC, copper SE cables, jacketed Type MC cable, rigid nonmetallic conduit, liquidtight flexible nonmetallic conduit, or other cables or raceways suitable for the location, with termination fittings approved by the authority having jurisdiction, can be installed in agricultural buildings or structures. The wiring methods of Article 502, Part II are permitted for areas described in 547.1(A).

FPN: See 300.7 and 352.44 for the installation requirements for raceways exposed to widely different temperatures.

(B) Mounting. All cables must be secured within 8 in. of terminations at cabinets, boxes, or fittings.

Author's Comment: See Article 100 for the definition of "Cabinet."

(C) Equipment Enclosures, Boxes, Conduit Bodies, and Fittings.

(1) Excessive Dust. Equipment enclosures, boxes, conduit bodies, and fittings in areas of agricultural buildings where excessive dust may be present must be designed to minimize the entrance of dust and have no openings, such as holes for attachment screws, through which dust could enter the enclosure.

Author's Comment: Weatherproof boxes and covers can be used to meet this requirement. **Figure 547-2**

(2) Damp or Wet Locations. In damp or wet locations of agricultural buildings, equipment enclosures, boxes, conduit bodies, and fittings must be located or equipped to prevent moisture from entering or accumulating within the enclosure, box, conduit body, or fitting. In agricultural buildings where surfaces are periodically washed or sprayed with water, boxes, conduit bodies, and fittings must be listed for use in wet locations and the enclosures must be weatherproof.

Author's Comments:

- See Article 100 for the definitions of "Conduit Body," "Location (Damp)," and "Location (Wet)."
- See 406.8 for the enclosure requirement for receptacles installed in damp and/or wet locations.

(3) Corrosive Atmosphere. Where wet dust, excessive moisture, corrosive gases or vapors, or other corrosive conditions may be present in an agricultural building, equipment enclosures, boxes, conduit bodies, and fittings must have corrosion-resistant properties suitable for the conditions.

(G) GFCI-Protected Receptacles. GFCI protection is required for all 15 and 20A, 125V general-purpose receptacles located: **Figure 547-3**

(1) In areas having an equipotential (stray voltage) plane in accordance with 547.10(A).

(2) Outdoors.

(3) In damp or wet locations.

(4) In dirt confinement areas for livestock.

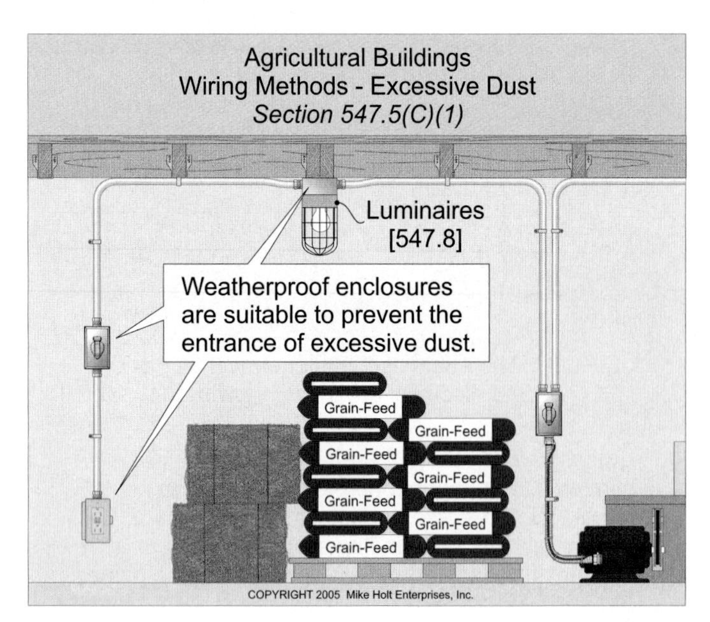

Figure 547–2

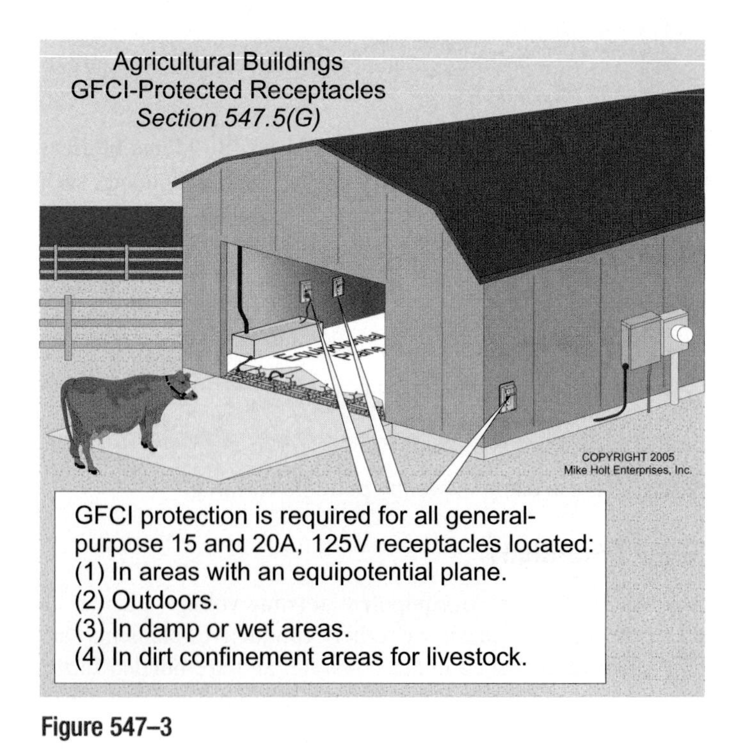

Figure 547–3

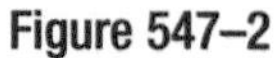

Author's Comment: The term "general-purpose receptacles" isn't defined in the *NEC*, but the intent is that GFCI protection isn't required for receptacles installed specifically for equipment such as brooders, incubators, feed mixers, feed grinders, feed conveyors, and the like.

547.10 Equipotential Planes and Bonding of Equipotential Planes.

The installation and bonding of equipotential (stray voltage) planes must comply with (A) and (B).

(A) Where Equipotential Plane is Required. An equipotential (stray voltage) plane must be installed in:

All concrete floor confinement areas of livestock buildings containing metallic equipment accessible to animals that may become energized. **Figure 547-4**

Outdoor concrete confinement areas, such as feedlots, must have an equipotential (stray voltage) plane around metallic equipment accessible to animals that may become energized.

The equipotential (stray voltage) plane must encompass the area around the equipment where the animal stands.

(B) Bonding of Equipotential Plane. The equipotential (stray voltage) plane must be bonded to the building or structure's electrical grounding system. The bonding conductor used for this purpose must be copper, insulated, covered, or bare, and not smaller than 8 AWG. The bonding conductor must terminate at pressure connectors or clamps of brass, copper, copper alloy, or an equally substantial means approved by the authority having jurisdiction.

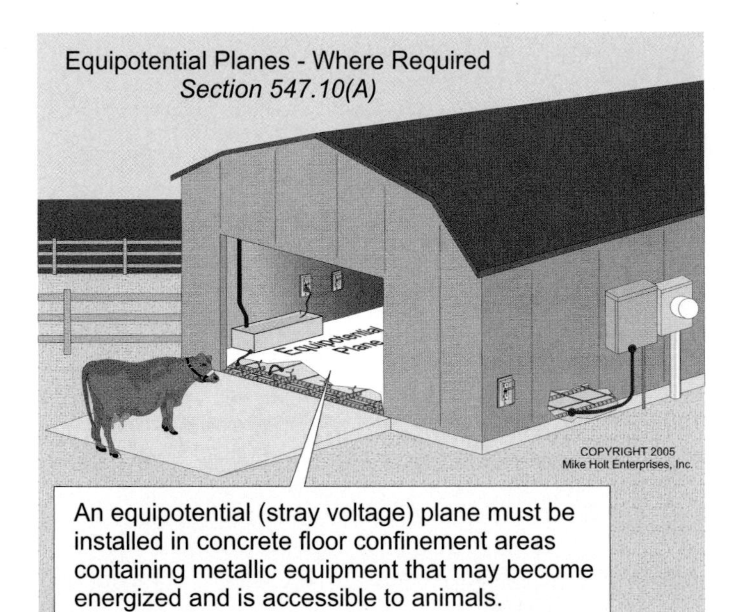

Equipotential Planes - Where Required
Section 547.10(A)
Equipotential Plane
COPYRIGHT 2005
Mike Holt Enterprises, Inc.
An equipotential (stray voltage) plane must be installed in concrete floor confinement areas containing metallic equipment that may become energized and is accessible to animals.

Figure 547–4

Author's Comment: See Article 100 for the definition of "Connector, Pressure."

FPN No. 1: Methods to establish equipotential (stray voltage) planes are described in American Society of Agricultural Engineers standard EP473, *Equipotential Planes in Animal Containment Areas.*

FPN No. 2: Low grounding electrode (earthing) system resistances may reduce potential differences in livestock facilities.

Author's Comments:

- The statement in Fine Print Note No. 2 that low grounding electrode resistance can be used to reduce potential difference in livestock facilities is false. However, this is beyond the scope of this textbook.
- Grounding (earthing) and bonding requirements contained in Article 547 are unique because of the sensitivity of livestock to stray voltage/current, especially in wet or damp concrete animal confinement areas. According to the IEEE 142 (Green Book), if the grounded neutral conductor is connected to earth at more than one location, part of the load current flows through the earth because it's in parallel with the grounded neutral conductor. Since there's impedance in both the conductor and the earth, a voltage drop will occur both along the earth and the conductor. Most of the voltage drop in the earth will occur in the vicinity of the point of connection to earth. Because of this nonlinear voltage drop in the earth, most of the earth will be at a different potential than the grounded neutral conductor due to the load current that flows from this conductor to earth.

  In most instances the potential difference between metal parts and the earth will be too low to present a shock hazard to persons. However, livestock might detect the potential difference if they come in contact with the metal parts, or if there is sufficient difference in potential between the earth contacts of their hoofs. Although potential differences may not be life threatening to the livestock, it has been reported that as little as 0.50V RMS can affect milk production.

  The topic of stray voltage/current is beyond the scope of this textbook. For more information, visit www.MikeHolt.com. Click on the "Technical" link, then the "Stray Voltage" link for more details.

Article 547 Questions

1. The distribution point is also known as the _____.

   (a) center yard pole (b) meter pole (c) common distribution point (d) all of these

2. All cables installed in agricultural buildings must be secured within _____ of each cabinet, box, or fitting.

   (a) 8 in. (b) 12 in. (c) 10 in. (d) 18 in.

3. Where _____ may be present in an agricultural building, enclosures and fittings must have corrosion-resistance properties suitable for the conditions.

   (a) wet dust (b) corrosive gases or vapors (c) other corrosive conditions (d) any of these

4. Where livestock is housed, that portion of the equipment grounding conductor run underground to the building or structure from a distribution point must be insulated or covered _____.

   (a) aluminum (b) copper (c) copper-clad aluminum (d) none of these

5. The equipotential planes in an agricultural building must be bonded to the electrical grounding system. The bonding conductor must be copper, insulated, covered, or bare and not smaller than _____.

   (a) 6 AWG (b) 8 AWG (c) 4 AWG (d) 10 AWG

Mobile Homes, Manufactured Homes, and Mobile Home Parks

Introduction

Among dwelling types, mobile homes have the highest rate of fire. Article 550 addresses some of the causes of those fires with the intent of reducing these statistics.

Article 550 recognizes that the same structures used for mobile or manufactured homes are also used for nondwelling purposes, such as construction offices or clinics [550.4(A)]. Thus, it excludes those structures from the 100A minimum service requirement.

Many people don't know there's a difference between a mobile home and a manufactured home, but there is—and Article 550 has different requirements for each. For example, you cannot locate service equipment on a mobile home. However, you can install service equipment on a manufactured home (provided you meet seven conditions). Pay close attention to the definitions in 550.2 so you don't get confused.

PART I. GENERAL

550.1 Scope. The provisions of Article 550 cover electrical conductors and equipment within or on mobile and manufactured homes, conductors that connect mobile and manufactured homes to the electric supply, and the installation of electrical wiring, luminaires, and electrical equipment within or on mobile and manufactured homes.

In addition, Article 550 applies to electrical equipment related to the mobile home feeder/service-entrance conductors and service equipment as covered by Part III of this article.

550.2 Definitions.

Manufactured Home. A structure that is 320 sq ft or more when assembled, transportable (minimum of 8 ft x 40 ft) in one or more sections that is built on a chassis and designed to be used as a dwelling unit with or without a permanent foundation. For the purpose of this article, the term "mobile home" includes manufactured homes. **Figure 550-1**

> **Author's Comment:** See Article 100 for the definition of "Dwelling Unit."

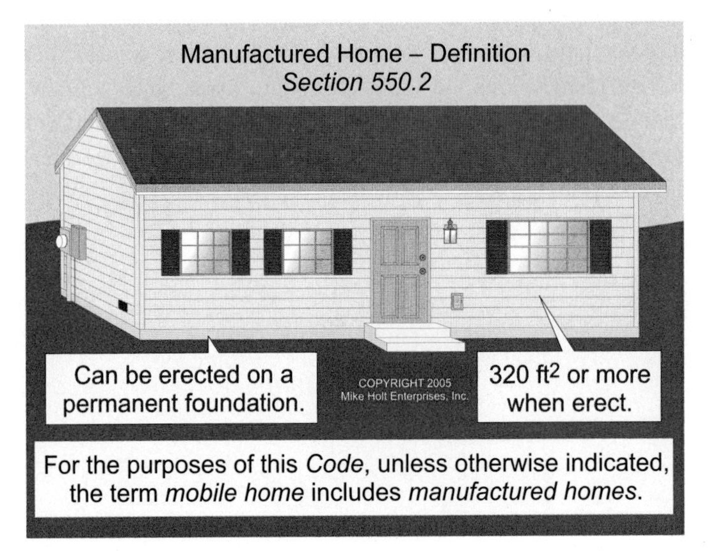

Manufactured Home – Definition
Section 550.2
Can be erected on a permanent foundation.
COPYRIGHT 2005 Mike Holt Enterprises, Inc.
320 ft2 or more when erect.
For the purposes of this Code, unless otherwise indicated, the term mobile home includes manufactured homes.

Figure 550–1

Mobile Home. A transportable structure built on a permanent chassis and designed to be used as a dwelling unit without a permanent foundation. In this article, the term "mobile home" includes manufactured homes.

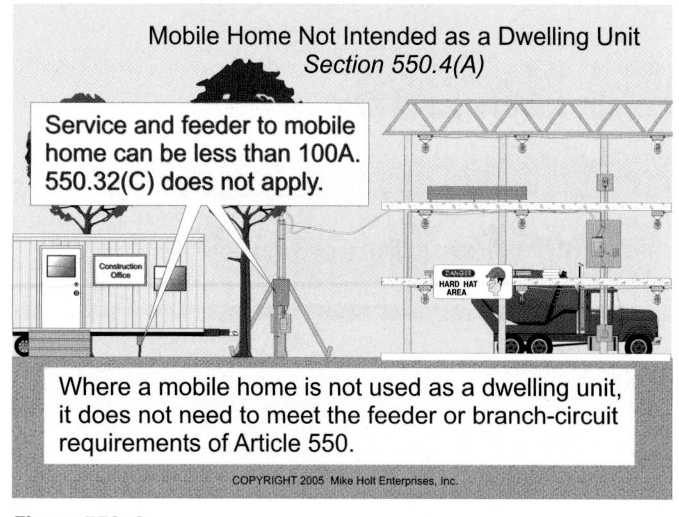

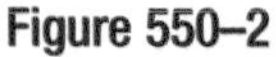
Figure 550–2

Author's Comment: One of the key differences between a manufactured home and a mobile home is that the manufactured home can go on or off a permanent foundation, while a mobile home can never be installed on a permanent foundation. Also, the *NEC* states that a manufactured home is 8 ft or more wide and 40 ft or more long in the traveling mode. The *Code* doesn't specify a minimum size for a mobile home.

550.4 General Requirements.

(A) Mobile Home Not Intended as a Dwelling Unit. A mobile home not intended as a dwelling unit, such as those equipped for sleeping purposes only, contractor's on-site offices, construction job dormitories, mobile studio dressing rooms, banks, clinics, mobile stores, or intended for the display or demonstration of merchandise or machinery, isn't required to meet the 100A minimum circuit capacity requirements of 550.32(C) for a dwelling unit. **Figure 550-2**

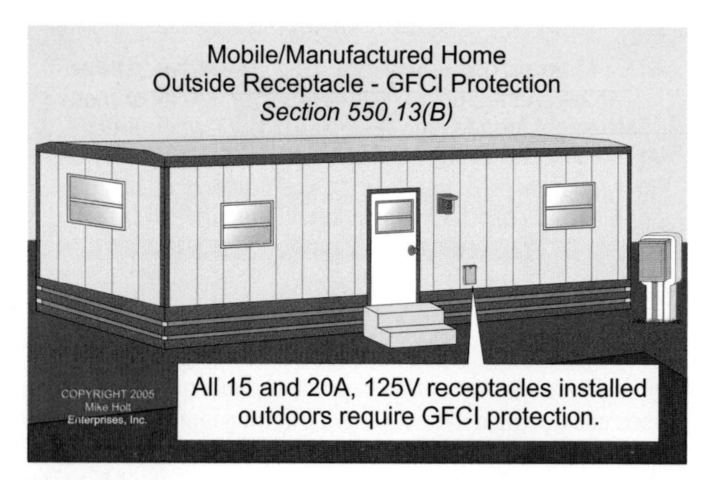

Mobile/Manufactured Home
Outside Receptacle - GFCI Protection
Section 550.13(B)
COPYRIGHT 2005 Mike Holt Enterprises, Inc.
All 15 and 20A, 125V receptacles installed outdoors require GFCI protection.

Figure 550–3

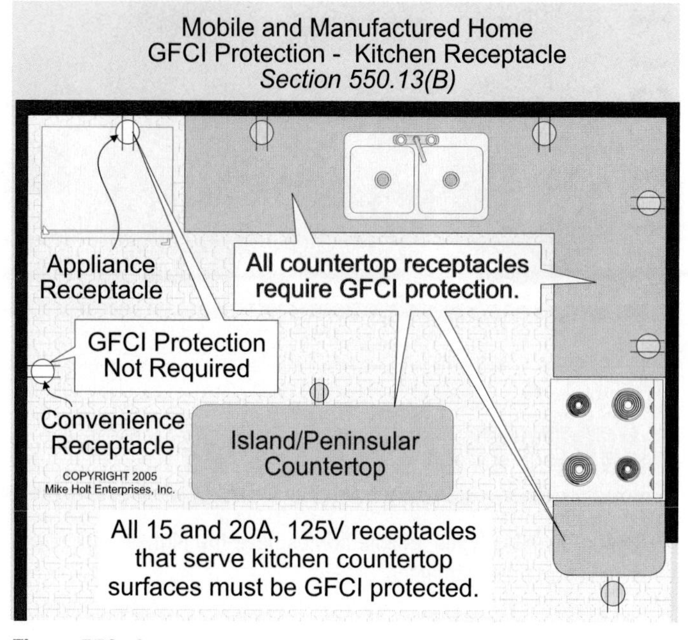

Figure 550–4

550.13 Receptacle Outlets.

(B) GFCI-Protected Receptacles. All 15 and 20A, 125V receptacles installed outdoors and in bathrooms must be GFCI protected. **Figure 550-3**

In addition, GFCI protection is required for all receptacles that serve countertops in kitchens and receptacles within 6 ft of a wet bar sink. **Figures 550-4** and **550-5**

(E) Pipe Heating Cable Receptacle Outlet. Mobile and manufactured home pipe heat tape receptacle outlets, if installed, must

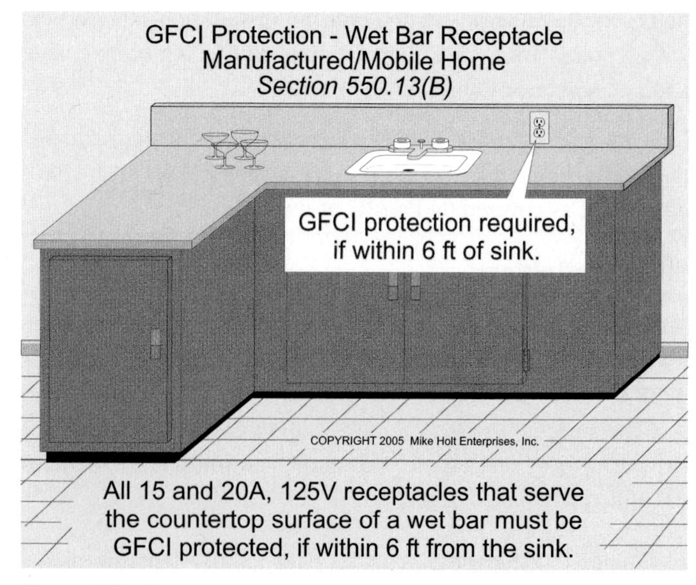

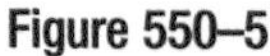
Figure 550–5

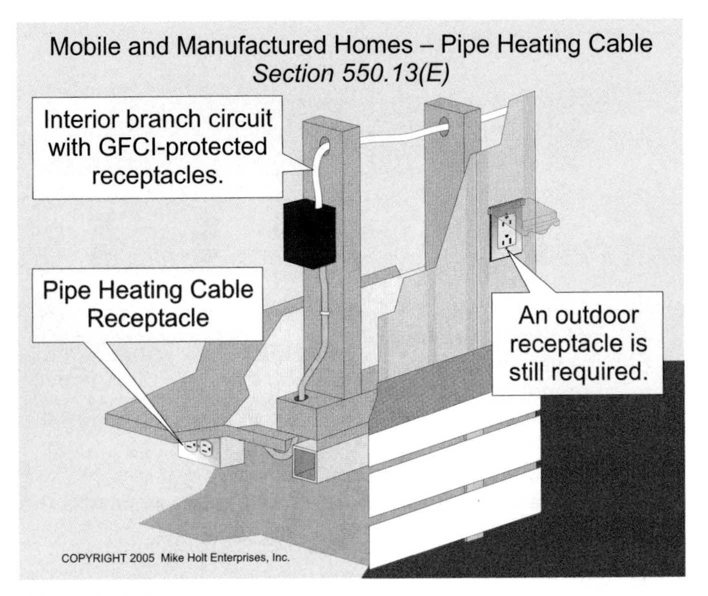
Mobile and Manufactured Homes – Pipe Heating Cable
Section 550.13(E)
Interior branch circuit with GFCI-protected receptacles.
Pipe Heating Cable Receptacle
An outdoor receptacle is still required.
COPYRIGHT 2005 Mike Holt Enterprises, Inc.

Figure 550–6

be connected to an interior GFCI-protected branch circuit. In addition, the pipe heating cable receptacle outlet must not substitute for the required outdoor receptacle outlet specified in 550.13(D)(8). **Figure 550-6**

> **Author's Comment:** The purpose of connecting the pipe heating cable receptacle to an interior GFCI-protected circuit is to alert the occupants if the GFCI-protection device has opened and the pipe heating cable is no longer energized, therefore no longer protecting the water pipes from freezing.

(F) Receptacle Outlets Not Permitted. Receptacle outlets aren't permitted in the following locations:

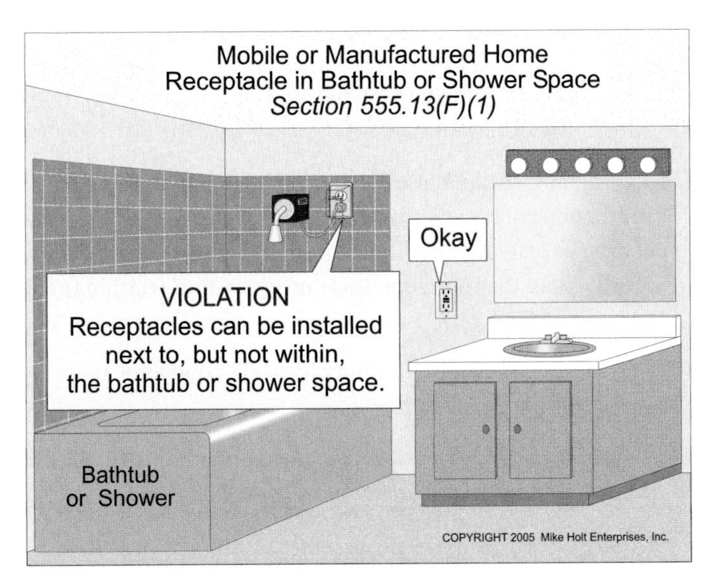
Mobile or Manufactured Home
Receptacle in Bathtub or Shower Space
Section 555.13(F)(1)
Okay
VIOLATION
Receptacles can be installed next to, but not within, the bathtub or shower space.
Bathtub or Shower
COPYRIGHT 2005 Mike Holt Enterprises, Inc.

Figure 550–7

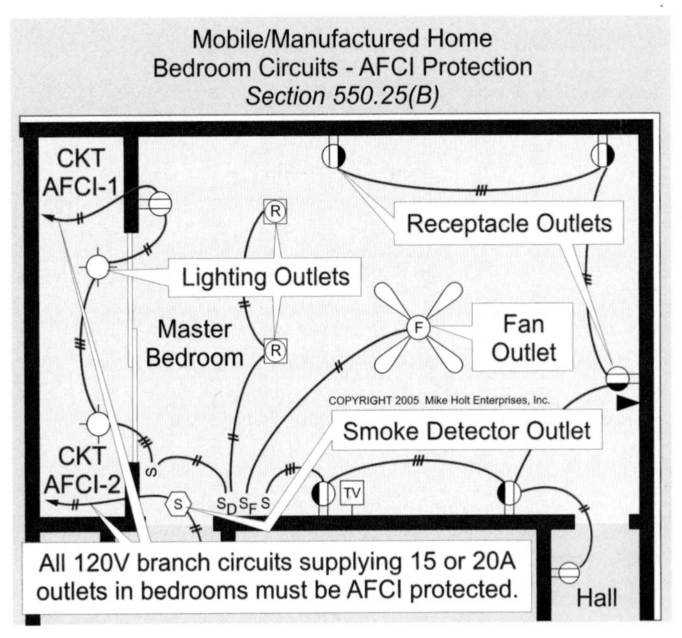
Mobile/Manufactured Home
Bedroom Circuits - AFCI Protection
Section 550.25(B)
CKT AFCI-1
Receptacle Outlets
Lighting Outlets
Master Bedroom
Fan Outlet
COPYRIGHT 2005 Mike Holt Enterprises, Inc.
Smoke Detector Outlet
CKT AFCI-2
TV
All 120V branch circuits supplying 15 or 20A outlets in bedrooms must be AFCI protected.
Hall

Figure 550–8

(1) Receptacle outlets must not be installed within a bathtub or shower space. **Figure 550-7**

(2) A receptacle must not be installed in a face-up position in any countertop.

550.25 Arc-Fault Circuit-Interrupter (AFCI) Protection.

(A) Definition. Arc-fault circuit interrupters are defined in 210.12(A).

(B) Bedrooms of Mobile Homes and Manufactured Homes. All 15 or 20A, 120V branch circuits that supply outlets in dwelling unit bedrooms must be AFCI protected. **Figure 550-8**

PART III. SERVICES AND FEEDERS

550.30 Distribution Systems. The electrical distribution system to mobile home lots must be 120/240V single-phase.

550.31 Allowable Demand Factors. Mobile park feeder/service loads can be calculated (at 120/240V) in accordance with Table 550.31, based on the larger of: **Figure 550-9**

(1) 16,000 VA for each mobile home lot, or

(2) In accordance with 550.18 for the largest mobile home each lot will accept.

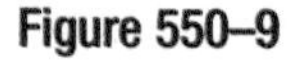

Mobile/Manufactured Home Park Site Demand Load
Section 550.31
Park Contains 35 Sites
COPYRIGHT 2005 Mike Holt Enterprises, Inc.
Determine the service demand load for 35 sites.
550.31, minimum site size is 16,000 VA.
Table 550.31 demand factor for 35 sites is 24%.
16,000 VA x 35 sites x 0.24 = 134,400 VA demand load
I = VA/E = 134,400 VA/240V = 560A

Figure 550–9

The feeder or service conductors to a single mobile or manufactured home can be sized in accordance with Table 310.15(B)(6).

> **Author's Comment:** See Article 100 for the definitions of "Feeder" and "Service Conductor."

550.32 Disconnect.

(A) Mobile Home Disconnect. A disconnecting means for a mobile home must be located adjacent to, within sight from, but not more than 30 ft from the exterior wall of the mobile home, and it must be rated at not less than 100A [550.32(C)]. **Figure 550-10**

> **Author's Comment:** The disconnecting means must be installed so the bottom of the enclosure isn't less than 2 ft above the finished grade or working platform [550.32(F)].

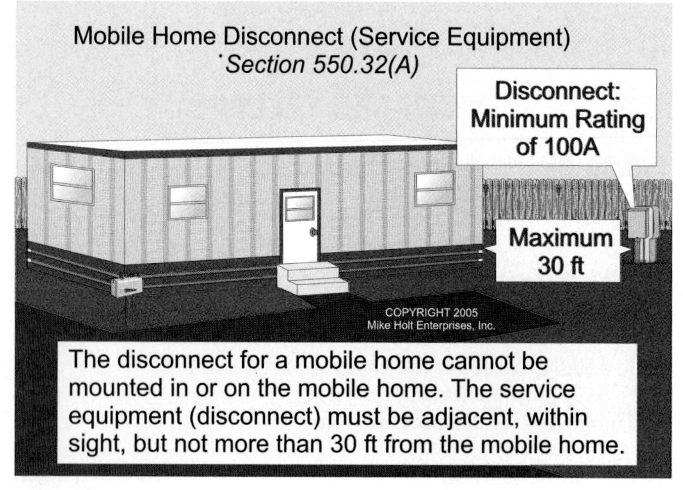

Mobile Home Disconnect (Service Equipment)
Section 550.32(A)
Disconnect: Minimum Rating of 100A
Maximum 30 ft
COPYRIGHT 2005 Mike Holt Enterprises, Inc.
The disconnect for a mobile home cannot be mounted in or on the mobile home. The service equipment (disconnect) must be adjacent, within sight, but not more than 30 ft from the mobile home.

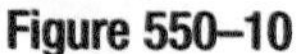

Figure 550–10

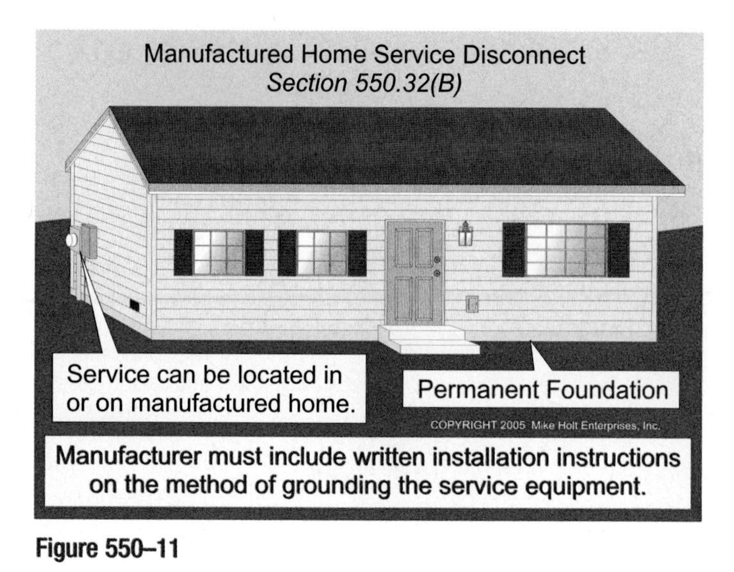

Figure 550–11

(B) Manufactured Home Service Equipment. The service equipment for a manufactured home is permitted in or on a manufactured home, provided all the conditions contained in (1) through (7) are met: **Figure 550-11**

(1) The manufacturer must include in its written installation instructions information indicating that the home must be secured in place by an anchoring system or installed on and secured to a permanent foundation.

(2) The installation of the service equipment complies with Article 230.

(3) Means are provided for the connection of a grounding electrode conductor to the service equipment and routing it outside the structure.

> **Author's Comment:** See Article 100 for the definition of "Grounding Electrode Conductor."

(4) Grounding and bonding of the service must comply with Article 250.

(5) The manufacturer must include in its written installation instructions one method of grounding (earthing) the service equipment at the installation site. The instructions must clearly state that other methods of grounding (earthing) are found in Article 250.

(6) The minimum size grounding electrode conductor must be specified in the instructions.

(7) A red warning label must be mounted on, or adjacent to, the service equipment. The label must state the following:

> **WARNING:** ***DON'T PROVIDE ELECTRICAL POWER UNTIL THE GROUNDING ELECTRODE IS INSTALLED AND CONNECTED.***

Where the service equipment isn't installed in or on the unit, the installation must comply with 550.32(A).

(C) Rating. Service equipment for a mobile or manufactured home must be rated at not less than 100A at 120/240V.

(F) Mounting Height. The disconnecting means for a mobile home must be installed so the bottom of the enclosure isn't less than 2 ft above the finished grade or working platform. The disconnecting means must be installed so that the center of the grip of the operating handle, when in its highest position, will not be more than 6 ft 7 in. above the finished grade or working platform. **Figure 550-12**

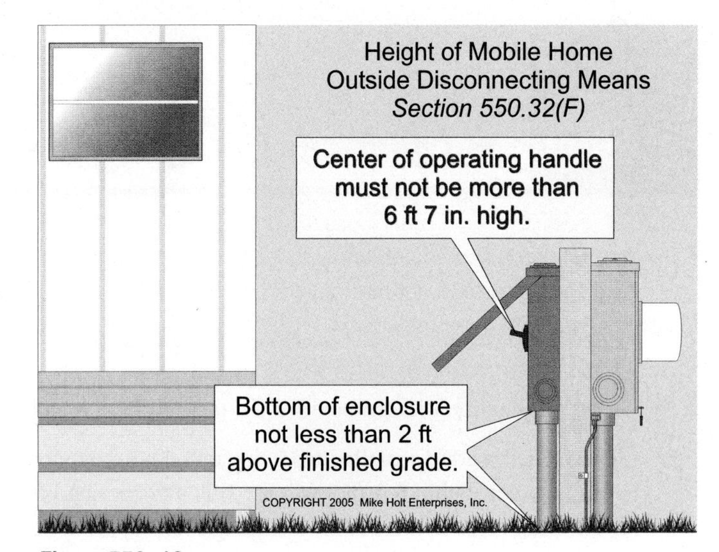

Figure 550–12

550.33 Feeder.

(A) Feeder Conductors. Feeder conductors to a mobile home must consist of either a listed cord that was factory-installed in accordance with 550.10(B), or a permanently installed feeder that consists of four insulated, color-coded conductors.

(B) Feeder Capacity. The feeder circuit for a mobile or manufactured home lot must be rated at not less than 100A at 120/240V.

Article 550 Questions

1. In reference to mobile/manufactured homes, examples of portable appliances could be _____, but only if these appliances are cord connected and not hard wired.

   (a) refrigerators (b) range equipment (c) clothes washers (d) all of these

2. Ground-fault circuit-interrupter (GFCI) protection in a mobile home is required for _____.

   (a) receptacle outlets installed outdoors and in compartments accessible from outside
   (b) receptacles within 6 ft of a wet bar sink and serving kitchen countertops
   (c) all receptacles in bathrooms including receptacles in luminaires (light fixtures)
   (d) all of these

3. The mobile home park secondary electrical distribution system to mobile home lots must be _____.

   (a) 120/240V, 1Ø, 3-wire (b) 208/208V, 3Ø, 4-wire (c) either a or b (d) none of these

4. An outdoor disconnecting means for a mobile home must be installed so the bottom of the enclosure is not less than _____ above the finished grade or working platform.

   (a) 1 ft (b) 2 ft (c) 3 ft (d) 6 ft

5. Mobile home and manufactured home lot feeder circuit conductors must have adequate capacity for the loads supplied and must be rated at not less than _____ at 120/240V.

   (a) 50A (b) 60A (c) 100A (d) 200A

Recreational Vehicles and Recreational Vehicle Parks

Introduction

Article 551 is similar to Article 550. After all, RVs are similar to mobile homes. While mobile homes are essentially trailers, RVs are essentially vehicles. Thus, RVs have their own article. RVs also have voltage converters, which mobile homes don't have. Other differences emerge as you look more closely.

While most of the requirements in Parts II and III apply to the RV manufacturer, an electrician doing work in an RV must also comply with these requirements.

If you're going to install an engine generator for an RV, you'll need to understand Part IV. Part V is primarily for RV manufacturers.

The typical electrician starts to get involved in Article 551 with Part VI, which provides the requirements for power distribution systems in RV parks. Perhaps the most important aspect of Part VI is understanding how to calculate loads and apply the demand factor, based on the number of RV sites in the park.

PART I. GENERAL

551.1 Scope. Article 551 covers electrical conductors and equipment installed within or on recreational vehicles, the conductors that connect recreational vehicles to a supply of electricity, and electrical installations within a recreational vehicle park.

551.2 Definitions.

Recreational Vehicle. A vehicular-type unit designed as temporary living quarters for recreational, camping, or travel use, which either has its own motive power or is mounted on or drawn by another vehicle. The basic entities are travel trailer, camping trailer, truck-camper, and motor home.

Recreational Vehicle Park. A plot of land upon which two or more recreational vehicle sites are located, established, or maintained for occupancy by recreational vehicles of the general public as temporary living quarters for recreation or vacation purposes.

551.3 Other Articles. Wherever the provisions of Article 551 differ from Chapters 1 through 4, the provisions of Article 551 apply.

PART VI. RECREATIONAL VEHICLE PARKS

551.71 Type of Receptacles Required. Every recreational vehicle site with electrical supply must have at least one 20A, 125V receptacle. A minimum of 20 percent of all recreational vehicle sites with electrical supply must have a 50A, 125/250V receptacle. A minimum of 70 percent of all recreational vehicle sites with electrical supply must have a 30A, 125V receptacle. **Figure 551-1**

All 15 and 20A, 125V receptacles located at a recreational vehicle park must be GFCI protected.

> **FPN:** The percentage of 50A sites required by 551.71 may be inadequate for seasonal recreational vehicle sites that serve a higher percentage of recreational vehicles with 50A electrical systems. In that type of recreational vehicle park, the percentage of 50A sites could approach 100 percent.

551.73 Calculated Load.

(A) Basis of Calculations. Electrical service and feeders are calculated on the basis of not less than 9,600 VA per site equipped with 50A, 120/240V supply facilities; 3,600 VA per site

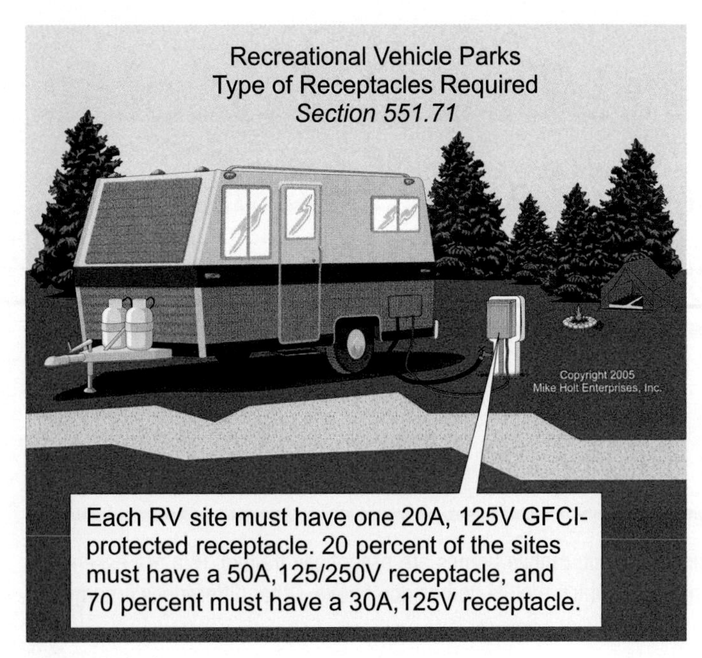

Recreational Vehicle Parks
Type of Receptacles Required
Section 551.71
Copyright 2005
Mike Holt Enterprises, Inc.
Each RV site must have one 20A, 125V GFCI-protected receptacle. 20 percent of the sites must have a 50A,125/250V receptacle, and 70 percent must have a 30A,125V receptacle.

Figure 551–1

equipped with both 20A and 30A supply facilities; 2,400 VA per site equipped with only 20A supply facilities; and 600 VA per site equipped with only 20A supply facilities that are dedicated to tent sites.

The demand factors in Table 551.73 are the minimum permitted in calculating the service and feeder loads. Where the electrical supply for a recreational vehicle site has more than one receptacle, the calculated load is based on the highest-rated receptacle.

(C) Demand Factors. The demand factors for a given number of sites must apply to all of the sites equally.

FPN: These demand factors may be inadequate in areas of extreme hot or cold temperatures with loaded circuits for heating or air conditioning.

(D) Feeder-Circuit Capacity. Recreational vehicle site feeder-circuit conductors must have sufficient ampacity for the loads supplied. They must be rated at least 30A each with full size grounded neutral conductors.

FPN: Due to the long circuit lengths typical in most recreational vehicle parks, feeder conductor sizes found in the ampacity tables of Article 310 may be inadequate to maintain satisfactory voltage regulation.

551.79 Clearance for Overhead Conductors. Open conductors must have a vertical clearance not less than 18 ft and a horizontal clearance not less than 3 ft in all areas subject to recreational vehicle movement. In all other areas, clearances must conform to 225.18 and 225.19.

Article 551 Questions

(• Indicates that 75% or fewer exam takers get the question correct.)

1. A minimum of 20 percent of all recreational vehicle sites with electrical supply must each be equipped with a _____,125/250V receptacle.

   (a) 15A (b) 20A (c) 30A (d) 50A

2. A minimum of 70 percent of all recreational vehicle sites with electrical supply must each be equipped with a _____,125V receptacle.

   (a) 15A (b) 20A (c) 30A (d) 50A

3. •Electrical service and feeders of a recreational vehicle park must be calculated at a minimum of _____ per site equipped with only 20A supply facilities (not including tent sites).

   (a) 1,200 VA (b) 2,400 VA (c) 3,600 VA (d) 9,600 VA

Notes

ARTICLE 555

Marinas and Boatyards

Introduction

Water level isn't constant. As the earth and the moon play their eons-old game of tug-of-war, oceans, lakes, rivers, and other bodies of water rise and fall at the shoreline. Other forces also cause the water level to change. For example, lakes and rivers vary in depth in response to rain. The variations can sometimes be dramatic.

To provide power to a marina or boatyard, you must allow for the variations in water level between the point of use and the power source. Article 555 addresses this issue.

But that's not the only issue involved with marinas and boatyards. As you might expect, Article 555 also presents requirements for accommodating the high levels of moisture inherent in these installations. Boatyard and marina installations pose further challenges as well. For example, sunlight reflected off the water is much more intense than it would otherwise be—and this has implications for insulation. Other factors to consider include temperature extremes, abrasion caused by movement, oil, gasoline, diesel fuel, ozone, acids, and chemicals.

Then, of course, docking a boat isn't as easy as pulling into a shopping center parking spot with your automobile. Electrical equipment must meet certain spatial requirements, such as not interfering with mooring lines or masts.

Article 555 begins with the concept of the electrical datum plane. You might think of it as the border of a "demilitarized zone" for electrical equipment. Or, you can think of it as a line that marks the beginning of a "no man's land" where you simply don't place electrical equipment. Once you determine where this plane is, don't locate transformers, connections, or receptacles below that line.

555.1 Scope. Article 555 covers the installation of wiring and equipment for fixed or floating piers, wharfs, docks, and other areas in marinas, boatyards, boat basins, boathouses, and similar occupancies. This article doesn't apply to docking facilities or boathouses used for the owners of single-family dwellings.

> **Author's Comment:** GFCI protection is required for all 15 or 20A, 125V receptacles at single-family dwelling boathouses [210.8(A)(8)].

555.2 Definitions.

Electrical Datum Plane.

(1) *Land Area Subject to Tidal Fluctuation.* The horizontal plane 2 ft above the highest high tide that occurs under normal circumstances. **Figure 555-1**

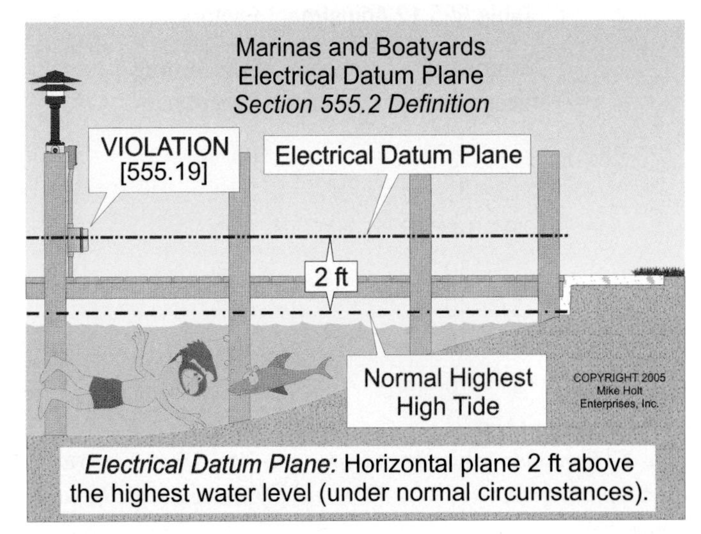
Marinas and Boatyards
Electrical Datum Plane
Section 555.2 Definition
VIOLATION
[555.19]
Electrical Datum Plane
2 ft
Normal Highest
High Tide
COPYRIGHT 2005
Mike Holt
Enterprises, Inc.
Electrical Datum Plane: Horizontal plane 2 ft above
the highest water level (under normal circumstances).

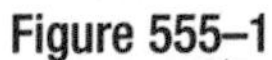

Figure 555–1

(2) Land Areas not Subject to Tidal Fluctuation. The horizontal plane 2 ft above the highest water level that occurs under normal circumstances.

(3) Floating Piers. The horizontal plane 30 in. above the water level at the floating pier and a minimum of 12 in. above the level of the deck.

Author's Comment: This definition is necessary for the location of transformers [555.5], electrical connections [555.9], and receptacles [555.19] near water.

Marine Power Outlet. An assembly that can include receptacles, circuit breakers, fused switches, fuses, and watt-hour meters, approved by the authority having jurisdiction for marine use.

Author's Comment: This definition is necessary for the application of shore power receptacles [555.19(A)(1)] and disconnecting means [555.17(B)].

555.5 Transformers. Transformers must be approved by the authority having jurisdiction for the location, and the bottom must not be located below the electrical datum plane.

555.9 Electrical Connections. All electrical connections must be located not less than 12 in. above the deck of a floating pier, and not less than 12 in. above the deck of a fixed pier, but not below the electrical datum plane.

555.12 Load Calculations for Service and Feeder Conductors. The calculated service or feeder load for shore power receptacles can be calculated using the adjustment factors contained in Table 555.12.

Table 555.12 Adjustment Factors

Number of Receptacles	Sum of the Rating of the Receptacles
1 – 4	100%
5 – 8	90%
9 – 14	80%
15 – 30	70%
31 – 40	60%
41 – 50	50%
51 – 70	40%
Over 71	30%

Note 1. Where shore power provides two receptacles for an individual boat slip, the receptacle with the larger kW demand must be used. **Figure 555-2**

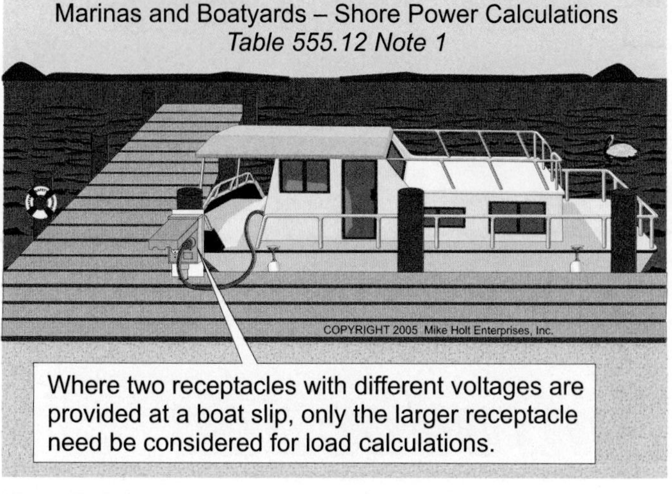

Marinas and Boatyards – Shore Power Calculations
Table 555.12 Note 1
COPYRIGHT 2005 Mike Holt Enterprises, Inc.
Where two receptacles with different voltages are provided at a boat slip, only the larger receptacle need be considered for load calculations.

Figure 555–2

Question: *What size 120/240V single-phase service is required for a marina with twenty 20A, 125V receptacles, and twenty 30A, 250V receptacles?* **Figure 555-3**

(a) 200A (b) 400A (c) 600A (d) 800A

Answer: *(c) 600A*

	Line 1	Line 2
Twenty 20A, 125V	200A	200A (ten on each line)
Twenty 30A, 250V	600A	600A (twenty on each line)
	800A	800A

The calculated load per line for the marina is based on the demand factor listed in Table 555.12 for 30 receptacles (per line). 800A x 0.7 = 560A

Author's Comment: Circuits are rated 120V or 240V and receptacles are rated 125V and 250V, but this has no effect on the actual load calculations.

555.17 Boat Receptacle Disconnecting Means. A disconnecting means must isolate each boat from its shore power receptacle.

(A) Type of Disconnect. A circuit breaker or switch can be used to serve as the required shore power receptacle disconnecting means. Each disconnect must be identified as to which receptacle it controls.

(B) Location. The disconnecting means for shore power receptacles must be readily accessible and located not more than 30 in. from the receptacle. Circuit breakers or switches located in marine power outlets can be used for the boat receptacle disconnecting means. **Figure 555-4**

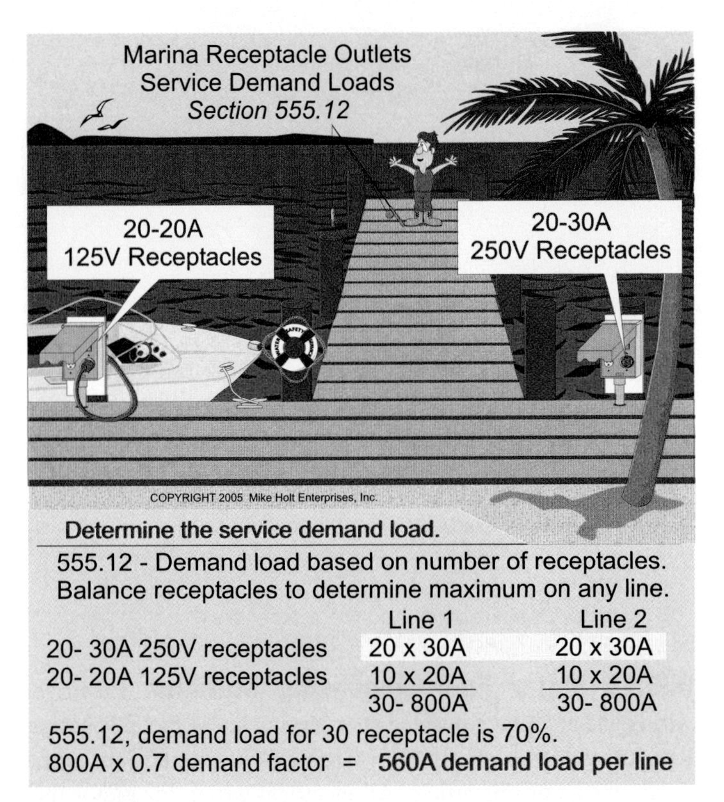

Marina Receptacle Outlets
Service Demand Loads
Section 555.12
20-20A
125V Receptacles
20-30A
250V Receptacles
COPYRIGHT 2005 Mike Holt Enterprises, Inc.
Determine the service demand load.
555.12 - Demand load based on number of receptacles.
Balance receptacles to determine maximum on any line.
Line 1 Line 2
20- 30A 250V receptacles 20 x 30A 20 x 30A
20- 20A 125V receptacles 10 x 20A 10 x 20A
30- 800A 30- 800A
555.12, demand load for 30 receptacle is 70%.
800A x 0.7 demand factor = 560A demand load per line

Figure 555–3

Author's Comment: This shore power receptacle disconnecting means is intended to eliminate the hazard of someone engaging or disengaging the boat's shore power attachment plug with wet, slippery hands, and possibly contacting energized blades. The "30-in. requirement" helps someone not familiar with the marina layout to quickly reach the disconnecting means in an emergency.

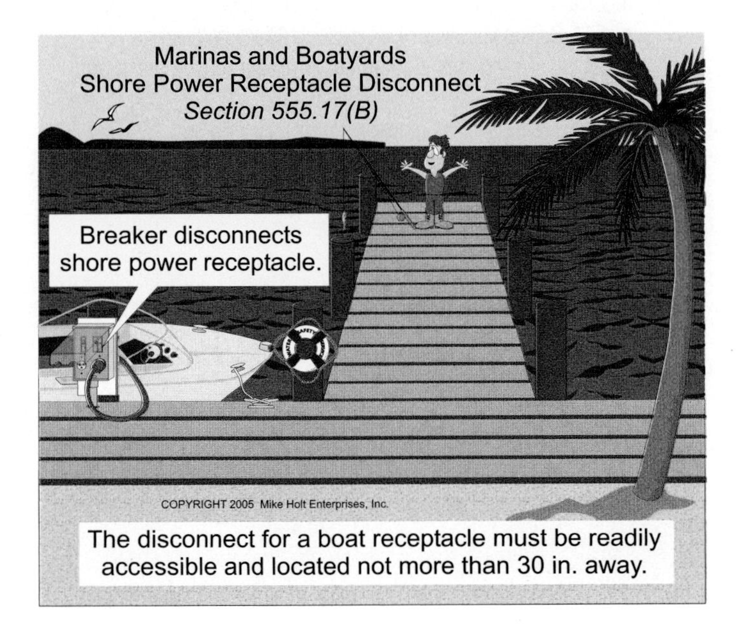

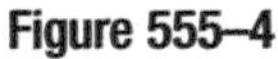
Figure 555–4

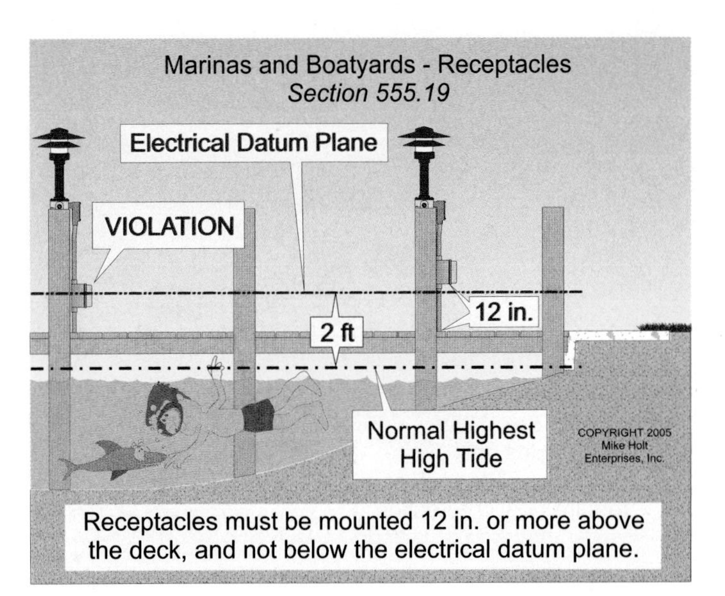

Figure 555–5

555.19 Receptacles. Receptacles must be mounted not less than 12 in. above the deck surface, and not below the electrical datum plane. **Figure 555-5**

(A) Shore Power Receptacles.

(4) Ratings. Receptacles that provide shore power for boats must be rated not less than 30A.

(a) Receptacles rated 50A and less must be of the locking and grounding (bonding) type. **Figure 555-6**

(b) Receptacles rated for 60A or 100A must be of the pin-and-sleeve type.

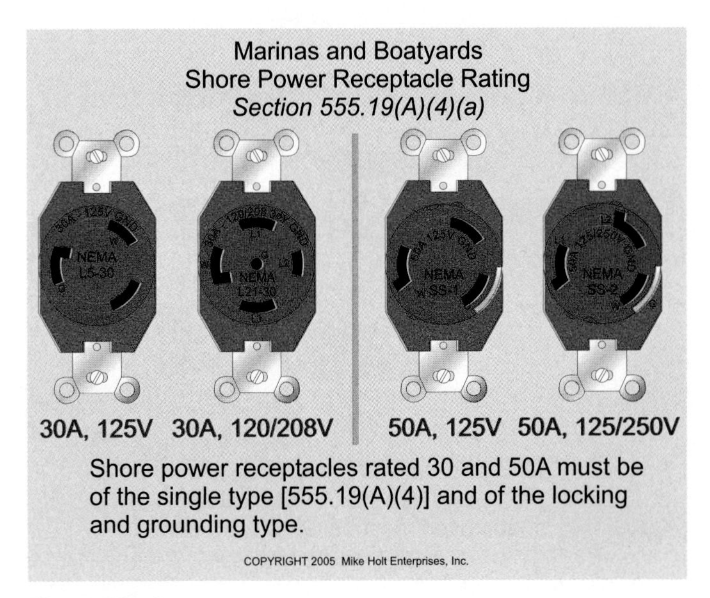

Marinas and Boatyards
Shore Power Receptacle Rating
Section 555.19(A)(4)(a)
NEMA
L5-30
NEMA
L21-30
NEMA
SS-1
NEMA
SS-2
30A, 125V
30A, 120/208V
50A, 125V
50A, 125/250V
Shore power receptacles rated 30 and 50A must be
of the single type [555.19(A)(4)] and of the locking
and grounding type.
COPYRIGHT 2005 Mike Holt Enterprises, Inc.

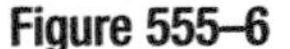
Figure 555–6

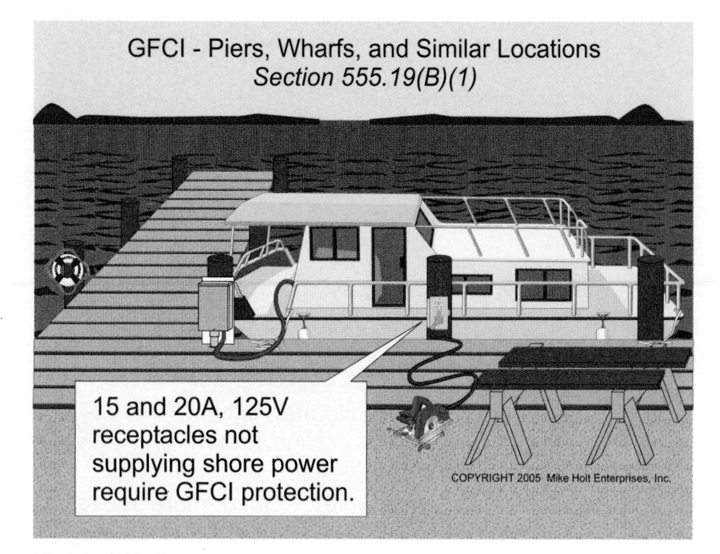
GFCI - Piers, Wharfs, and Similar Locations
Section 555.19(B)(1)
15 and 20A, 125V receptacles not supplying shore power require GFCI protection.
COPYRIGHT 2005 Mike Holt Enterprises, Inc.

Figure 555–7

Author's Comment: The rating of the shore power receptacle isn't dependent upon the length of the boat. The *Code* simply sets a minimum rating of 30A and leaves it up to the designer and/or owner to provide the receptacles they deem necessary, based on projected usage of the slips.

(B) Other Than Shore Power.

(1) GFCI Protection. All 15 and 20A, 125V receptacles installed outdoors for portable electric hand tools, electrical diagnostic equipment, or portable lighting equipment must be GFCI protected. **Figure 555-7**

Author's Comment: All 15 and 20A, 125V receptacles in boathouses of dwelling units must be GFCI protected [210.8(A)(8)]. **Figure 555-8**

GFCI Protection - Dwelling Boathouse Receptacles
Section 210.8(A)(8)
COPYRIGHT 2005 Mike Holt Enterprises, Inc.
All 15 and 20A, 125V receptacles in boathouses must be GFCI protected.

Figure 555–8

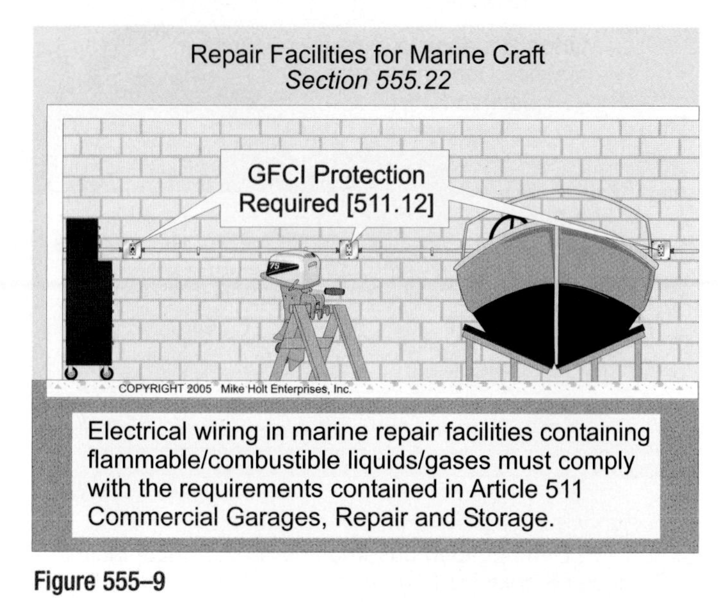
Repair Facilities for Marine Craft
Section 555.22
GFCI Protection Required [511.12]
COPYRIGHT 2005 Mike Holt Enterprises, Inc.
Electrical wiring in marine repair facilities containing flammable/combustible liquids/gases must comply with the requirements contained in Article 511 Commercial Garages, Repair and Storage.

Figure 555–9

555.21 Motor Fuel Dispensing Stations. Electrical wiring and equipment located at or serving motor fuel dispensing equipment must comply with Article 514, in addition to the requirements of this article. All electrical wiring for motor fuel dispensing equipment must be installed on the side of the wharf, pier, or dock opposite from the liquid motor fuel piping system.

555.22 Repair Facilities. Electrical wiring and equipment at marine craft repair facilities containing flammable or combustible liquids or gases must comply with the requirements contained in Article 511. **Figure 555-9**

Author's Comments:

- Important rules in Article 511 to consider include:
  - 511.3 Classification of Hazardous (Classified) Areas
  - 511.4 Wiring and Equipment In Hazardous (Classified) Locations
  - 511.7 Wiring and Equipment Above Hazardous (Classified) Locations
  - 511.9 Explosionproof Seals
  - 511.12 GFCI-Protected Receptacles

Article 555 Questions

1. In marinas and boatyards, transformers and enclosures must be specifically approved for the intended location. The bottom of enclosures for transformers must not be located below _____.

   (a) 2 ft above the dock
   (b) 18 in. above the electrical datum plane
   (c) the electrical datum plane
   (d) a dock

2. The demand used to calculate 45 receptacles in a boatyard feeder is _____ percent.

   (a) 90 (b) 80 (c) 70 (d) 50

3. The disconnecting means for a boat must be readily accessible, not more than _____ from the receptacle it controls and must be in the supply circuit ahead of the receptacle.

   (a) 12 in. (b) 24 in. (c) 30 in. (d) none of these

4. Electrical wiring and equipment at marine craft repair facilities containing flammable or combustible liquids or gases must comply with the requirements contained in _____

   (a) Article 511 (b) Article 555 (c) Article 513 (d) a and b

Notes

ARTICLE 590 Temporary Installations

Introduction

It's a common misconception that temporary wiring represents a lower standard than that of other wiring. In truth, it merely meets a *different* standard. The same rules of workmanship, ampacity, and circuit protection apply to temporary installations as to other installations.

So, how is a temporary installation different? In one sense, it does represent a lower standard. For example, you can use Types NM and NMC rather than the normally required raceway-enclosed wiring—without height limitation. And you do not have to put splices in boxes.

You must remove a temporary installation upon completion of the purpose for which it was installed. If the temporary installation is for holiday displays, it cannot last more than 90 days.

Article 590 addresses practicality and execution issues inherent in temporary installations, thereby making the installation less time consuming to install.

590.1 Scope. The requirements of Article 590 apply to temporary power and lighting installations, including power for construction, remodeling, maintenance, repair, demolitions, and decorative lighting. This article also applies when temporary installations are necessary during emergencies or for tests and experiments.

Author's Comment: Temporary installations for trade shows must comply with the requirements contained in Article 518, and temporary installations for carnivals, circuses, fairs, and similar events must be installed in accordance with Article 525.

590.2 All Installations.

(A) Other Articles. All requirements of the *NEC* apply to temporary installations unless specifically modified in this article.

CAUTION: *Just because an installation is temporary doesn't mean you can rig up anything you want. General contractors, insurance companies, and the Occupational Safety and Health Administration (OSHA) enforce job safety. All trades use temporary electrical installations, and they must be kept in a safe working condition.*

(B) Approval. Temporary wiring methods are acceptable only if approved by the authority having jurisdiction based on the conditions of use and any special requirements of the temporary installation.

Author's Comment: "Approved" means acceptable to the authority having jurisdiction; usually the electrical inspector.

590.3 Time Constraints.

(A) Construction Period. Temporary electrical power and lighting installations are permitted during the period of construction, remodeling, maintenance, repair, or demolition of buildings, structures, equipment, or similar activities.

(B) Decorative Lighting. Temporary electrical power for decorative lighting is permitted for a period of up to 90 days. **Figure 590-1**

Author's Comment: Decorative lighting must be listed as required by 410.110.

(C) Emergencies and Tests. Temporary electrical power and lighting installations are permitted for the duration necessary for emergencies or for tests and experiments.

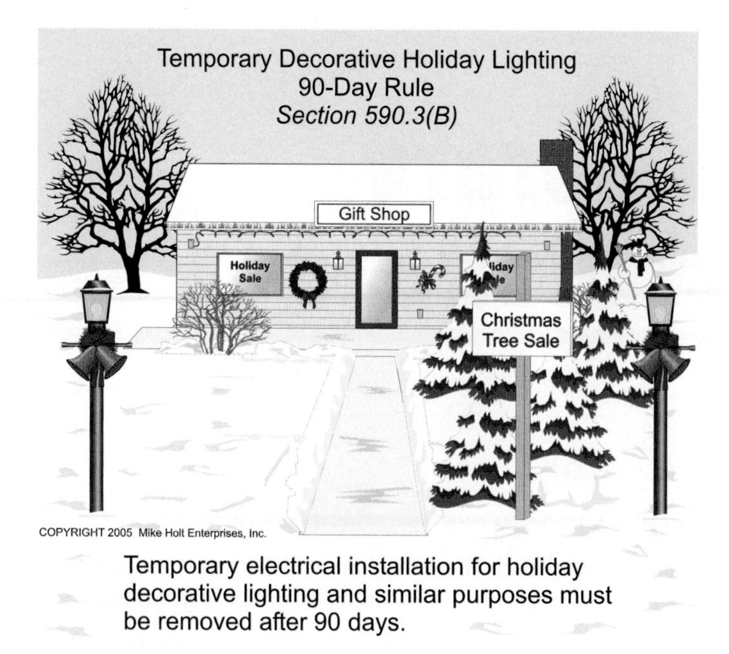

Temporary Decorative Holiday Lighting
90-Day Rule
Section 590.3(B)
Gift Shop
Holiday Sale
Christmas Tree Sale
COPYRIGHT 2005 Mike Holt Enterprises, Inc.
Temporary electrical installation for holiday decorative lighting and similar purposes must be removed after 90 days.

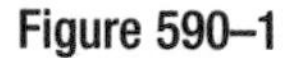
Figure 590–1

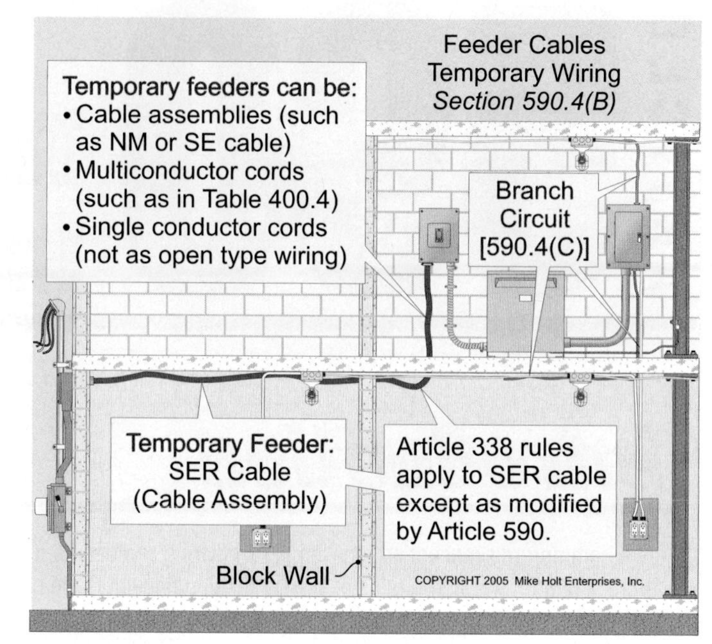

Figure 590–2

(D) Removal. Temporary installations must be removed immediately upon the completion of the purpose for which they were installed.

590.4 General.

(A) Services. The installation of services for temporary installations must conform to Article 230.

(B) Feeders. Open conductors are not permitted for temporary installations, however, cable assemblies, and hard-usage and extra-hard usage cords are permitted. **Figure 590-2**

Type NM cable, exposed or concealed, can be used in any building or structure, without any height limitation. **Figure 590-3**

(C) Branch Circuits. Open conductors are not permitted for temporary installations, however, cable assemblies, and hard-usage and extra-hard usage cords are permitted. See **Figure 590-2**.

Type NM cable, exposed or concealed, can be used in any building or structure, without any height limitation. See **Figure 590-3**.

(D) Receptacles. The receptacle grounding terminal must be bonded to an effective ground-fault current path in accordance with 250.146 and 406.3.

On a construction site, luminaires and receptacles must not be placed on the same branch circuit. **Figure 590-4**

> **Author's Comment:** This requirement is necessary so that illumination is maintained, even when the receptacle's GFCI-protection device opens.

(E) Disconnecting Means. Ungrounded conductors of a multiwire branch circuit must have a disconnecting means at the panelboard that opens all of the ungrounded conductors simultaneously. Where single-pole circuit breakers are used for a multiwire circuit, an approved handle tie must be used to secure the trip handles together.

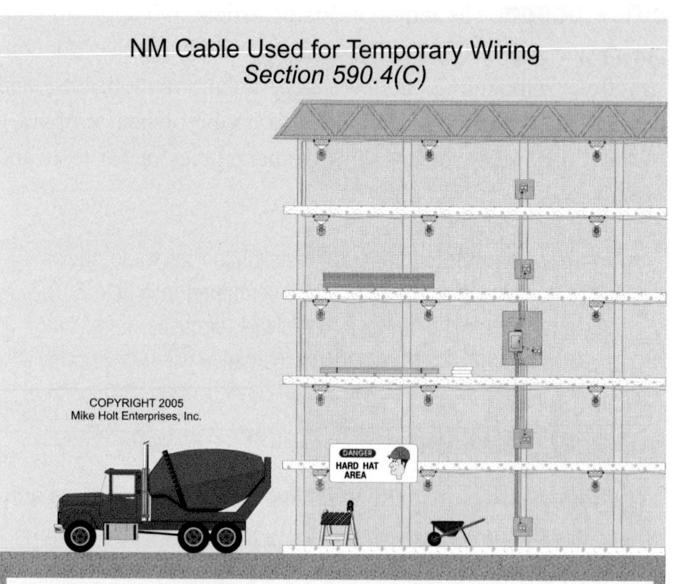

NM Cable Used for Temporary Wiring
Section 590.4(C)
COPYRIGHT 2005
Mike Holt Enterprises, Inc.
DANGER
HARD HAT AREA
NM Cable is permitted for feeders and branch circuits [590.4(C)] for temporary wiring in dwellings and buildings without any height limitations or building construction type [334.12(2)], and without having to be concealed within walls, floors, or ceilings.

Figure 590–3

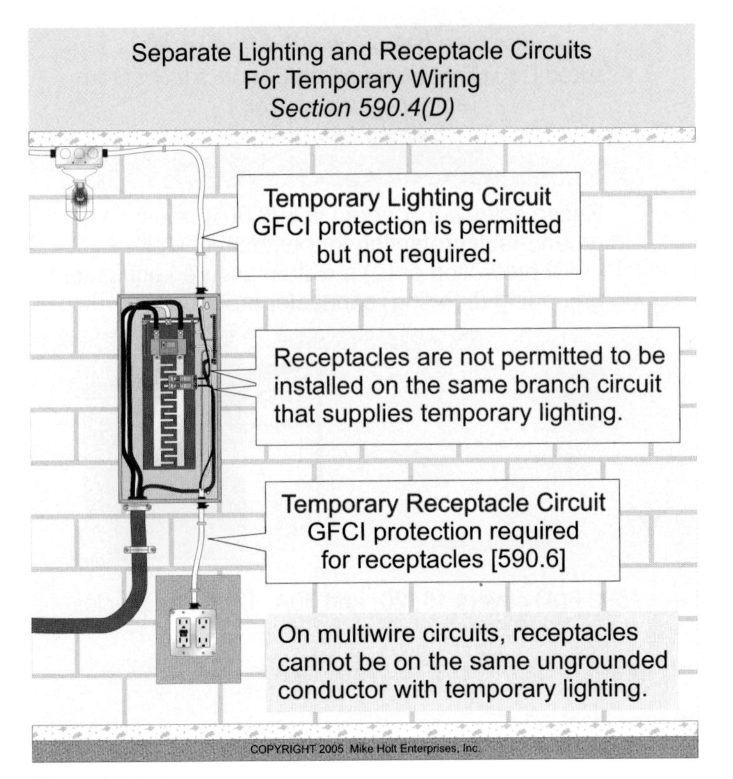

Separate Lighting and Receptacle Circuits
For Temporary Wiring
Section 590.4(D)
Temporary Lighting Circuit
GFCI protection is permitted
but not required.
Receptacles are not permitted to be
installed on the same branch circuit
that supplies temporary lighting.
Temporary Receptacle Circuit
GFCI protection required
for receptacles [590.6]
On multiwire circuits, receptacles
cannot be on the same ungrounded
conductor with temporary lighting.
COPYRIGHT 2005 Mike Holt Enterprises, Inc.

Figure 590–4

Author's Comments:

- See Article 100 for the definition of "Panelboard."
- For additional multiwire branch-circuit requirements, see 210.4 and 300.13(B).

(F) Lamp Protection. Lamps (bulbs) must be protected from accidental contact by a suitable luminaire or by the use of a lampholder with a guard.

(H) Protection from Accidental Damage. Cables and flexible cords must be protected from accidental damage and from sharp corners and projections. Protection must also be provided when cables and flexible cords pass through doorways or other pinch points.

(J) Support. Cable assemblies and flexible cords must be supported at intervals that ensure protection from physical damage. Support must be in the form of staples, cable ties, straps, or other similar means designed not to damage the cable or cord assembly.

Author's Comment: The support requirement for temporary cables is determined by the authority having jurisdiction, based on the jobsite conditions [590.2(B)].

Vegetation cannot be used for the support of overhead branch circuit or feeder conductors. **Figure 590–5**

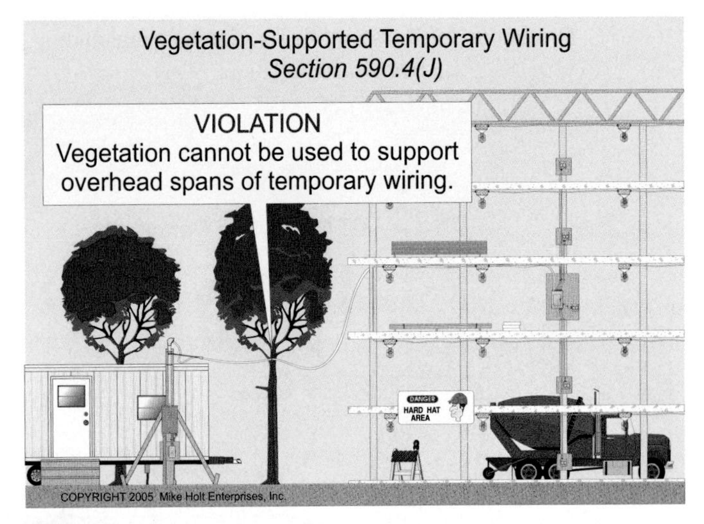

Vegetation-Supported Temporary Wiring
Section 590.4(J)
VIOLATION
Vegetation cannot be used to support
overhead spans of temporary wiring.
COPYRIGHT 2005 Mike Holt Enterprises, Inc.

Figure 590–5

Exception: Vegetation can be used to support decorative lighting, where strain-relief devices, tension take-up devices, or other approved means prevent damage to the conductors from the movement of the live vegetation.

590.6 Ground-Fault Protection for Personnel. Ground-fault protection for personnel is required for all temporary wiring used for construction, remodeling, maintenance, repair, or demolition of buildings, structures, or equipment.

(A) 15A, 20A, and 30A, 125V Receptacles. GFCI protection is required for all receptacles used by personnel during construction, remodeling, maintenance, repair, or demolition of buildings, structures, equipment, or similar activities. **Figure 590-6**

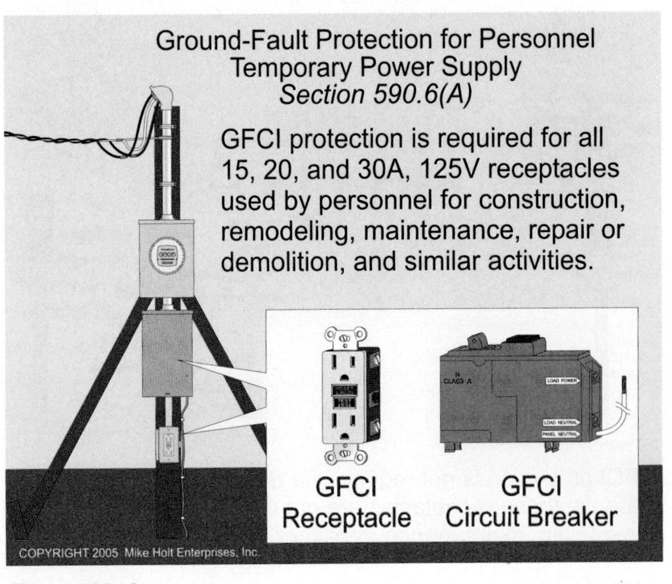

Figure 590–6

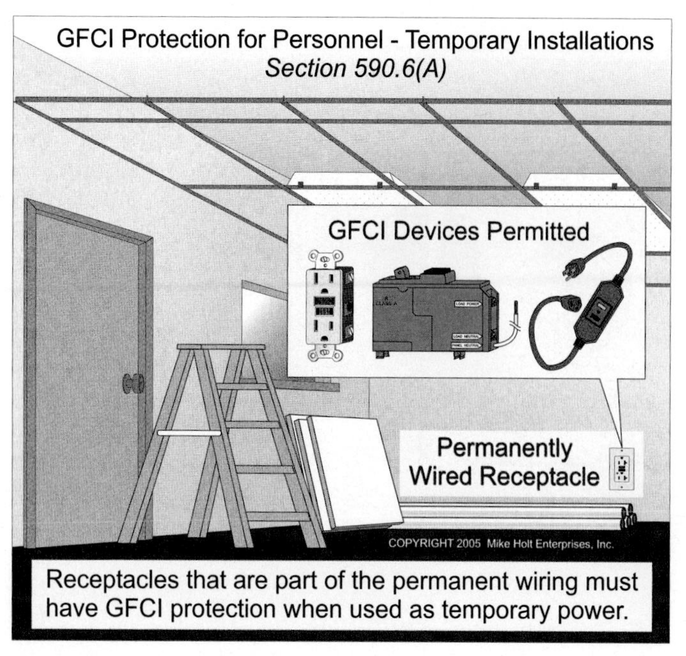
GFCI Protection for Personnel - Temporary Installations
Section 590.6(A)
GFCI Devices Permitted
Permanently
Wired Receptacle
COPYRIGHT 2005 Mike Holt Enterprises, Inc.
Receptacles that are part of the permanent wiring must have GFCI protection when used as temporary power.

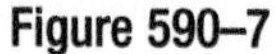
Figure 590–7

GFCI protection can be provided by circuit breakers, receptacles, cord sets, or adapters that incorporate listed GFCI protection. **Figure 590-7**

Author's Comment: GFCI protection isn't required for receptacles that supply power for decorative holiday lighting [590.3(B)]! This is because these receptacles aren't used for "construction, remodeling, maintenance, repair, or demolition of buildings or structures" [590.6]. **Figure 590-8**

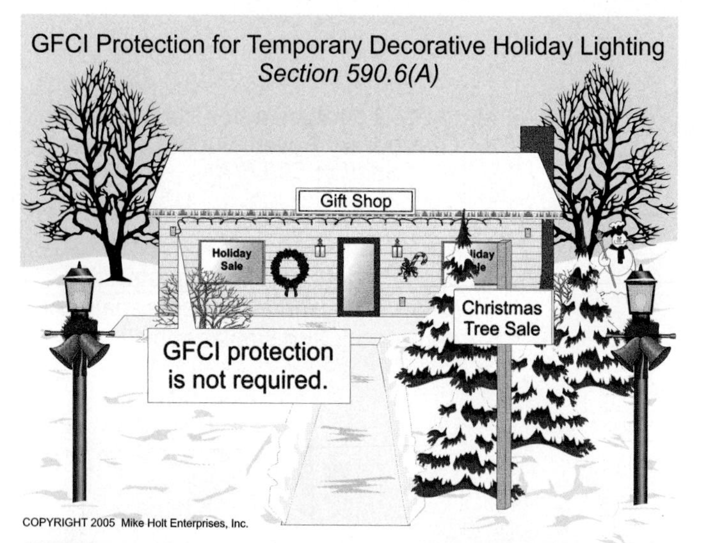
GFCI Protection for Temporary Decorative Holiday Lighting
Section 590.6(A)
Gift Shop
Holiday Sale
Christmas Tree Sale
GFCI protection is not required.
COPYRIGHT 2005 Mike Holt Enterprises, Inc.
GFCI protection is not required for decorative holiday lighting because these receptacles are not used for construction, remodeling, maintenance, repair, or demolition [590.3].

Figure 590–8

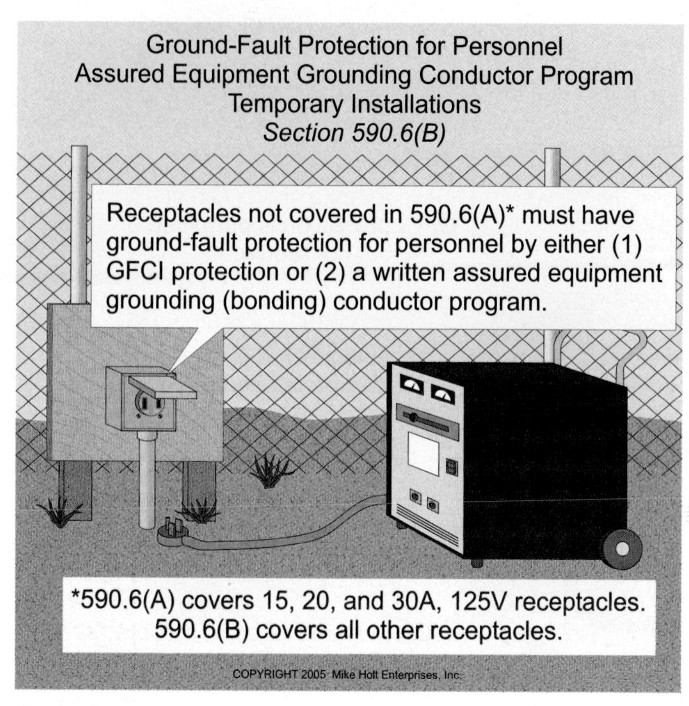
Ground-Fault Protection for Personnel
Assured Equipment Grounding Conductor Program
Temporary Installations
Section 590.6(B)
Receptacles not covered in 590.6(A)* must have ground-fault protection for personnel by either (1) GFCI protection or (2) a written assured equipment grounding (bonding) conductor program.
*590.6(A) covers 15, 20, and 30A, 125V receptacles.
590.6(B) covers all other receptacles.
COPYRIGHT 2005 Mike Holt Enterprises, Inc.

Figure 590–9

(B) Other Receptacles. Receptacles rated other than 15A, 20A, or 30A, 125V that supply temporary power used by personnel during construction, remodeling, maintenance, repair, or demolition of buildings, structures, equipment, or similar activities must comply with either (1) or (2): **Figure 590-9**

(1) GFCI Protection. Receptacles must be GFCI protected.

(2) Assured Equipment Grounding Conductor Program. GFCI protection is not required of receptacles if a written assured equipment grounding conductor program continuously enforced at the site to ensure that equipment grounding (bonding) conductors for all cord sets, receptacles that aren't a part of the permanent wiring of the building or structure, and equipment connected by cord and plug, are installed and maintained in accordance with the applicable requirements of 250.114, 250.138, 406.3(C), and 590.4(D).

(a) The assured equipment grounding conductor program requires the following tests of cord sets, receptacles, and cord-and-plug connected equipment.

(1) All equipment grounding (bonding) conductors must be tested for continuity.

(2) Each receptacle and attachment plug must be tested for the correct attachment of the equipment grounding (bonding) conductor.

(3) The tests must be performed before:

a. The first use on-site

b. When there's evidence of damage

c. Before equipment is returned to service following repairs, and

d. At intervals not exceeding 3 months.

(b) The results of the tests required in 590.6(B)(2)(a) must be recorded and made available to the authority having jurisdiction.

Article 590 Questions

1. Except as specifically modified by Article 590, all other requirements of the *Code* for permanent wiring apply to temporary wiring installations.

   (a) True (b) False

2. Services for temporary installations are not required to comply with the requirements of Article 230.

   (a) True (b) False

3. All receptacles for temporary branch circuits are required to be electrically connected to the _____ conductor.

   (a) grounded (b) grounding (c) equipment grounding (d) grounding electrode

4. For temporary installations, cable assemblies, as well as flexible cords and flexible cables, must be supported at intervals that ensure protection from physical damage. Support must be in the form of _____ or similar type fittings installed so as not to cause damage.

   (a) staples (b) cable ties (c) straps (d) any of these

5. Receptacles rated other than 125V single-phase 15, 20, and 30A for temporary installations must be protected by _____.

   (a) a GFCI device
   (b) the Assured Equipment Grounding Conductor Program
   (c) an AFCI device
   (d) a or b

CHAPTER 6

Special Equipment

Introduction

Chapter 6, which covers special equipment, is the second of four chapters that deal with special topics. Chapters 5, 7, and 8 deal with special occupancies, special conditions, and communications systems, respectively. Remember, the first four Chapters of the *NEC* are sequential and form a foundation for each of the subsequent four Chapters.

What exactly is "special equipment?" It's equipment that, by nature of its use, construction, or unique nature creates a need for additional measures to ensure the "safeguarding of people and property" mission of the *NEC,* as stated in Article 90.

The *NEC* groups these logically, as you might expect. Here are the general groupings:

- Pre-fab things assembled in the field. Articles 600—605. These are signs, manufactured wiring systems, and office furnishings.
- Lifting equipment. Articles 610 and 620. These include cranes, hoists, elevators, dumbwaiters, wheelchair lifts, and escalators.
- Electric vehicle charging systems. Article 625.
- Electric welders. Article 630.
- Equipment for creating or processing information. Articles 640—650. These include amplifiers, computers (information technology), and pipe organs.
- X-ray equipment. Article 660.
- Process and production equipment. Articles 665—675. These include induction heaters, electrolytic cells, electroplating, and irrigation machines.
- Swimming pools. Article 680.
- Integrated electrical systems. Article 685.
- "New fuel" technologies. Article 690 covers solar photovoltaic systems, and Article 692 covers fuel cells. Perhaps a future Article 694 will cover wind power systems or some other energy source.
- Fire pumps. Article 695.

Author's Comment: The *NFPA* also produces a fire pump standard. It is NFPA 20, *Standard for the Installation of Stationary Pumps for Fire Protection.*

It's important to remember that Chapter 6 covers only the minimum safety requirements for the special equipment it covers and that, in most cases, other standards also apply. The *NEC* isn't meant as a substitute for those other standards. So before you begin work on any of this special equipment, first obtain the standards written for that equipment. If you're not sure what standards apply, contact the manufacturer for assistance.

Article 600. Electric Signs and Outline Lighting. This article covers the installation of conductors and equipment for electric signs and outline lighting as defined in Article 100. Electric signs and outline lighting include all products and installations that utilize neon tubing, such as signs, decorative elements, skeleton tubing, or art forms.

Article 604. Manufactured Wiring Systems. The provisions of Article 604 apply to field-installed manufactured wiring systems used for branch circuits, remote-control circuits, signaling circuits, and communications circuits in accessible areas. The components of a listed manufactured wiring system can be assembled together at the jobsite.

Article 605. Office Furnishings (Wired Partitions). This article covers electrical equipment, lighting accessories, and wiring systems used to connect, or contained in or on, relocatable partitions. Partitions can be fixed or freestanding and can have communications, signaling, and optical fiber cable wiring in addition to wiring for receptacles and lighting.

Article 620. Elevators, Escalators, and Moving Walks. Article 620 covers the installation of electrical equipment and wiring used in connection with elevators, dumbwaiters, escalators, moving walks, wheelchair lifts, and stairway chair lifts.

Article 625. Electric Vehicle Charging Systems. Article 625 covers conductors and equipment external to electric vehicles that are used for electric vehicle charging. This only applies to automotive-type vehicles for highway use.

Article 630. Electric Welders. Article 630 covers the wiring of arc welders, resistance welders, and other welding equipment connected to an electric supply system.

Article 640. Audio Signal Processing, Amplification, and Reproduction Equipment. This article covers equipment and wiring for audio signal generation, recording, processing, amplification and reproduction, distribution of sound, public address, speech input systems, temporary audio system installations, and electronic musical instruments such as electric organs, electric guitars, and electronic drums/percussion instruments.

Article 645. Information Technology Equipment. Article 645 covers equipment, power-supply wiring, equipment interconnecting wiring, grounding, and bonding of information technology equipment and systems, including terminal units in an information technology equipment room.

Article 647. Sensitive Electronic Equipment. A technical power system (called "balanced power" by some) is a separately derived, 120V line-to-line, single-phase, 3-wire system with 60V-to-ground used for sensitive electronic equipment.

Article 680. Swimming Pools, Spas, Hot Tubs, Fountains, and Similar Installations. Article 680 covers the installation of electric wiring and equipment that supplies swimming, wading, therapeutic and decorative pools, fountains, hot tubs, spas, and hydromassage bathtubs, whether permanently installed or storable.

Article 690. Solar Photovoltaic Systems. The provisions of Article 690 apply to solar photovoltaic electrical energy systems, including the array circuit(s), inverter(s), and controller(s) for such systems. Solar photovoltaic systems within the scope of Article 690 may be interactive with other electrical power production sources or may stand alone, with or without electrical energy storage, such as batteries.

Article 692. Fuel Cell Systems. This article covers the installation of fuel cell power systems, which may be stand-alone or interactive with other electrical power production sources, and may be with or without electrical energy storage, such as batteries. A fuel cell is an electrochemical device that combines hydrogen, or hydrogen-rich fuel, with oxygen to produce electricity, heat, and water.

Article 695. Fire Pumps. Article 695 covers the electric power sources and interconnecting circuits for electric motor-driven fire pumps. It also covers switching and control equipment dedicated to fire pump drivers. Article 695 doesn't apply to sprinkler system pumps in one- and two-family dwellings or to pressure-maintenance (jockey) pumps.

ARTICLE

600 Electric Signs and Outline Lighting

Introduction

One of the first things you notice when entering a strip mall is that there's a sign for every store. Every commercial occupancy needs a form of identification, and the standard method is the electric sign. Thus, 600.5 requires a sign outlet for the entrance of each tenant location. Article 600 requires a disconnecting means within the line of sight of a sign unless the disconnecting means can be locked in the open position. The lamps in signs burn out and tenants or owners change the lamps. But the signs themselves also change—tenants relocate all the time.

Author's Comment: Article 100 defines an electric sign as any "fixed, stationary, or portable self-contained, electrically illuminated utilization equipment with words or symbols designed to convey information or attract attention."

Another requirement is height. Freestanding signs, such as those that might be erected in a parking lot, must be located at least 14 ft above areas accessible to vehicles unless they're protected from physical damage.

As you studied Chapters 1 through 4, you saw requirements based on the laws of physics. A major theme in those chapters is how to prevent wires from melting. But, the physics approach isn't the only one the *NEC* takes to safety. From bird droppings to bus traffic to bulb burnouts, you'll find Article 600 requirements compensate for the realities of commercial establishments.

Neon art forms or decorative elements are subsets of electric signs and outline lighting. If installed and not attached to an enclosure or sign body, they are considered skeleton tubing for the purpose of applying the requirements of Article 600. However, if that neon tubing is attached to an enclosure or sign body, which may be a simple support frame, it is considered a sign or outline lighting subject to all the provisions that apply to signs and outline lighting, such as 600.3, which requires listing of this product.

Author's Comment: Outline lighting is an arrangement of incandescent lamps or electric-discharge lighting to outline or call attention to certain features, such as the shape of a building or the decoration of a window [Article 100].

PART I. GENERAL

600.1 Scope. Article 600 covers the installation of conductors and equipment for electric signs and outline lighting, including all products and installations that use neon tubing, decorative elements, skeleton tubing, or art forms.

Author's Comment: See Article 100 for the definition of "Outline Lighting."

600.2 Definitions.

Section Sign. A sign or outline lighting system, shipped as subassemblies that require field-installed wiring between the subassemblies to complete the overall sign.

600.3 Listing. Electric signs, section signs, and outline lighting—fixed, mobile, or portable—must be listed and installed in accordance with the listing instructions. Field-installed outline lighting isn't required to be listed if wired in accordance with Chapter 3.

600.4 Markings.

(A) Signs and Outline Lighting Systems. Signs and outline lighting systems must be marked with the manufacturer's name, trademark, or other means of identification.

600.5 Branch Circuits.

(A) Required Branch Circuit. Each commercial building or occupancy accessible to pedestrians must be provided with an outlet at an accessible location for a sign. The sign outlet must be located at the entrance of each tenant space, and it must be supplied with an individual branch circuit rated not less than 20A. **Figure 600-1**

(C) Wiring Methods.

(1) The wiring for a sign must terminate at the sign or outline lighting enclosure, box, or conduit body.

(3) Metal poles used to support signs are permitted to contain the sign circuit conductors, provided the installation complies with 410.15(B).

600.6 Disconnects. Each circuit that supplies a sign or outline lighting system must be controlled by an externally operable switch or circuit breaker that opens all ungrounded conductors. **Figure 600-2**

(A) Location.

(1) Within Sight of Sign. The disconnecting means must be within sight of the sign or outline lighting system. Where the disconnecting means is out of the line of sight from any section of the sign or outline lighting that is able to be energized, the disconnecting means must be capable of being locked in the open position.

Signs for Commercial Buildings
With Pedestrian Access
Section 600.5(A)
Accessible sign outlet located at each entrance to each tenant space.
Last Bank & Trust
COPYRIGHT 2005
Mike Holt Enterprises, Inc.
Each commercial building and each commercial occupancy accessible to pedestrians must have at least one sign or outline outlet supplied by a 20A branch circuit.

Figure 600–1

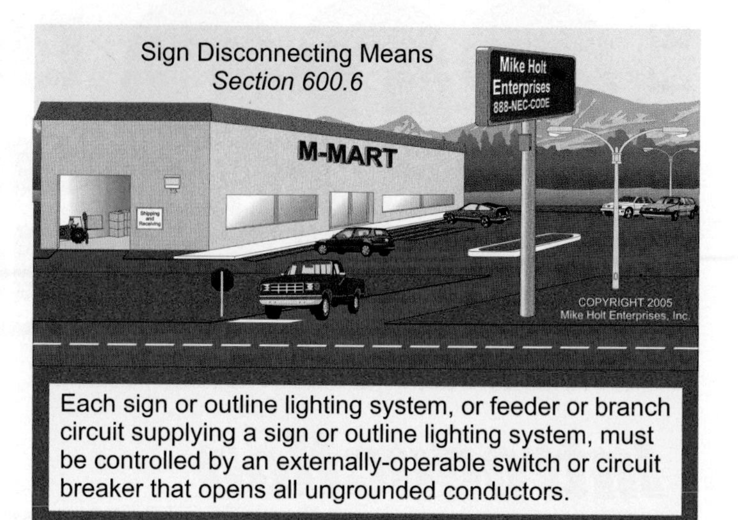

Figure 600–2

> **Author's Comment:** See Article 100 for the definition of "Within Sight."

(2) Within Sight of the Controller. Signs or outline lighting systems operated by electronic or electromechanical controllers located external to the sign or outline lighting system must have the disconnecting means installed in accordance with (1) through (3):

(1) Located within sight of or in the same enclosure with the controller.

(2) Be capable of disconnecting the sign or outline lighting and the controller from all ungrounded supply conductors.

(3) Be capable of being locked in the open position.

600.9 Location.

(A) Vehicles. Unless protected from physical damage, a sign or outline lighting system must be located not less than 14 ft above areas accessible to vehicles. **Figure 600-3**

> **Author's Comment:** A minimum of 18 ft of overhead conductor clearance is required over public streets, alleys, roads, parking areas subject to truck traffic, and driveways on other than residential property [225.18]. This may be an important consideration if a sign is fed by overhead conductors.

(B) Pedestrians. Neon tubing, other than dry-location portable signs accessible to pedestrians, must be protected from physical damage.

(C) Adjacent to Combustible Materials. Signs and outline lighting systems must be installed so that adjacent combustible materials aren't subjected to temperatures that exceed 194°F. An incandescent or HID lamp or lampholder must have not less than 2 in. of spacing from wood or other combustible materials.

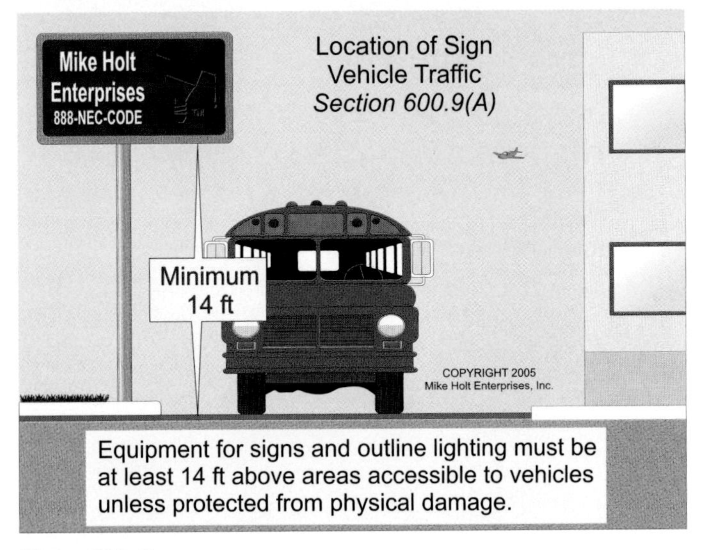

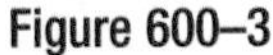

Figure 600–3

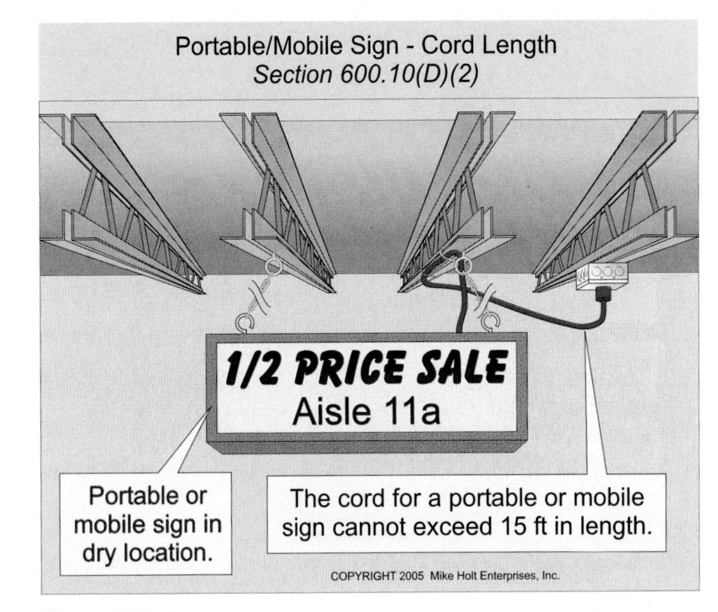

Figure 600–4

(D) Wet Location. Signs and outline lighting systems installed at wet locations must be weatherproof and must have drain holes.

600.10 Portable or Mobile Signs.

(A) Support. Portable or mobile signs must be adequately supported and readily movable without the use of tools.

(B) Attachment Plug. An attachment plug is required for each portable or mobile sign.

(C) Wet or Damp Location. Portable or mobile signs in wet or damp locations must comply with the following:

(1) Cords. Cords must be junior hard-service or hard-service types with an equipment grounding (bonding) conductor.

(2) GFCI. Portable or mobile signs must be GFCI protected by a factory-installed GFCI-protection device that is an integral part of the attachment plug or is located in the power-supply cord located within 1 ft of the attachment plug.

(D) Dry Location.

(2) To prevent the use of extension cords, the cord on a portable or mobile sign installed in a dry location can be up to 15 ft long. **Figure 600-4**

Author's Comments:

- See Article 100 for the definition of "Location (Dry)."
- 400.8(2) and (5) prohibit flexible cords from being run through, or above, a suspended ceiling.

600.21 Ballasts, Transformers, and Electronic Power Supplies.

(A) Accessibility. Ballasts, transformers, and electronic power supplies must be accessible and must be securely fastened in place.

(B) Location. The secondary conductors from ballasts, transformers, and electronic power must be as short as possible.

(D) Working Space. A working space not less than 3 ft high by 3 ft wide by 3 ft deep is required for each ballast, transformer, and electronic power supply where not installed in a sign.

(E) Attic Locations. Ballasts, transformers, and electronic power supplies located in attics and soffits are permitted, provided there's an access door and a passageway not less than 3 ft high by 2 ft wide with a suitable permanent walkway at least 1 ft wide to the point of entry for each component.

(F) Suspended Ceilings. Ballasts, transformers, and electronic power supplies can be above a suspended ceiling, provided the enclosures are securely fastened in place and don't use the suspended-ceiling grid for support.

Ballasts, transformers, and electronic power supplies installed above a suspended ceiling cannot be connected to the branch-circuit wiring by a flexible cord. **Figure 600-5**

Author's Comment: 400.8(2) and (5) prohibit flexible cords from being run through, or above, a suspended ceiling.

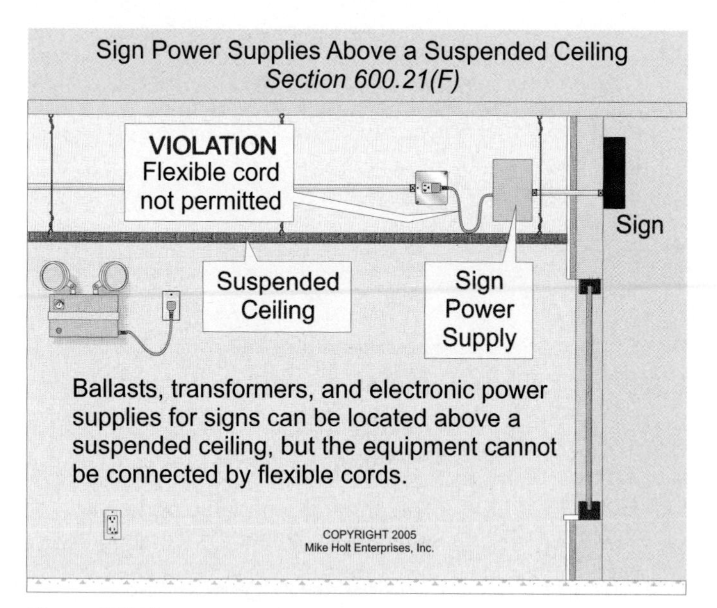
Sign Power Supplies Above a Suspended Ceiling
Section 600.21(F)
VIOLATION
Flexible cord
not permitted
Sign
Suspended
Ceiling
Sign
Power
Supply
Ballasts, transformers, and electronic power supplies for signs can be located above a suspended ceiling, but the equipment cannot be connected by flexible cords.
COPYRIGHT 2005
Mike Holt Enterprises, Inc.

Figure 600–5

Article 600 Questions

1. Electric signs and outline lighting—fixed, mobile, or portable—are not required to be listed.

   (a) True (b) False

2. Branch circuits that supply signs and outline lighting systems containing incandescent and fluorescent forms of illumination must be rated not to exceed _____.

   (a) 20A (b) 30A (c) 40A (d) 50A

3. The disconnecting means for each circuit leading to a sign located within a fountain must be located in accordance with _____.

   (a) 430.102 (b) 440.14 (c) 680.12 (d) any of these

4. Signs and outline lighting systems must be installed so that adjacent combustible materials are not subjected to temperatures in excess of _____.

   (a) 90°C (b) 60°C (c) 75°C (d) 40°C

5. Ballasts, transformers, and electronic power supplies for signs installed in suspended ceilings can be connected to the branch circuit by a _____.

   (a) fixed wiring method (b) flexible wiring method (c) flexible cord (d) a or b

Get Ready Now!

2005 NEC

A 15 or 20A, 125V recept requires GFCI protection

125V receptacles installed at on for the servicing of hea equipment in acco

NEC 2005
NFPA 70: National Electrical Code®
International Electrical Code® Series

Order Mike's Code Change Library and SAVE OVER $450

2005 Code® Change Library Order Form

COMPANY

NAME TITLE

MAILING ADDRESS

CITY STATE ZIP

SHIPPING ADDRESS

CITY STATE ZIP

PHONE FAX

E-MAIL ADDRESS WEB SITE

	Item	Description	Price
❑	05BK	Illustrated Changes to the NEC® Textbook	$44
❑	05CCOLP1	Part 1 8 hours Online Code Change Program	$89
❑	05CCOLP2	Part 2 8 hours Online Code Change Program	$89
❑	05CCV1	Video 1 (4.5 hours) - Articles 90-314	$109
❑	05CCV2	Video 2 (4.5 hours) - Articles 320-830	$109
❑	05CCD1	DVD 1 (4.5 hours) - Articles 90-314	$109
❑	05CCD2	DVD 2 (4.5 hours) - Articles 320-830	$109
❑	05TB	Code Tabs	$10
❑	05CCMP	MP3 Audio CD Articles 90-830 NEW	$59
❑	**05CCP**	**Code® Change Library** (includes everything above) List ~~$727~~ You Pay	**$249**
	Also Available		
❑	05SB	*NEC* Spiral Code book	$68
❑	05ECB	2005 NEC Code book (electronic version)	$100
❑	05HB	2005 NEC Handbook Available February, 2005	$120
❑	05EHB	2005 NEC Handbook (electronic version) Available February, 2005	$175

❑ CHECK ❑ VISA ❑ MASTER CARD ❑ DISCOVER ❑ AMEX ❑ MONEY ORDER

CREDIT CARD # : ______________________ EXP. DATE: ________

Sub-Total $ ________

Sales Tax **FLORIDA RESIDENTS ONLY** add 6% $ ________

Shipping: 4% of Total Price (or minimum $7.50) $ ________

TOTAL $ ________

1.888.NEC®CODE • 1.954.720.1999 • FAX 1.954.720.7944 • E-mail-Info@MikeHolt.com

www.NECcode.com

Mike Holt Enterprises, Inc. • 7310 W. McNab Rd., Suite 201 • Tamarac, FL 33321

All prices and availability are subject to change. • If you are not 100% satisfied with your order, return it within 10 days and receive a full refund.

ARTICLE

604 Manufactured Wiring Systems

Introduction

Article 604 applies to field-installed manufactured wiring systems, while Article 605 applies to relocatable partitions. In both cases, you're installing and/or assembling prewired components or subassemblies. These normally come with manufacturer's instructions, so *Code* compliance is almost automatic if you follow the instructions. However, you still need to:

- Compare the instructions against the provisions of Article 604 or Article 605, as appropriate.
- Use the proper connection cords.
- Apply the correct branch-circuit protection.
- Apply proper mechanical support to the electrical components, cords, and connectors.

The specific requirements contained in this article aren't difficult to understand, but you should be familiar with them before installing manufactured wiring systems or wired partitions.

604.1 Scope. The provisions of Article 604 apply to field-installed manufactured wiring systems. **Figure 604-1**

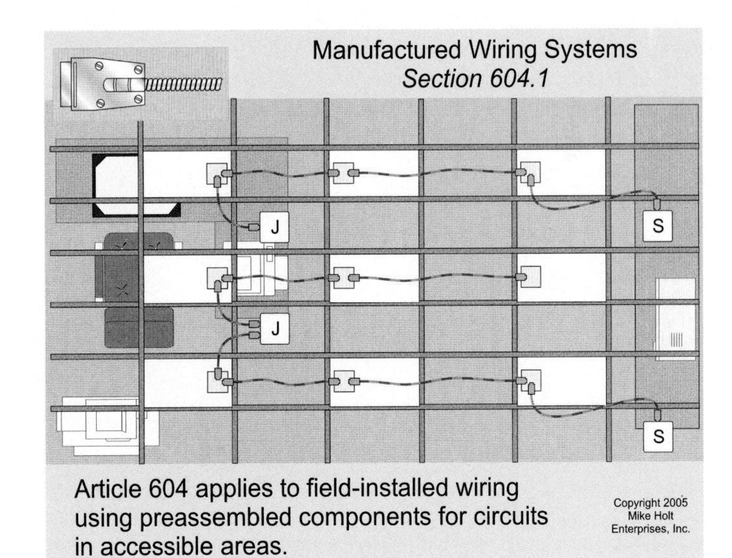

Manufactured Wiring Systems
Section 604.1
J
J
S
S
Article 604 applies to field-installed wiring using preassembled components for circuits in accessible areas.
Copyright 2005
Mike Holt
Enterprises, Inc.

Figure 604–1

604.2 Definition.

Manufactured Wiring System. A system assembled by a manufacturer that cannot be inspected at the building site without damage or destruction to the assembly.

> **Author's Comment:** Manufactured wiring systems are typically installed in commercial and institutional occupancies because of their ability to be easily relocated.

604.3 Other Articles. Except as modified by the requirements of this article, all other applicable articles of the *NEC* apply to manufactured wiring systems.

604.4 Uses Permitted. Manufactured wiring systems can be installed in accessible and dry locations, and in other environmental air spaces, where installed in accordance with 300.22(C).

Exception 1: In concealed spaces, one end of the cable can extend into hollow walls for direct termination to an outlet.

604.5 Uses Not Permitted. Manufactured wiring systems constructed with Type AC cable must be installed in accordance with Article 320. Where constructed with Type MC cable, the installations must comply with Article 330.

604.6 Construction.

(A) Cable or Conduit Types.

(1) Cable. Manufactured wiring systems can be constructed with Type AC or Type MC cable containing 12 AWG or 10 AWG conductors.

(2) Conduits. Manufactured wiring systems can be constructed with listed flexible metal conduit or listed liquidtight flexible conduit that contains insulated copper conductors with a bare or insulated copper equipment grounding (bonding) conductor equivalent in size to the ungrounded (neutral) conductor.

Exception 1: Flexible raceways no longer than 6 ft for a single luminaire are permitted to contain conductors smaller than 12 AWG but not smaller than 18 AWG.

(3) Flexible Cord. Flexible cord with minimum 12 AWG is permitted for the connection of utilization equipment, other than luminaires, where the cord is part of a listed factory-made manufactured wiring assembly no longer than 6 ft. The cord must not be permanently secured to the building structure, it must be visible for its entire length, and it must not be subject to strain or physical damage.

Author's Comment: The minimum size conductor requirements and the maximum 6 ft cord length don't apply when used for luminaires. See 604.6(F).

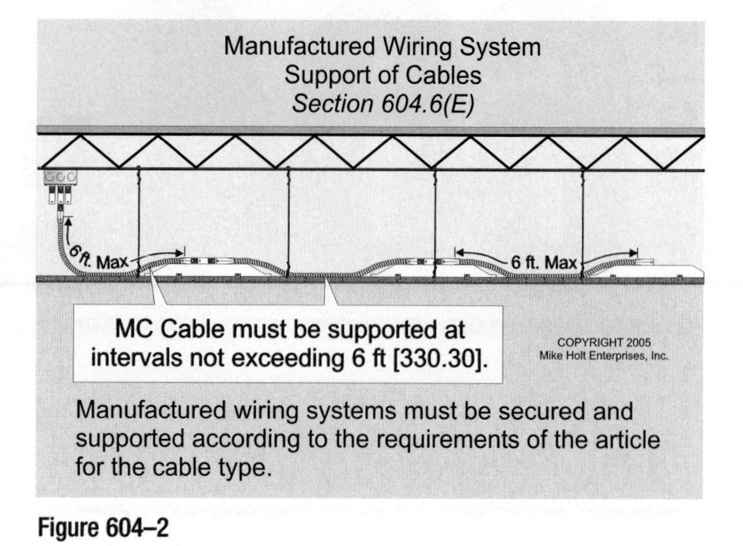

Manufactured Wiring System
Support of Cables
Section 604.6(E)
6 ft. Max
6 ft. Max
MC Cable must be supported at intervals not exceeding 6 ft [330.30].
COPYRIGHT 2005
Mike Holt Enterprises, Inc.
Manufactured wiring systems must be secured and supported according to the requirements of the article for the cable type.

Figure 604–2

(E) Securing and Supporting. Manufactured wiring systems must be secured and supported in accordance with the applicable article for the cable.

Author's Comment: Type AC cable must be secured at intervals not exceeding 4½ ft, and Type MC cable can be run unsupported or secured at intervals not exceeding 6 ft [330.30]. **Figure 604-2**

(F) Luminaires. Listed electric discharge luminaires can be connected via a flexible cord in accordance with 410.30(C).

Article 604 Questions

1. The provisions of Article 604 apply to field-installed manufactured wiring systems using off-site manufactured subassemblies for branch circuits, remote-control circuits, signaling circuits, and communications circuits in _____ areas.

   (a) accessible
   (b) only patient care
   (c) hazardous (classified)
   (d) concealed

2. A manufactured wiring system is a system assembled by a manufacturer which cannot be inspected at the building site without _____.

   (a) a permit
   (b) a manufacturer's representative present
   (c) damage or destruction to the assembly
   (d) an engineer's supervision

3. Article 604 contains all requirements for the installation of manufactured wiring systems, no other *Code* articles will apply.

   (a) True
   (b) False

4. Manufactured wiring systems are permitted in _____ locations and in plenums and spaces used for environmental air, where installed in accordance with 300.22.

   (a) accessible
   (b) dry
   (c) wet
   (d) both a and b

5. Manufactured wiring systems must be constructed with _____.

   (a) listed Type AC or Type MC cable
   (b) 10 or 12 AWG copper-insulated conductors
   (c) conductors that are suitable for nominal 600V
   (d) all of these

Notes

ARTICLE

605 Office Furnishings (Wired Partitions)

Introduction

Article 604 applied to field-installed manufactured wiring systems, while Article 605 applies to relocatable partitions. In both cases, you're installing and/or assembling prewired components or subassemblies. These normally come with the manufacturer's instructions, so *Code* compliance is almost automatic if you follow the instructions. However, you still need to:

- Compare the instructions against the provisions of Article 604 or Article 605, as appropriate.
- Use the proper connection cords.
- Apply the correct branch-circuit protection.
- Apply proper mechanical support to the electrical components, cords, and connectors.

The specific requirements contained in this article aren't difficult to understand, but you should be familiar with them before installing manufactured wiring systems or wired partitions.

605.1 Scope. Article 605 contains the requirements for electrical equipment, lighting accessories, and wiring systems for relocatable wired partitions.

605.2 General.

(B) Other Articles. Except as modified by the requirements of this article, all other articles of the *NEC* apply.

605.4 Partition Interconnections.

The electrical connection between wired partitions must be a flexible assembly identified for use with wired partitions. A flexible cord is acceptable, provided all of the following conditions are met:

(1) The cord is of the extra-hard usage type with 12 AWG or larger conductors with an insulated equipment grounding (bonding) conductor.

(2) The partitions are mechanically contiguous.

> **Author's Comment:** "Contiguous" means being in actual contact, or touching or connected throughout in an unbroken sequence.

(3) The cord isn't longer than necessary, but in no case can it exceed 2 ft.

(4) The cord must terminate at an attachment plug and cord connector with strain relief.

605.6 Fixed-Type Partitions. Branch circuits for fixed-type wired partitions use a Chapter 3 wiring method. Multiwire branch circuits that supply power to the wired partition must be provided with a means to disconnect simultaneously all ungrounded conductors where the branch circuit originates.

Author's Comments:

- Individual single-pole circuit breakers with handle ties identified for the purpose, or a breaker with common internal trip, can be used for this purpose [240.20(B)(1)]. **Figure 605-1**
- Providing a multipole breaker with a common internal trip mechanism will cause all the branch circuits tied to the same breaker to take out all of the circuits, resulting in unnecessary downtime. Now when John or Jane Doe decide to take a 1,500W space heater to work and accidentally trip the breaker, instead of 4 personal computers (PCs) going down, it will be 12!

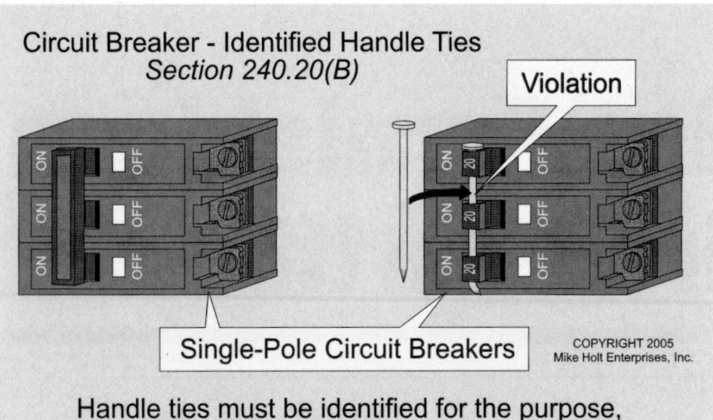
Circuit Breaker - Identified Handle Ties
Section 240.20(B)
Violation
Single-Pole Circuit Breakers
COPYRIGHT 2005
Mike Holt Enterprises, Inc.
Handle ties must be identified for the purpose, which means that handle ties made from nails, screws, wires, or other nonconforming material are not permitted.

Figure 605–1

605.7 Freestanding-Type Partitions. Branch circuits for wired partitions of the freestanding type must be of a type contained in Chapter 3. Multiwire branch circuits that supply power to the wired partition must be provided with a means to simultaneously disconnect all ungrounded conductors where the branch circuit originates.

Article 605 Questions

1. Wiring systems for the wiring of office furnishings must be identified as suitable for providing power for lighting accessories and appliances in wired partitions. These partitions are allowed to extend from the floor to above the ceiling.

   (a) True (b) False

2. The electrical connection between mechanically contiguous wired partitions (for office furnishings) is permitted to be a flexible cord if the cord _____.

   (a) is extra hard-usage with 12 AWG or larger conductors
   (b) has an insulated equipment grounding conductor
   (c) is no longer than 2 ft and terminates at an attachment plug and connector with strain relief
   (d) all of these

3. Wired partitions for office furnishings that are fixed (secured to building surfaces) must be permanently connected to the building electrical system by a Chapter 3 wiring method.

   (a) True (b) False

4. Multiwire branch circuits that supply power to the wired partitions of office furnishings for _____ must be provided with a means to simultaneously disconnect all ungrounded conductors where the branch circuit originates.

   (a) fixed-type partitions (b) free-standing type partitions
   (c) both a and b (d) none of these

Notes

ARTICLE

620 Elevators, Escalators, and Moving Walks

Introduction

With the exception of dumbwaiters, the equipment covered by Article 620 moves people. Thus, a major concept in Article 620 is that of keeping people separate from electrical power. That's why, for example, 620.3 requires live parts to be enclosed.

Article 620 consists of the following Parts:

- Part I. General. This provides scope, definitions, and voltage limitations.
- Part II. Conductors. The single-line diagram of Figure 620.13 shown in the *NEC* book illustrates how the requirements of Part II work together.
- Part III. Wiring. This addresses wiring method and branch-circuit requirements for different equipment.
- Part IV. Installation of Conductors. This covers wire fill, supports, and related items.
- Part V. Traveling Cables. This covers installation, suspension, location, and protection of cables that move with the motion of the elevator or lift.
- Part VI. Disconnecting Means and Control. The requirements vary with the application.
- Part VII. Overcurrent Protection. While most of it refers to Article 430, Part VII does add additional requirements, such as providing selective coordination.
- Part VIII. Machine and Control Rooms and Spaces. The primary goal here is the prevention of unauthorized access.
- Part IX. Grounding (Bonding). While most of this refers to Article 250, Part IX adds additional requirements. For example, all 15 and 20A, 125V receptacles in certain locations must be GFCI protected.
- Part X. Emergency and Standby Systems. This deals with regenerative power and with the need for a disconnecting means that can disconnect the elevator from both the normal power system and the emergency or standby system.

PART I. GENERAL

620.1 Scope. Article 620 contains the requirements for the installation of electrical equipment and wiring in connection with elevators, escalators, and moving walks.

PART III. WIRING

620.23 Branch Circuit for Machine Room/Machinery Space.

(A) Branch Circuit. A separate branch circuit must supply machine room/machinery space lighting and receptacle(s). The required lighting must not be GFCI protected. **Figure 620-1**

(B) Light Switch. The machine room lighting switch must be located at the point of entry to the machine room or machinery space.

(C) Receptacle. At least one 15 or 20A, 125V GFCI-protected duplex receptacle is required in each machine room and machinery space [620.85]. **Figure 620-2**

620.24 Branch Circuit for Hoistway Pit.

(A) Branch Circuit. A separate branch circuit must supply the hoistway pit lighting and receptacle(s). The required lighting must not be connected to the load side of a GFCI [680.85].

(B) Light Switch. The lighting switch must be readily accessible from the pit access door.

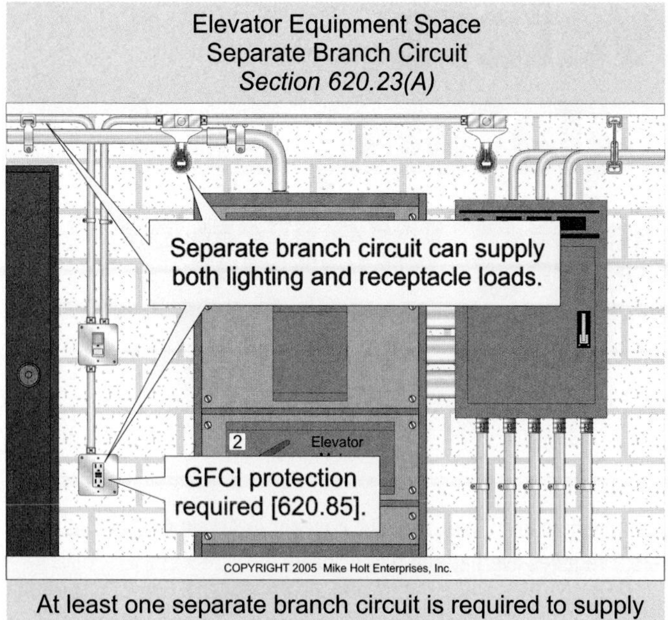

At least one separate branch circuit is required to supply machine room/space lighting and receptacle(s). Lighting cannot be connected to the load side of a GFCI device.

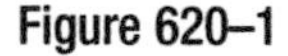

Figure 620–1

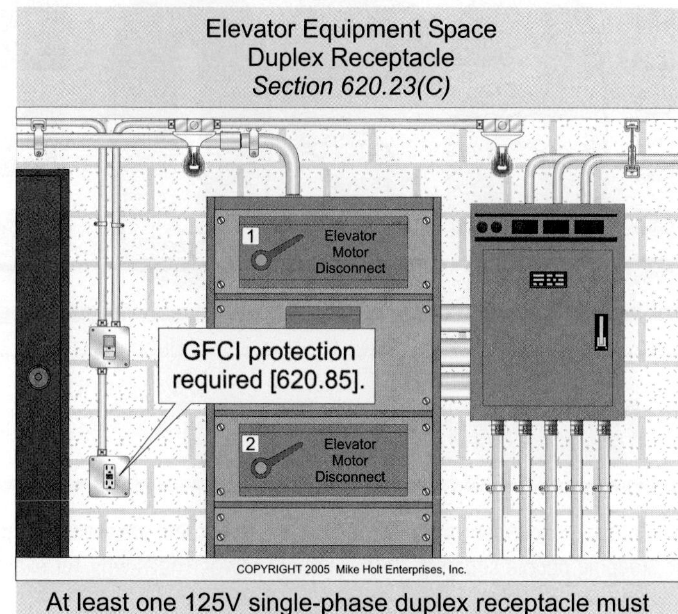

At least one 125V single-phase duplex receptacle must be installed in each machine room or control room, and each machinery space or control space.

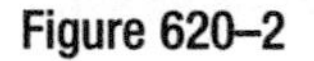

Figure 620–2

(C) Receptacle. At least one 15 or 20A, 125V GFCI-protected duplex receptacle is required in the hoistway pit.

PART IV. INSTALLATION OF CONDUCTORS

620.37 Wiring in Elevator Hoistways and Machine Rooms.

(A) Uses Permitted. Only wiring, raceways, and cables, including communications, fire detection, power, lighting, heating, ventilating, and air conditioning used directly in connection with the elevator, are permitted inside the elevator hoistway, machine room, or control room.

PART VI. DISCONNECTING MEANS AND CONTROL

620.51 Disconnecting Means. A disconnect must be provided with a means to simultaneously disconnect all ungrounded main power-supply conductors for each unit.

(A) Type. The disconnecting means must be an externally operable fused motor-circuit switch or circuit breaker capable of being locked in the open position.

(B) Operation. If sprinklers are installed in hoistways, machine rooms, control rooms, machinery spaces, or control spaces, the disconnecting means is permitted to automatically open the power supply to the affected elevator(s) prior to the application of water.

(C) Location. The disconnecting means must be located so that it's only readily accessible to qualified persons.

(D) Identification and Signs. Where there's more than one driving machine in a machine room, the disconnecting means must be numbered to correspond to the identifying number of the driving machine that they control.

PART VIII. MACHINE ROOMS, CONTROL ROOMS, MACHINERY SPACES, AND CONTROL SPACES

620.85 GFCI-Protected Receptacles. All 15 and 20A, 125V receptacles located in pits, hoistways, elevator car tops, or escalator and moving walk wellways must be of the GFCI type.

All 15 and 20A, 125V receptacles installed in machine rooms and machinery spaces must be GFCI protected. **Figure 620-3**

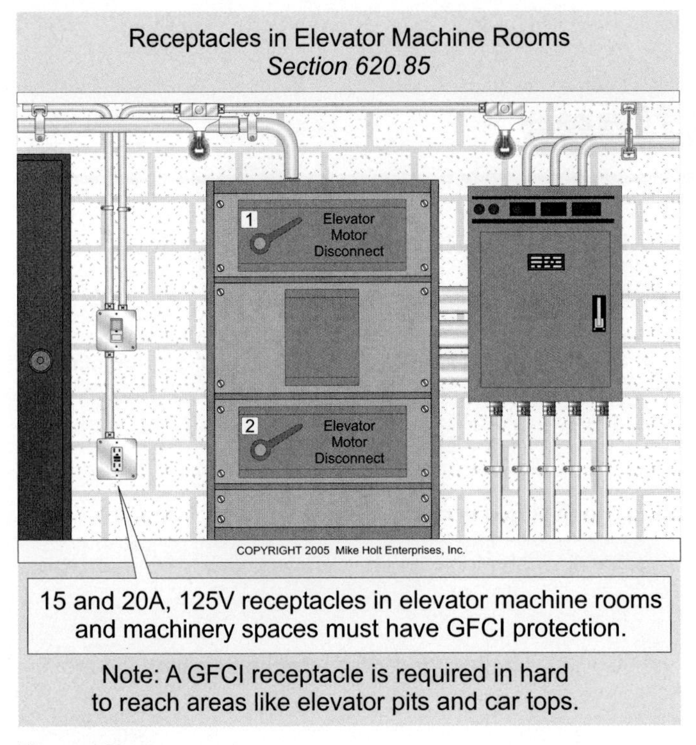
Receptacles in Elevator Machine Rooms
Section 620.85
1
Elevator
Motor
Disconnect
2
Elevator
Motor
Disconnect
COPYRIGHT 2005 Mike Holt Enterprises, Inc.
15 and 20A, 125V receptacles in elevator machine rooms and machinery spaces must have GFCI protection.
Note: A GFCI receptacle is required in hard to reach areas like elevator pits and car tops.

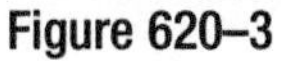
Figure 620–3

Article 620 Questions

1. At least _____ 125V, single-phase, duplex receptacle(s) must be provided in each machine room and machinery space.

   (a) one (b) two (c) three (d) four

2. Only wiring, raceways, and cables used directly in connection with the elevator must be inside the hoistway and the machine room.

   (a) True (b) False

3. The location of the disconnecting means for an elevator must be _____ to qualified persons.

   (a) accessible
   (b) readily accessible
   (c) disclosed only
   (d) accessible only with a key

4. All 15 and 20A, 125V single-phase receptacles installed in pits, in hoistways, on elevator car tops, and in escalator and moving walk wellways must be _____.

   (a) on a GFCI-protected circuit
   (b) of the GFCI type
   (c) a or b
   (d) none of these

5. Each elevator must have a single means for disconnecting all ungrounded main power supply conductors for each unit _____.

   (a) excluding the emergency power system
   (b) including the emergency or standby power system
   (c) excluding the emergency power system if it is automatic
   (d) and the power supply may not be an emergency power system

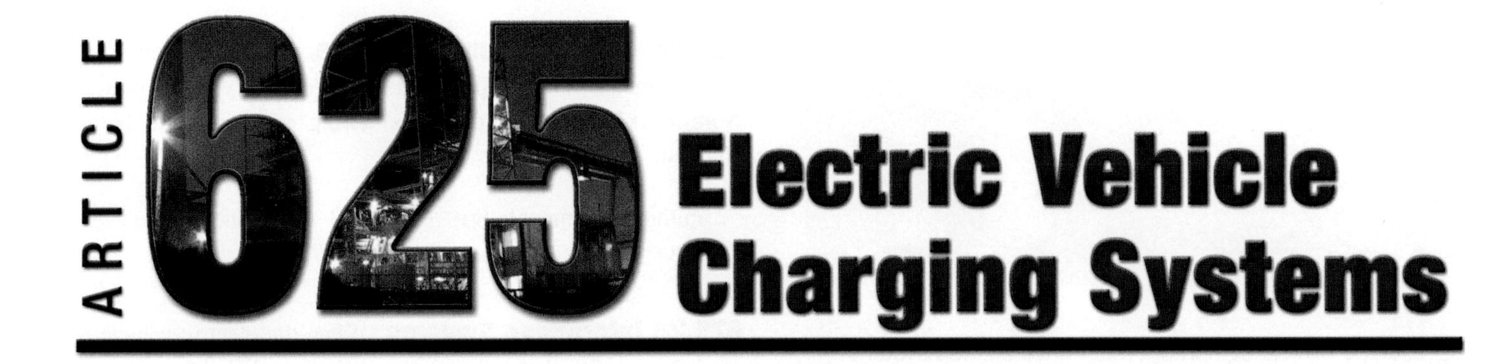

ARTICLE 625 Electric Vehicle Charging Systems

Introduction

Article 625 covers electrical conductors and equipment used to connect an electric vehicle to a source of electricity, and the equipment and devices related to electric vehicle charging.

Vehicle charging connectors must be configured such that they aren't interchangeable with other electrical systems. They must prevent inadvertent contact of live parts with people, and they must prevent unintentional disconnection. In fact, Article 625 requires automatic de-energization of charging cables under certain conditions.

PART I. GENERAL

625.1 Scope. Article 625 covers conductors and equipment external to electric vehicles and used for electric vehicle charging. **Figure 625-1**

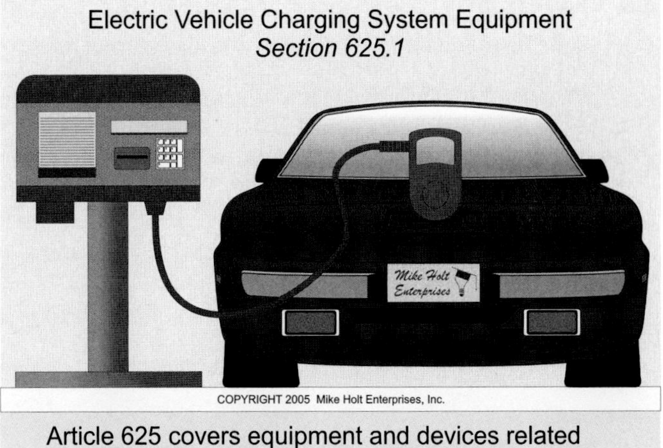

Article 625 covers equipment and devices related to electric vehicle charging equipment that connects the vehicle to a supply of electricity by conductive or inductive means.

Figure 625–1

625.2 Definitions.

Electric Vehicle. An automotive-type vehicle for highway use, such as passenger automobiles, buses, trucks, vans, or neighborhood electric vehicles primarily powered by an electric motor. The term electric vehicle doesn't apply to electric motorcycles, industrial trucks, hoists, lifts, transports, golf carts, airline ground support equipment, tractors, or boats.

> **Author's Comment:** A neighborhood electric vehicle has automotive grade headlights, seat belts, windshields, brakes, and other safety equipment, and a top speed of 25 MPH. These vehicles are often used in retirement or other planned communities for short trips for shopping, social, and recreational purposes.

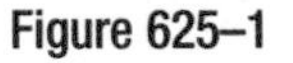

PART IV. CONTROL AND PROTECTION

625.22 Personnel Protection System. Electric vehicle supply equipment must provide personnel protection against electric shock by incorporating a listed personnel protection device. For plug-connected electric vehicle supply equipment, personnel protection can be either an integral part of the attachment plug or located in the power-supply cable. **Figure 625-2**

> **Author's Comment:** GFCI protection doesn't meet this requirement. For more information, see UL Standard 2231, *Personnel Protection Systems for Electrical Vehicle Charging System Equipment.*

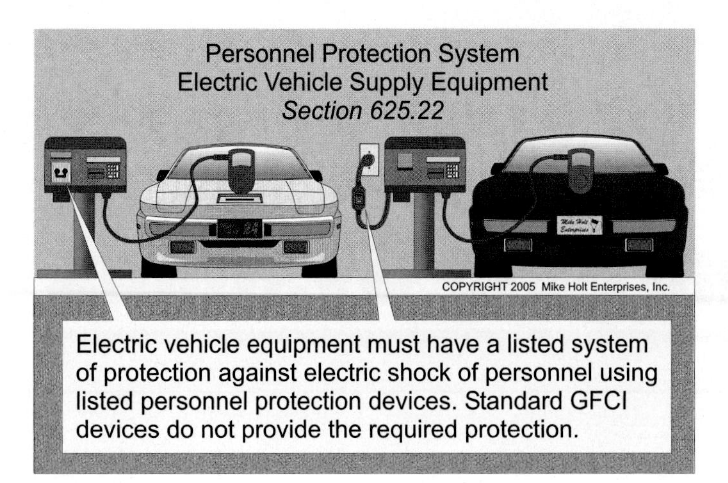
Personnel Protection System
Electric Vehicle Supply Equipment
Section 625.22
COPYRIGHT 2005 Mike Holt Enterprises, Inc.
Electric vehicle equipment must have a listed system of protection against electric shock of personnel using listed personnel protection devices. Standard GFCI devices do not provide the required protection.

Figure 625–2

625.25 Loss of Primary Source. Unless permitted by 625.26, means must be provided to prevent the electric vehicle from backfeeding the premises wiring system while being charged.

> **Author's Comment:** See Article 100 for the definition of "Premises Wiring."

625.26 Interactive Systems. When the electric vehicle is used as an optional standby system, it must be listed for this purpose and installed in accordance with Article 702.

> **Author's Comment:** This means that a transfer switch must be installed on the premises wiring system [702.6].

Article 625 Questions

1. An electric vehicle that falls within the scope of Article 625 could be _____.

    (a) an automotive-type vehicle for highway use
    (b) an automotive-type vehicle for off-road use
    (c) an electric golf cart
    (d) any of these

2. According to Article 625, automotive-type vehicles for highway use include _____.

    (a) passenger automobiles
    (b) trucks
    (c) neighborhood electric vehicles
    (d) all of these

3. An electric vehicle connector is a device that, by insertion into an electric vehicle inlet, establishes an electrical connection to the electric vehicle for the purpose of charging and information exchange.

    (a) True
    (b) False

4. For plug-connected electric vehicle supply equipment, the listed system of personnel protection can be _____.

    (a) an integral part of the attachment plug
    (b) in the power supply cable not more than 12 in. from the attachment plug
    (c) a or b
    (d) none of these

5. Electric vehicle supply equipment that is identified for and intended to be interconnected to a vehicle, and also serve _____, must be listed as suitable for that purpose.

    (a) as an optional standby system
    (b) as an electric power production source
    (c) to provide bidirectional power feed
    (d) all of these

Notes

ARTICLE

630 Electric Welders

Introduction

Electric welding equipment does its job either by creating an electric arc between two surfaces or by heating a rod that melts from overcurrent. Either way results in a hefty momentary current draw. Welding machines come in many shapes and sizes. On the smaller end of the scale are portable welding units used for manual welding, such as in a fabrication shop. At the larger end of the scale are robotic welding machines the size of a house, used for making everything from automobile bodies to refrigerator panels. All of these must comply with Article 630.

The primary concern of Article 630 is adequately sizing the conductors and circuit protection to handle this type of load. Fortunately for the design engineer and the field electrician, this article requires certain information to be provided on the nameplate of the equipment. Article 630 explains how to use this information for the proper sizing of conductors and circuit protection.

Welding cable has requirements other conductors do not have. For example, it must be supported at not less than 6 in. intervals. Also, the insulation on these cables must be flame-retardant.

PART I. GENERAL

630.1 Scope. Article 630 covers electric arc welding, resistance welding apparatus, and other similar welding equipment connected to an electric supply system.

PART II. ARC WELDERS

630.11 Ampacity of Supply Conductors.

(A) Individual Welders. The supply conductors for arc welders must have an ampacity not less than the welder nameplate rating.

If the nameplate rating isn't available, the supply conductors must have an ampacity not less than the rated primary current as adjusted by the multiplier in Table 630.11(A), based on the duty cycle of the welder.

Table 630.11(A) Duty Cycle Multiplication Factors for Arc Welders

Duty Cycle	Nonmotor Generator	Motor Generator
100	1.00	1.00
90	0.95	0.96
80	0.89	0.91
70	0.84	0.86
60	0.78	0.81
50	0.71	0.75
40	0.63	0.69
30	0.55	0.62
20 or less	0.45	0.55

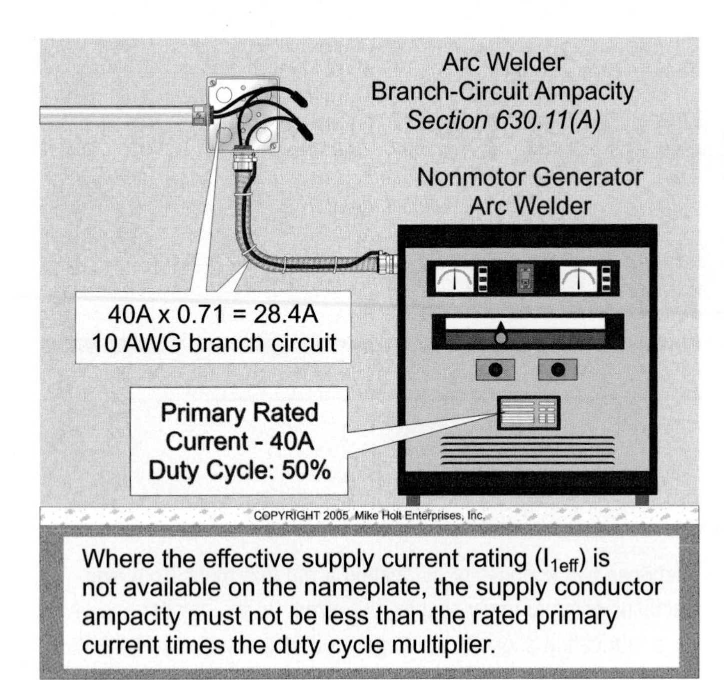
Arc Welder
Branch-Circuit Ampacity
Section 630.11(A)
Nonmotor Generator
Arc Welder
40A x 0.71 = 28.4A
10 AWG branch circuit
Primary Rated
Current - 40A
Duty Cycle: 50%
COPYRIGHT 2005 Mike Holt Enterprises, Inc.
Where the effective supply current rating (I_{1eff}) is not available on the nameplate, the supply conductor ampacity must not be less than the rated primary current times the duty cycle multiplier.

Figure 630–1

Example: *A nonmotor-generator arc welder has a primary current rating of 40A with a duty cycle of 50 percent. The branch-circuit conductor for the welder must not be sized less than:* **Figure 630-1**

(a) 15A (b) 20A (c) 25A (d) 30A

Answer: *(d) 30A*

40A x 0.71 = 28.40A, 10 AWG [Table 310.16].

(B) Group of Welders. Feeder conductors that supply a group of welders must have a minimum ampacity not less than the sum of the currents as determined in 630.11(A), based on 100 percent of the two largest welders, 85 percent for the third largest welder, 70 percent for the fourth largest welder, and 60 percent for all remaining welders.

Author's Comment: This calculation method provides an ample margin of safety under high-production conditions.

Example: *What would the feeder conductor be for five 50A non-motor-generator welders with a duty cycle of 50 percent?*

(a) 1 AWG (b) 2 AWG (c) 3 AWG (d) 4 AWG

Answer: *(a) 1 AWG*

Welder 1: 50A x 0.71 [630.11(A)]	=	
35.5A x 100% [630.11(B)]	=	*35.5A*
Welder 2: 50A x 0.71 [630.11(A)]	=	
35.5 x 100% [630.11(B)]	=	*35.5A*
Welder 3: 50A x 0.71 [630.11(A)]	=	
35.5A x 85% [630.11(B)]	=	*30.18A*
Welder 4: 50A x 0.71 [630.11(A)]	=	
35.5A x 70% [630.11(B)]	=	*24.85A*
Welder 5: 50A x 0.71 [630.11(A)]	=	
35.5A x 60% [630.11(B)]	=	*21.30A*
Total		*147.33A*

Conductor size from Table 310.16 is 1 AWG.

630.12 Overcurrent Protection. Where the calculated overcurrent protection value does not correspond to the standard protection device ratings in 240.6(A), the next higher standard rating is permitted.

(A) Welders. Each welder must have overcurrent protection rated at not more than 200 percent of the maximum rated supply current at maximum rated output nameplate, or not more than 200 percent of the rated primary current of the welder.

(B) Conductors. The conductors must be protected by an overcurrent device rated at not more than 200 percent of the conductor rating.

Question: *What is the maximum overcurrent protection device rating for 10 THHN?*

(a) 30A (b) 40A (c) 50A (d) 70A

Answer: *(d) 70A*

10 AWG at 75°C is rated 35A [Table 310.16]

35A x 2 = 70A maximum overcurrent protection device size

630.13 Disconnecting Means. A disconnecting means is required for each arc welder that isn't equipped with an integral disconnect.

PART III. RESISTANCE WELDERS

630.31 Ampacity of Supply Conductors.

(A) Individual Welders.

(1) The ampacity of the supply conductors for varied duty-cycle welders must not be less than 70 percent of the rated primary current for seam and automatically fed welders, and 50 percent of the rated primary current for manually operated welders.

(2) The ampacity of the supply conductors for welders with a specific duty cycle and nonvarying current levels must have an ampacity not less than the rated primary current as adjusted by the multipliers in Table 630.31(A)(2), based on the duty cycle of the welder.

Table 630.31(A)(2) Duty Cycle Multiplication Factors for Resistance Welders

Duty Cycle	Multiplier
50	0.71
40	0.63
30	0.55
25	0.50
20	0.45
15	0.39
10	0.32
7.5	0.27
5 or less	0.22

(B) Groups of Welders. Feeder conductors that supply a group of resistance welders must have an ampacity not less than the sum of the value determined using 630.31(A)(2) for the largest welder in the group, plus 60 percent of the values determined for all remaining welders.

630.32 Overcurrent Protection.

Where the calculated overcurrent protection value does not correspond with the standard protection device ratings in 240.6(A), the next higher standard rating is permitted.

(A) Welders. Each welder must have overcurrent protection set at not more than 300 percent of the rated primary current.

(B) Conductors. Branch circuit conductors must be protected by an overcurrent device rated at not more than 300 percent of the conductor rating.

630.33 Disconnecting Means.

A switch or circuit breaker is required to disconnect each resistance welder and its control equipment from the supply circuit.

Article 630 Questions

1. The ampacity of the supply conductors to an individual electric welder must not be less than the effective current value on the rating plate.

   (a) True (b) False.

2. Feeder conductors that supply a group of welders must have an ampacity not less than the sum of the currents, as determined in accordance with 630.11(A) based on _____ percent of the two largest welders, 85 percent for the third largest welder, 70 percent for the fourth largest welder, and 60 percent for all remaining welders.

   (a) 90 (b) 100 (c) 125 (d) 250

3. A disconnecting means must be provided in the supply circuit for each arc welder that is not equipped with _____.

   (a) a governor
   (b) a shunt trip breaker
   (c) an integral disconnect
   (d) ground-fault circuit-interrupter protection

4. The ampacity for the supply conductors for a resistance welder with a duty cycle of 15 percent and a primary current of 21A is _____.

   (a) 9.45A (b) 8.19A (c) 6.72A (d) 5.67A

5. A _____ must be provided to disconnect each resistance welder and its control equipment from the supply circuit.

   (a) switch (b) circuit breaker (c) magnetic starter (d) a or b

Audio Signal Processing, Amplification, and Reproduction Equipment

Introduction

If you understand three major goals of Article 640, you will be able to better understand and apply the requirements. These three goals are:

- Reduce the spread of fire and smoke.
- Comply with other articles.
- Prevent shock. Article 640 includes several requirements, such as specifics in the mechanical execution of work and requirements when audio equipment is located near bodies of water, to reduce shock hazards peculiar to audio equipment installations.

In addition, Article 640 distinguishes between permanent and temporary audio installations. Part II provides requirements for permanent installations, and Part III provides requirements for temporary installations.

PART I. GENERAL

640.1 Scope. Article 640 covers equipment and wiring for audio distribution of sound and public address systems, including temporary audio system installations.

FPN No. 1: Permanently installed distributed audio system locations include, but aren't limited to, restaurants, hotels, business offices, commercial and retail sales environments, churches, and schools. Temporary installations include, but aren't limited to, auditoriums, theaters, stadiums, and outdoor events such as fairs, festivals, circuses, public events, and concerts.

640.2 Definitions.

Abandoned Cable. A cable that isn't terminated and isn't identified for future use with a tag.

Author's Comment: 640.3(A) requires accessible abandoned audio distribution cables to be removed.

Audio System. Within this article, the totality of all equipment and interconnecting wiring used to fabricate a fully functional audio signal processing, amplification, and reproduction system.

640.3 Locations and Other Articles.

(A) Spread of Fire or Products of Combustion. The accessible portion of abandoned audio distribution cables must be removed.

Audio circuits installed through fire-resistant rated walls, partitions, floors, or ceilings must be firestopped using methods approved by the authority having jurisdiction to maintain the fire-resistance rating. Cables installed through fire-rated assemblies must be firestopped with a fire-stop material approved by the authority having jurisdiction in accordance with the specific instructions supplied by the manufacturer for the specific type of cable and construction material (drywall, brick, etc.) [300.21]. Figure 640-1

(B) Air-Handling Spaces. Audio circuits installed in air-handling spaces must be installed in accordance with 300.22.

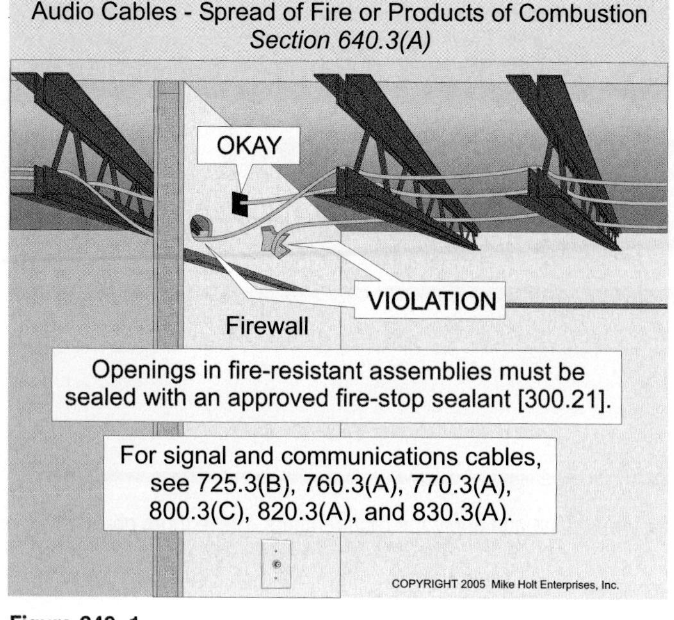
Audio Cables - Spread of Fire or Products of Combustion
Section 640.3(A)
OKAY
VIOLATION
Firewall
Openings in fire-resistant assemblies must be sealed with an approved fire-stop sealant [300.21].
For signal and communications cables, see 725.3(B), 760.3(A), 770.3(A), 800.3(C), 820.3(A), and 830.3(A).
COPYRIGHT 2005 Mike Holt Enterprises, Inc.

Figure 640–1

Author's Comment: Audio cables may be installed in other spaces used for environmental air if the cable is plenum rated [725.3(C) and 725.82(A)]. **Figure 640-2**

(C) Cable Trays. Audio cables installed in cable trays must comply with Article 392.

FPN: See 725.61(C) for the use of Class 2, Class 3, and Type PLTC cable in cable trays. **Figure 640-3**

(D) Hazardous (Classified) Locations. Audio equipment in hazardous (classified) locations must comply with Chapter 5.

640.4 Protection of Electrical Equipment.

Amplifiers, loudspeakers, and other audio equipment must be located or protected against environmental or physical damage that might cause a fire, shock, or personal hazard.

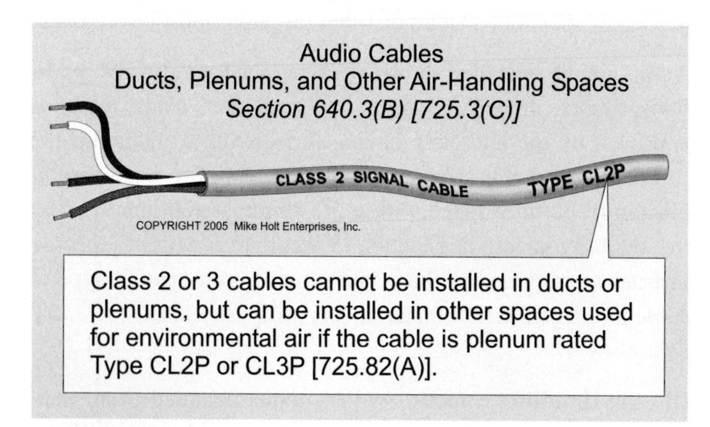
Audio Cables
Ducts, Plenums, and Other Air-Handling Spaces
Section 640.3(B) [725.3(C)]
CLASS 2 SIGNAL CABLE
TYPE CL2P
COPYRIGHT 2005 Mike Holt Enterprises, Inc.
Class 2 or 3 cables cannot be installed in ducts or plenums, but can be installed in other spaces used for environmental air if the cable is plenum rated Type CL2P or CL3P [725.82(A)].

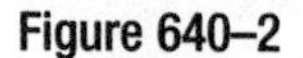

Figure 640–2

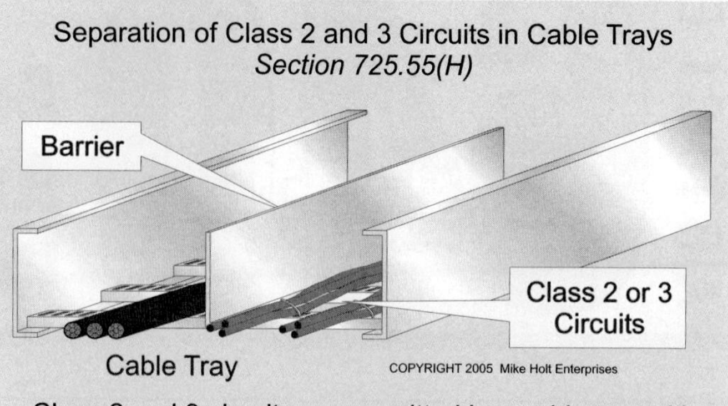
Separation of Class 2 and 3 Circuits in Cable Trays
Section 725.55(H)
Barrier
Class 2 or 3 Circuits
Cable Tray
COPYRIGHT 2005 Mike Holt Enterprises
Class 2 and 3 circuits are permitted in a cable tray with conductors of light, power, Class 1, NPL fire alarm, and medium power network-powered broadband where separated by a barrier, see 725.61(C).

Figure 640–3

640.5 Access to Electrical Equipment.

Equipment access must not be prohibited by an accumulation of cables that prevents the removal of suspended-ceiling panels. **Figure 640-4**

640.6 Mechanical Execution of Work.

Equipment and cabling must be installed in a neat and workmanlike manner.

Exposed audio cables must be supported so that the cable will not be damaged by normal building use. In addition, straps, staples, hangers, or similar fittings must secure audio cables so that the cable will not be damaged. **Figure 640-05**

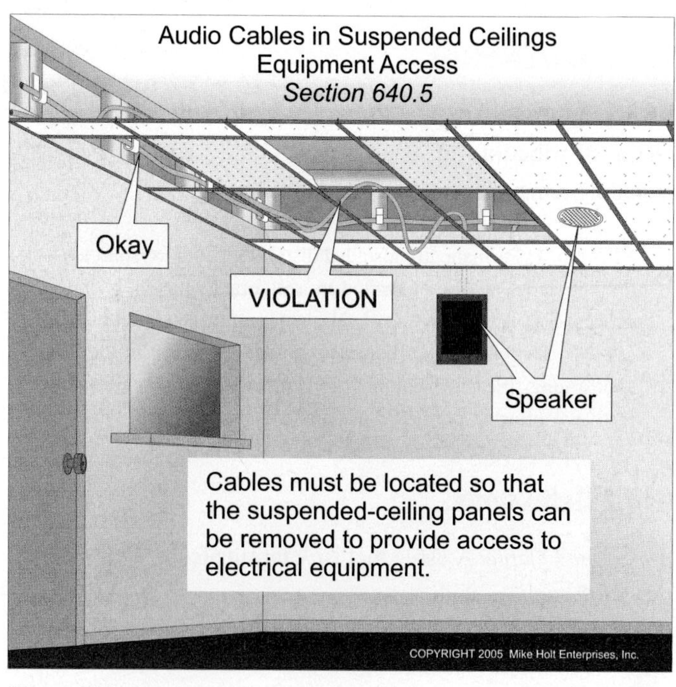
Audio Cables in Suspended Ceilings
Equipment Access
Section 640.5
Okay
VIOLATION
Speaker
Cables must be located so that the suspended-ceiling panels can be removed to provide access to electrical equipment.
COPYRIGHT 2005 Mike Holt Enterprises, Inc.

Figure 640–4

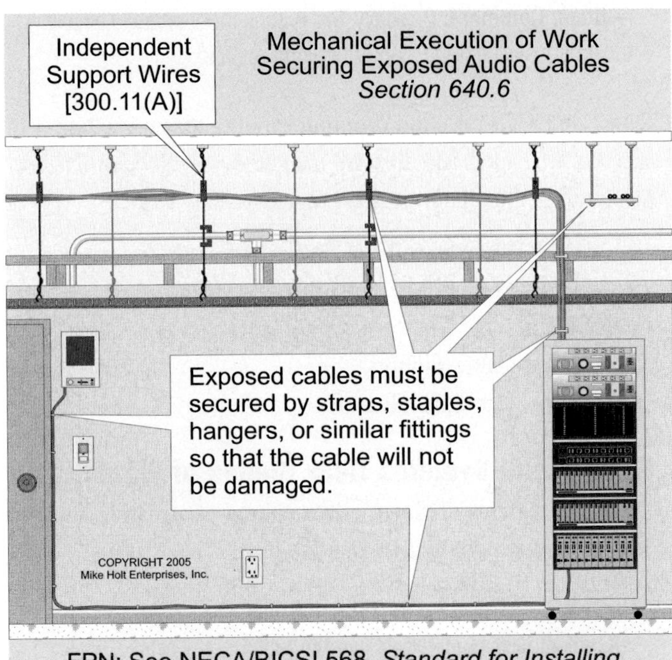

Figure 640–5

Author's Comment: See Article 100 for the definition of "Exposed."

Cables installed parallel to framing members must be protected against physical damage from penetration by screws or nails by 1¼ in. separation from the face of the framing member or by a suitable metal plate, in accordance with 300.4(D). **Figure 640-6**

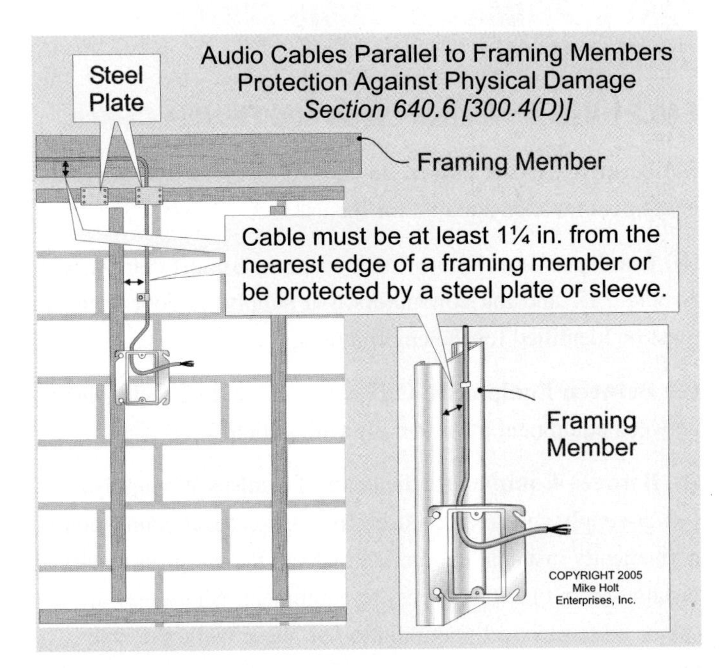

Figure 640–6

FPN: Accepted industry practices are described in ANSI/NECA/BICSI 568, *Standard for Installing Commercial Building Telecommunications Cabling*.

Author's Comment: For more information about this standard, visit http://www.necaneis.org/.

640.7 Grounding (Earthing) and Bonding.

(B) 60V-to-Ground Systems. Separately derived 60/120V systems must be grounded and bonded in accordance with 250.30 and 647.6. **Figure 640-7**

(C) Isolated Ground Receptacles. Isolated ground receptacles, where installed, must comply with 250.146(D). **Figure 640-8**

FPN: See 406.2(D) for the identification requirements for isolated ground-type receptacles.

640.9 Wiring Methods.

(A) Wiring to and Between Audio Equipment.

(1) Input Wiring. Branch-circuit wiring for audio equipment must comply with Chapters 1 through 4.

(2) Separately Derived Power Systems. A separately derived 60V-to-ground system installed in accordance with Article 647 is permitted.

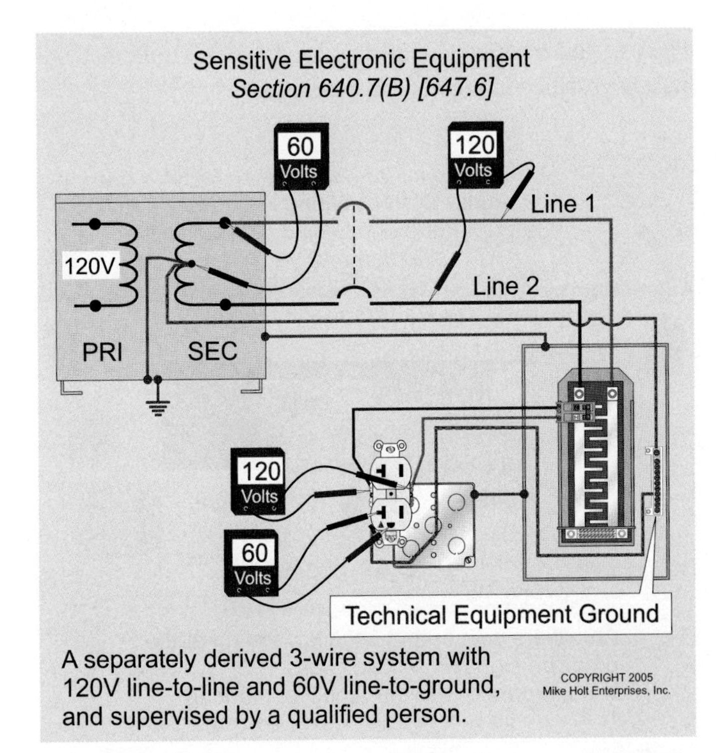

Figure 640–7

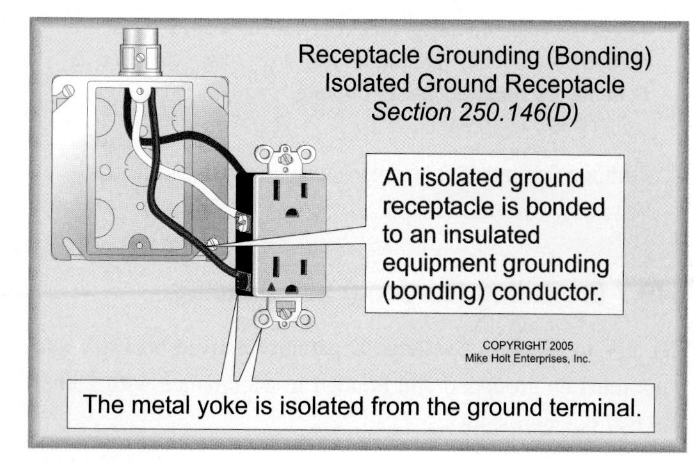

Figure 640–8

(3) Other Wiring. Output wiring not connected to the premises wiring system must be installed in accordance with Article 725.

(C) Output Wiring. Output circuits are permitted to use Class 1, Class 2, or Class 3 wiring methods where the amplifier is marked for use with the specific class of wiring method. Overcurrent protection for the output conductors is required, and it can be inherently limited within the amplifier. **Figure 640-9**

Author's Comment: The term "inherently limited" means that the design of the system (amplifier) is such that overcurrent on the load side of the amplifier will cause the internal windings of the amplifier to burn apart, which in turn opens the circuit.

Class 1. Audio output circuits that use Class 1 wiring methods must be installed in accordance with 725.25.

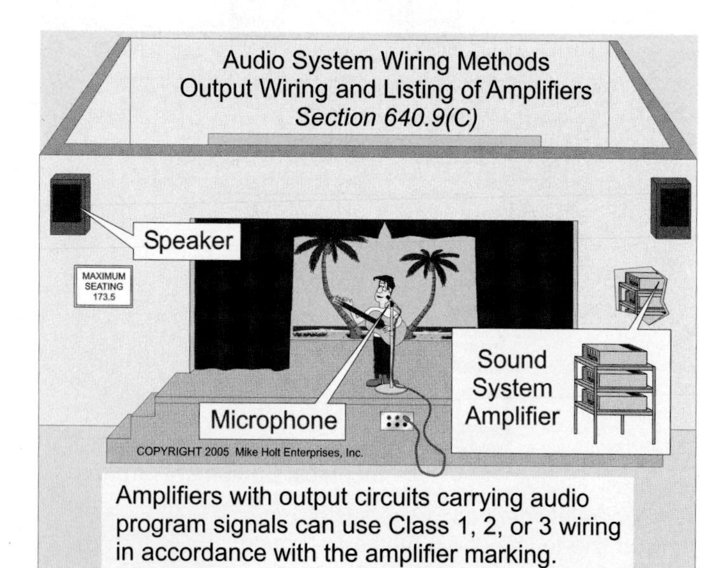

Figure 640–9

Author's Comment: Basically, this means that the audio circuit must be installed in a Chapter 3 wiring method.

Class 2 and Class 3. Audio output circuits that use Class 2 or Class 3 wiring methods must be insulated and marked in accordance with 725.82, and installed in accordance with:

- 725.55—Separation from other conductors
- 725.56—Different systems in same raceway or enclosure
- 725.57—Cable support
- 725.61—Cable installation

640.10 Audio Systems Near Bodies of Water. Audio systems near bodies of water, either natural or artificial, are subject to the restrictions in (A) and (B).

FPN: See 680.27(A) for the requirements of underwater audio equipment near pools, spas, and hot tubs.

(A) Branch-Circuit Power. Audio system equipment must not be located within 5 ft of the inside wall of a pool, spa, hot tub, fountain, or tidal high-water mark.

(B) Class 2 Power Supply. Audio system speakers powered by a listed Class 2 power supply, or by an amplifier listed for use with Class 2 wiring, are restricted in their placement by the manufacturer's recommendations.

PART II. PERMANENT AUDIO SYSTEM INSTALLATIONS

640.21 Use of Flexible Cords and Flexible Cables.

(A) Branch-Circuit Power. Power-supply cords used for audio equipment must be suitable for the use.

(B) Loudspeakers. Cables for loudspeakers must comply with Article 725, and the conductors that supply outdoor speakers must be identified for the environment.

(C) Between Equipment. Cables for distributing audio signals between equipment must comply with Article 725.

(E) Between Equipment Racks and Premises Wiring System. Power-supply cords can be used for the electrical connection of permanently installed equipment racks to the premises electrical installation to facilitate access to equipment, where they are not subjected to physical manipulation or abuse while the rack is in use. **Figure 640-10**

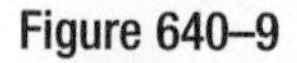

Figure 640–10

640.22 Wiring of Equipment Racks. Metal equipment racks and enclosures must be grounded (bonded) to an effective ground-fault current path in accordance with Part VI of Article 250 with an equipment grounding (bonding) conductor of a type specified in 250.118 [250.4(A)(3)].

640.23 Raceway Fill.

(A) Number of Conductors. The number of conductors permitted in a single conduit or tubing must not exceed the percentage fill specified in Table 1, Chapter 9 [300.17].

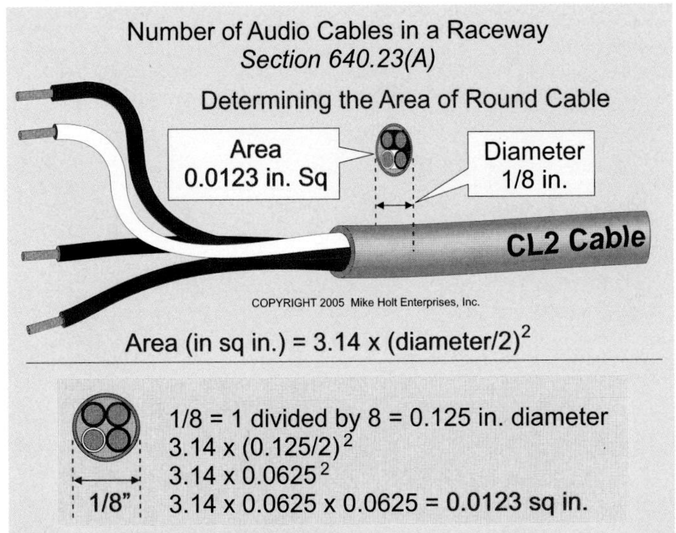

Figure 640–11

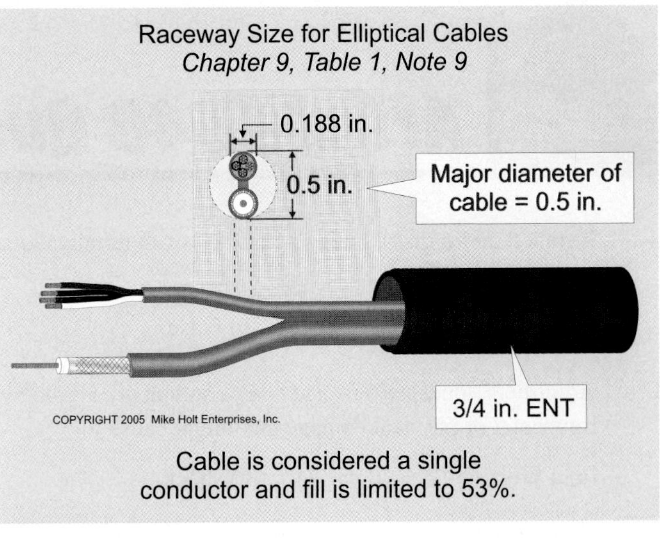

Figure 640–12

Author's Comment: A multiconductor cable is treated as a single conductor for calculating percentage conduit fill area. **Figure 640-11**

For cables that have elliptical cross sections, the cross-sectional area calculation is based on using the major diameter of the ellipse as a circle diameter [Note (9) of Table 1, Chapter 9]. **Figure 640-12**

Author's Comment: Elliptical cables have a cross section that is taller than it is wide. The major (larger) diameter must be used for calculations. Examples of elliptical cables include two-wire (flat) NM cable, or a duplex or Siamese-type low voltage cable as shown in **Figure 640-12**.

640.25 Loudspeakers in Fire-Resistance Rated Partitions, Walls, and Ceilings. Loudspeakers installed in fire-resistance rated partitions, walls, or ceilings must be listed for the purpose or installed in an enclosure or recess that maintains the fire-resistance rating.

Author's Comment: Few (if any) enclosures are listed for this purpose, so "tenting" with fire-rated ceiling panel tiles is required for speakers. Since this requirement is contained in the *National Electrical Code*, the responsibility falls within the scope of the electrical installer.

Article 640 Questions

1. Installed audio distribution cable that is not terminated at equipment and not identified for future use with a tag is considered abandoned.

   (a) True (b) False

2. Amplifiers, loudspeakers, and other equipment covered by Article 640 must be so located or protected so as to guard against environmental or physical damage that might cause _____.

   (a) a fire (b) shock (c) personal hazard (d) all of these

3. Audio system equipment supplied by branch circuit power must not be located within _____ of the inside wall of a pool, spa, hot tub, fountain, or tidal high-water mark.

   (a) 2 ft (b) 10 ft (c) 5 ft (d) 18 in.

4. Flexible cords and flexible cables are not allowed for the electrical connection of permanently installed equipment racks of audio systems to the premises wiring to facilitate access to equipment.

   (a) True (b) False

5. Metal equipment racks and enclosures for permanent audio system installations must be grounded (bonded).

   (a) True (b) False

ARTICLE

645 Information Technology Equipment

Introduction

One of the unique things about Article 645 is the requirement for a shutoff switch readily accessible from the exit doors of information technology equipment rooms [645.10]. This requirement would seem to be wrong on its face because it allows someone to shut off the IT room from a single point. So despite having a UPS and taking every precaution against a power outage, the IT system is still vulnerable to a shutdown from a switch readily accessible at the principal exit doors. What was the *Code*-making panel thinking of when they added this requirement?

They were thinking of fire and rescue teams. Having a means to shut down the power and disconnect the batteries before entering the IT room during a fire allows the rescue team to use fire hoses and other equipment without risking contact with energized equipment. Yes, there's loss of IT function during the shutdown. But if the room needs fire and rescue teams, loss of the IT function is the least of its problems at that time. The shutdown allows the rescue of people and property. A breakaway lock can protect the IT room from inadvertent shutdown via this switch.

What about the rest of Article 645? The major goal is to reduce the spread of fire and smoke. The raised floors common in IT rooms pose special additional challenges to achieving this goal, so this article devotes a fair percentage of its text to raised floor requirements. Fire-resistant walls, separate HVAC systems, and other requirements further help to achieve this goal.

645.1 Scope. Article 645 covers equipment, power-supply wiring, and interconnecting wiring of information technology equipment and systems in an information technology equipment room that meets all of the requirements of 645.4.

Author's Comment: Article 645 applies only to wiring and equipment located within an information technology equipment room, not to all wiring associated with information technology equipment.

645.4 Information Technology Equipment Room. The requirements contained in this article only apply if all of the following conditions are met:

(1) A disconnecting means that complies with 645.10 is provided.

(2) A dedicated heating/ventilating/air-conditioning (HVAC) system is provided for information technology equipment and is separated from other areas of the occupancy.

(3) Only listed information technology equipment is installed in the room.

(4) The information technology equipment room is occupied only by persons needed for the maintenance and operation of information technology equipment.

(5) The information technology equipment room is separated from other occupancies by fire-resistant rated walls, floors, and ceilings with protected openings.

Author's Comment: An information technology equipment room is an enclosed area specifically designed to comply with the construction and fire protection provisions of NFPA 75, *Standard for the Protection of Electronic Computer/Data-Processing Equipment.*

645.5 Supply Circuits and Interconnecting Cables.

(A) Branch-Circuit Conductors. Branch-circuit conductors for data-processing equipment must have an ampacity not less than 125 percent of the total connected load.

> **Author's Comment:** See Article 100 for the definition of "Ampacity."

(B) Connecting Cables. Data-processing equipment can be connected to a branch circuit by:

(1) A flexible cord no longer than 15 ft, with an attachment plug.

(2) A cord set assembly. Where run on the surface of the floor, this cord set assembly must be protected against physical damage.

(C) Interconnecting Cables. Cables listed for information technology equipment can interconnect data-processing units. Where exposed to physical damage, the interconnecting cables must be protected by a means approved by the authority having jurisdiction.

(D) Under Raised Floors. Power cables, communications cables, connecting cables, interconnecting cables, and receptacles associated with the information technology equipment are permitted under a raised floor, provided the following conditions are met:

(1) The raised floor is of suitable construction, and the area under the floor is accessible.

(2) The branch-circuit supply conductors are installed in rigid metal conduit, rigid nonmetallic conduit, intermediate metal conduit, electrical metallic tubing, electrical nonmetallic tubing, metal wireway, nonmetallic wireway, surface metal raceway with metal cover, nonmetallic surface raceway, flexible metal conduit, liquidtight flexible metal conduit, liquidtight flexible nonmetallic conduit, Type MI cable, Type MC cable, or Type AC cable. **Figure 645-1**

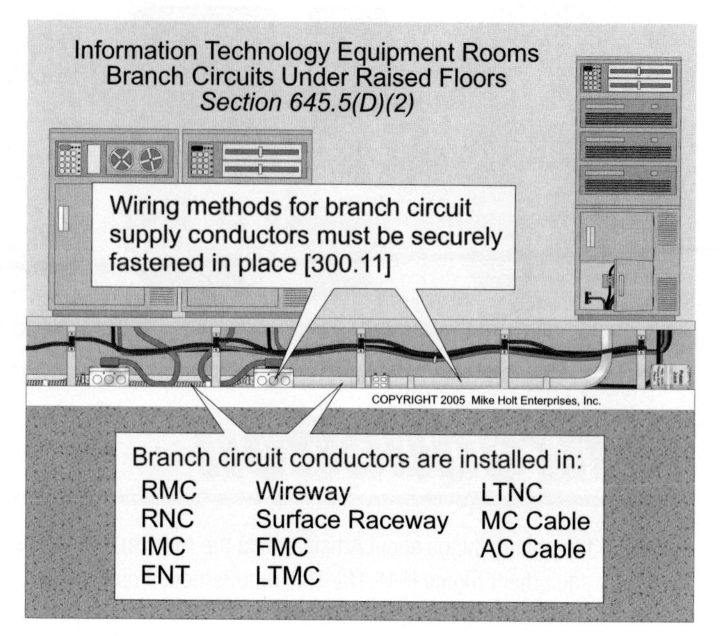

Figure 645–1

> **Author's Comment:** Nonmetallic raceways are permitted within the raised floor area because this space is neither subject to physical damage, nor required to comply with the environmental airspace requirements of 300.22(D).

Branch-circuit wiring methods must be securely fastened in place in accordance with 300.11.

(3) Ventilation in the underfloor area is for the information equipment room only, and it's arranged so that air circulation will cease upon the detection of fire or products of combustion.

(4) Openings in raised floors protect the cables against abrasions.

(5) DP-rated cables except:

(a) Cables enclosed in a raceway.

(b) Interconnecting cables listed with equipment manufactured prior to July 1, 1994, and installed with that equipment.

(c) Signal or communications cables of the following types: Type TC [Article 336]; Types CL2, CL3, and PLTC [Article 725]; Types NPLF and FPL [Article 760]; Types OFC and OFN [Article 770]; Type CM [Article 800]; Type CATV [Article 820].

> **Author's Comment:** Signaling and communications cables installed within the raised floor area in an information technology equipment room aren't required to be plenum rated. **Figure 645-2**

Conductors with green or green with one or more yellow stripes, insulated 4 AWG and larger, and marked "for use in cable trays" or "CT use" are permitted for equipment grounding (bonding) of the raised floor parts.

(6) Abandoned cables must be removed unless they are contained in metal raceways.

(E) Securing in Place. Power, communications, connecting, and interconnecting cables that are part of listed information technology equipment aren't required to be secured in place. **Figure 645-3**

> **Author's Comment:** The wiring methods for the branch-circuit supply conductors must be securely fastened in place in accordance with 300.11 [645.5(D)(2)].

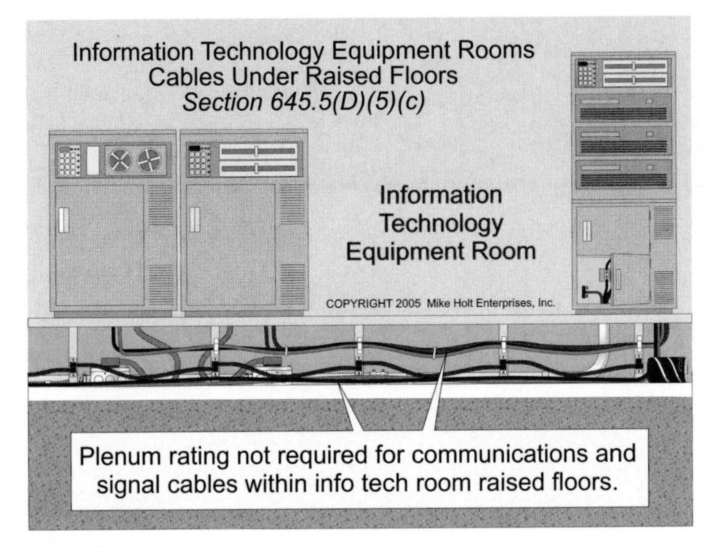
Information Technology Equipment Rooms
Cables Under Raised Floors
Section 645.5(D)(5)(c)
Information Technology Equipment Room
COPYRIGHT 2005 Mike Holt Enterprises, Inc.
Plenum rating not required for communications and signal cables within info tech room raised floors.

Figure 645–2

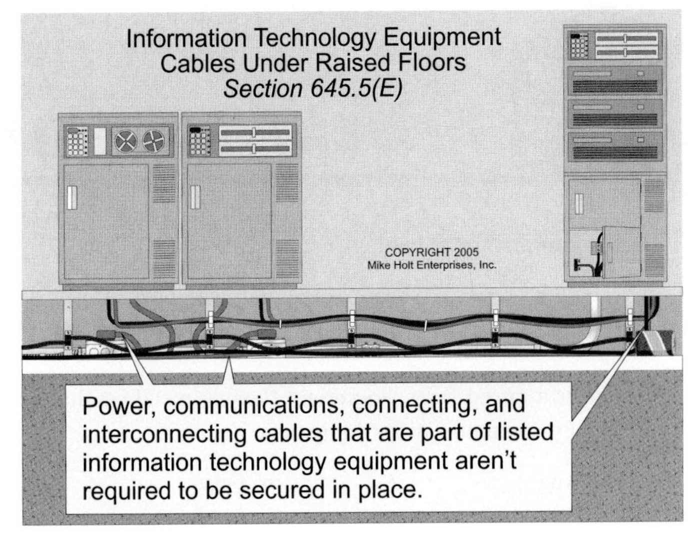
Information Technology Equipment
Cables Under Raised Floors
Section 645.5(E)
COPYRIGHT 2005
Mike Holt Enterprises, Inc.
Power, communications, connecting, and interconnecting cables that are part of listed information technology equipment aren't required to be secured in place.

Figure 645–3

645.6 Cables Not in Information Technology Equipment Room. Signaling and communications cables that extend beyond the information technology equipment room must be installed in accordance with the applicable article.

> **FPN**: For signaling circuits, see Article 725. For fiber optic cable, see Article 770. For communications circuits, see Articles 800 through 820. For fire alarm circuits, see Article 760.

645.7 Penetrations. Wiring systems that penetrate the fire-resistant room boundary must be sealed in accordance with 300.21.

645.10 Disconnecting Means. A disconnecting means must disconnect power to all electronic equipment in the information technology equipment room and dedicated HVAC systems that serve the room.

The control(s) for the disconnecting means must be grouped, identified, and located so as to be readily accessible at the principal exit doors. A single means to control both the electronic equipment and HVAC systems is permitted. If a button is used as a means to disconnect power, pushing the button "in" must disconnect the power.

> **Author's Comment:** The "control" for the disconnect is generally accomplished by the use of a normally open (N.O.) momentary pushbutton located at all principal exit doors. When the emergency button is pressed, the closing contacts open the shunt-trip circuit breaker(s), thereby turning off all power.

645.11 Uninterruptible Power Supplies (UPS). A disconnecting means installed in accordance with 645.10 must disconnect power to all UPS systems within the information technology room. This disconnecting means must disconnect all supply and output circuits, and the battery, from its load.

645.15 Grounding (Bonding). Metal parts of an information technology equipment room must be grounded (bonded) to an effective ground-fault current path, in accordance with Part VI of Article 250, with an equipment grounding (bonding) conductor of a type specified in 250.118 [250.4(A)(3)].

> **FPN No. 2**: Where isolated ground receptacles are installed in the information technology equipment room, they must be grounded (bonded) to an effective ground-fault current path in accordance with 250.146(D) and 406.2(D).

Article 645 Questions

1. An information technology equipment room must have _____.

   (a) a disconnecting means complying with 645.10
   (b) a separate heating/ventilating/air conditioning system
   (c) separation by fire-resistance rated walls, floors, and ceiling
   (d) all of these

2. Under a raised floor, liquidtight flexible metal conduit is permitted to enclose branch-circuit conductors for information technology communications equipment.

   (a) True
   (b) False

3. _____ cables such as CL2, CM, or CATV are permitted within the raised floor area of an information technology equipment room.

   (a) Control
   (b) Signal
   (c) Communications
   (d) all of these

4. Signal and communications cables that extend beyond the information technology equipment room are required to comply only with Article 645.

   (a) True
   (b) False

5. All exposed noncurrent-carrying metal parts of an information technology system must be _____.

   (a) grounded in accordance with Article 250
   (b) double insulated
   (c) fed from GFCI-protected circuits
   (d) a or b

ARTICLE 647 Sensitive Electronic Equipment

Introduction

Power quality problems cause misoperation of electronic equipment in some cases and outright loss in others. Attempts to protect equipment from power quality problems have often taken approaches that endanger people and property. For example, a designer might consider the grounding system to be a source of dirty power and specify that the equipment have a floating ground. Article 647 provides some direction for designers and installers to resolve power quality issues without creating a dangerous situation.

Some people have criticized the title of Article 647, arguing that all electronic equipment is sensitive. The intent of that title is to convey the idea that Article 647 will apply to electronic equipment that the designer, installer, or user considers in need of extra protection from power quality issues. If the title simply said "electronic equipment," it would imply a much broader coverage than what was intended.

People often refer to the voltage-drop "requirement" in 310.15, 210.19, and 215.2. These are actually Fine Print Notes (FPN) that suggest a given circuit voltage-drop limit. The voltage-drop specification in 647.4(D), though, is a requirement, and it's twice as stringent as the suggestion in the FPNs. However, it applies only to sensitive electronic equipment.

Pay attention to the grounding, bonding, and labeling requirements of this article. Power quality problems disappear when an installation is brought into compliance with Articles 250 and 647.

647.1 Scope. Article 647 covers the installation and wiring of a separately derived 60/120V system for sensitive electronic equipment. **Figure 647-1**

> **Author's Comment:** The receptacle configuration in **Figure 647-1** shows a typical NEMA 20A, 125V receptacle. However, all receptacles used for 60/120V technical power must have a unique configuration [647.7].

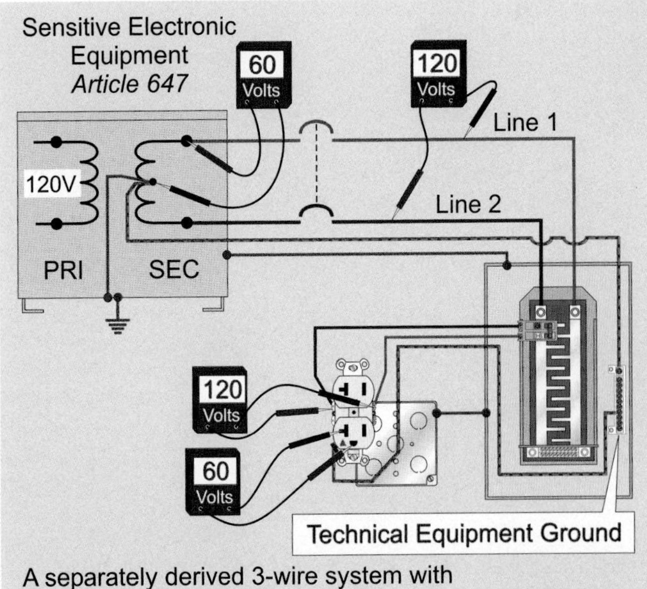

Figure 647–1

647.3 General. The use of a 60/120V system (120V line-to-line) is permitted for the purpose of reducing objectionable noise in sensitive electronic equipment provided:

(1) The system is only permitted in commercial or industrial occupancies.

(2) The system is under close supervision by qualified personnel.

647.4 Wiring Methods.

(A) Panelboards and Overcurrent Protection. Standard panelboards and distribution equipment with higher voltage ratings are permitted. The system must be clearly marked on, or inside, the panelboard cover, and common-trip 2-pole circuit breakers are required to disconnect all ungrounded conductors simultaneously.

(B) Junction Boxes. Junction boxes must be clearly marked to indicate the system voltage (60/120V).

(C) Color Coding. Feeder and branch-circuit conductors for 60/120V systems must be identified by color, marking, tagging, or other effective means, and the means of identification must be posted at each panelboard and disconnecting means.

(D) Voltage Drop. Voltage drop must not exceed 1.5 percent for branch circuits, and the combined voltage drop of feeder and branch-circuit conductors must not exceed 2.5 percent.

647.6 Grounding (Bonding).

(A) System. The power supply must be grounded and bonded as a separately derived system in accordance with 250.30.

(B) Grounding (Bonding) Conductor. Permanently wired utilization equipment and receptacles must be grounded (bonded) to an effective ground-fault current path with a separate conductor run with the circuit conductors to an equipment grounding bus prominently marked "Technical Equipment Ground."

> **FPN No. 1**: When the circuit conductors are increased in size for voltage drop, the size of the equipment grounding (bonding) conductor must also be increased. See 250.122(B).

647.7 Receptacles.

(A) General.

(1) All 15 and 20A, 125V receptacles must be GFCI protected.

(4) Receptacles used for 60/120V technical power must have a unique configuration.

Article 647 Questions

1. Power for sensitive electronic equipment, called "Technical Power" is a separately derived 1Ø, 3-wire system with _____ volts to a grounded neutral conductor on each of two ungrounded conductors. The line-to-line voltage is _____.

   (a) 30, 60
   (b) 60, 120
   (c) 120, 120
   (d) none of these

2. Panelboards for sensitive electronic equipment systems are allowed to be standard single-phase panelboards with the requirements that _____.

   (a) 2 pole common-trip circuit breakers are used
   (b) circuit breakers are identified for operation at the system voltage
   (c) the system is clearly marked on the face of the panel or inside cover
   (d) all of these

3. Junction boxes used in sensitive electronic equipment systems must be clearly marked to indicate _____.

   (a) the installer's name
   (b) the system voltage
   (c) the distribution panel
   (d) b and c

4. Voltage drop on sensitive electronic equipment systems must not exceed _____ percent for branch circuits.

   (a) 1.5
   (b) 3
   (c) 2.5
   (d) 5

5. Permanently-wired utilization equipment and receptacles in sensitive electronic equipment systems must be grounded with a separate equipment grounding conductor run with the circuit conductors to an equipment grounding bus prominently marked _____.

   (a) "Technical Equipment Ground"
   (b) "Equipment Ground Bar"
   (c) "Green"
   (d) "Isolation Bonding"

Notes

Swimming Pools, Spas, Hot Tubs, Fountains, and Similar Installations

Introduction

The overriding concern of Article 680 is to keep people and electricity separated. Some ways in which Article 680 accomplishes this include:

- References to other articles.
- Equipment Requirements. Any equipment that goes into a pool, spa, hot tub, fountain, or similar installation must meet the appropriate requirements of Article 680.
- Equipotential Bonding. This prevents stray currents from taking a path through people.
- Conductor Control. For example, flexible cords cannot be 6 ft long—3 ft is the limit. Article 680 contains many other conductor-related requirements, such as minimum clearances from pool, spa, hot tub, and fountain structures.

PART I. GENERAL REQUIREMENTS FOR PERMANENTLY INSTALLED POOLS, STORABLE POOLS, OUTDOOR SPAS, OUTDOOR HOT TUBS, OR FOUNTAINS

Author's Comment: The requirements contained in Part I of Article 680 apply to permanently installed pools [680.20], storable pools [680.30], outdoor spas and hot tubs [680.42], and fountains [680.50].

680.1 Scope. The requirements contained in Article 680 apply to the installation of electric wiring and equipment for swimming pools, hot tubs, spas, fountains, and hydromassage bathtubs.

680.2 Definitions.

Forming Shell. A structure mounted in the wall of permanently installed pools, storable pools, outdoor spas, outdoor hot tubs, or fountains designed to support a wet-niche luminaire.

Hydromassage Bathtub. A permanently installed bathtub with a recirculating piping system designed to accept, circulate, and discharge water after each use.

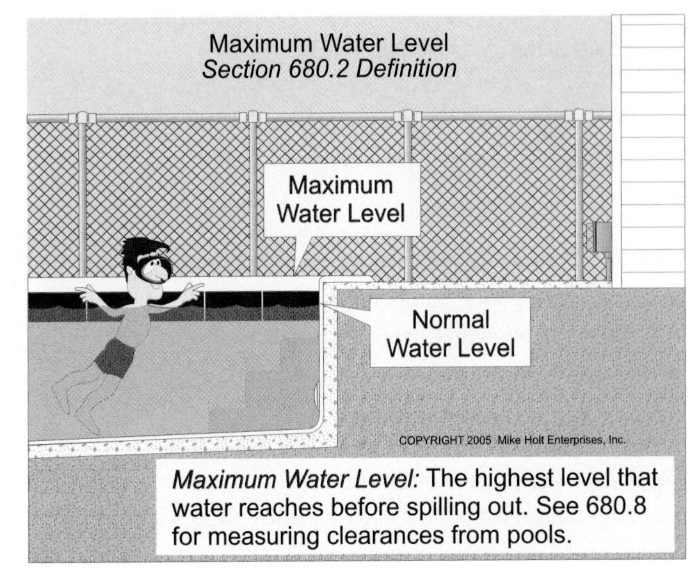

Maximum Water Level
Section 680.2 Definition
Maximum
Water Level
Normal
Water Level
COPYRIGHT 2005 Mike Holt Enterprises, Inc.
Maximum Water Level: The highest level that water reaches before spilling out. See 680.8 for measuring clearances from pools.

Figure 680–1

Maximum Water Level. The highest level that water reaches before it spills out. Figure 680-1

Spa or Hot Tub. A hydromassage pool or tub designed for recreational use that is typically not drained after each use.

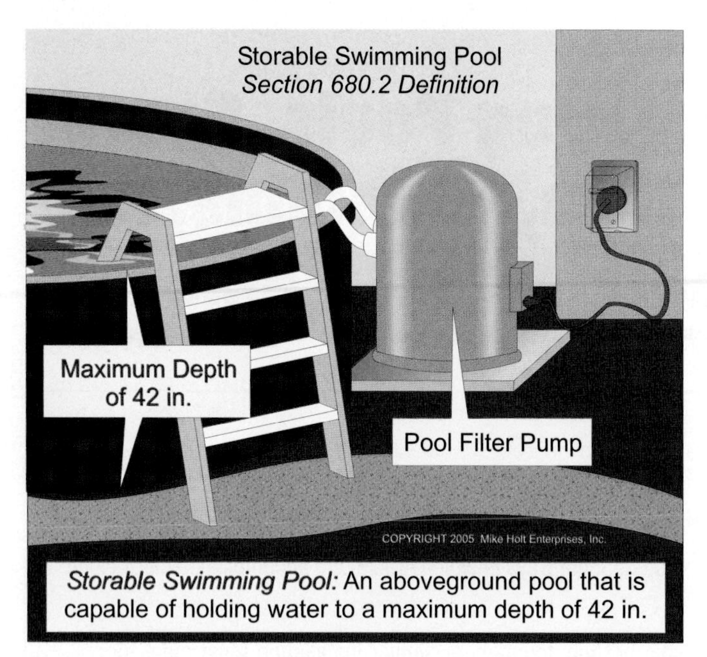

Figure 680–2

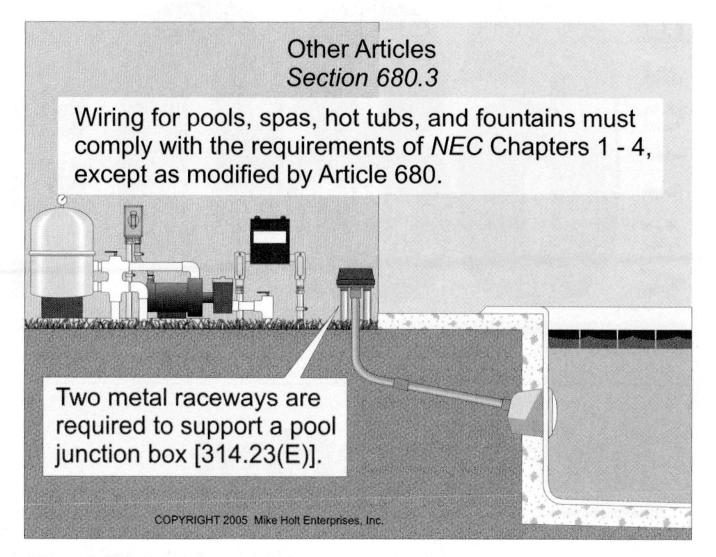

Other Articles
Section 680.3
Wiring for pools, spas, hot tubs, and fountains must comply with the requirements of NEC Chapters 1 - 4, except as modified by Article 680.
Two metal raceways are required to support a pool junction box [314.23(E)].
COPYRIGHT 2005 Mike Holt Enterprises, Inc.

Figure 680–3

Storable Swimming Pool. An aboveground pool with a maximum water depth of 42 in. **Figure 680-2**

> **Author's Comment:** Storable pools are sold as a complete package that consists of the pool walls, vinyl liner, plumbing kit, and pump/filter device. Underwriters Laboratories Inc. (UL) requires the pump/filter units to have a minimum 25-ft cord to discourage the use of extension cords.

Wet-Niche Luminaire. A luminaire intended to be installed in a forming shell where the luminaire will be completely surrounded by water.

680.3 Other Articles. The wiring of permanently installed pools, storable pools, outdoor spas, outdoor hot tubs, or fountains must comply with Chapters 1 through 4, except as modified by this article. **Figure 680-3**

680.7 Cord-and-Plug Connected Equipment. Fixed or stationary equipment other than an underwater luminaire (lighting fixture) for permanently installed pools, storable pools, outdoor spas, outdoor hot tubs, or fountains can be cord-and-plug connected to facilitate the removal or disconnection for maintenance or repair.

(A) Length. Except for storable pools, the cord must not exceed 3 ft.

> **Author's Comment:** There is no maximum cord length specified in the *NEC* for a storable pool pump motor.

(B) Equipment Grounding (Bonding). The cord must have a copper equipment grounding (bonding) conductor not smaller than 12 AWG that terminates at a grounding-type attachment plug.

680.8 Overhead Conductor Clearance. Overhead conductors must meet the clearance requirements contained in Table 680.8. The clearance measurement is to be taken from the maximum water level.

(A) Overhead Power Conductors. Permanently installed pools, storable pools, outdoor spas, outdoor hot tubs, fountains, diving structures, observation stands, towers, or platforms must not be placed under or within 22½ ft of service-drop conductors or open overhead wiring.

> **Author's Comment:** This rule doesn't prohibit utility-owned overhead service-drop conductors from being installed over a permanently installed pool, storable pool, outdoor spa, outdoor hot tub, or fountain [90.2(B)(4)]. It does prohibit a permanently installed pool, storable pool, outdoor spa, outdoor hot tub, or fountain from being installed under an existing service drop.

(B) Communications Systems. Permanently installed pools, storable pools, outdoor spas, outdoor hot tubs, fountains, diving structures, observation stands, towers, or platforms must not be placed under, or within, 10 ft of communications cables. **Figure 680-4**

> **Author's Comment:** This rule doesn't prohibit a utility-owned communications overhead cable from being installed over a permanently installed pool, storable pool, outdoor spa, outdoor hot tub, or fountain [90.2(B)(4)]. It does prohibit a permanently installed pool, storable pool, outdoor spa, outdoor hot tub, or fountain from being installed under an existing communications utility overhead supply.

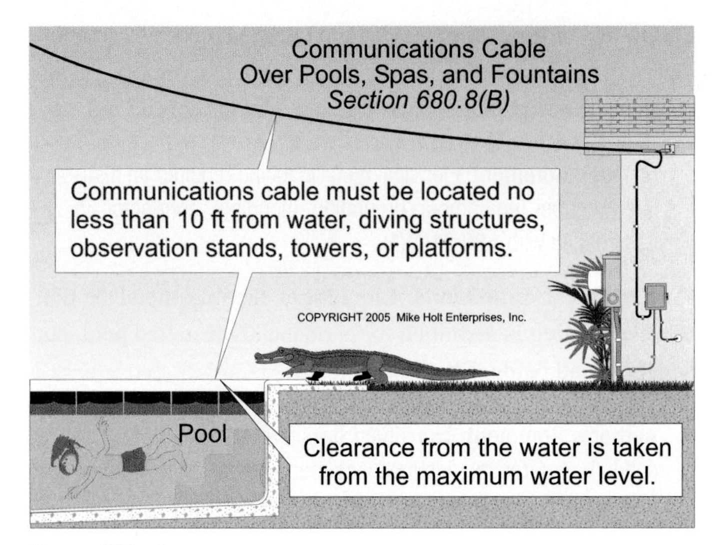
Communications Cable
Over Pools, Spas, and Fountains
Section 680.8(B)
Communications cable must be located no less than 10 ft from water, diving structures, observation stands, towers, or platforms.
COPYRIGHT 2005 Mike Holt Enterprises, Inc.
Pool
Clearance from the water is taken from the maximum water level.

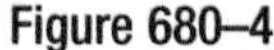
Figure 680–4

(C) Network-Powered Broadband Communications Systems. Permanently installed pools, storable pools, outdoor spas, outdoor hot tubs, or fountains must not be placed under, or within, 22½ ft of network-powered broadband communications cables. Observation stands, towers, or diving platforms must not be placed under, or within, 14½ ft of network-powered broadband communications cables. **Figure 680-5**

> **Author's Comment:** This rule doesn't prohibit a utility-owned broadband overhead cable from being installed over a permanently installed pool, storable pool, outdoor spa, outdoor hot tub, or fountain [90.2(B)(4)]. It does prohibit a permanently installed pool, storable pool, outdoor spa, outdoor hot tub, or fountain from being installed under an existing communications utility overhead supply.

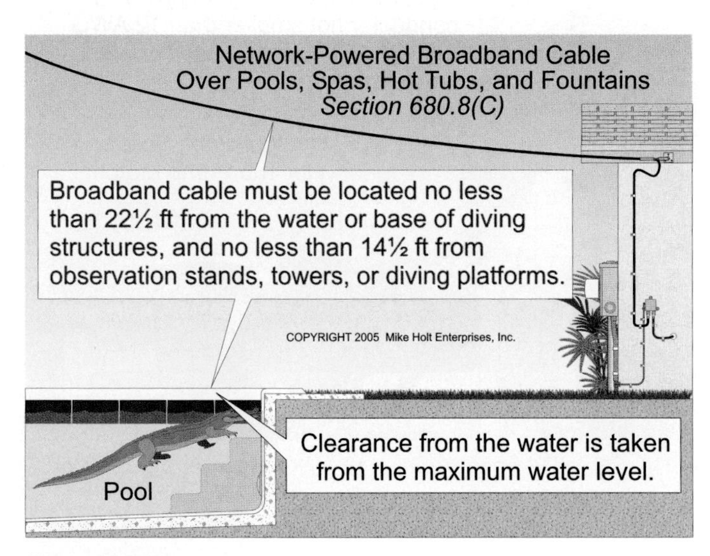
Network-Powered Broadband Cable
Over Pools, Spas, Hot Tubs, and Fountains
Section 680.8(C)
Broadband cable must be located no less than 22½ ft from the water or base of diving structures, and no less than 14½ ft from observation stands, towers, or diving platforms.
COPYRIGHT 2005 Mike Holt Enterprises, Inc.
Pool
Clearance from the water is taken from the maximum water level.

Figure 680–5

680.9 Electric Water Heater. The ampacity of branch-circuit conductors and overcurrent protective devices for pool or outdoor spa and hot tub water heaters cannot be less than 125 percent of the total nameplate rating.

680.10 Underground Wiring Location. Underground wiring isn't permitted under permanently installed pools, storable pools, outdoor spas, outdoor hot tubs, or fountains. Nor is it permitted within 5 ft horizontally from the inside wall of the pool, spa, hot tub, or fountain.

When necessary to supply permanently installed pools, storable pools, outdoor spas, outdoor hot tubs, or fountain equipment, or where space limitations prevent wiring from being at least 5 ft away, underground wiring must be installed in rigid metal conduit, intermediate metal conduit, or a nonmetallic raceway system that is listed for direct burial. The minimum burial depth is 6 in. for metal raceways and 18 in. for nonmetallic raceways.

680.11 Equipment Rooms and Pits. Permanently installed pools, storable pools, outdoor spas, outdoor hot tubs, or fountain equipment cannot be located in rooms or pits that do not have adequate drainage to prevent water accumulation during normal operation or filter maintenance.

680.12 Maintenance Disconnecting Means. A maintenance disconnecting means is required for a permanently installed pool, storable pool, outdoor spa, outdoor hot tub, or fountain equipment, other than lighting for these water bodies. The maintenance disconnecting means must be readily accessible and located within sight from the permanently installed pool, storable pool, outdoor spa, outdoor hot tub, or fountain equipment. **Figure 680-6**

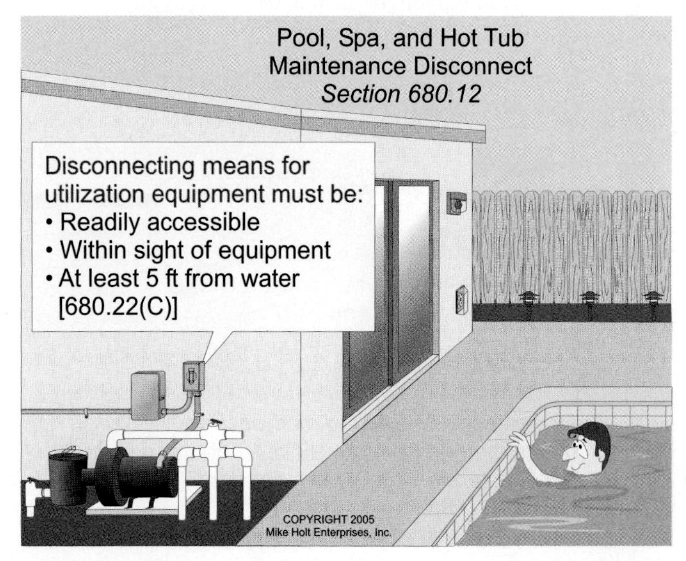
Pool, Spa, and Hot Tub
Maintenance Disconnect
Section 680.12
Disconnecting means for utilization equipment must be:
• Readily accessible
• Within sight of equipment
• At least 5 ft from water [680.22(C)]
COPYRIGHT 2005 Mike Holt Enterprises, Inc.

Figure 680–6

Author's Comment: Disconnects must be located not less than 5 ft horizontally from the inside walls of a permanently installed pool, storable pool, outdoor spa, outdoor hot tub, or fountain, unless separated by a solid fence, wall, or other permanent barrier [680.22(C)].

PART II. PERMANENTLY INSTALLED POOLS, OUTDOOR SPAS, AND OUTDOOR HOT TUBS

Author's Comment: The requirements contained in Part I of Article 680 also apply to permanently installed pools [680.20], outdoor spas, and outdoor hot tubs [680.42].

680.21 Motors.

(A) Wiring Methods.

(1) General. Branch-circuit conductors for permanently installed pool, outdoor spa, and outdoor hot tub motors must be installed in rigid metal conduit, intermediate metal conduit, rigid nonmetallic conduit, or Type MC cable listed for the location (sunlight-resistant or for direct burial). The wiring methods must contain an insulated copper equipment grounding (bonding) conductor sized in accordance with 250.122, but in no case can it be smaller than 12 AWG. **Figure 680-7**

Author's Comment: For interior wiring in one-family dwellings, Types UF, SE, and NM cable with an uninsulated equipment grounding (bonding) conductor are permitted [680.21(A)(4)].

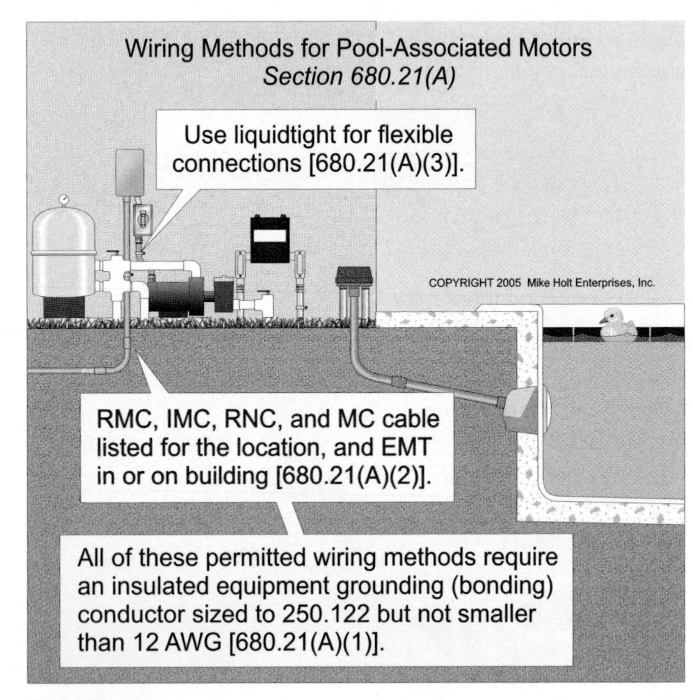

Figure 680–7

(2) On or Within Buildings. Where installed on or within buildings, electrical metallic tubing is permitted to supply permanently installed pool, outdoor spa, and outdoor hot tub motors.

Author's Comment: Electrical metallic tubing requires an insulated copper equipment grounding (bonding) conductor in accordance with 680.21(A)(1).

(3) Flexible Connections. Liquidtight flexible metal or nonmetallic conduit is permitted for permanently installed pool, outdoor spa, and outdoor hot tub motors.

Author's Comment: Liquidtight flexible metal or nonmetallic conduit requires an insulated copper equipment grounding (bonding) conductor in accordance with 680.21(A)(1).

(4) One-Family Dwellings Interior Wiring Method. Any Chapter 3 wiring method is permitted in the interior of one-family dwellings, or in the interior of accessory buildings associated with a one-family dwelling.

Where run in a raceway, the wiring method requires an insulated copper equipment grounding (bonding) conductor [680.21(A)(1)]. **Figure 680-8**

Where run in a cable, Types NM, SE, or UF cable with an uninsulated equipment grounding (bonding) conductor are permitted.

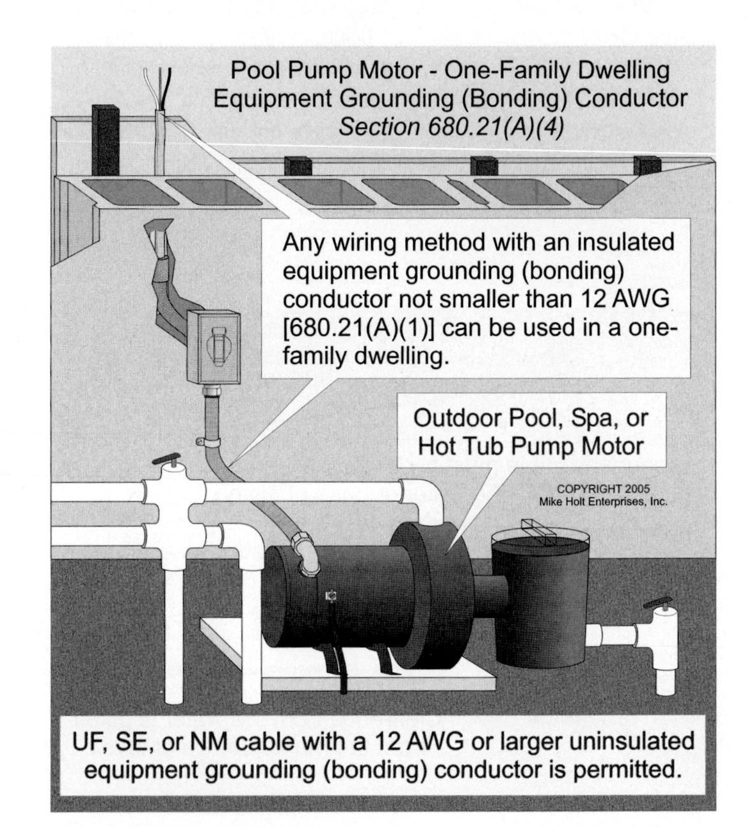

Figure 680–8

Author's Comment: Underground or exterior wiring of a one-family dwelling for permanently installed pool, outdoor spa, and outdoor hot tub motors must be installed in rigid metal conduit, intermediate metal conduit, or rigid nonmetallic conduit with an insulated copper equipment grounding (bonding) conductor in accordance with 680.21(A)(1).

(5) Cord-and-Plug Connections. A cord no longer than 3 ft, with an attachment plug and containing an equipment grounding (bonding) conductor, is permitted for permanently installed pool, outdoor spa, and outdoor hot tub motors.

Author's Comment: For outdoor spas and hot tubs, the cord must be GFCI protected and it can have a length of up to 15 ft [680.42(A)(2)].

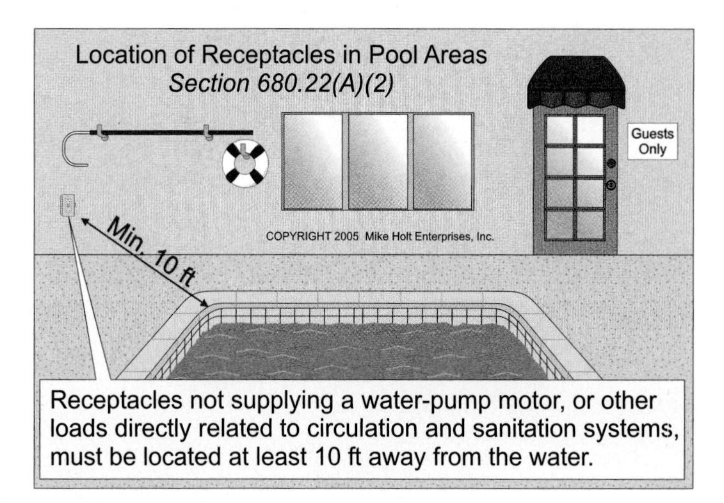

Location of Receptacles in Pool Areas
Section 680.22(A)(2)
Guests Only
Min. 10 ft
COPYRIGHT 2005 Mike Holt Enterprises, Inc.
Receptacles not supplying a water-pump motor, or other loads directly related to circulation and sanitation systems, must be located at least 10 ft away from the water.

Figure 680–10

680.22 Area Lighting, Receptacles, and Equipment.

(A) Receptacles.

(1) Circulation System. Receptacles for permanently installed pool, outdoor spa, and outdoor hot tub motors, or other loads directly related to the circulation system must be located not less than 10 ft from the inside walls, or not less than 5 ft from the water, if the receptacle is a single twist-locking type that is GFCI protected. **Figure 680-9**

(2) Other Receptacles. Receptacles not for motors or other loads directly related to the circulation system must be not less than 10 ft from the water. **Figure 680-10**

(3) Dwelling Unit. At a dwelling unit, one 15 or 20A, 125V receptacle must be located not less than 10 ft and not more than 20 ft from the water from a permanently installed pool, outdoor spa, or outdoor hot tub. This receptacle must be located not more than 6½ ft above the floor, platform, or grade level serving the permanently installed pool, outdoor spa, or outdoor hot tub. **Figure 680-11**

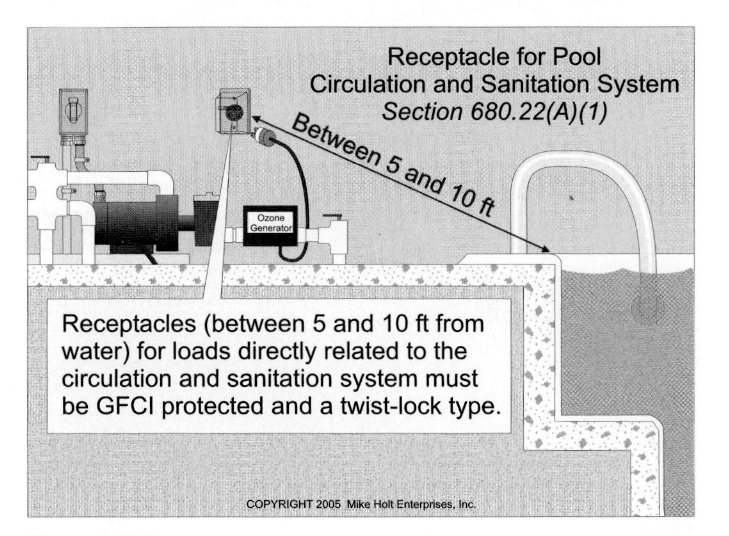

Receptacle for Pool
Circulation and Sanitation System
Section 680.22(A)(1)
Between 5 and 10 ft
Ozone Generator
Receptacles (between 5 and 10 ft from water) for loads directly related to the circulation and sanitation system must be GFCI protected and a twist-lock type.
COPYRIGHT 2005 Mike Holt Enterprises, Inc.

Figure 680–9

(4) Restricted Dwelling Space. Where a permanently installed pool, outdoor spa, or outdoor hot tub is within 10 ft of a dwelling, the receptacle required by 680.22(A)(3) can be located less than 10 ft, but not less than 5 ft measured horizontally from the water. **Figure 680-12**

(5) GFCI-Protected Receptacles. All 15 and 20A, 125V receptacles located within 20 ft of the inside walls of a permanently installed pool, outdoor spa, or outdoor hot tub must be GFCI protected. **Figure 680-13**

Author's Comment: All outdoor dwelling-unit receptacles must be GFCI protected, regardless of the distance from a permanently installed pool, outdoor spa, or outdoor hot tub [210.8(A)(3)]. In addition, all 15 and 20A, 125V receptacles for nondwelling units located outdoors with public access, or accessible to the public, require GFCI protection [210.8(B)(4)].

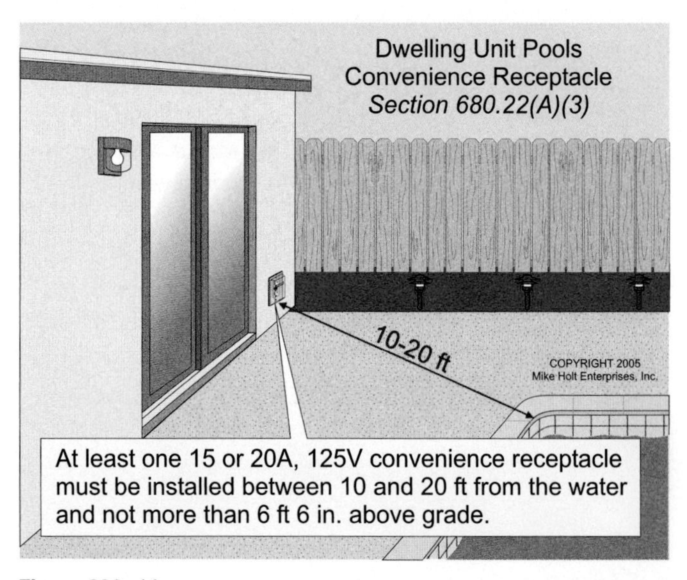

Dwelling Unit Pools
Convenience Receptacle
Section 680.22(A)(3)
10-20 ft
COPYRIGHT 2005 Mike Holt Enterprises, Inc.
At least one 15 or 20A, 125V convenience receptacle must be installed between 10 and 20 ft from the water and not more than 6 ft 6 in. above grade.

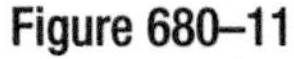

Figure 680–11

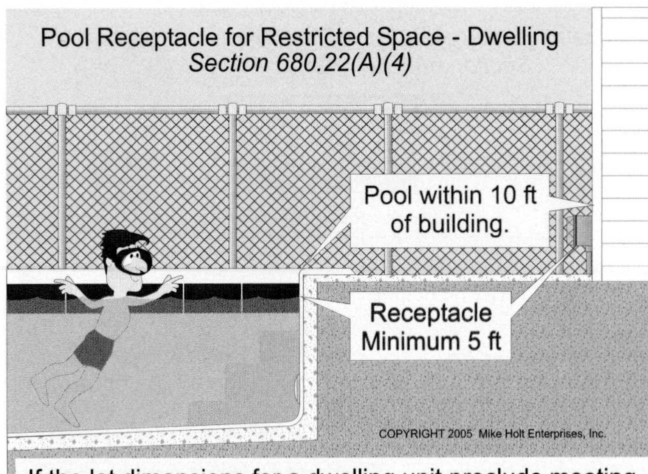

Figure 680–12

All receptacles rated 15 or 20A, 125V through 250V for a permanently installed pool, outdoor spa, or outdoor hot tub supplying cord-and-plug connected pump motors must be GFCI protected. **Figure 680-14**

(6) Measurements. In determining the above dimensions, the distance must be the shortest path a supply cord connected to the receptacle would follow without piercing a floor, wall, ceiling, doorway with hinged or sliding door, window opening, or other effective permanent barrier.

(B) Luminaires and Ceiling Fans.

(1) New Outdoor Installations. Luminaires and ceiling fans installed above the water, or the area extending within 5 ft horizontally from the water, must not be less than 12 ft above the maximum water level.

(3) Existing Installations. Existing luminaires located less than 5 ft horizontally from the water must be not less than 5 ft above the surface of the maximum water level and must be GFCI protected. **Figure 680-15**

(4) Adjacent Areas. New luminaires and ceiling fans installed between 5 ft and 10 ft horizontally, and not more that 5 ft above the maximum water level of a permanently installed pool, outdoor spa, or outdoor hot tub must be GFCI protected.

Author's Comment: Low-voltage lighting systems must not be located within 10 ft of a pool, spa, or hot tub, even if GFCI protected [411.4]. **Figure 680-16**

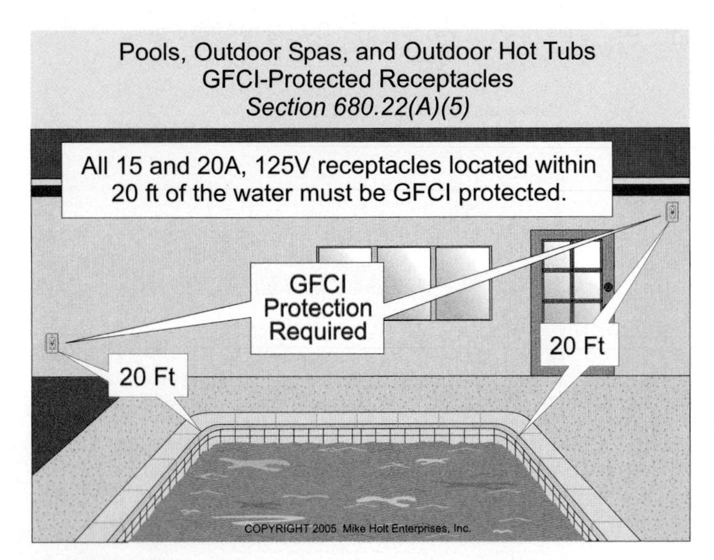

Figure 680–13

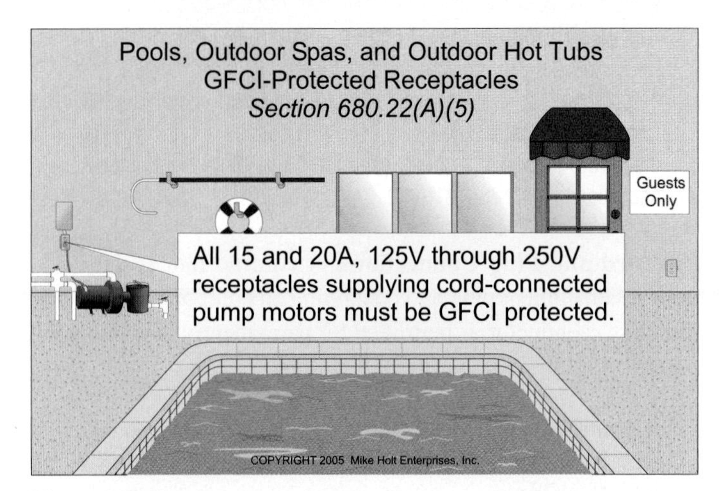

Pools, Outdoor Spas, and Outdoor Hot Tubs
GFCI-Protected Receptacles
Section 680.22(A)(5)
Guests Only
All 15 and 20A, 125V through 250V receptacles supplying cord-connected pump motors must be GFCI protected.
COPYRIGHT 2005 Mike Holt Enterprises, Inc.

Figure 680–14

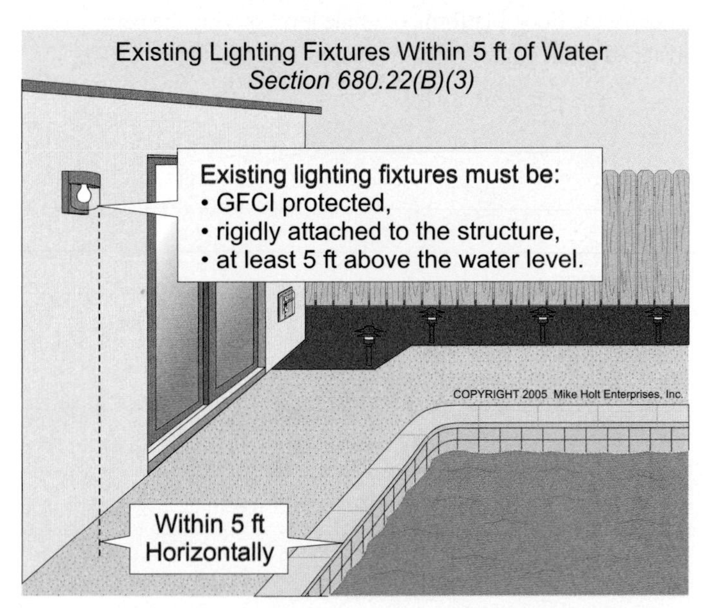

Figure 680–15

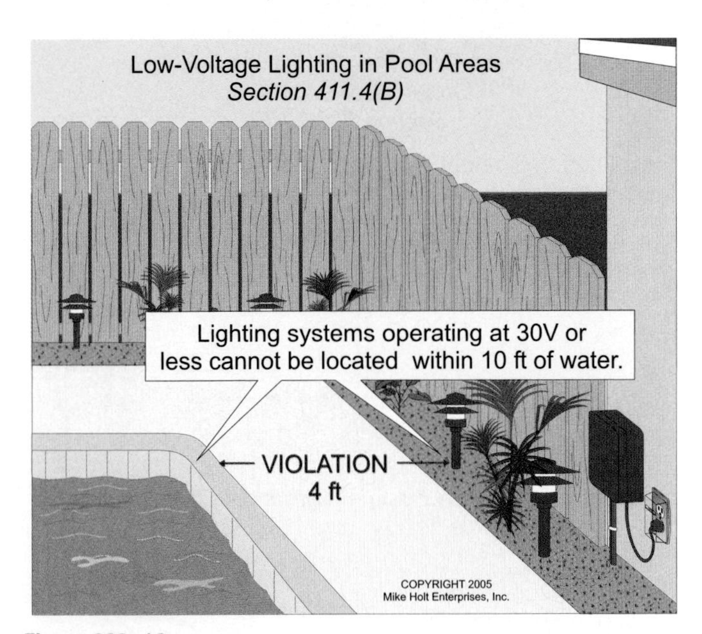
Low-Voltage Lighting in Pool Areas
Section 411.4(B)
Lighting systems operating at 30V or less cannot be located within 10 ft of water.
VIOLATION
4 ft
COPYRIGHT 2005
Mike Holt Enterprises, Inc.

Figure 680–16

(C) **Switching Devices.** Circuit breakers, time clocks, pool light switches, and other switching devices must be located not less than 5 ft horizontally from the inside walls of a permanently installed pool, outdoor spa, or outdoor hot tub unless separated by a solid fence, wall, or other permanent barrier. **Figure 680-17**

680.23 Underwater Luminaires.

(A) General.

(2) Transformers. Transformers for underwater luminaires must be listed as a swimming pool transformer of the isolating winding type and have a grounded metal barrier between the primary and secondary windings.

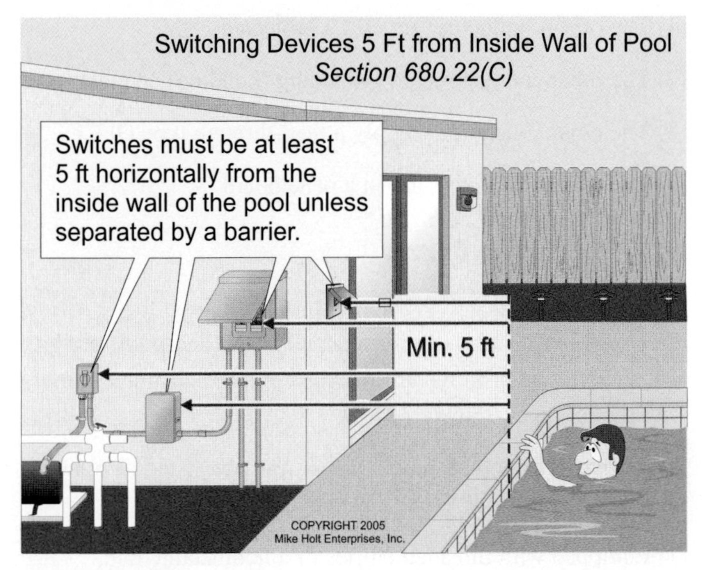
Switching Devices 5 Ft from Inside Wall of Pool
Section 680.22(C)
Switches must be at least 5 ft horizontally from the inside wall of the pool unless separated by a barrier.
Min. 5 ft
COPYRIGHT 2005
Mike Holt Enterprises, Inc.

Figure 680–17

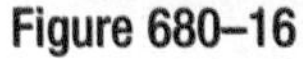

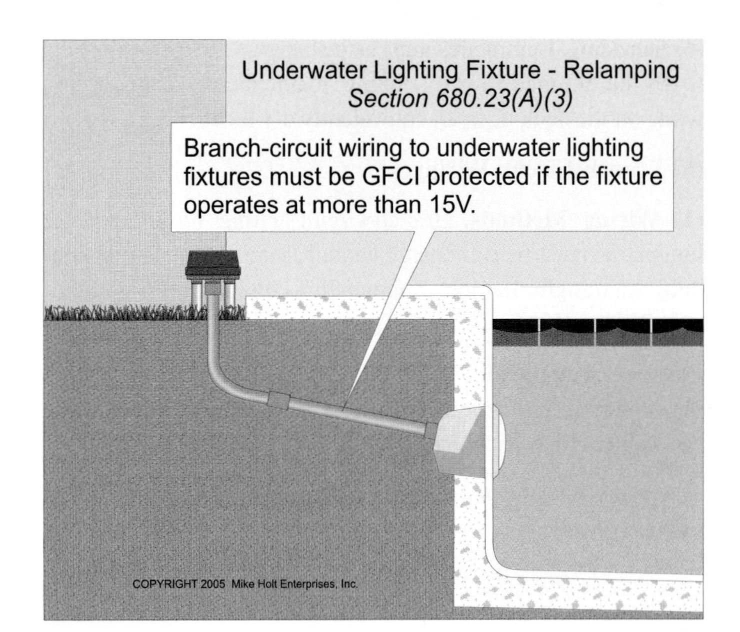
Underwater Lighting Fixture - Relamping
Section 680.23(A)(3)
Branch-circuit wiring to underwater lighting fixtures must be GFCI protected if the fixture operates at more than 15V.
COPYRIGHT 2005 Mike Holt Enterprises, Inc.

Figure 680–18

(3) GFCI Protection of Underwater Luminaires. Branch circuits that supply underwater luminaires operating at more than 15V must be GFCI protected. **Figure 680-18**

(5) Wall-Mounted Luminaires. Underwater luminaires must be installed so that the top of the luminaire lens isn't less than 18 in. below the normal water level.

> **Author's Comment:** The 18 in. requirement reduces the likelihood that persons hanging on the side of the pool will have their chest cavity in line with the underwater luminaire.

(B) Wet-Niche Underwater Luminaires.

(1) Forming Shells. Forming shells for wet-niche underwater luminaires must be equipped with provisions for conduit entries. All forming shells used with nonmetallic conduit systems must include provisions for terminating an 8 AWG copper conductor.

(2) Wiring to the Forming Shell. The conduit that extends directly to the underwater pool wet-niche forming shell must comply with (a) or (b).

(a) Metal Conduit. Brass or corrosion-resistant rigid metal approved by the authority having jurisdiction.

(b) Nonmetallic Conduit. Nonmetallic conduit containing an 8 AWG insulated (solid or stranded) copper bonding jumper, which must terminate in the forming shell and junction box. The termination of the 8 AWG bonding jumper in the forming shell must be covered with, or encapsulated in, a listed potting compound to protect the connection from the possible deteriorating effect of pool water.

(6) Servicing. Luminaires must be installed so that personnel can reach the luminaire for relamping, maintenance, or inspection while on the deck or in an equivalently dry location.

(F) Branch-Circuit Wiring.

(1) Wiring Methods. Branch-circuit wiring for underwater luminaires must be rigid metal conduit, intermediate metal conduit, liquidtight flexible nonmetallic conduit, or rigid nonmetallic conduit. See 680.23(B)(2).

Electrical metallic tubing can be installed on a building. Where installed within a building, electrical nonmetallic tubing, Type MC cable, or electrical metallic tubing is permitted.

Exception: Where connecting to transformers for pool lights, liquidtight flexible metal conduit or liquidtight flexible nonmetallic conduit is permitted in individual lengths not exceeding 6 ft.

(2) Equipment Grounding (Bonding). Branch-circuit conductors for an underwater luminaire must contain an insulated copper equipment grounding (bonding) conductor sized in accordance with Table 250.122, but not smaller than 12 AWG. **Figure 680-19**

The equipment grounding (bonding) conductor for the underwater luminaire must not be spliced, except as permitted in (a) or (b). **Figure 680-20**

(a) Where more than one underwater luminaire is supplied by the same branch circuit, the equipment grounding (bonding) conductor can terminate at a listed pool junction box that meets the requirements of 680.24(A).

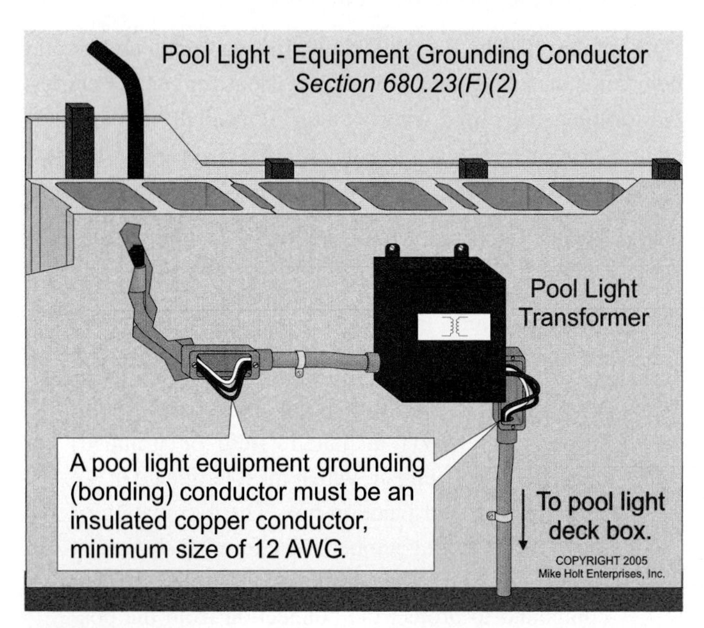
Pool Light - Equipment Grounding Conductor
Section 680.23(F)(2)
Pool Light Transformer
A pool light equipment grounding (bonding) conductor must be an insulated copper conductor, minimum size of 12 AWG.
To pool light deck box.
COPYRIGHT 2005 Mike Holt Enterprises, Inc.

Figure 680–19

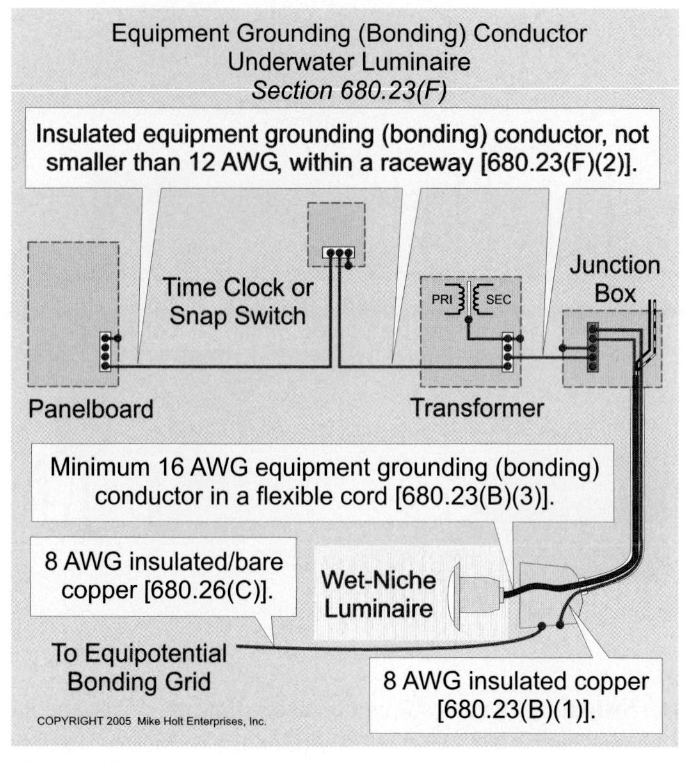
Equipment Grounding (Bonding) Conductor
Underwater Luminaire
Section 680.23(F)
Insulated equipment grounding (bonding) conductor, not smaller than 12 AWG, within a raceway [680.23(F)(2)].
Time Clock or Snap Switch
PRI
SEC
Junction Box
Panelboard
Transformer
Minimum 16 AWG equipment grounding (bonding) conductor in a flexible cord [680.23(B)(3)].
8 AWG insulated/bare copper [680.26(C)].
Wet-Niche Luminaire
To Equipotential Bonding Grid
8 AWG insulated copper [680.23(B)(1)].
COPYRIGHT 2005 Mike Holt Enterprises, Inc.

Figure 680–20

(b) The equipment grounding (bonding) conductor can terminate at the grounding terminal of a listed pool transformer that meets the requirements of 680.23(A)(2).

(3) Conductors. The branch-circuit conductors for the underwater luminaire must not occupy raceways, boxes, or enclosures containing other conductors on the load side of a GFCI or transformer, unless one of the following conditions applies:

(1) The other conductors are GFCI protected.

(2) The other conductors are grounding (bonding) conductors.

(3) The other conductors supply a feed-through type GFCI.

(4) The other conductors are in a panelboard.

680.24 Junction Boxes.

(A) Junction Boxes. The junction box (deck box) that connects directly to an underwater permanently installed pool, outdoor spa, or outdoor hot tub luminaire forming shell must comply with the following: **Figure 680-21**

(1) Construction. The junction box must be listed as a swimming pool junction box and must be:

(1) Equipped with threaded entries or a nonmetallic hub,

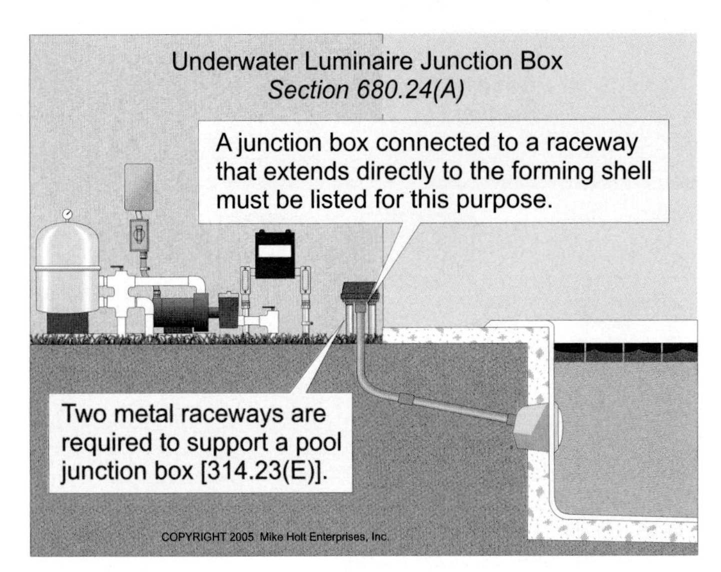

Figure 680–21

(2) Constructed of copper, brass, or corrosion-resistant material approved by the authority having jurisdiction, and

(3) Provided with electrical continuity between all metal conduit and the grounding (bonding) terminals within the junction box.

Author's Comment: In addition, the junction box must be provided with at least one more grounding (bonding) terminal than the number of conduit entries [680.24(D)], and it must be provided with a strain relief [680.24(E)].

(2) Installation. Where the luminaire operates at over 15 volts, the junction box location must comply with (a) and (b).

(a) Vertical Spacing. The junction box must be located not less than 4 in. above the ground or permanently installed pool, outdoor spa, or outdoor hot tub deck, or not less than 8 in. above the maximum water level.

(b) Horizontal Spacing. The junction box must be located not less than 4 ft from the inside wall of the permanently installed pool, outdoor spa, or outdoor hot tub, unless separated by a solid fence, wall, or other permanent barrier.

Author's Comment: The underwater luminaire junction box must be supported by two metal conduits threaded wrenchtight into the enclosure [314.23(E)].

(B) Transformer or Ground-Fault Circuit-Interrupter Enclosure. Where the enclosure for a transformer or GFCI is connected to a conduit that extends directly to an underwater permanently installed pool, outdoor spa, or outdoor hot tub luminaire forming shell, the enclosure must comply with the following:

(1) Construction. The enclosure must be listed and labeled for the purpose, and be:

(1) Equipped with threaded entries or a nonmetallic hub,

(2) Constructed of copper, brass, or corrosion-resistant material approved by the authority having jurisdiction, and

(4) Provide electrical continuity between all metal conduit and the grounding (bonding) terminals of the enclosure.

Author's Comment: See Article 100 for the definitions of "Labeled" and "Listed."

(C) Physical Protection. Junction boxes for underwater pool, spa, or hot tub luminaires must not be located in the walkway unless afforded protection by being located under diving boards or adjacent to fixed structures.

(D) Grounding (Bonding) Terminals. The junction box for an underwater permanently installed pool, outdoor spa, or outdoor hot tub luminaire must be provided with at least one more grounding (bonding) terminal than the number of conduit entries.

Author's Comment: Typically, there are four grounding (bonding) terminals in the junction box and three conduit entries.

(E) Strain Relief. The termination of a flexible cord that supplies an underwater permanently installed pool, outdoor spa, or outdoor hot tub luminaire must be provided with a strain relief.

680.25 Feeders

(A) Wiring Methods. Feeder conductors to panelboards containing permanently installed pool, outdoor spa, or outdoor hot tub equipment circuits must be installed in rigid metal conduit, intermediate metal conduit, liquidtight flexible nonmetallic conduit, or rigid nonmetallic conduit. Electrical metallic tubing is permitted where installed on or within a building, and electrical nonmetallic tubing is permitted where installed within a building.

Exception: Branch circuits for permanently installed pool, outdoor spa, or outdoor hot tub equipment can originate from an existing panelboard supplied by a cable assembly that includes an equipment grounding (bonding) conductor within its outer sheath.

(B) Grounding (Bonding). Except for existing feeders [680.25(A) Ex], an insulated equipment grounding (bonding) conductor must be installed with the feeder conductors between the grounding terminal of the permanently installed pool, outdoor spa, or outdoor hot tub equipment panelboard and the grounding terminal of the service equipment.

(1) Size. This feeder equipment grounding (bonding) conductor must be sized in accordance with 250.122, but not smaller than 12 AWG.

(2) Separate Buildings. Where a feeder is run to a separate building or structure to supply permanently installed swimming pool, outdoor spa, or outdoor hot tub equipment, an insulated equipment grounding (bonding) conductor must be installed with the feeder conductors to the disconnecting means in the separate building or structure [250.32(B)(1)].

680.26 Equipotential Bonding.

(A) Performance. Equipotential (stray voltage) bonding is intended to reduce voltage gradients in a permanently installed pool, outdoor spa, or outdoor hot tub area by forming a common bonding grid.

Author's Comment: Equipotential (stray voltage) bonding isn't intended to provide a low-impedance ground-fault current path to help assist in clearing a ground fault. The topic of stray voltage is beyond the scope of this textbook. For more information, visit www.MikeHolt.com, click on the Technical link, then the Stray Voltage link.

FPN: The 8 AWG or larger solid copper equipotential (stray voltage) bonding conductor [680.26(C)] isn't required to extend to or be attached to any panelboard, service equipment, or an electrode.

(B) Bonded Parts. The following parts of a permanently installed pool, outdoor spa, or outdoor hot tub must be bonded together.

Author's Comment: See 680.42(B) for the bonding methods permitted for outdoor spas and hot tubs.

(1) Metallic Parts of Structure. All metallic parts of the water structure, including the reinforcing metal of the permanently installed pool, outdoor spa, or outdoor hot tub shell and deck. The usual steel tie-wires are considered suitable for bonding the reinforcing steel together. Welding or special clamping is not required, but the tie-wires must be made tight. **Figure 680-22**

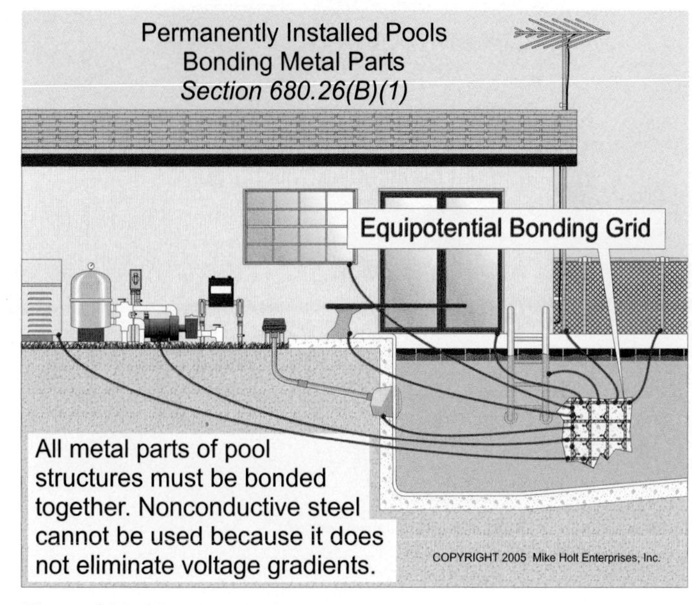

Permanently Installed Pools
Bonding Metal Parts
Section 680.26(B)(1)
Equipotential Bonding Grid
All metal parts of pool structures must be bonded together. Nonconductive steel cannot be used because it does not eliminate voltage gradients.
COPYRIGHT 2005 Mike Holt Enterprises, Inc.

Figure 680–22

Where the reinforcing steel of the permanently installed pool, outdoor spa, or outdoor hot tub shell and deck is encapsulated with a nonconductive compound, or if it's not available, provisions must be made for an alternative means to eliminate voltage gradients that would otherwise be provided by unencapsulated, bonded reinforcing steel

Author's Comment: This means that an equipotential (stray voltage) grid constructed in accordance with 680.26(C) must be installed.

(2) Underwater Lighting. All metal forming shells for underwater permanently installed pool, outdoor spa, or outdoor hot tub luminaires and speakers.

(3) Metal Fittings. Metal fittings within or attached to the permanently installed pool, outdoor spa, or outdoor hot tub structure, such as ladders and handrails.

(4) Electrical Equipment. Metal parts of electrical equipment associated with the permanently installed pool, outdoor spa, or outdoor hot tub water circulating system, such as water heaters and pump motors.

Where a double-insulated water-pump motor is installed, a solid 8 AWG copper conductor from the bonding grid must be provided for a replacement motor.

(5) Metal Wiring Methods and Equipment. Metal-sheathed cables and raceways, metal piping, and all fixed metal parts, as well as metallic surfaces of electrical equipment if located:

(1) Within 5 ft horizontally of the inside walls of a permanently installed pool, outdoor spa, or outdoor hot tub, and

(2) Within 12 ft measured vertically above the maximum water level of a permanently installed pool, outdoor spa, or outdoor hot tub, or any observation stands, towers, platforms, or any diving structures.

(C) Equipotential Grid. A solid copper conductor not smaller than 8 AWG must be used to bond the metallic parts of a permanently installed pool, outdoor spa, or outdoor hot tub as specified in 680.26(B) to an equipotential (stray voltage) grid. The termination of the bonding conductor must be made by exothermic

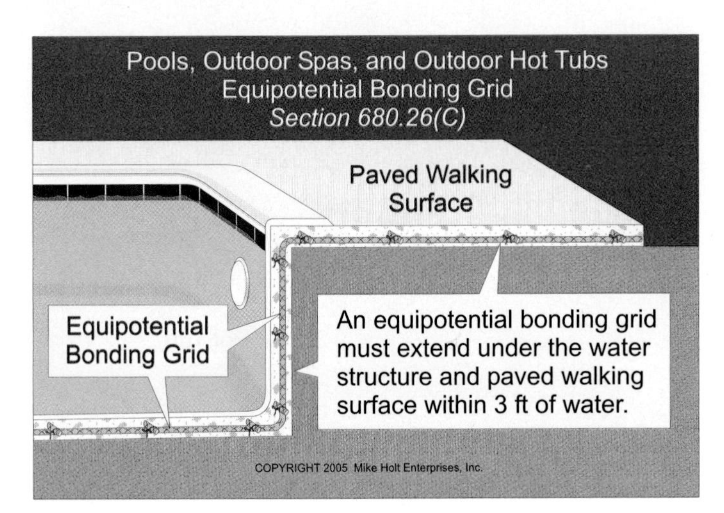
Pools, Outdoor Spas, and Outdoor Hot Tubs
Equipotential Bonding Grid
Section 680.26(C)
Paved Walking Surface
Equipotential Bonding Grid
An equipotential bonding grid must extend under the water structure and paved walking surface within 3 ft of water.
COPYRIGHT 2005 Mike Holt Enterprises, Inc.

Figure 680–23

welding, listed pressure connectors, or listed clamps that are suitable for the purpose.

To properly mask stray voltage, an equipotential (stray voltage) grid must extend under walking surfaces for 3 ft horizontally from the water, **Figure 680-23**. The equipotential (stray voltage) grid must be formed from one or more of the following:

(1) Structural Reinforcing Steel. Structural reinforcing steel of the concrete permanently installed pool, outdoor spa, or outdoor hot tub.

(2) Bolted or Welded Metal Pools. The walls of a bolted or welded metal permanently installed pool, outdoor spa, or outdoor hot tub.

(3) Other Methods. The equipotential (stray voltage) grid can be constructed as specified in (a) through (c) below.

(a) Materials and Connections. The equipotential (stray voltage) grid can be constructed with 8 AWG bare solid copper conductors that are bonded to each other at all points of crossing.

(b) Grid. The equipotential (stray voltage) grid must cover the contour of the permanently installed pool, outdoor spa, or outdoor hot tub, and deck extending 3 ft horizontally from the water. The equipotential (stray voltage) grid must be arranged in a 1 ft x 1 ft network of conductors in a uniformly spaced perpendicular grid pattern with a tolerance of 4 in.

(c) Securing. The equipotential (stray voltage) grid must be secured.

(D) Connections. Where structural reinforcing steel or the walls of bolted or welded metal permanently installed pool, outdoor spa, or outdoor hot tub structures are used as an equipotential (stray voltage) grid for nonelectrical parts, the connections must be made in accordance with 250.8.

680.27 Specialized Equipment.

(B) Electrically Operated Covers.

(1) Motors and Controllers. The electric motors, controllers, and wiring for an electrically operated cover must be located not less than 5 ft from the inside wall of a permanently installed pool, outdoor spa, or outdoor hot tub, unless separated by a permanent barrier.

(2) Wiring Methods. The electric motor and controller circuit must be GFCI protected.

PART III. STORABLE SWIMMING POOL

680.30 General. Electrical installations for storable pools must also comply with Part I of Article 680.

> **Author's Comment:** The requirements contained in Part I of Article 680 include the location of switches, receptacles, and luminaires.

680.32 GFCI-Protected Receptacles. GFCI protection is required for all electrical equipment, including power-supply cords, used with storable pools.

GFCI protection is required for all 15 or 20A, 125V receptacles located within 20 ft of the inside walls of a storable pool or that supply storable pool pump motors. **Figure 680-24**

Storable Pools
GFCI-Protected Receptacles
Section 680.32
COPYRIGHT 2005 Mike Holt Enterprises, Inc.
GFCI protection is required for all 15 and 20A, 125V receptacles located within 20 ft of a storable swimming pool.

Figure 680–24

The measured distance is the shortest path a supply cord connected to the receptacle would follow without piercing a floor, wall, ceiling, doorway with hinged or sliding door, window opening, or other effective permanent barrier.

Author's Comment: This requirement mirrors the requirements contained in 680.25(A)(5) and (6) for permanently installed pools.

680.34 Receptacle Locations. Receptacles must not be located less than 10 ft from the inside walls of a storable pool. **Figure 680-25**

The measured distance is the shortest path a supply cord connected to the receptacle would follow without piercing a floor, wall, ceiling, doorway with hinged or sliding door, window opening, or other effective permanent barrier.

Author's Comment: This rule mirrors the requirements contained in 680.22(A)(1) for permanently installed pools.

PART IV. SPAS AND HOT TUBS

680.40 General. Electrical installations for spas and hot tubs must comply with Part I as well.

680.41 Emergency Switch for Spas and Hot Tubs. In other than a single-family dwelling, a clearly labeled emergency spa or hot tub water recirculation and jet system shutoff must be supplied. The emergency shutoff must be readily accessible to the users and located not less than 5 ft away, but adjacent to and within sight of the spa or hot tub. **Figure 680-26**

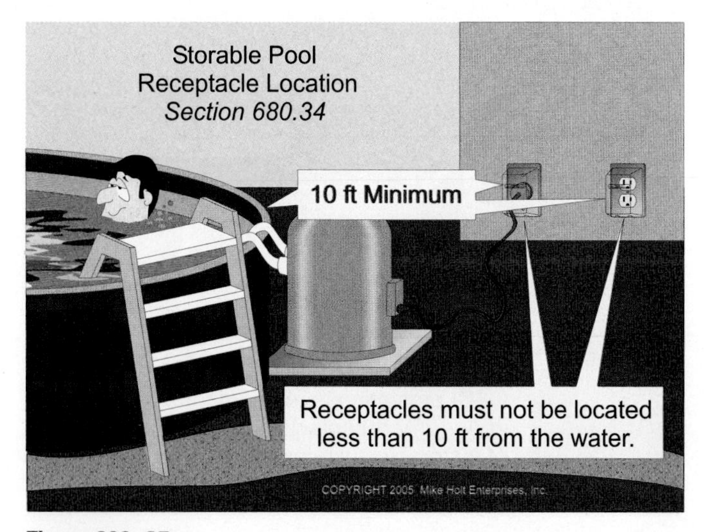

Figure 680–25

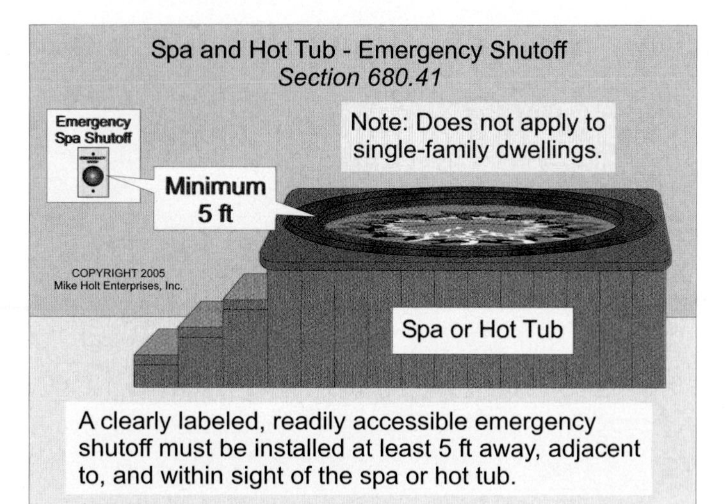

Figure 680–26

Author's Comments:

- Either the maintenance disconnecting means required by 680.12 or a pushbutton that controls a relay located in accordance with this section can be used to meet the emergency shutoff requirement.
- The purpose of the emergency shutoff is to protect users. Deaths and injuries have occurred in less than 3 ft of water because individuals became stuck to the water intake opening. This requirement applies to spas and hot tubs installed indoors as well as outdoors.

680.42 Outdoor Installations. Electrical installations for outdoor spas or hot tubs must comply with Parts I and II of this article, except as permitted in the following:

(A) Flexible Connections. Listed packaged spa or hot tub equipment assemblies or self-contained spas or hot tubs are permitted to use flexible connections as follows:

(1) Flexible Conduit. Liquidtight flexible metal conduit or liquidtight flexible nonmetallic conduit in lengths of not more than 6 ft.

(2) Cord-and-Plug Connections. Cord-and-plug connections with a GFCI-protected cord that is not longer than 15 ft.

(B) Bonding. Bonding is permitted by mounting equipment to a metal frame or base. Metal bands that secure wooden staves aren't required to be bonded.

(C) Interior Wiring for Outdoor Spas or Hot Tubs. Any Chapter 3 wiring method containing a copper equipment grounding (bonding) conductor that is insulated or enclosed within the outer sheath of the wiring method and not smaller than 12 AWG is permitted for the connection to motor, heating, and control loads that are part of a self-contained spa or hot tub, or a packaged spa or hot tub equipment assembly.

Wiring to an underwater light must comply with 680.23 or 680.33.

680.43 Indoor Installations. Electrical installations for an indoor spa or hot tub must comply with Parts I and II of Article 680, except as modified by this section. Indoor installations of spas or hot tubs can be connected by any of the wiring methods contained in Chapter 3.

Exception: Listed packaged units rated 20A or less can be cord-and-plug connected.

(A) Receptacles. At least one 15 or 20A, 125V receptacle must be located at least 5 ft, but not more than 10 ft, from the inside wall of the spa or hot tub.

(1) Location. Other receptacles must be located not less than 5 ft, measured horizontally, from the inside walls of the indoor spa or hot tub.

(2) GFCI-Protected Receptacles. Receptacles rated 30A or less at 125V, located within 10 ft of the inside walls of an indoor spa or hot tub, must be GFCI protected. **Figure 680-27**

(3) Spa or Hot Tub Receptacle. Receptacles that provide power for an indoor spa or hot tub must be GFCI protected.

(4) Measurements. In determining the above dimensions, the distance to be measured must be the shortest path that the supply cord of an appliance connected to the receptacle would follow without piercing a floor, wall, ceiling, doorway with hinged or sliding door, window opening, or other effective permanent barrier.

(B) Luminaires and Ceiling Fans.

(1) Elevation. Luminaires and ceiling fans within 5 ft, measured horizontally, from the inside walls of the indoor spa or hot tub must be:

(a) Not less than 12 ft above an indoor spa or hot tub where no GFCI protection is provided.

(b) Not less than 7½ ft above an indoor spa or hot tub where GFCI protection is provided.

(c) If GFCI protection is provided and the installation meets the following, luminaires and ceiling fans can be mounted less than 7½ ft above an indoor spa or hot tub:

(1) Recessed luminaires with a glass or plastic lens, nonmetallic or electrically isolated metal trim, and suitable for use in damp locations.

Author's Comment: See Article 100 for the definition of "Location, Damp."

(2) Surface-mounted luminaires with a glass or plastic globe, a nonmetallic body, or a metallic body isolated from contact, and suitable for use in damp locations.

(C) Switches. Switches must be located not less than 5 ft, measured horizontally, from the inside walls of the indoor spa or hot tub. **Figure 680-28**

(D) Bonding. The following parts of an indoor spa or hot tub must be bonded together:

(1) Metal fittings within or attached to the indoor spa or hot tub structure.

(2) Metal parts of electrical equipment associated with the indoor spa or hot tub water circulating system.

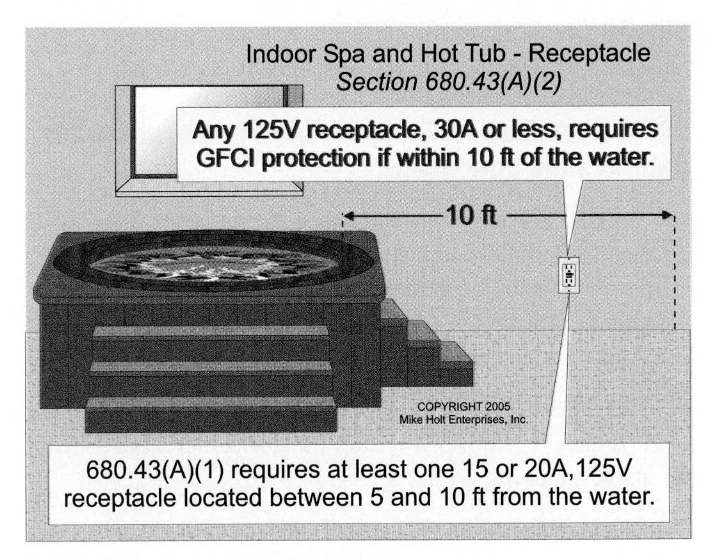

Indoor Spa and Hot Tub - Receptacle
Section 680.43(A)(2)
Any 125V receptacle, 30A or less, requires GFCI protection if within 10 ft of the water.
10 ft
COPYRIGHT 2005
Mike Holt Enterprises, Inc.
680.43(A)(1) requires at least one 15 or 20A,125V receptacle located between 5 and 10 ft from the water.

Figure 680–27

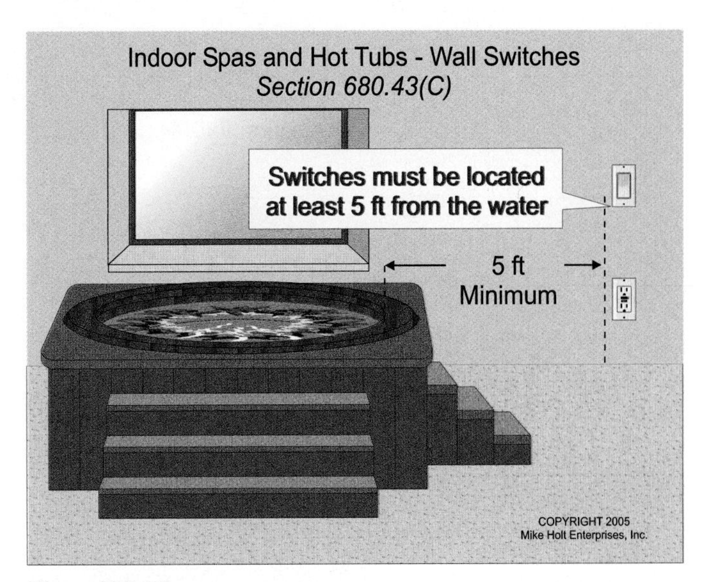

Figure 680–28

(3) Metal conduit, and metal piping within 5 ft of the inside walls of the indoor spa or hot tub, and not separated from the indoor spa or hot tub by a permanent barrier.

(4) Metal surfaces within 5 ft of the inside walls of an indoor spa or hot tub not separated from the indoor spa or hot tub area by a permanent barrier.

Exception: Small conductive surfaces, such as air and water jets, not likely to become energized, aren't required to be bonded. Other nonelectrical equipment, such as towel bars or mirror frames, which aren't connected to metallic piping, aren't required to be bonded.

(E) Methods of Bonding. All metal parts required by 680.43(D) to be bonded must be bonded together to create an equipotential (stray voltage) plane by any of the following methods:

(1) The interconnection of threaded metal piping and fittings.

(2) Metal-to-metal mounting on a common frame or base.

(3) A solid copper bonding jumper not smaller than 8 AWG.

680.44 GFCI Protection. The outlet that supplies a self-contained indoor spa or hot tub, a packaged spa or hot tub equipment assembly, or a field-assembled spa or hot tub must be GFCI protected. **Figure 680-29**

Author's Comment: A self-contained spa or hot tub is a factory-fabricated unit that consists of a spa or hot tub vessel with all water-circulating, heating, and control equipment integral to the unit. A packaged spa or hot tub equipment assembly is a factory-fabricated unit that consists of water circulating, heating, and control equipment mounted on a common base intended to operate a spa or hot tub [680.2].

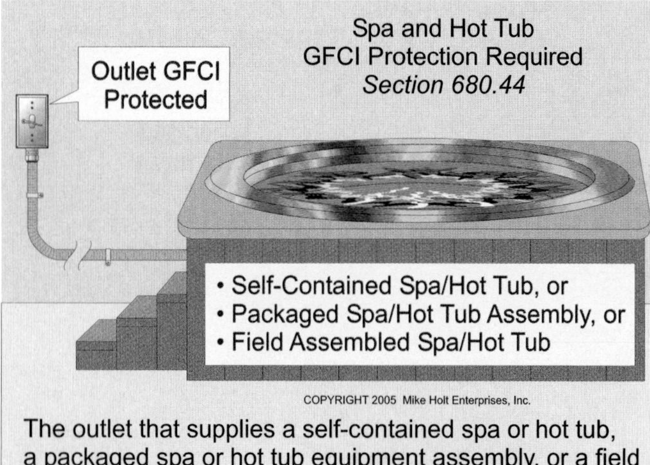
Outlet GFCI Protected
Spa and Hot Tub
GFCI Protection Required
Section 680.44
• Self-Contained Spa/Hot Tub, or
• Packaged Spa/Hot Tub Assembly, or
• Field Assembled Spa/Hot Tub
COPYRIGHT 2005 Mike Holt Enterprises, Inc.

The outlet that supplies a self-contained spa or hot tub, a packaged spa or hot tub equipment assembly, or a field assembled spa or hot tub must be GFCI protected.

Figure 680–29

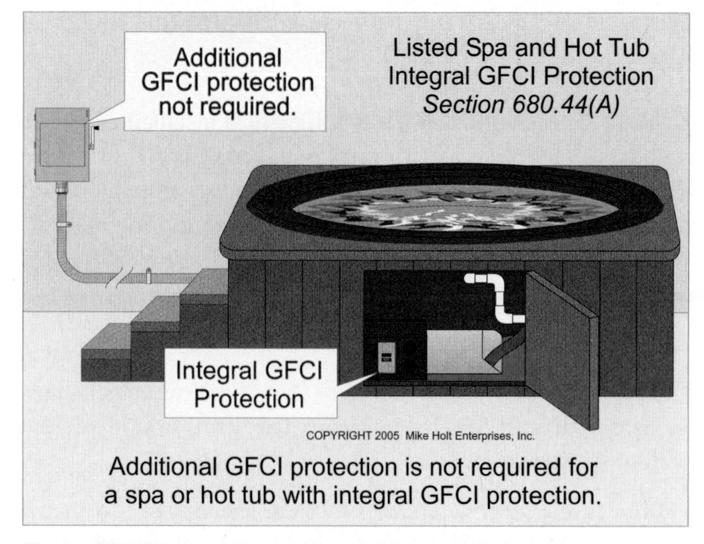
Additional GFCI protection not required.
Listed Spa and Hot Tub
Integral GFCI Protection
Section 680.44(A)
Integral GFCI Protection
COPYRIGHT 2005 Mike Holt Enterprises, Inc.

Additional GFCI protection is not required for a spa or hot tub with integral GFCI protection.

Figure 680–30

(A) Listed Units. Additional GFCI protection isn't required for a listed self-contained spa or hot tub unit or listed packaged spa or hot tub assembly marked to indicate that integral GFCI protection has been provided for all electrical parts within the unit or assembly. **Figure 680-30**

(B) Other Units. GFCI protection isn't required for a field-assembled spa or hot tub that is three-phase or that has a voltage rating over 250V, or has a heater load above 50A.

(C) Combination Pool and Spa or Hot Tub. GFCI protection isn't required for equipment that supplies a combination pool/hot tub or spa assembly.

PART V. FOUNTAINS

680.50 General. Fountains that have water common to a pool must comply with the pool requirements of Parts I and II.

680.51 Luminaires, Submersible Pumps, and Other Submersible Equipment.

(A) GFCI Protection for Fountain Equipment. The branch circuit that supplies luminaires, submersible pumps, and other submersible equipment must be GFCI protected, unless the equipment is listed for not more than 15V and is supplied by a listed pool transformer that complies with 680.23(A)(2).

(C) Luminaire Lenses. Luminaires must be installed so the top of the luminaire lens is below the normal water level unless listed for above-water use.

(E) Cords. The maximum length of exposed cord in the fountain is 10 ft. Power supply cords that extend beyond the fountain perimeter must be enclosed in a wiring enclosure approved by the authority having jurisdiction.

(F) Servicing. Equipment must be capable of being removed from the water for relamping or for normal maintenance.

(G) Stability. Equipment must be inherently stable or securely fastened in place.

680.53 Bonding. All metal piping systems associated with the fountain must be bonded to the equipment grounding (bonding) conductor of the branch circuit that supplies the fountain.

680.55 Methods of Grounding (Bonding).

(B) Supplied by a Flexible Cord. Equipment supplied by a flexible cord must have all exposed metal parts grounded (bonded) to an effective ground-fault current path by an insulated copper equipment grounding (bonding) conductor that is an integral part of the cord.

680.56 Cord-and-Plug Connected Equipment.

(A) GFCI Protection of Cord-and-Plug Equipment. All cord-and-plug connected fountain equipment must be GFCI protected.

(B) Cord Type. Flexible cords immersed in or exposed to water must be of the hard-service type, as designated in Table 400.4, and must be marked "Water-Resistant."

680.57 Signs in or Adjacent to Fountains.

(B) GFCI Protection of Sign Equipment. Each circuit that supplies a sign installed within a fountain, or within 10 ft of the fountain edge, must be GFCI protected [680.57(A)]. **Figure 680-31**

(C) Sign Location.

(1) Fixed or Stationary. A fixed or stationary electric sign installed within a fountain must be not less than 5 ft from the outside edge of the fountain.

(2) Portable. A portable electric sign must not be placed in or within 5 ft from the inside walls of a fountain.

> **Author's Comment:** Because portable electric signs pose a greater hazard from electric shock than a fixed or stationary sign, they cannot be installed within 5 ft of the inside walls of the fountain.

(D) Disconnect. Either the sign disconnect must be within sight of the sign or the disconnecting means must be capable of being locked in the open position, in accordance with 600.6 and 680.12.

680.58 GFCI-Protected Receptacles. GFCI protection is required for all 15 and 20A, 125V through 250V, receptacles located within 20 ft of the inside walls of a fountain. **Figure 680-32**

PART VII. HYDROMASSAGE BATHTUBS

680.70 Protection. A hydromassage bathtub, which is a permanently installed bathtub equipped with a recirculating piping system designed to accept, circulate, and discharge water upon each use, isn't required to comply with the other parts of this article.

Electric Signs
Installed in Fountains
Section 680.57
Mike Holt
Enterprises
888-NEC-CODE
A sign in or within 10 ft of a fountain must:
• Be GFCI protected [680.57(B)].
• If in the fountain, be at least 5 ft from the inside edge of the fountain [680.57(C)].
• Have a disconnect within sight or capable of being locked in the open position [680.57(D)].
COPYRIGHT 2005
Mike Holt Enterprises, Inc.
Fountains
GFCI-Protected Receptacles
Section 680.58
GFCI protection is required for all 15 and 20A, 125V to 250V 1-phase receptacles located within 20 ft of the fountain.
20 ft
or Less
GFCI
Protection
Required
COPYRIGHT 2005
Mike Holt Enterprises, Inc.

Figure 680–31

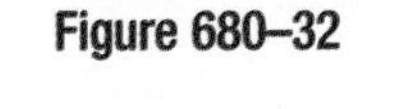
Figure 680–32

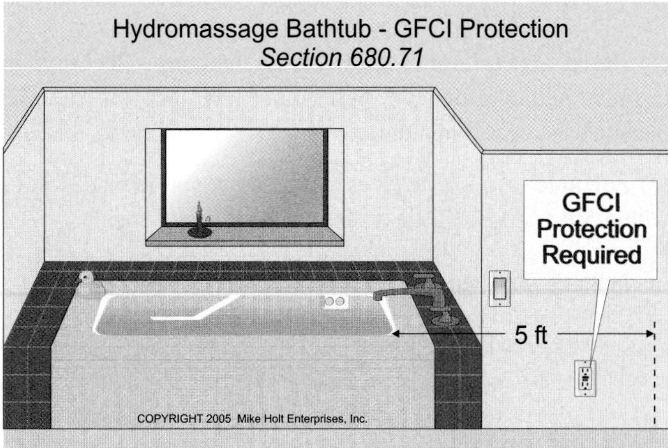

The electrical components of a hydromassage bathtub and all 125V receptacles within 5 ft must be GFCI protected.

Figure 680–33

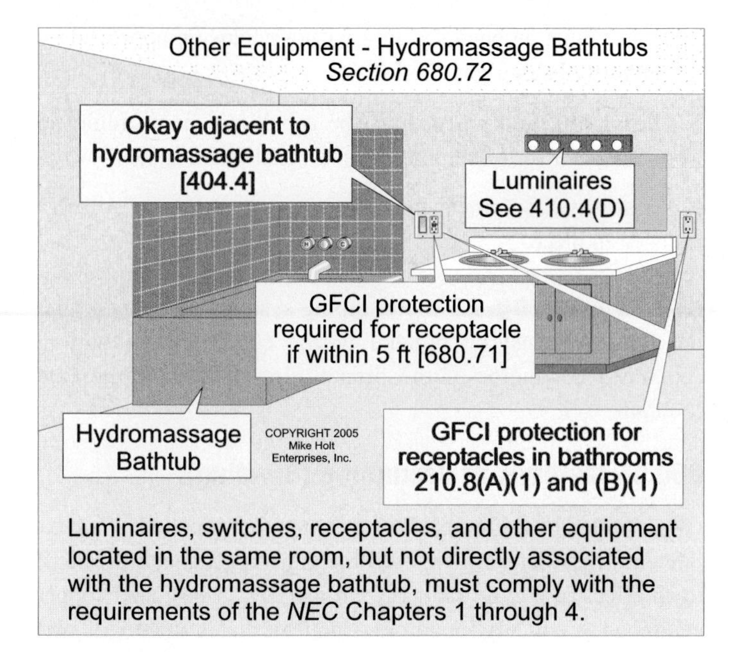

Luminaires, switches, receptacles, and other equipment located in the same room, but not directly associated with the hydromassage bathtub, must comply with the requirements of the *NEC* Chapters 1 through 4.

Figure 680–34

680.71 Other Electric Equipment. Hydromassage bathtubs and their associated electrical components must be GFCI protected. In addition, GFCI protection is required for all receptacles rated 30A or less at 125V located within 5 ft of the inside walls of a hydromassage tub. **Figure 680-33**

680.72 Other Electrical Equipment. Luminaires, switches, receptacles, and other electrical equipment located in the same room and not directly associated with a hydromassage bathtub must be installed in accordance with Chapters 1 through 4.

Author's Comment: A hydromassage bathtub is treated like a regular bathtub. For example, a 5 ft clearance isn't required for switches or receptacles, and the fixtures must be installed in accordance with 410.4(D). **Figure 680-34**

680.73 Accessibility. Electrical equipment for hydromassage bathtubs must be capable of being removed or exposed without damaging the building structure or finish.

680.74 Bonding. All metal piping systems and grounded metal parts in contact with the circulating water must be bonded together using a solid copper bonding jumper not smaller than 8 AWG.

Article 680 Questions

1. Fixed or stationary pool, outdoor spa, and hot tub equipment is permitted to be cord-and-plug connected to facilitate the removal or disconnection for maintenance or repair. The flexible cord must _____.

   (a) not exceed 3 ft except for storable pools
   (b) have a copper equipment grounding conductor not smaller than 12 AWG
   (c) terminate in a grounding-type attachment plug
   (d) all of these

2. Underground rigid nonmetallic wiring located less than 5 ft from the inside wall of a pool or spa must be buried not less than _____

   (a) 6 in. (b) 10 in. (c) 12 in. (d) 18 in.

3. In dwelling units, a 125V receptacle is required to be installed a minimum of 10 ft and a maximum of 20 ft from the inside wall of the pool.

   (a) True (b) False

4. Switching devices must be at least 5 ft horizontally from the inside walls of a pool unless the switch is listed as being acceptable for use within 5 ft. An example of a switch that meets this requirement would be a pneumatic switch listed for this purpose.

   (a) True (b) False

5. Branch-circuit wiring for underwater luminaires must be installed in _____.

   (a) rigid metal conduit or intermediate metal conduit
   (b) liquidtight flexible nonmetallic conduit or rigid nonmetallic conduit
   (c) any raceway wiring method
   (d) a or b

6. A pool light junction box that has a raceway that extends directly to underwater pool light forming shells must be located not less than _____ from the outdoor pool or spa.

   (a) 2 ft (b) 3 ft (c) 4 ft (d) 6 ft

7. The pool structure, including the reinforcing metal of the pool shell and deck, must be bonded together.

   (a) True (b) False

8. The _____ pool bonding conductor must be connected to the equipotential bonding grid either by exothermic welding or by pressure connectors that are labeled as being suitable for the purpose.

   (a) 8 AWG (b) insulated or bare (c) copper (d) all of these

9. In spas or hot tubs, a clearly labeled emergency shutoff or control switch for the purpose of stopping the motors(s) that provide power to the recirculation system and jet system must be installed. The emergency shutoff control switch must be _____ to the users and located not less than 5 ft away, and within sight of, the spa or hot tub. This requirement does not apply to single-family dwelling units.

   (a) accessible (b) readily accessible (c) available (d) none of these

10. At least one 15 or 20A, 125V receptacle must be located a minimum of _____ (and a maximum of 10 ft) from the inside wall of a spa or hot tub installed indoors.

(a) 2 ft (b) 5 ft (c) 18 in. (d) no minimum

11. Surface-mounted luminaires _____ located over or within 5 ft, measured horizontally, from the inside walls of an indoor spa or hot tub are permitted to be installed at less than 7 ft 6 in. above the maximum water level when GFCI protection is provided.

(a) with a glass or plastic globe
(b) with a nonmetallic body or a metallic body isolated from contact
(c) suitable for use in a damp location
(d) all of these

12. Small conductive surfaces of an indoor spa or hot tub, such as air and water jets not likely to become energized, are not required to be bonded. Other nonelectric equipment, such as towel bars or mirror frames, which are not connected to metallic piping, are not required to be bonded.

(a) True (b) False

13. The maximum length of exposed cord in a fountain must be _____.

(a) 3 ft (b) 4 ft (c) 6 ft (d) 10 ft

14. Each circuit supplying a sign within or adjacent to a fountain must _____.

(a) have ground-fault circuit-interrupter protection
(b) be equipped with a lock out
(c) operate at less than 50V
(d) be an intrinsically safe circuit

15. A portable electric sign cannot be placed in or within _____ from the inside walls of a fountain.

(a) 3 ft (b) 5 ft (c) 10 ft (d) none of these

16. GFCI protection is required for all 125V, single-phase receptacles not exceeding 30A and located within 5 ft measured _____ from the inside walls of a hydromassage bathtub.

(a) vertically (b) horizontally (c) across (d) none of these

17. Hydromassage bathtub equipment must be _____ without damaging the building structure or building finish.

(a) readily accessible (b) accessible (c) within sight (d) none of these

18. All metal piping systems, metal parts of electrical equipment, and pump motors associated with a hydromassage tub must be bonded together with a(n) _____ solid copper bonding jumper not smaller than 8 AWG.

(a) insulated (b) covered (c) bare (d) any of these

ARTICLE

690 Solar Photovoltaic Systems

Introduction

In the typical application, solar systems complement, rather than replace, one or more power sources. Additionally, solar power systems include some sort of energy storage (usually batteries) or connections to the utility grid. Both of these factors complicate the picture. When you have another power source, you must prevent backfeed to the solar cells, or backfeed from the solar cells to the utility grid if the utility power source fails.

Even a stand-alone system needs an inverter to allow running normal equipment from the batteries. And whether the system is stand-alone or supplemental, you still need to meet requirements for grounding, bonding, overcurrent protection, labeling, and disconnecting means. Plus, the charging systems pose additional challenges to ensure a safe installation. Article 690 addresses all of these issues.

PART I. GENERAL

690.1 Scope. The provisions of Article 690 apply to solar photovoltaic electrical energy systems including the array circuit(s), inverter(s), and controller(s) for such systems. Solar photovoltaic systems within the scope of Article 690 may be interactive with other electrical power production sources or stand-alone, with or without electrical energy storage such as batteries. These systems can be either all ac or all dc, or they can be a combination of both ac and dc. **Figure 690-1**

> **Author's Comment:** A solar photovoltaic system can be a stand-alone system but it's frequently installed with other power supplies. When a solar photovoltaic system is interconnected with other power sources, such as the electric utility or an on-site generator system, Article 705 Interconnected Electrical Power Production Sources also applies [705.3 Exception 1].

Article 690 is very detailed and contains specific installation requirements for solar photovoltaic systems, which are beyond the scope of this textbook.

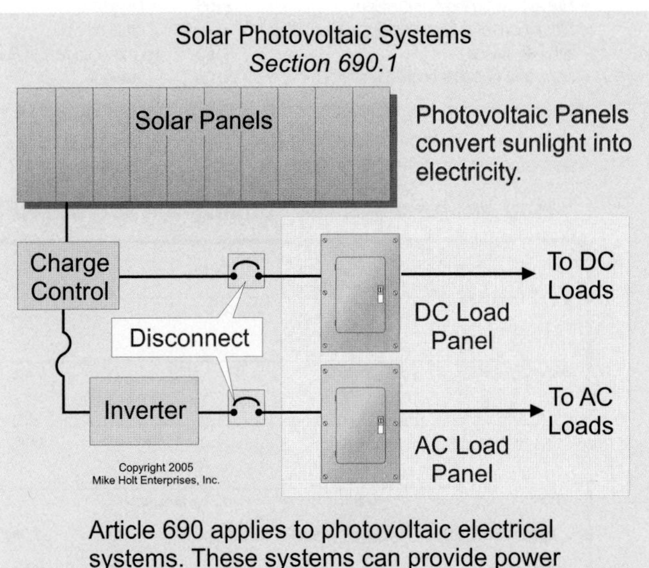
Solar Photovoltaic Systems
Section 690.1
Solar Panels
Photovoltaic Panels convert sunlight into electricity.
Charge Control
Disconnect
Inverter
To DC Loads
DC Load Panel
To AC Loads
AC Load Panel
Copyright 2005
Mike Holt Enterprises, Inc.
Article 690 applies to photovoltaic electrical systems. These systems can provide power to ac loads, dc loads, or both.

Figure 690–1

Order our Ultimate Training Library and save almost $1,600!

A $3,352 value—you pay only $1,795!

As an added bonus, the first 500 orders will receive Mike's Deluxe Estimating Library—a $295 Value!

ELECTRICAL THEORY

- *Basic Electrical Theory* textbook/workbook $50
- *Electrical Fundamentals and Basic Electricity* video or DVD $109
- *Electrical Circuits, Systems and Protection* video or DVD $109
- *Alternating Current, Motors, Generators, and Transformers* video or DVD $109
- *Fire Alarms* textbook and video $109
- *Motor Control & Signaling Circuits* textbook and two videos $169
- *Harmonic Currents* booklet and video $109

UNDERSTANDING THE NEC®

- *Understanding the NEC*—Volume 1 textbook $59
- *Understanding the NEC*—Volume 2 textbook $45
- *NEC Exam Practice Questions* book $40
- *General Requirements* two videos or DVDs $198
- *Grounding and Bonding* two videos or DVDs $198
- *Grounding and Bonding* MP3 audio CD $59
- *Wiring Methods* videos or DVDs $198
- *Equipment for General Use* video or DVD $99
- *Special Occupancies* video or DVD $109
- *Special Equipment* video or DVD $109
- *Limited Energy and Communication Systems* video or DVD $109
- *Low-Voltage and Power-Limited Systems* textbook $35

CODE® CHANGES TO THE NEC

- *Changes to the NEC* textbook $44
- *Changes to the NEC* video or DVD $198
- *Changes to the NEC* MP3 audio CD $59
- Code Change tabs $10
- 16-Hour Online Code Change program $178

ELECTRICAL CALCULATIONS

- *Exam Prep* textbook/workbook $59
- *Raceway and Box Calculations* video or DVD $99
- *Conductor Sizing and Protection* video or DVD $109
- *Motor Calculations* video or DVD $79
- *Voltage-Drop Calculations* video or DVD $79
- *Dwelling Unit Calculations* video or DVD $99
- *Multifamily Dwelling Calculations* video $99
- *Commercial Calculations* video $109
- *Delta/Delta and Delta/Wye Transformers* video $109

Find out how you may qualify to receive FREE PowerPoint presentations—each valued at $495!

NAME _____ COMPANY TITLE _____

MAILING ADDRESS _____ CITY _____ STATE _____ ZIP _____

SHIPPING ADDRESS _____ CITY _____ STATE _____ ZIP _____

PHONE _____ FAX _____ E-MAIL ADDRESS _____ WEB SITE _____

❑ CHECK ❑ VISA ❑ MASTER CARD ❑ DISCOVER ❑ AMEX ❑ MONEY ORDER

CREDIT CARD # : _____ EXP. DATE: _____

Mike Holt's 2005 Ultimate Training Library

❑ 05UTP	2005 Ultimate Training Library w/Videos	$1,795
❑ 05UTPD	2005 Ultimate Training Library w/DVDs	$1,795
❑ *FREE Deluxe Estimating Library (available only to the first 500 orders)		

Subtotal $ _____

Sales Tax **FLORIDA RESIDENTS ONLY** add 6% _____

Total Price $ _____

Shipping: 4% of Total Price (or Minimum $7.50) $ _____

TOTAL DUE $ _____

Mike Holt Enterprises, Inc. • 7310 W. McNab Rd., Suite 201 • Tamarac, FL 33321 • FAX 1.954.720.7944 • www.NECcode.com

Contact Freddy@MikeHolt.com • 1.888.NEC®CODE

ARTICLE 692 Fuel Cell Systems

Introduction

Article 692, covering requirements for the installation and use of fuel cells, was added to the 2002 *NEC*. Fuel cells convert a gas into electrical energy that can then be used to power a building or residential dwelling.

692.1 Scope. This article identifies the requirements for the installation of fuel cell power systems. They may be stand-alone or interactive with other electrical power production sources, and may be with or without electrical energy storage, such as in the case of batteries. These systems may have ac or dc output for utilization.

692.2 Definitions.

Fuel Cell. An electrochemical system that consumes fuel to produce an electrical current. The main chemical reaction used in a fuel cell for producing electrical power isn't combustion. There may, however, be sources of combustion used within the overall fuel cell system, such as reformers/fuel processors.

Fuel Cell System. The complete aggregate of equipment used to convert chemical fuel into usable electricity. A fuel cell system typically consists of a reformer, stack, power inverter, and auxiliary equipment.

692.6 Listing Requirement. The fuel cell system must be evaluated and listed for its intended application prior to installation.

Article 692 Questions

1. A fuel cell is an electrochemical system that consumes fuel to produce an electric current. The main chemical reaction used in a fuel cell to produce electrical power is not combustion.

   (a) True (b) False

2. A fuel cell system typically consists of a reformer, stack, power inverter, and auxiliary equipment.

   (a) True (b) False

3. The fuel cell system must be evaluated and _____ for its intended application prior to installation.

   (a) approved (b) identified (c) listed (d) marked

ARTICLE

695 Fire Pumps

Introduction

The general philosophy behind *Code* articles is that circuit protection will shut down equipment before letting the supply conductors melt from overload. Article 695 departs from this philosophy. The idea is that the fire pump motor must run, no matter what; it supplies water to a facility's fire protection piping, which in turn supplies water to the sprinkler system and fire hoses. Article 695 contains many requirements to keep that supply of water uninterrupted.

Some of these requirements are intuitively obvious. For example, locating the pump so as to minimize its exposure to fire. Or, ensuring that the fire pump and its jockey (pressure maintenance) pump have a reliable source of power. And, of course, it makes sense to keep fire pump wiring independent of all other wiring.

Other requirements seem wrong at first glance, until you remember why that fire pump is there in the first place. For example, the disconnect must be lockable in the closed position. You would normally expect these to be lockable in the open position because other articles require that for the safety of maintenance personnel. But the fire pump runs to ensure the safety of an entire facility and everyone in it. For the same reason, fire pump power circuits cannot have automatic protection against overloads.

Remember, the fire pump must be kept in service, even if doing so damages or destroys the pump. It's better to run the pump until its windings melt, than to save the fire pump and lose the facility. And the intent of Article 695 is to save the facility.

695.1 Scope.

(A) Covered. Article 695 covers the installation of:

(1) Electric power sources and interconnecting circuits.

(2) Switching and control equipment dedicated to fire pump drivers.

(B) Not Covered. The article doesn't cover:

(1) Performance, maintenance, and testing.

(2) Pressure maintenance (jockey or makeup) pumps.

> **FPN**: See NFPA 20, *Standard for the Installation of Stationary Pumps for Fire Protection*, for further information.

695.3 Power Sources.

(A) Individual Source. Power to fire pump motors must be supplied by a reliable source of power that has the capacity to carry the locked-rotor current of the fire pump motor(s), pressure maintenance pump motors, and the full-load current of any associated fire pump equipment. Reliable sources of power include:

(1) Electric Utility Service. A separate service from a connection located ahead of but not within the service disconnecting means.

(2) On-Site Power. An on-site power supply, such as a generator, located and protected to minimize damage by fire is permitted to supply a fire pump.

> **Author's Comment:** The determination of a reliable source of power is subject to the approval of the authority having jurisdiction.

695.4 Continuity of Power. Circuits that supply electric motor-driven fire pumps must be supervised from inadvertent disconnection as covered in (A) or (B).

(A) Direct Connection. The supply conductors must directly connect the power source either to a listed fire pump controller or to a listed combination fire pump controller and power transfer switch. [NFPA 20, 6.3.2.2.1]

(B) Supervised Connection. A single disconnecting means and associated overcurrent protective device(s) is permitted between a remote power source and one of the following:

(1) A listed fire pump controller

(2) A listed fire pump power transfer switch

(3) A listed combination fire pump controller and power transfer switch

(1) Overcurrent Device Selection. The overcurrent protective device(s) must be selected or set to carry indefinitely the sum of the locked-rotor current of the fire pump and pressure maintenance pump motor(s), and 100 percent of the ampere rating of the fire pump's accessory equipment. The requirement to carry the locked-rotor currents indefinitely does not apply to fire pump motor conductors.

(2) Disconnecting Means. The disconnecting means must comply with all of the following:

(1) Be identified as suitable for use as service equipment

(2) Be lockable in the closed position

(3) Not be located within equipment that feeds loads other than the fire pump

(4) Be located sufficiently remote from other building or other fire pump source disconnecting means

(3) Disconnect Marking. The disconnecting means must be marked "Fire Pump Disconnecting Means." The letters must be at least 1 in. in height and be visible without opening enclosure doors or covers.

695.5 Transformers.

(A) Size. If a transformer supplies an electric fire pump motor, it must be sized no less than 125 percent of the sum of the fire pump and pressure maintenance pump(s) motor loads, and 100 percent of the ampere rating of the fire pump's accessory equipment.

(B) Overcurrent Protection. The primary overcurrent protective device must be sized to carry indefinitely the sum of the locked-rotor current of the fire pump and pressure maintenance pump motor(s), and 100 percent of the ampere rating of the fire pump's accessory equipment. Secondary overcurrent protection is not permitted. The requirement to carry the locked-rotor currents indefinitely does not apply to fire pump motor conductors.

695.6 Power Wiring.

(A) Service and Feeder Conductors. Supply conductors must be physically routed outside buildings and must be installed in accordance with Article 230. Where supply conductors cannot be routed outside buildings, they must be encased in 2 in. of concrete or brick [230.6(1) or (2)].

(B) Circuit Conductors. Fire pump supply conductors on the load side of the final disconnecting means and overcurrent device(s) [695.4(B)] must be kept entirely independent of all other wiring. They can be routed through a building using one of the following methods:

(1) Be encased in a minimum 2 in. of concrete

(2) Be within an enclosed construction dedicated to the fire pump circuit(s) and having a minimum of a 1-hour fire-resistive rating

(3) Be a listed electrical circuit protective system with a minimum 1-hour fire rating

(C) Conductor Size.

(1) Fire Pump Motors and Other Equipment. Conductors supplying fire pump motors and accessory equipment must be sized no less than 125 percent of the sum of the motor full-load current as listed in Table 430.248 or 430.250, plus 100 percent of the ampere rating of the fire pump's accessory equipment.

(2) Fire Pump Motors Only. Conductors supplying a single fire pump motor must be sized in accordance with the requirements of 430.22.

Author's Comment: This means that the branch-circuit conductors to a single fire pump motor must have an ampere rating of not less than 125 percent of the fire pump motor full-load current (FLC), as listed in Table 430.248 or 430.250.

Question: *What size conductor is required for a 25 hp, 208V three-phase fire pump motor?* **Figure 695-1**

(a) 4 AWG *(b) 3 AWG* *(c) 2 AWG* *(d) 1 AWG*

Answer: *(b) 3 AWG*

Conductors are sized at 125 percent of the motor's FLC as listed in Table 430.250

FLC of 25 hp = 74.8A, Table 430.250
Conductor = 74.8A x 1.25
Conductor = 93.5A
3 AWG at 75°C is rated 100A

Note: *The fire pump motor circuit protective device size must be set to carry indefinitely the sum of the locked-rotor current of the fire pump motor. According to Table 430.251(B), the locked-rotor current of a 25 hp, 208V three-phase motor is 404A.*

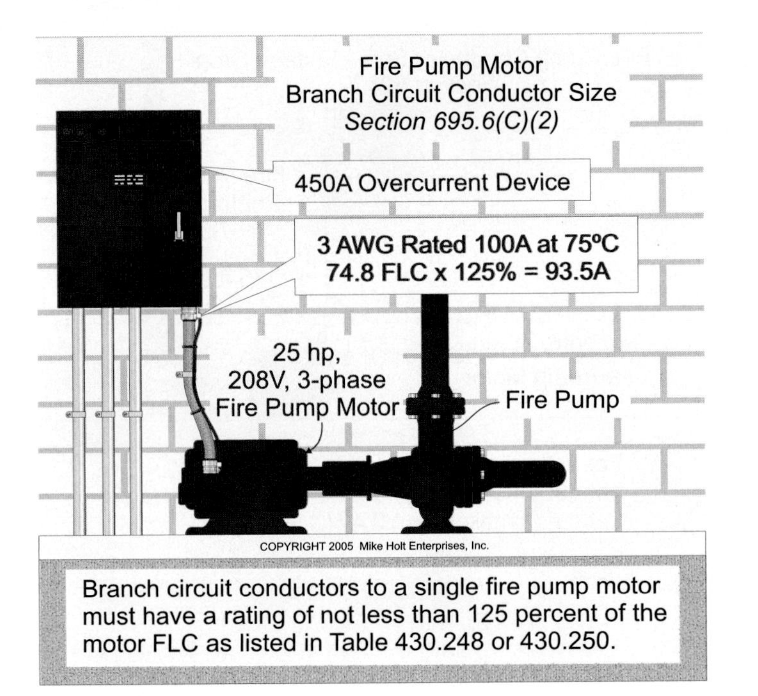

Figure 695–1

In addition, branch-circuit conductors for a fire pump motor must be sized to accommodate the voltage-drop requirements of 695.7.

(D) Overload Protection. Branch-circuit and feeder conductors must be protected against short circuit, not an overload.

(E) Pump Wiring. Wiring from the fire pump controller to the fire pump motor [not run through a building – 695.6(B)] must be in rigid metal conduit, intermediate metal conduit, liquidtight flexible metal conduit, liquidtight flexible nonmetallic conduit Type B, listed Type MC cable with an impervious covering, or Type MI cable.

(H) Ground-Fault Protection of Equipment. Ground-fault protection of equipment is not permitted for fire pumps.

Author's Comment: See Article 100 for the definition of "Ground-Fault Protection of Equipment."

695.7 Voltage Drop.

Controller. The voltage at the line terminals of the controller, when the motor starts (locked-rotor current), must not drop more than 15 percent below the controller's rated voltage. In addition, the voltage at the motor terminals must not drop more than 5 percent below the voltage rating of the motor when the motor operates at 115 percent of the fire pump full-load current rating.

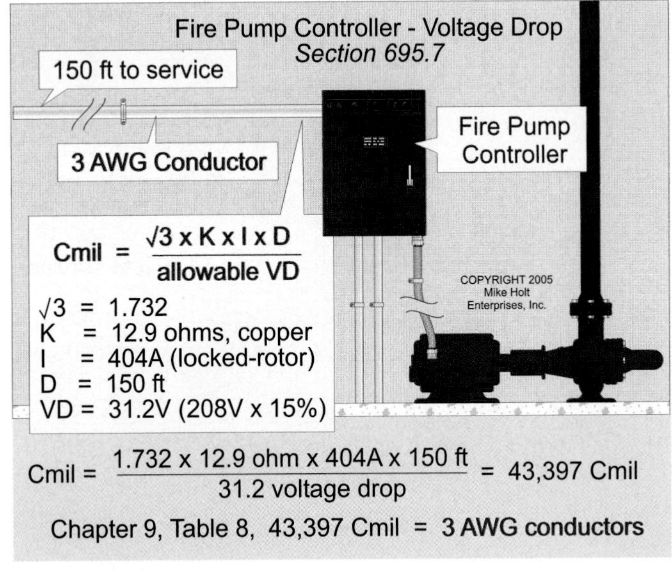

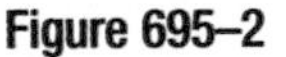
Figure 695–2

Example—Three-Phase

Question: *A 25 hp, 208V three-phase fire pump motor is located 175 ft from the service. The fire pump motor controller is located 150 ft from the service. What size conductor must be installed to the fire pump motor controller?* **Figure 695-2**
Note: Equipment terminals are rated 75°C.

(a) 4 THHN (b) 3 THHN (c) 2 THHN (d) 1 THHN

Answer: *(b) 3 THHN*

Cmil = (1.732 x K x I x D)/VD
Cmil = Wire size, Chapter 9, Table 8
K = 12.9 ohms, copper
I = 404A (locked-rotor, Table 430.251B)
D = 150 ft
VD = 31.2V (208V x 15%)
Cmil = (1.732 x 12.9 ohms x 404A x 150 ft)/31.2V
Cmil = 43,397, Chapter 9, Table 8 = 3 AWG

Motor. Where ungrounded conductors are increased in size, equipment grounding (bonding) conductors (where installed) must be proportionally increased in size according to the circular mil area of the ungrounded conductors [250.122(B)].

Example—Three-Phase

Question: *A 25 hp, 208V three-phase fire pump motor is located 175 ft from the service. The fire pump motor controller is located 150 ft from the service. What size conductor must be installed to the fire pump motor if it's located 25 ft from the controller? Note: Equipment terminals are rated 75°C.* **Figure 695-3**

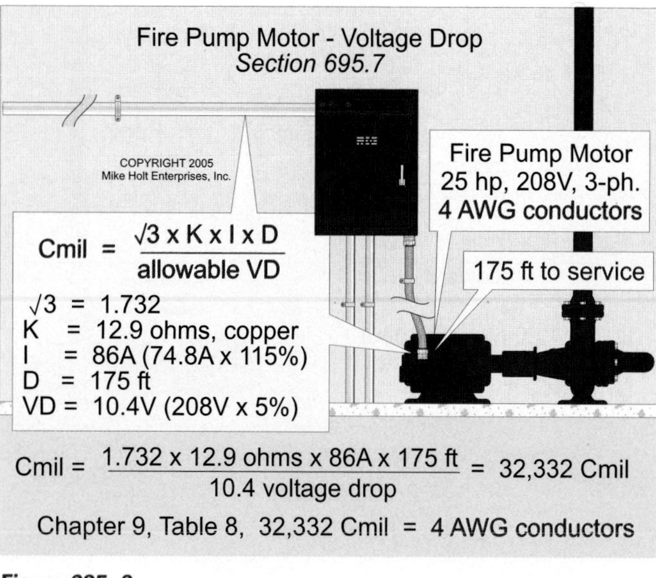

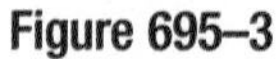

Figure 695–3

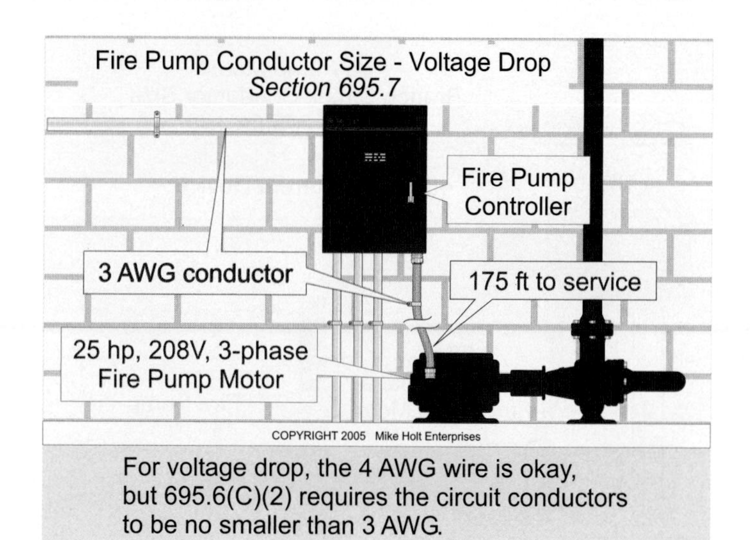

Figure 695–4

(a) 4 THHN (b) 3 THHN (c) 2 THHN (d) 1 THHN

Answer: *(b) 3 THHN*

The operating voltage at the terminals of the motor must not drop more than 5 percent below the voltage rating of the motor while the motor is operating at 115 percent of the full-load current rating of the motor.

Cmil = (1.732 x K x I x D)/VD
Cmil = Wire size, Chapter 9, Table 8
K = 12.9 ohms, copper
I = 86A (74.8A at 115%), Table 430.250
D = 175 ft
VD 5 % = 10.4V (208V x 5%)
Cmil = (1.732 x 12.9 ohms x 86A x 175 ft)/10.4V
Cmil = 32,332
Chapter 9, Table 8 = 4 AWG

CAUTION: ***For voltage drop, the 4 AWG wire is okay from the controller to the motor, but 695.6(C)(2) requires the branch-circuit conductors to be sized not smaller than 3 AWG.*** **Figure 695-4**

695.14 Control Wiring.

(E) Wiring Methods. Control wiring must be in rigid metal conduit, intermediate metal conduit, liquidtight flexible metal conduit, liquidtight flexible nonmetallic conduit Type B, listed Type MC cable with an impervious covering, or Type MI cable.

Article 695 Questions

1. Power to fire pump motors must be supplied by a reliable source. This source must have the capacity to carry the locked-rotor current of the fire pump motor(s), the pressure maintenance pump motors, and the full-load current of any associated fire pump equipment. This source can be _____.

   (a) a separate service or a tap located ahead of but not within the utility service disconnecting means
   (b) an on-site power supply, such as a generator
   (c) a or b
   (d) a and b

2. Where a transformer supplies an electric fire pump motor, it must be sized no less than _____ percent of the sum of the fire pump motor(s) and pressure maintenance pump motors, and 100 percent of any associated fire pump accessory equipment supplied by the transformer.

   (a) 100 (b) 125 (c) 250 (d) 300

3. Secondary overcurrent protection is permitted for transformers supplying fire pumps.

   (a) True (b) False

4. Feeder conductors supplying fire-pump motors and accessory equipment must be sized no less than _____ percent of the sum of the motor full-load currents as listed in Article 430, plus 100 percent of the ampere rating of the fire-pump accessory equipment.

   (a) 100 (b) 125 (c) 250 (d) 600

5. Ground-fault protection of equipment _____ for fire pumps.

   (a) is not permitted (b) is permitted
   (c) is allowed (d) must be approved by the AHJ

Notes

CHAPTER 7
Special Conditions

Introduction

Chapter 7, which covers special conditions, is the third of four chapters that deal with special topics. Chapters 5, 6, and 8 cover special occupancies, special equipment, and communications systems, respectively. Remember, the first four Chapters of the *NEC* are sequential and form a foundation for each of the subsequent four Chapters.

What exactly is a "special condition?" It's a situation that doesn't fall under the category of Special Occupancies or Special Equipment, but creates a need for additional measures to ensure the "safeguarding of people and property" mission of the *NEC*, as stated in Article 90.

The *NEC* groups these logically, as you might expect. Here are the general groupings:

- Emergency and standby power systems. Articles 700, 701, and 702. Article 700 addresses emergency standby systems, Article 701 addresses legally required standby systems, and Article 702 addresses optional ones.
- Interconnected power sources. Article 705. This primarily has to do with generators or photovoltaic systems used for on-site power generation.
- Low-voltage, low-power wiring. Articles 720–780. Examples include control, signaling, instrumentation, fire alarm systems, and optical fiber installations.

Article 700. Emergency Power Systems. The requirements of Article 700 apply only to the wiring methods for "emergency systems" that are essential for safety to human life and required by federal, state, municipal, or other regulatory codes. When normal power is lost, emergency systems must be capable of supplying emergency power in 10 seconds or less.

Article 701. Legally Required Standby Power Systems. Legally required standby systems provide electric power to aid in firefighting, rescue operations, control of health hazards, and similar operations, and are required by federal, state, municipal, or other regulatory codes. When normal power is lost, legally required systems must be capable of automatically supplying standby power in 60 seconds or less, instead of the 10 seconds or less required of emergency systems.

Article 702. Optional Standby Power Systems. Optional standby systems are intended to protect public or private facilities or property where life safety doesn't depend on the performance of the system. These systems are typically installed to provide an alternate source of electric power for such facilities as industrial and commercial buildings, farms, and residences, and to serve loads that, when stopped during any power outage, could cause discomfort, serious interruption of a process, or damage to a product or process. Optional standby systems are intended to supply on-site generated power to loads selected by the customer either automatically or manually.

Article 720. Circuits and Equipment Operating at Less than 50 Volts. Article 720 applies to electrical installations that operate below 50V, either direct current or alternating current. Other installations that operate below 50V and are covered in Articles 411, 551, 650, 669, 725, and 760, aren't required to comply with Article 720.

Article 725. Class 1, Class 2, and Class 3 Remote-Control, Signaling, and Power-Limited Circuits. Article 725 contains the requirements for remote-control, signaling, and power-limited circuits that aren't an integral part of a device or appliance.

- Remote-Control Circuit—A circuit that controls other circuits through a relay or solid-state device. For example, a circuit that controls the coil of a motor starter or lighting contactor.

- Signaling Circuit—A circuit that supplies energy to an appliance or device that gives a visual and/or audible signal. For example, a circuit for doorbells, buzzers, code-calling systems, signal lights, annunciators, burglar alarms, and other indication or alarm devices.

Article 760. Fire Alarm Systems. Article 760 covers the installation of wiring and equipment for fire alarm systems. Fire alarm systems include fire detection and alarm notification, voice communications, guard's tour, sprinkler waterflow, and sprinkler supervisory systems.

Article 770. Optical Fiber Cables and Raceways. Article 770 covers the installation of optical fiber cables, which transmit signals using light for control, signaling, and communication. This article also contains the installation requirements for raceways that contain and support the optical fiber cables, and requirements for composite cables (often called "hybrid" in the field) that combine optical fibers with current-carrying metallic conductors.

ARTICLE

Emergency Power Systems

Introduction

Emergency systems are legally required, often as a condition of an operating permit for a given facility based on its use. The authority having jurisdiction makes the determination as to whether an emergency system is necessary for a given facility and what it must entail. Sometimes, an emergency system simply provides power for exit lighting or the illumination of exit signs upon loss of main power or in the case of fire. Its purpose isn't to provide power for normal business operations, but rather to provide lighting and controls essential for human life.

This background information will help you understand that not all emergency actions to save human life fall under Article 700. The general goal is to keep the emergency operation as reliable as possible. One way to do that is to use inherently-safe actuation devices, such as valves that "fail safe" to a predetermined position upon loss of power. Another is to limit what needs to be an emergency load in the first place so the emergency system powers only what is needed to save human life.

In an emergency, it's difficult to administratively control loads. Thus, the emergency system must be able to supply all emergency loads simultaneously. When the emergency power source also supplies power for load shedding or other nonemergency loads, the emergency loads take priority over the other loads, and those other loads may be dropped to support the emergency loads.

As you study Article 700, keep in mind that emergency systems are essentially lifelines for people. The entire article is based on keeping those lifelines from breaking.

PART I. GENERAL

700.1 Scope. Article 700 applies to the installation, operation, and maintenance of emergency systems consisting of circuits and equipment intended to supply illumination or power within 10 seconds [700.12] when the normal electrical supply is interrupted. Emergency power systems are those systems legally required and classed as emergency by a governmental agency having jurisdiction. These systems are intended to automatically supply illumination and/or power essential for safety to human life. **Figure 700-1**

FPN No. 3: Emergency power systems are generally installed where artificial illumination is required for safe exiting and for panic control in buildings subject to occupancy by large numbers of persons, such as hotels, theaters, sports arenas, health care facilities, and similar institutions.

Emergency power systems may also provide power to maintain life, fire detection and alarm systems, elevators, fire pumps, public safety communications systems, industrial processes where current interruption would produce serious life safety or health hazards, and similar functions.

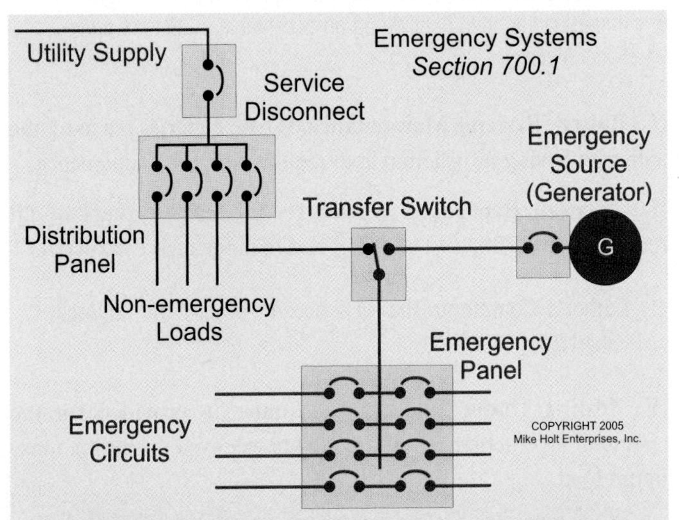
Emergency Systems
Section 700.1
Utility Supply
Service
Disconnect
Emergency
Source
(Generator)
Transfer Switch
Distribution
Panel
G
Non-emergency
Loads
Emergency
Panel
Emergency
Circuits
COPYRIGHT 2005
Mike Holt Enterprises, Inc.
Article 700 applies to the installation, operation, and maintenance of emergency systems for illumination and/or power within 10 seconds of the interruption of the normal electrical supply.

Figure 700–1

FPN No. 4: For specific locations where emergency lighting is required, see NFPA 101, *Life Safety Code*.

700.2 Application of Other Articles. Except as modified by this article, all other requirements contained in Chapters 1 through 4 of the *NEC* apply.

700.3 Equipment Approval. The authority having jurisdiction must approve all equipment used for the emergency system.

Author's Comment: The authority having jurisdiction determines the approval of equipment. This means that the AHJ can reject an installation of listed equipment and can approve the use of unlisted equipment. Given our highly litigious society, approval of unlisted equipment is becoming increasingly difficult to obtain.

700.4 Tests and Maintenance. Emergency power system testing consists of acceptance testing and operational testing.

(A) Conduct or Witness Test. To ensure that the emergency power system meets or exceeds the original installation specification, the authority having jurisdiction must conduct or witness an acceptance test of the emergency system upon completion and periodically afterward [700.4(B)].

(B) Tested Periodically. Emergency power systems must be periodically tested to ensure that adequate maintenance has been performed and that the systems are in proper operating condition.

Author's Comment: Running the emergency power system under load is often considered an acceptable method of operational testing.

(C) Battery Systems Maintenance. Where batteries are used, the authority having jurisdiction is to require periodic maintenance.

(D) Written Record. A written record must be kept of all required tests [700.4(A) and (B)] and maintenance [700.4(C)].

Author's Comment: The *NEC* doesn't specify the required record retention period.

(E) Testing Under Load. Means must be provided for the testing of all emergency lighting and power systems during maximum load.

700.5 Capacity.

(A) Capacity and Rating. An emergency system power source must have adequate capacity to carry safely all emergency loads that are expected to operate simultaneously.

(B) Selective Load Pickup, Load Shedding, and Peak Load Shaving. If an alternate power source has adequate capacity, it is permitted to supply emergency loads [Article 700], legally required standby loads [Article 701], and optional standby system loads [Article 702]. If the alternate power source does not have adequate capacity to carry the entire load, it must have automatic selective load pickup and load shedding to ensure adequate power in the following order of priority:

(1) The emergency circuits,

(2) The legally required standby circuits, and

(3) The optional standby circuits.

A temporary alternate source of power must be available whenever the emergency generator is out of service for more than a few hours for maintenance or repair.

700.6 Transfer Equipment.

(A) General. Transfer equipment must be automatic, identified for emergency use, and approved by the authority having jurisdiction.

(D) Use. Transfer equipment must supply only emergency loads. **Figure 700-2**

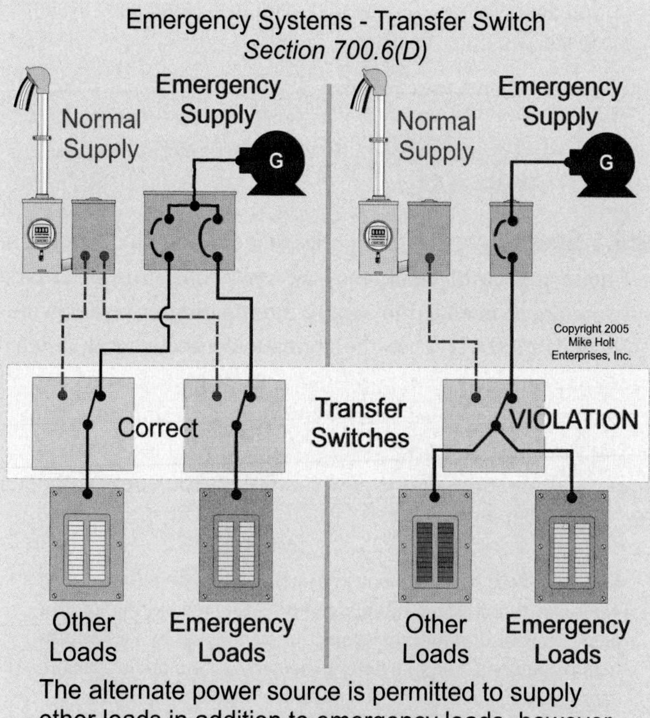

Figure 700–2

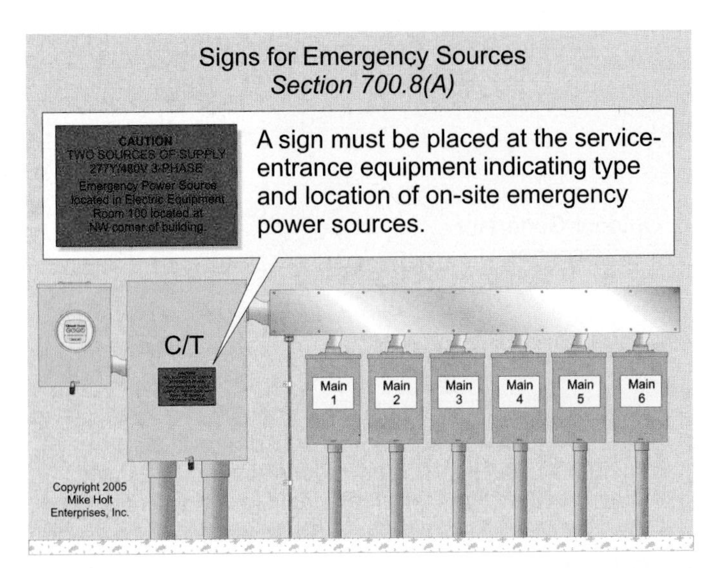
Signs for Emergency Sources
Section 700.8(A)
CAUTION
TWO SOURCES OF SUPPLY
277Y/480V 3-PHASE
Emergency Power Source located in Electric Equipment Room 100 located at NW corner of building.
A sign must be placed at the service-entrance equipment indicating type and location of on-site emergency power sources.
C/T
Main 1
Main 2
Main 3
Main 4
Main 5
Main 6
Copyright 2005 Mike Holt Enterprises, Inc.

Figure 700–3

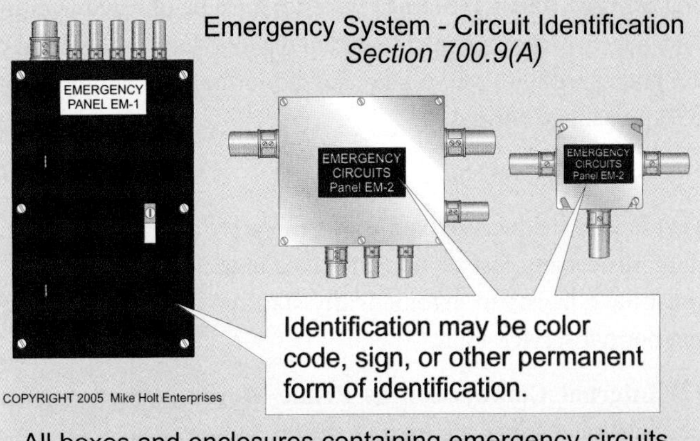
Emergency System - Circuit Identification
Section 700.9(A)
EMERGENCY PANEL EM-1
EMERGENCY CIRCUITS Panel EM-2
EMERGENCY CIRCUITS Panel EM-2
Identification may be color code, sign, or other permanent form of identification.
COPYRIGHT 2005 Mike Holt Enterprises
All boxes and enclosures containing emergency circuits must be permanently marked so they will be readily identifiable as a part of the emergency circuit or system.

Figure 700–4

Author's Comment: Multiple transfer switches are required where a single generator is used to supply both emergency loads and other loads.

700.8 Signs.

(A) Emergency Sources. A sign must be placed at the service-entrance equipment indicating the type and location of on-site emergency power sources. **Figure 700-3**

PART II. CIRCUIT WIRING

700.9 Wiring.

(A) Identification. All boxes and enclosures, including transfer switches, generators, and power panels for emergency circuits must be permanently marked as components of an emergency system. **Figure 700-4**

(B) Wiring. To ensure that a fault on the normal wiring circuits will not affect the performance of emergency wiring or equipment, all wiring from an emergency source, or emergency source distribution overcurrent protection, to emergency loads must be kept entirely independent of all other wiring and equipment, except as permitted for:

(1) Wiring in transfer equipment enclosures. **Figure 700-5**

(2) Luminaires supplied from two sources of power.

(3) A common junction box attached to luminaires supplied from two sources of power.

Two or more emergency circuits can be installed in the same raceway, cable, box, or cabinet.

(C) Wiring Design and Location. Emergency wiring circuits must be designed and located to minimize the hazards that might cause failure due to flooding, fire, icing, vandalism, and other adverse conditions.

PART III. SOURCES OF POWER

700.12 General Requirements. In the event of failure of the normal supply to the building or structure, emergency power must be available within 10 seconds. Emergency equipment must be designed and located so as to minimize the hazards that might cause complete failure due to flooding, fires, icing, and vandalism. The emergency power source must be one of the following:

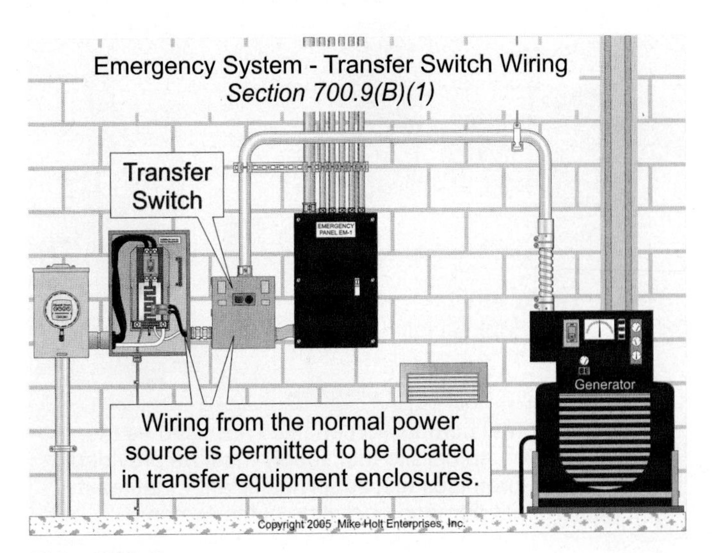
Emergency System - Transfer Switch Wiring
Section 700.9(B)(1)
Transfer Switch
Generator
Wiring from the normal power source is permitted to be located in transfer equipment enclosures.
Copyright 2005 Mike Holt Enterprises, Inc.

Figure 700–5

(A) Storage Battery. Storage batteries must be of suitable rating and capacity to supply and maintain the total load for a period of 1½ hours, without the voltage applied to the load falling below 87½ percent of normal.

(B) Generator Set.

(1) Prime Mover-Driven. A generator acceptable to the authority having jurisdiction and sized in accordance with 700.5 must have means to automatically start the prime mover when the normal service fails.

(2) Internal Combustion as Prime Movers. Where internal combustion engines are used as the prime mover, an on-site fuel supply must be provided for not less than two of hours full-demand operation of the system. **Figure 700-6**

(6) Outdoor Generator Sets. Where an outdoor-housed generator is equipped with a readily accessible disconnecting means located within sight (within 50 ft) of the building or structure, an additional disconnecting means isn't required on or at the building or structure for the generator feeder conductors that serve or pass through the building or structure. **Figure 700-7**

(C) Uninterruptible Power Supplies. Uninterruptible power supplies must comply with the applicable requirements of 700.12(A) and (B).

(D) Separate Service. An additional service installed in accordance with Article 230, where acceptable to the authority having jurisdiction, is permitted to serve as an emergency source of power. **Figure 700-8**

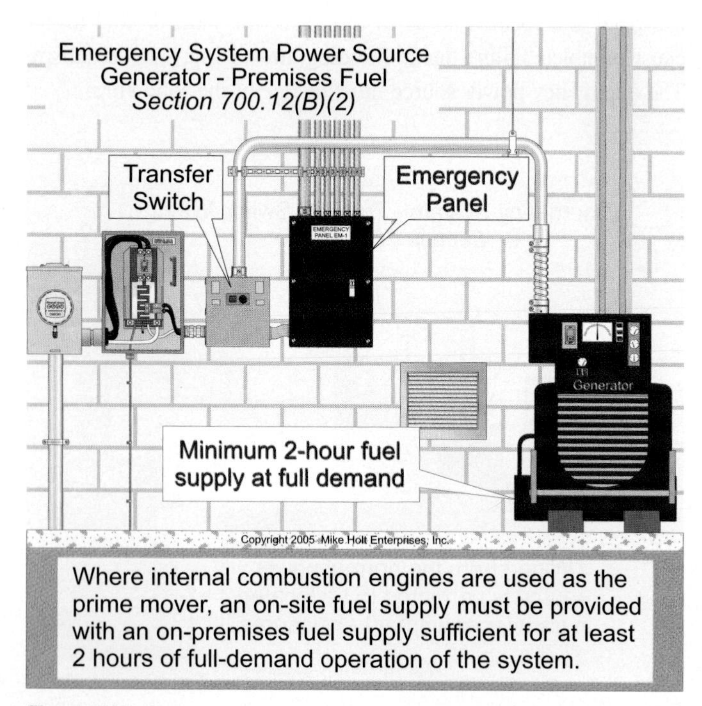
Emergency System Power Source
Generator - Premises Fuel
Section 700.12(B)(2)
Transfer Switch
Emergency Panel
Generator
Minimum 2-hour fuel supply at full demand
Copyright 2005 Mike Holt Enterprises, Inc.
Where internal combustion engines are used as the prime mover, an on-site fuel supply must be provided with an on-premises fuel supply sufficient for at least 2 hours of full-demand operation of the system.

Figure 700–6

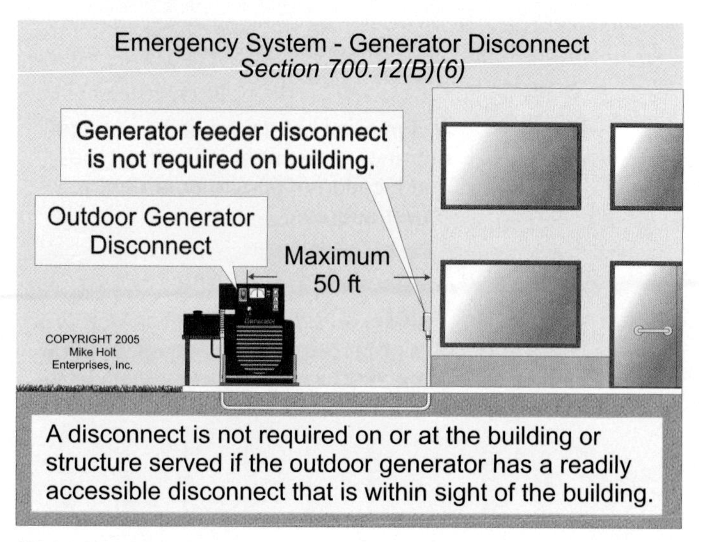
Emergency System - Generator Disconnect
Section 700.12(B)(6)
Generator feeder disconnect is not required on building.
Outdoor Generator Disconnect
Maximum 50 ft
COPYRIGHT 2005 Mike Holt Enterprises, Inc.
A disconnect is not required on or at the building or structure served if the outdoor generator has a readily accessible disconnect that is within sight of the building.

Figure 700–7

To minimize the possibility of simultaneous interruption of emergency supply, the additional service must:

(1) Be served by a separate service drop or lateral

(2) Be electrically and physically separated from all other service conductors

Author's Comment: Tapping ahead of the normal service equipment to provide power for the emergency system isn't permitted to serve as the required emergency source of power. **Figure 700-9**

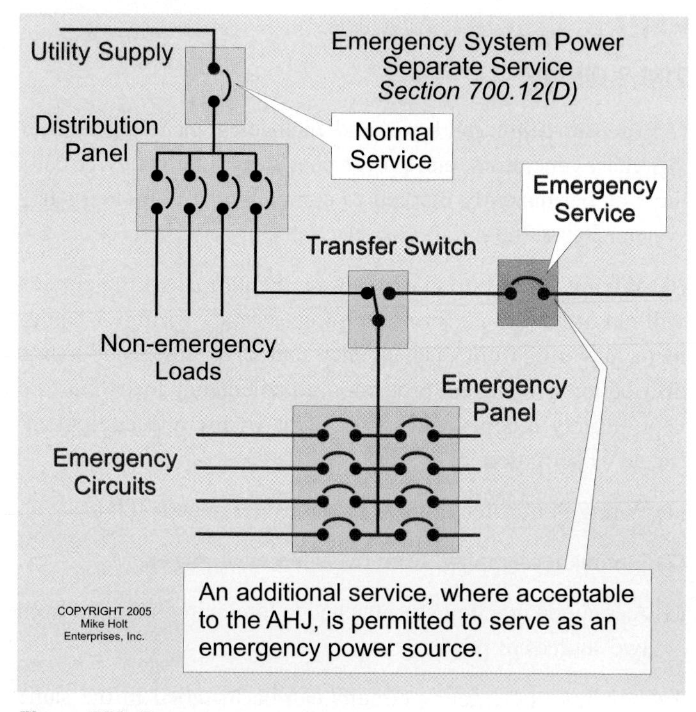
Utility Supply
Emergency System Power
Separate Service
Section 700.12(D)
Distribution Panel
Normal Service
Emergency Service
Transfer Switch
Non-emergency Loads
Emergency Panel
Emergency Circuits
COPYRIGHT 2005 Mike Holt Enterprises, Inc.
An additional service, where acceptable to the AHJ, is permitted to serve as an emergency power source.

Figure 700–8

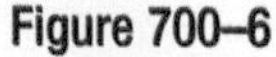
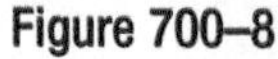

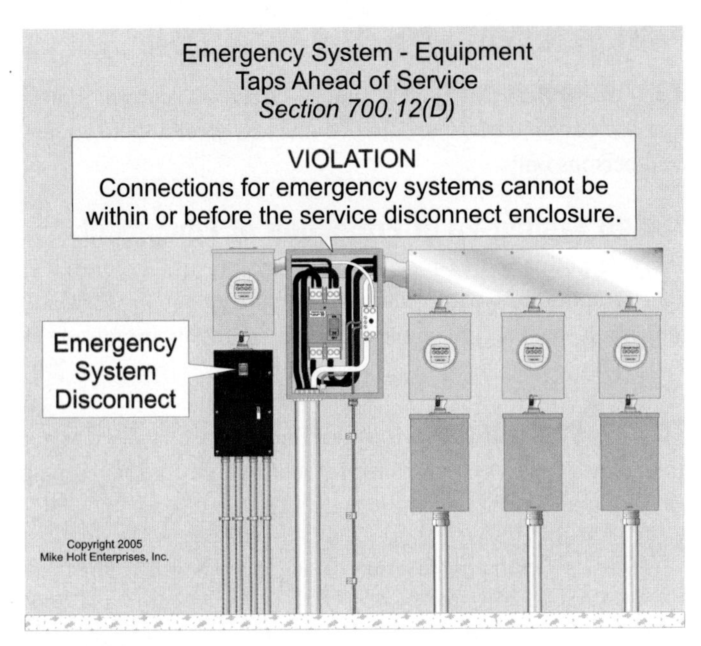

Emergency System - Equipment
Taps Ahead of Service
Section 700.12(D)
VIOLATION
Connections for emergency systems cannot be within or before the service disconnect enclosure.
Emergency System Disconnect
Copyright 2005
Mike Holt Enterprises, Inc.

Figure 700–9

(F) Unit Equipment. Individual unit equipment (emergency lighting battery pack) must consist of the following: **Figure 700-10**

(1) A rechargeable battery,

(2) A battery charging means,

(3) Provisions for one or more lamps mounted on the equipment, or terminals for remote lamps (or both), and

(4) A relaying device arranged to energize the lamps automatically upon failure of the supply to the unit equipment.

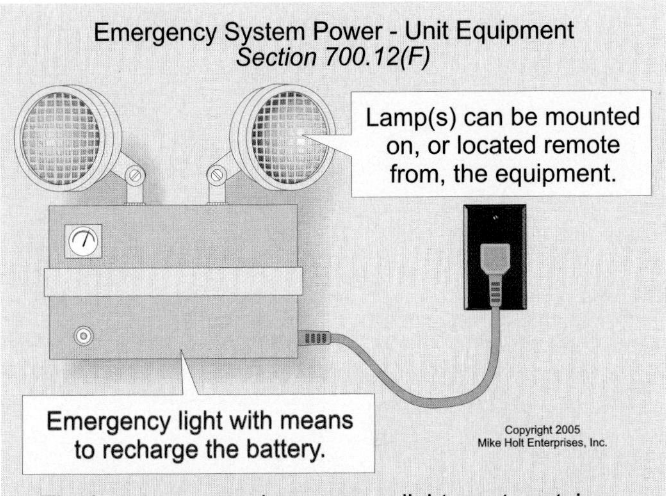

Emergency System Power - Unit Equipment
Section 700.12(F)
Lamp(s) can be mounted on, or located remote from, the equipment.
Emergency light with means to recharge the battery.
Copyright 2005
Mike Holt Enterprises, Inc.

The battery powered emergency light must contain a relay device to automatically energize the lamp when there is a power failure to the unit.

Figure 700–10

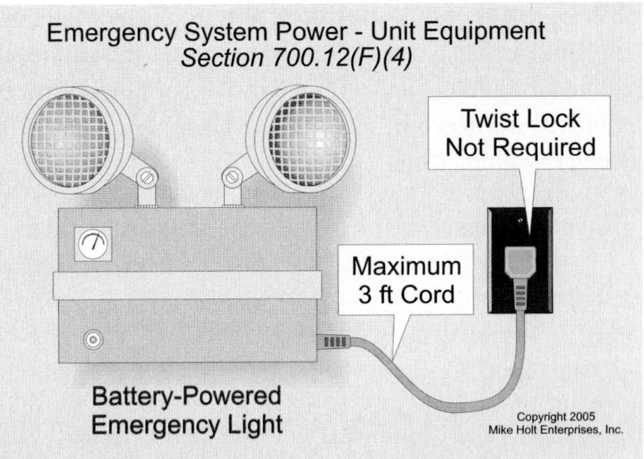

The branch circuit that supplies unit equipment (battery pack) must be the same branch circuit that supplies the normal lighting in the area, but the unit equipment must be connected ahead of any local switches.

Figure 700–11

Emergency lighting battery pack equipment must be permanently fixed in place. Flexible cord-and-plug connection (twist lock not required) is permitted for emergency lighting battery pack equipment that is designed for this purpose, provided the cord doesn't exceed 3 ft. **Figure 700-11**

The branch-circuit wiring that supplies emergency lighting battery pack equipment must be the same branch-circuit wiring that supplies the normal lighting in the area, but the emergency lighting battery pack equipment must be connected ahead of any local switches. The branch circuit that feeds the emergency lighting battery pack equipment must be clearly identified at the distribution panel in accordance with 110.21 and 408.4.

> **Author's Comment:** There are two reasons why the emergency lighting battery packs must be connected ahead of the switch controlling the normal area lighting: (1) in the event of a power loss to the lighting circuit, the emergency battery lighting packs will activate and provide emergency lighting for people to exit the building, and (2) the emergency lighting battery packs will not turn on when the switch controlling normal lighting is turned off!

> **CAUTION:** *Unit equipment must not be connected to the emergency circuit, because if it is, then it will not operate when normal power is lost!*

PART IV. EMERGENCY SYSTEM CIRCUITS FOR LIGHTING AND POWER

700.15 Loads on Emergency Branch Circuits. Emergency circuits must supply no loads other than those required for emergency use.

700.16 Emergency Illumination. Emergency lighting systems must be designed and installed so that the failure of any individual lighting element, such as the burning out of a light bulb, will not leave in total darkness any space that requires emergency illumination.

Author's Comment: This means that a single remote head is never sufficient for an area. A minimum of two lighting heads is always required.

PART VI. OVERCURRENT PROTECTION

700.25 Accessibility. The branch-circuit overcurrent protection devices for emergency circuits must be accessible to authorized persons only.

700.26 Ground-Fault Protection of Equipment. The alternate power source for emergency systems isn't required to have ground-fault protection of equipment, but ground-fault indication of the emergency power source is required by 700.7(D).

700.27 Coordination. Overcurrent protection devices for emergency power systems must be selectively coordinated with all supply-side overcurrent protective devices.

Author's Comment: See Article 100 for the definition of "Selective Coordination."

Article 700 Questions

1. To ensure that the emergency system meets or exceeds the original installation specification, the _____ must conduct or witness an acceptance test of the complete emergency system upon installation and periodically afterward.

   (a) electrical engineer
   (b) authority having jurisdiction
   (c) qualified person
   (d) manufacturer's representative

2. An emergency system must have adequate capacity to safely carry _____ that are expected to operate simultaneously on the emergency system.

   (a) all of the loads
   (b) 80 percent of the total loads
   (c) up to 200A of the loads
   (d) 300 percent of the total loads

3. A sign _____ be placed at the service-entrance equipment indicating the type and location of on-site emergency power sources.

   (a) must
   (b) should
   (c) is not required to
   (d) is not allowed to

4. Unit equipment (battery packs) must be on the same branch circuit that serves the normal lighting in the area and connected _____ any local switches.

   (a) with
   (b) ahead of
   (c) after
   (d) none of these

5. All manual switches for controlling emergency circuits must be in locations convenient to authorized persons responsible for their _____.

   (a) maintenance
   (b) actuation
   (c) inspection
   (d) evaluation

6. Overcurrent protection devices for emergency power systems _____ all supply-side overcurrent protective devices.

   (a) must be selectively coordinated with
   (b) are allowed to be selectively coordinated with
   (c) must be the same amperage as
   (d) must be a larger amperage than

Notes

ARTICLE

701 Legally Required Standby Power Systems

Introduction

In the hierarchy of electrical systems, Article 700 Emergency Power Systems get first priority. Taking the number two spot are legally required standby power systems, which fall under Article 701. But there are other differences. For example, legally required systems must supply standby power in 60 seconds or less after a power loss, instead of the 10 seconds or less required of emergency systems.

Article 701 systems do not serve the purpose of directly protecting human life. They supply specific loads that, if shut down, would create hazards or impede rescue operations. Thus, hospital communications systems fall under Article 701—evacuation instructions announced over the public address system are part of a rescue operation.

Article 700 basically applies to systems or equipment required to protect people who are in an emergency and trying to get out, while Article 701 basically applies to systems or equipment needed to aid the people *responding* to the emergency. For example, Article 700 lighting provides an exit path. But, Article 701 lighting might illuminate the fire hydrants and switchgear areas.

PART I. GENERAL

701.1 Scope. The provisions of Article 701 apply to the installation, operation, and maintenance of legally required standby systems consisting of circuits and equipment intended to supply illumination or power when the normal electrical supply or system is interrupted.

Author's Comment: Legally required standby systems provide electric power to aid in firefighting, rescue operations, control of health hazards, and similar operations. When normal power is lost, legally required systems must be capable of supplying standby power in 60 seconds or less [701.11], instead of the 10 seconds or less required of emergency systems [700.12].

701.2 Definitions.

Legally Required Standby Systems. Legally required standby systems are those systems classified as legally required by any governmental agency having jurisdiction. These systems are intended to automatically supply power to selected loads (other than those classed as emergency systems) in the event of failure of the normal power source.

FPN: Legally required standby systems typically supply loads such as heating and refrigeration systems, communications systems, ventilation and smoke removal systems, sewage disposal, lighting systems, and industrial processes that, when stopped, could create hazards, or hamper rescue or firefighting operations.

701.3 Application of Other Articles. Except as modified by this article, all other applicable requirements of the *NEC* apply.

701.4 Equipment Approval. The authority having jurisdiction must approve all equipment.

701.5 Tests and Maintenance. Legally required standby system testing consists of acceptance testing and operational testing, and written records of both types of testing and maintenance must be maintained [701.5(D)].

(A) Conduct or Witness Test. To ensure that the legally required standby system meets or exceeds the original installation specification, the authority having jurisdiction must conduct or witness an acceptance test of the emergency system upon completion of the installation, and periodically afterward [701.5(D)].

(B) Tested Periodically. Legally required standby systems must be periodically tested to ensure that adequate maintenance has been performed and that the systems are in proper operating condition.

> **Author's Comment:** Running the legally required standby system to power the loads of the facility is often considered an acceptable method of operational testing.

(C) Battery Systems Maintenance. Where batteries are used, the authority having jurisdiction is to require periodic maintenance.

(D) Written Record. A written record must be kept of all required tests [701.5(A) and (B)] and maintenance [701.5(C)].

> **Author's Comment:** The *NEC* doesn't specify the required record retention period.

(E) Testing Under Load. Means for testing all legally required standby systems, with the maximum anticipated load condition, must be provided.

701.6 Capacity and Rating.

A legally required standby system must have adequate capacity to carry safely all loads that are expected to operate simultaneously. Legally required standby system equipment must be suitable for the maximum available fault current at line terminals.

The legally required standby alternate power source is permitted to supply legally required standby and optional standby system loads under any of the following conditions:

(1) The alternate power source has adequate capacity to handle all connected loads.

(2) Where automatic selective load pickup and load shedding is provided to ensure adequate power to the legally required standby circuits.

701.7 Transfer Equipment.

(A) General. Transfer equipment must be identified for standby use, and be approved by the authority having jurisdiction.

701.9 Signs.

(A) Mandated Standby. A sign must be placed at the service-entrance equipment indicating type and location of on-site legally required standby power sources. **Figure 701-1**

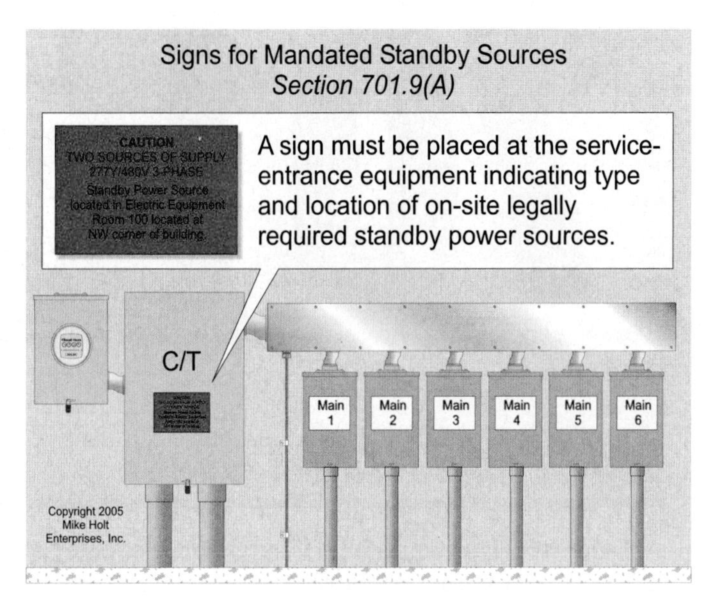

Signs for Mandated Standby Sources
Section 701.9(A)
CAUTION
TWO SOURCES OF SUPPLY
277Y/480V 3-PHASE
Standby Power Source
located in Electric Equipment
Room 100 located at
NW corner of building.
A sign must be placed at the service-entrance equipment indicating type and location of on-site legally required standby power sources.
C/T
Main 1
Main 2
Main 3
Main 4
Main 5
Main 6
Copyright 2005
Mike Holt
Enterprises, Inc.

Figure 701–1

PART II. CIRCUIT WIRING

701.10 Wiring.

Legally required standby system wiring is permitted to occupy the same raceways, cables, boxes, and cabinets with other general wiring.

PART III. SOURCES OF POWER

701.11 Legally Required Standby Systems.

If the normal supply fails, legally required power must be available within 60 seconds. The supply system for the legally required standby power source must be one of the following:

(A) Storage Battery. Storage batteries must be of suitable rating and capacity to supply and maintain the total load for a period of 1½ hours, without the voltage applied to the load falling below 87½ percent of normal.

(B) Generator Set.

(1) Prime Mover-Driven. A generator acceptable to the authority having jurisdiction and sized in accordance with 701.6 must have the means to automatically start the prime mover on failure of the normal service.

(2) Internal Combustion Engines as Prime Movers. Where internal combustion engines are used as the prime mover, an on-site fuel supply must be provided for not less than two hours of full-demand operation of the system. **Figure 701-2**

(5) Outdoor Generator Sets. Where an outdoor-housed generator is equipped with a readily accessible disconnecting means located within sight (within 50 ft) of the building or structure, an additional disconnecting means isn't required on or at the building or structure for the generator feeder conductors that serve or pass through the building or structure. **Figure 701-3**

(C) Uninterruptible Power Supplies. Uninterruptible power supplies must comply with 701.11(A) and (B).

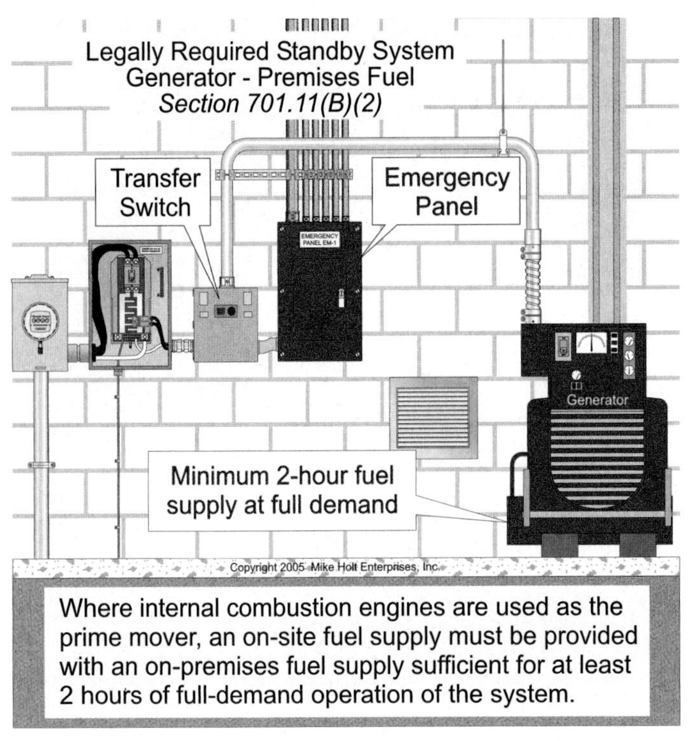

Legally Required Standby System
Generator - Premises Fuel
Section 701.11(B)(2)
Transfer Switch
Emergency Panel
Generator
Minimum 2-hour fuel supply at full demand
Copyright 2005 Mike Holt Enterprises, Inc.
Where internal combustion engines are used as the prime mover, an on-site fuel supply must be provided with an on-premises fuel supply sufficient for at least 2 hours of full-demand operation of the system.

Figure 701–2

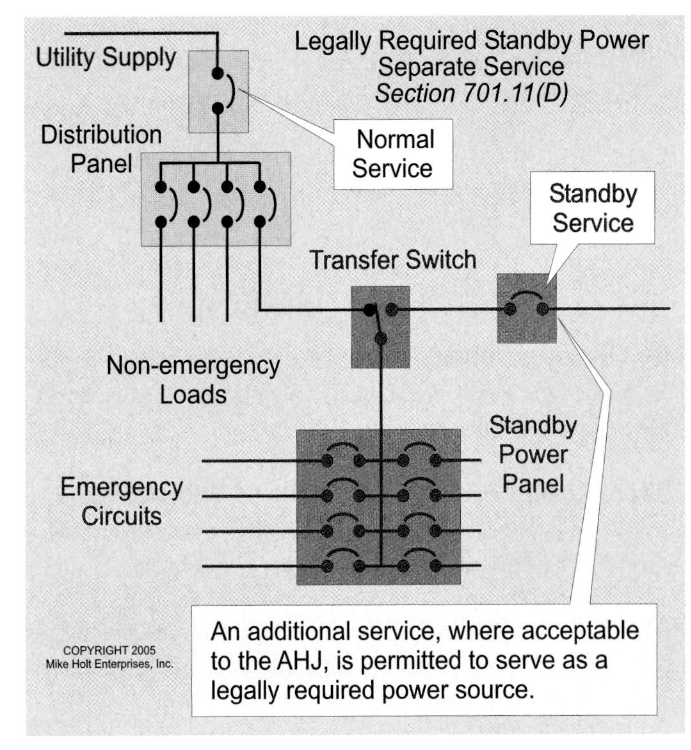

Figure 701–4

(D) Separate Service. An additional service installed in accordance with Article 230, where acceptable to the authority having jurisdiction, is permitted to serve as a legally required source of power. **Figure 701-4**

To minimize the possibility of simultaneous interruption of the legally required standby supply, a separate service drop or lateral must be electrically and physically separated from all other service conductors.

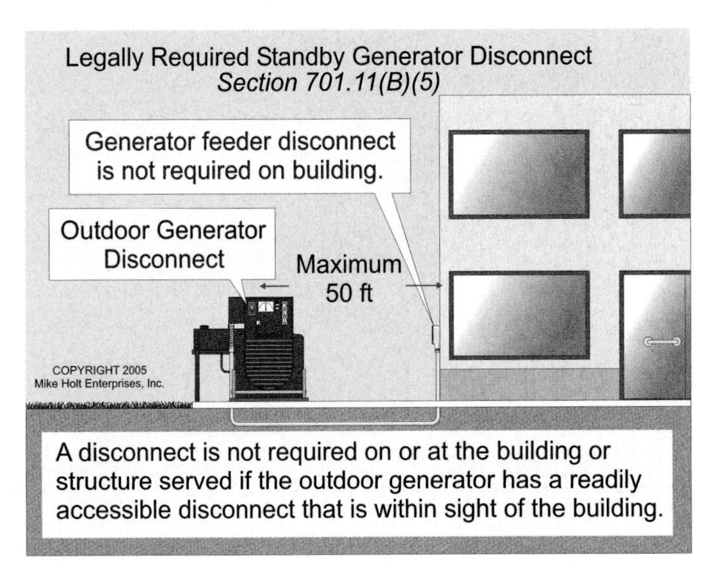

Figure 701–3

(E) Connection Ahead of Service Disconnecting Means. Where acceptable to the authority having jurisdiction, connection ahead of, but not within, the same cabinet, enclosure, or vertical switchboard section as the service disconnecting means is permitted. **Figure 701-5**

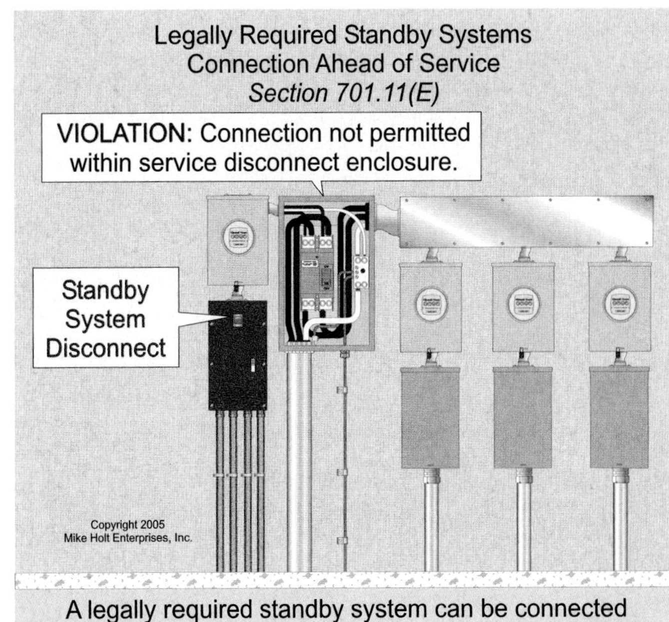

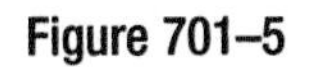

Figure 701–5

To prevent simultaneous interruption of supply, the legally required standby service disconnect must be sufficiently separated from the normal service disconnection means.

FPN: See 230.82 for equipment permitted on the supply side of a service disconnecting means.

PART IV. OVERCURRENT PROTECTION

701.15 Accessibility. The branch-circuit overcurrent protection devices for legally required standby circuits must be accessible to authorized persons only.

701.17 Ground-Fault Protection of Equipment. The alternate source for legally required standby power systems isn't required to have ground-fault protection of equipment.

701.18 Coordination. Overcurrent protection devices for legally required power systems must be selectively coordinated with all supply-side overcurrent protective devices.

Article 701 Questions

1. A legally required standby system is intended to automatically supply power to _____.

   (a) those systems classed as emergency systems
   (b) selected loads
   (c) a and b
   (d) none of these

2. A written record must be kept of required tests and maintenance on legally required standby systems.

   (a) True
   (b) False

3. The alternate power source (generator, UPS, etc.) is permitted to supply legally required standby and optional standby system loads where automatic selective load pickup and load shedding is provided as needed to ensure adequate power to the legally required standby systems.

   (a) True
   (b) False

4. Audible and visual signal devices must be provided on legally required standby systems, where practicable, to indicate _____.

   (a) derangement of the standby source
   (b) that the standby source is carrying load
   (c) that the battery charger is not functioning
   (d) all of these

5. Where acceptable to the authority having jurisdiction (AHJ), connections ahead of and not within the same cabinet, enclosure, or vertical switchboard section as the service disconnecting means are permitted for _____ standby service.

   (a) emergency
   (b) legally required
   (c) optional
   (d) all of these

Notes

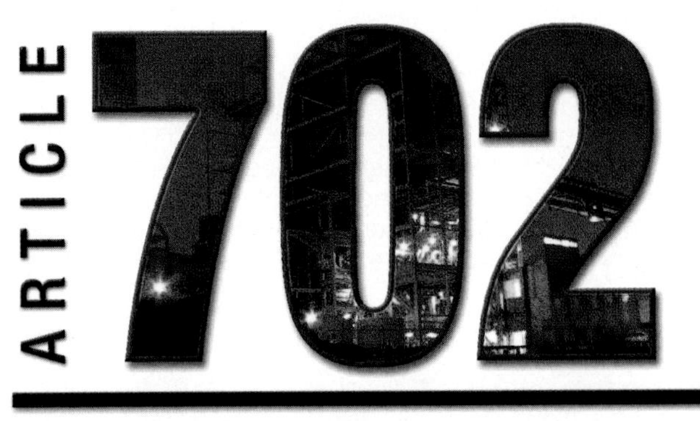

ARTICLE 702 Optional Standby Power Systems

Introduction

Taking third priority after emergency and then legally required systems, optional standby systems protect public or private facilities or property where life safety doesn't depend on the performance of the system. These systems are not required for rescue operations.

Suppose a glass plant loses power. Once glass hardens in the equipment—which it will do when process heat is lost—the plant is going to suffer a great deal of downtime and expense before it can resume operations. An optional standby system can prevent this loss.

You'll see these systems in facilities where loss of power could cause economic loss or business interruptions. Data centers can lose millions of dollars from a single minute of lost power. A chemical or pharmaceutical plant could lose an entire batch from a single momentary power glitch. In many cases, the lost revenue cannot be recouped.

When the power went out in Chicago in August a few years ago, restaurants lost millions of dollars in food inventory due to a loss of refrigeration. But many firms were using optional standby systems and didn't suffer such huge losses. In an extended outage, where the logistics of fuel delivery becomes a problem, optional standby systems would have to wait in line behind legally required standby systems, which would have to wait in line behind emergency systems.

PART I. GENERAL

702.1 Scope. The systems covered by Article 702 consist of those that are permanently installed, including prime movers, and those that are arranged for a connection to a premises wiring system from a portable alternate power supply. **Figure 702-1**

> **Author's Comment:** A portable generator that provides temporary power, like those used on construction sites, doesn't fall within the scope of Article 702 unless the generator is connected to the premises wiring. **Figure 702-2**

702.2 Definition.

Optional Standby Systems. Optional standby systems are intended to supply power to public or private facilities or property where life safety doesn't depend on the performance of the system. Optional standby systems are intended to supply on-site generated power to selected loads either automatically or manually.

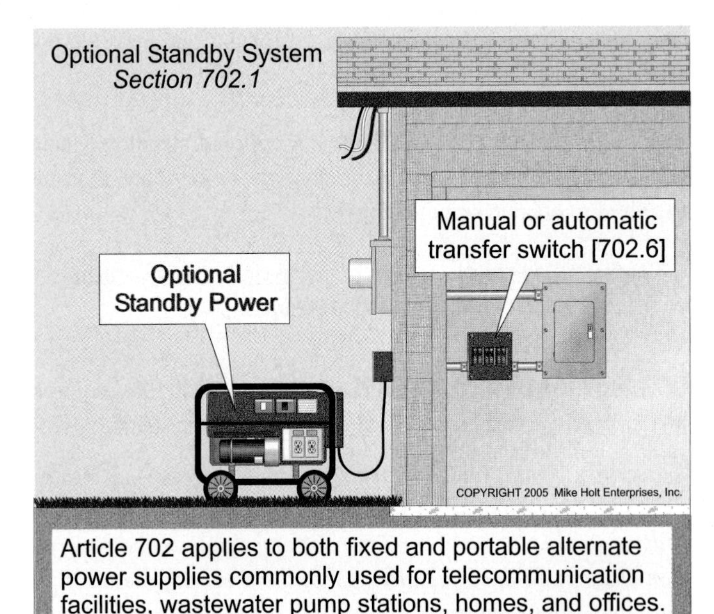
Optional Standby System
Section 702.1
Manual or automatic transfer switch [702.6]
Optional Standby Power
COPYRIGHT 2005 Mike Holt Enterprises, Inc.
Article 702 applies to both fixed and portable alternate power supplies commonly used for telecommunication facilities, wastewater pump stations, homes, and offices.

Figure 702–1

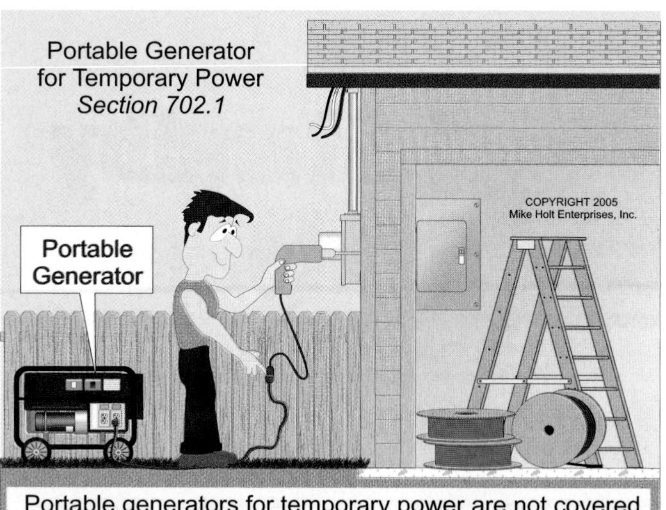
Portable Generator
for Temporary Power
Section 702.1
COPYRIGHT 2005
Mike Holt Enterprises, Inc.
Portable
Generator
Portable generators for temporary power are not covered
by Article 702 if not connected to the premises wiring.

Figure 702–2

FPN: Optional standby systems are typically installed to provide an alternate source of electric power for such facilities as industrial and commercial buildings, farms and residences, and to serve loads such as heating and refrigeration systems, data-processing and communications systems, and industrial processes that, when stopped during any power outage, could cause discomfort, economic loss, serious interruption of the process, damage to the product or process, or the like.

702.3 Application of Other Articles. Except as modified by this article, all other applicable *NEC* requirements apply.

702.4 Equipment Approval. The authority having jurisdiction must approve all equipment used for the optional standby power system.

702.5 Capacity and Rating. An optional standby system must have adequate capacity to carry safely all loads that are expected to operate simultaneously.

Author's Comment: The user of the optional standby system can select the load(s) connected to the system.

702.6 Transfer Equipment. A transfer switch is required for all fixed or portable optional standby power systems.

Exception: Temporary connection of a portable generator without transfer equipment is permitted where written safety procedures are in place and conditions of maintenance and supervision ensure that only qualified persons will service the installation, and where the normal supply is physically isolated by a lockable disconnecting means or by the disconnection of the normal supply conductors.

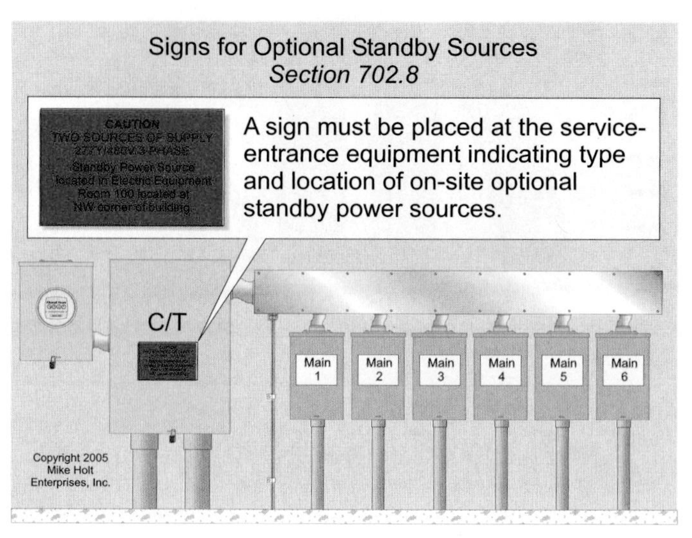
Signs for Optional Standby Sources
Section 702.8
CAUTION
TWO SOURCES OF SUPPLY
Standby Power Source
NW corner of building
A sign must be placed at the service-entrance equipment indicating type and location of on-site optional standby power sources.
C/T
Main 1
Main 2
Main 3
Main 4
Main 5
Main 6
Copyright 2005
Mike Holt
Enterprises, Inc.

Figure 702–3

702.8 Signs.

(A) Standby Power Sources. A sign that indicates type and location of on-site optional standby power sources must be placed at the service-entrance equipment. **Figure 702-3**

PART II. CIRCUIT WIRING

702.9 Wiring. The optional standby system wiring can occupy the same raceways, cables, boxes, and cabinets with other general wiring.

PART III. GROUNDING AND BONDING

702.10 Portable Generator Grounding and Bonding.

Author's Comment: Generators must be grounded and bonded in accordance with 250.30 as a separately derived system, or in accordance with 250.34 if the generator is a portable or vehicle-mounted unit. **Figure 702-4**

PART IV. SOURCES OF POWER

702.11 Outdoor Generator Sets. Where an outdoor-housed generator is equipped with a readily accessible disconnecting means located within sight (within 50 ft) of the building or structure, an additional disconnecting means isn't required on or at the building or structure for the generator feeder conductors that serve or pass through the building or structure.

Author's Comment: Similar requirements are contained in 700.12(B)(6) for emergency power systems and 701.11(B)(5) for legally required standby systems.

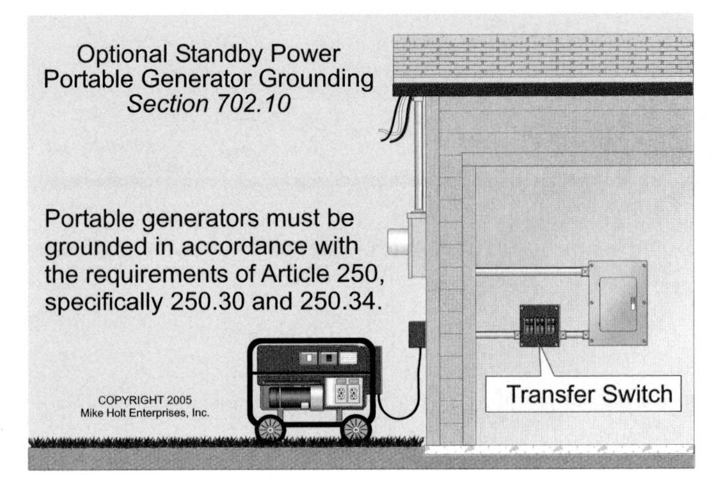
Optional Standby Power
Portable Generator Grounding
Section 702.10
Portable generators must be grounded in accordance with the requirements of Article 250, specifically 250.30 and 250.34.
COPYRIGHT 2005
Mike Holt Enterprises, Inc.
Transfer Switch

Figure 702–4

Article 702 Questions

1. Article 702 applies to _____ generators used for backup power to telecommunications facilities, water and wastewater pump stations, as well as homes and offices.

   (a) permanently installed (b) portable (c) a and b (d) none of these

2. Optional standby systems must have adequate capacity and rating for the supply of _____.

   (a) all emergency lighting and power loads
   (b) all equipment intended to be operated at one time
   (c) 100 percent of the appliance loads and 50 percent of the lighting loads
   (d) 100 percent of the lighting loads and 75 percent of the appliance loads

3. For optional standby systems, the temporary connection of a portable generator without transfer equipment is permitted, where written safety procedures are in place and conditions of maintenance and supervision ensure that only qualified persons will service the installation, and where the normal supply is physically isolated by _____.

   (a) a lockable disconnecting means
   (b) the disconnection of the normal supply conductors
   (c) an extended power outage
   (d) a or b

4. A sign must be placed at the service-entrance equipment indicating the _____ of on-site optional standby power sources.

   (a) type (b) location (c) manufacturer (d) a and b

5. Where a generator for an optional standby system is installed outdoors and equipped with a readily accessible disconnecting means located _____, an additional disconnecting means is not required where ungrounded conductors serve or pass through the building or structure.

   (a) inside the building or structure
   (b) within sight of the building or structure
   (c) inside the generator enclosure
   (d) a or c

Article 720 Circuits and Equipment Operating at Less Than 50 Volts

Introduction

Article 720 applies to electrical installations that operate below 50V, direct current or alternating current [720.1]. Article 720 was originally developed for installations known as "farm lighting plants," which operate at 32V (six 6V batteries connected in series, allowing for voltage drop). **Figure 720-1**

Article 720 doesn't apply to installations within the scope of Articles 411 Lighting Systems Operating at 30V or Less, 725 Remote-Control, Signaling, and Power-Limited Circuits, or 760 Fire Alarm Systems [720.2].

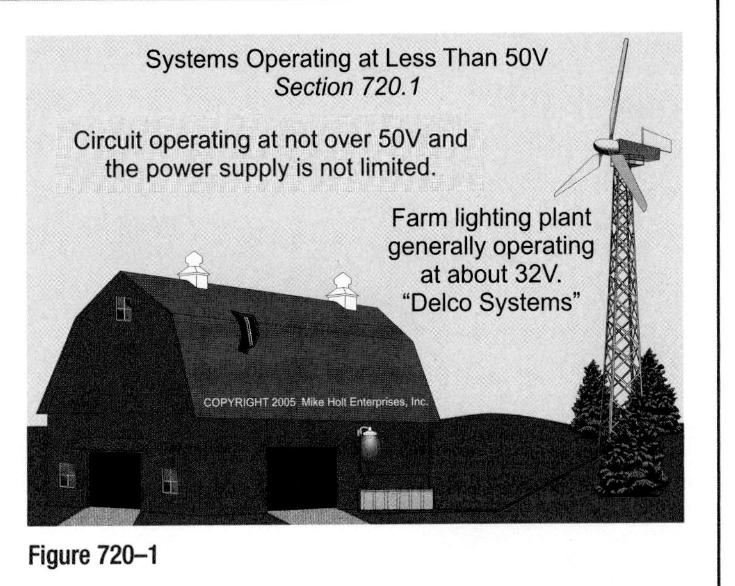

Systems Operating at Less Than 50V
Section 720.1
Circuit operating at not over 50V and the power supply is not limited.
Farm lighting plant generally operating at about 32V. "Delco Systems"
COPYRIGHT 2005 Mike Holt Enterprises, Inc.

Figure 720–1

Article 720 Questions

1. Conductors for an appliance circuit supplying more than one appliance or appliance receptacle in an installation operating at less than 50V must not be smaller than _____ AWG copper or equivalent.

   (a) 18 (b) 14 (c) 12 (d) 10

2. Circuits and equipment operating at less than 50V must use receptacles that are rated at not less than _____.

   (a) 10A (b) 15A (c) 20A (d) 30A

Class 1, Class 2, and Class 3 Remote-Control, Signaling, and Power-Limited Circuits

Introduction

Circuits that fall under Article 725 are remote-control, signaling, and power-limited circuits that aren't an integral part of a device or appliance. Article 725 includes circuits for burglar alarms, access control, sound, nurse call, intercoms, some computer networks, some lighting dimmer controls, and some low-voltage industrial controls.

Let's take a quick look at the types of circuits:

- A remote-control circuit controls other circuits through a relay or solid-state device, such as a motion-activated security lighting circuit.
- A signaling circuit that provides an output that is a signal or indicator, such as a buzzer, flashing light, or annunciator.
- A power-limited circuit operates at no more than 30V and 1,000 VA.

We'll cover these definitions in more depth later and explain the differences between Class 1, Class 2, and Class 3 circuits.

The purpose of Article 725 is to allow for the fact that these circuits "are characterized by usage and power limitations that differentiate them from electric light and power circuits." Article 725 provides alternative requirements for minimum wire sizes, derating factors, overcurrent protection, insulation requirements, wiring methods, and materials.

As with all other articles in Chapters 5 and 6, the wiring methods required by Chapters 1 through 4 apply [90.3]. But because of the inherently lower danger of fire risk with qualifying circuits, Article 725 specifies conditions where these methods aren't required. For example, a Class 2 splice isn't required to be in a box or enclosure. At the same time, additional requirements apply to ensure safety. While there's not a net compromise on safety, there's a net cost-savings.

Article 725 consists of four parts. Part I provides general information, Part II pertains to Class 1 cabling, and Part III pertains to Class 2 and Class 3 cabling. The key to understanding and applying each of the three parts is knowing the voltage and energy levels of the cabling involved and the purposes of that cabling. Article 725 allows you to save time and money when working with particular types of circuits.

PART I. GENERAL

725.1 Scope. Article 725 contains the requirements for remote-control, signaling, and power-limited circuits that aren't an integral part of a device or appliance.

> **FPN:** Class 2 and Class 3 circuits have electrical power and voltage limitations that differentiate them from electric light and power circuits. Alternative requirements are given with regard to minimum wire sizes, derating factors, overcurrent protection, insulation requirements, wiring methods, and materials.

Author's Comments:

- To understand when to apply the requirements of Article 725 for control, signaling, and power-limited circuits, you must understand the following Article 100 Definitions:
    - *Remote-Control Circuit.* Any electric circuit that controls another circuit through a relay or equivalent device. An example is the 120V circuit that operates the coil of a motor starter or lighting contactor, or the 24V circuit for a garage door opener. **Figure 725-1**
    - *Signaling Circuit.* Any electric circuit that energizes signaling equipment. Examples include doorbells, buzzers, signal lights, annunciators, burglar alarm, and other detection indication or alarm devices. **Figure 725-2**

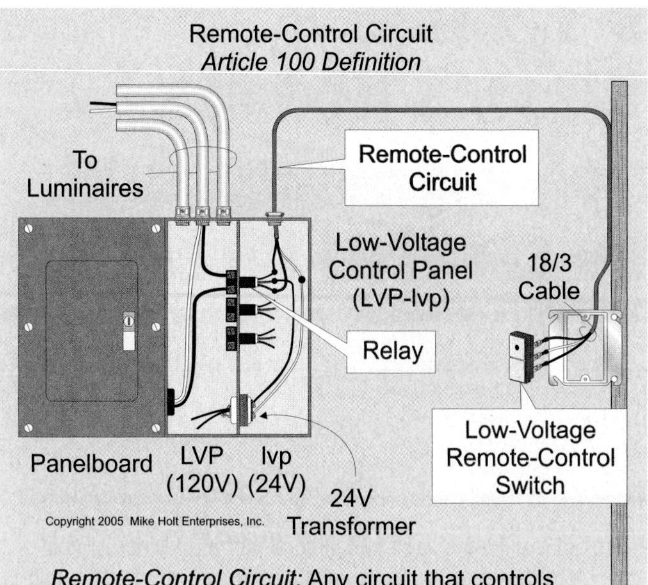

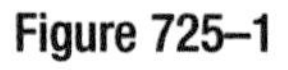
Figure 725–1

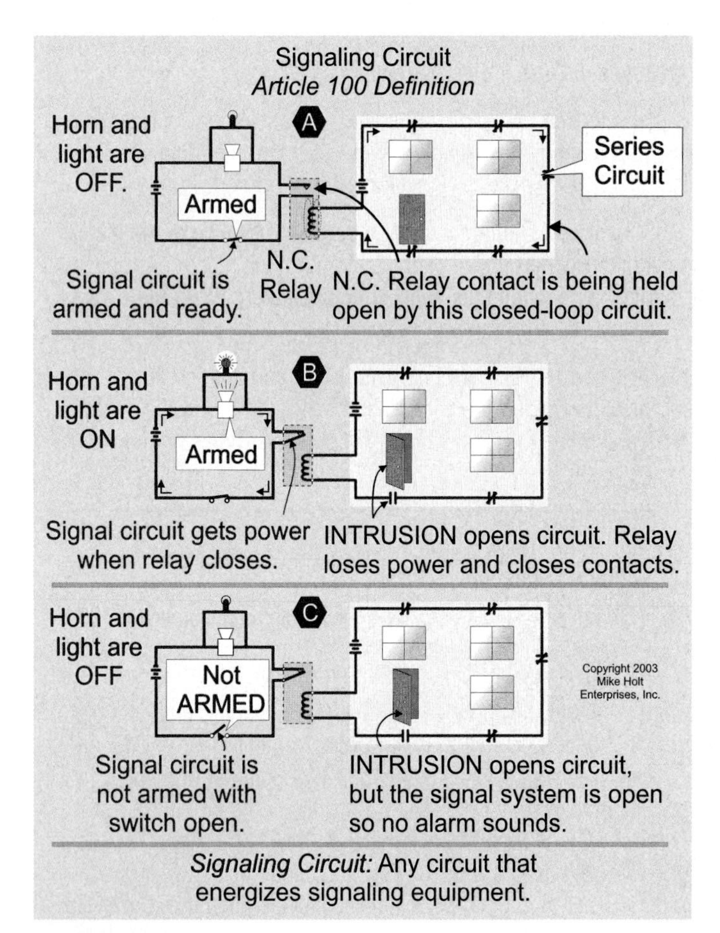

Figure 725–2

725.2 Definitions.

Abandoned Cable. Cable that isn't terminated at equipment, and not identified for future use with a tag, is considered as abandoned for the purpose of this article.

> **Author's Comment:** 725.3(B) requires abandoned cables to be removed.

Class 1 Circuit. That wiring system between the load side of the Class 1 circuit overcurrent device and the connected equipment such as relays, controllers, lights, audible devices, etc.

> **Author's Comment:** Class 1 remote-control circuits are commonly used to operate motor controllers in conjunction with moving equipment or mechanical processes, such as elevators, conveyors, and shunt-trip circuits for circuit breakers.

- Class 1 Nonpower-Limited Circuit. Class 1 nonpower-limited circuits can operate at up to 600V and the power output isn't limited [725.21(B)].
- Class 1 Power-Limited Circuit. Power-limited Class 1 circuits can be either ac or dc from a power source that limits the output to 30V and 1,000 VA [725.21(A)]. Power-limited Class 1 circuits are necessary when the energy demands of the system exceed the energy limitations of Class 2 or Class 3 circuits (energy limit of 100 VA or less) [Chapter 9 Table 11(A)]. **Figure 725-3**

Class 2 Circuit. The wiring system between the load side of a Class 2 power source and the connected Class 2 equipment.

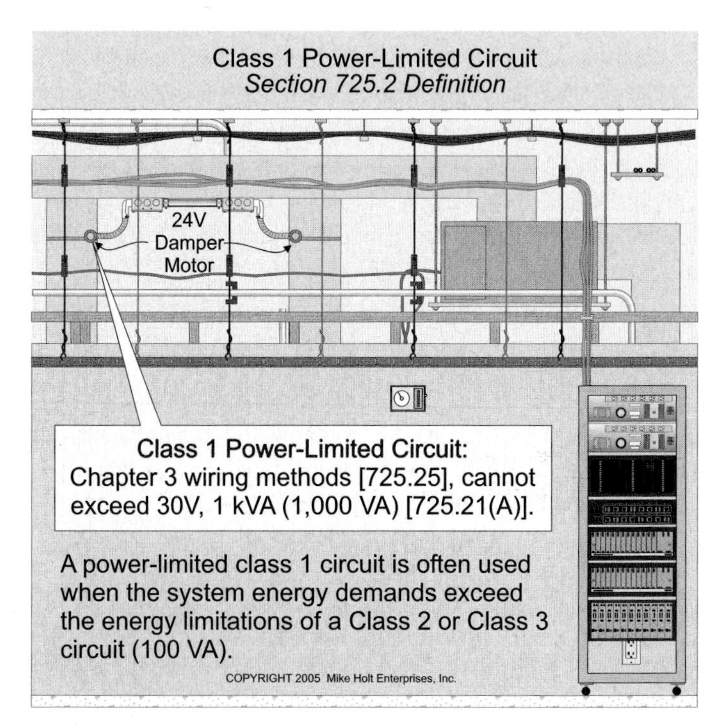

Class 1 Power-Limited Circuit
Section 725.2 Definition
24V Damper Motor
Class 1 Power-Limited Circuit:
Chapter 3 wiring methods [725.25], cannot exceed 30V, 1 kVA (1,000 VA) [725.21(A)].
A power-limited class 1 circuit is often used when the system energy demands exceed the energy limitations of a Class 2 or Class 3 circuit (100 VA).
COPYRIGHT 2005 Mike Holt Enterprises, Inc.

Figure 725–3

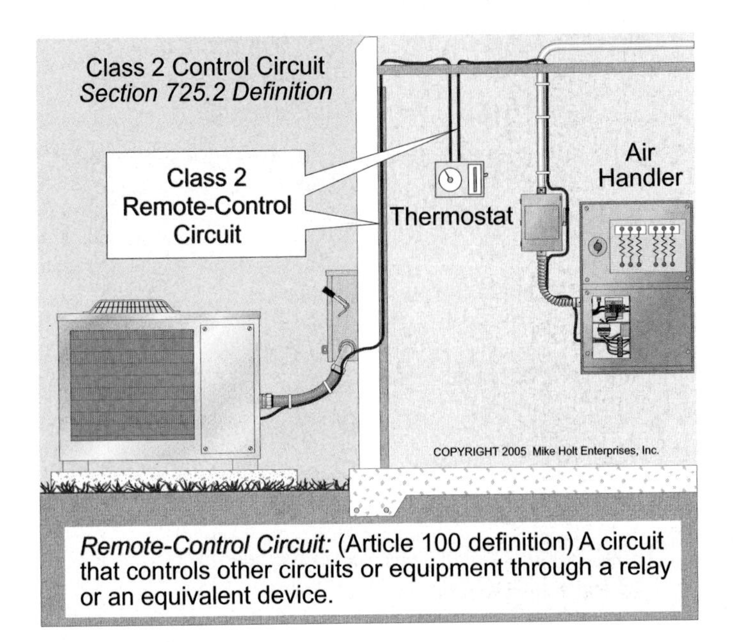

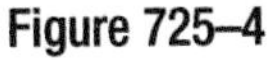
Figure 725–4

Author's Comment: Class 2 circuits typically include wiring for low-energy, low-voltage loads such as thermostats, programmable controllers, burglar alarms, and security systems. This type of circuit also includes twisted-pair or coaxial cable that interconnects computers for Local Area Networks (LANs) and programmable controller I/O circuits [725.41(A)(3)].

Class 2 circuits are rendered safe by limiting the power supply to 100 VA for circuits that operate at 30V or less and the current to 5 mA for circuits over 30V [Chapter 9, Table 11]. **Figures 725-4** and **725-5**

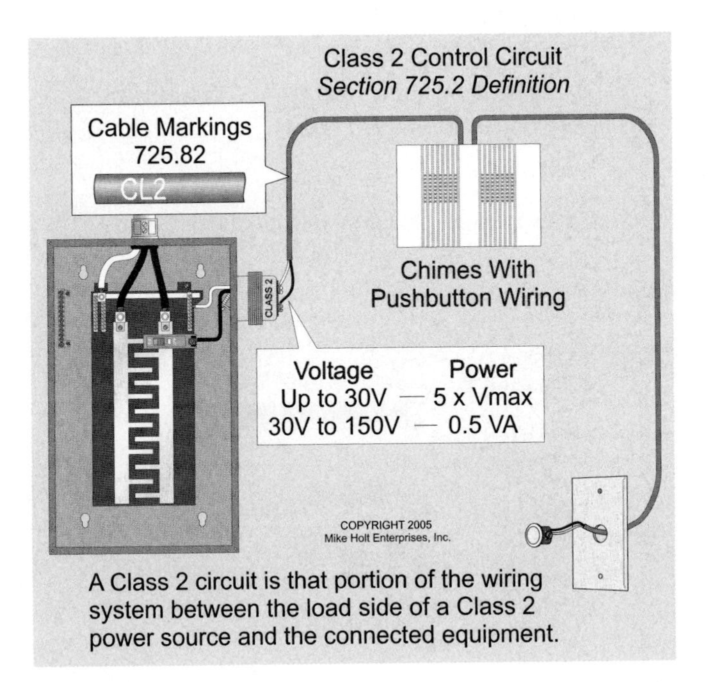

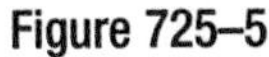
Figure 725–5

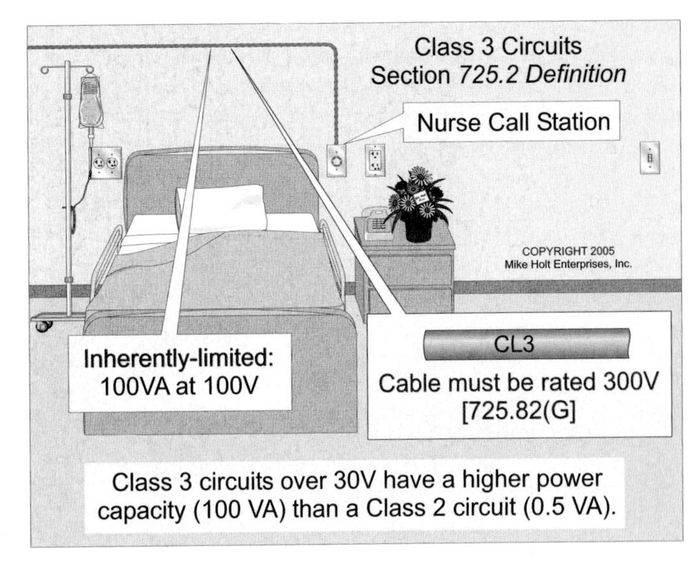

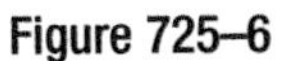
Figure 725–6

Class 3 Circuit. The wiring system between the load side of a Class 3 power source and the connected Class 3 equipment.

Author's Comment: Class 3 circuits are used when the power demand exceeds 0.50 VA, but not more than 100 VA, for circuits over 30V [Chapter 9, Table 11]. **Figure 725-6**

725.3 Other Articles. Only those sections contained in Article 300 specifically referenced in Article 725 apply to Class 1, 2, and 3 circuits.

Author's Comment: Boxes or other enclosures aren't required for Class 2 or Class 3 splices or terminations because Article 725 doesn't reference 300.15. **Figure 725-7**

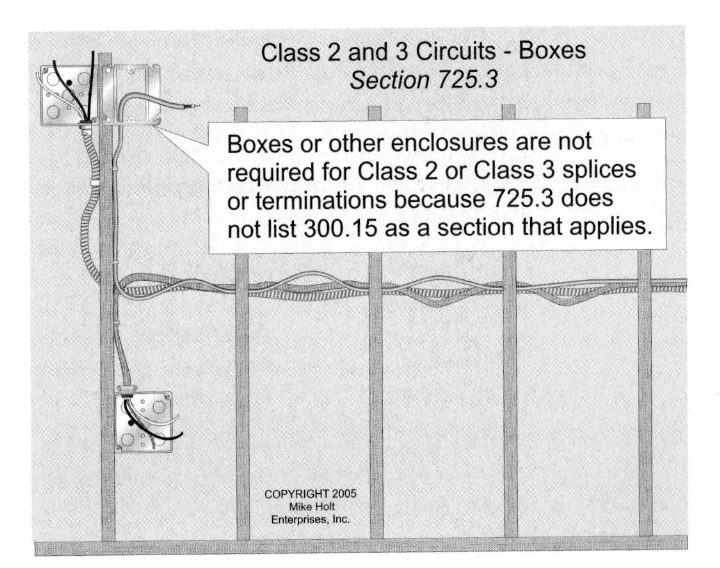

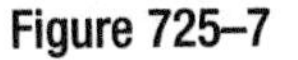
Figure 725–7

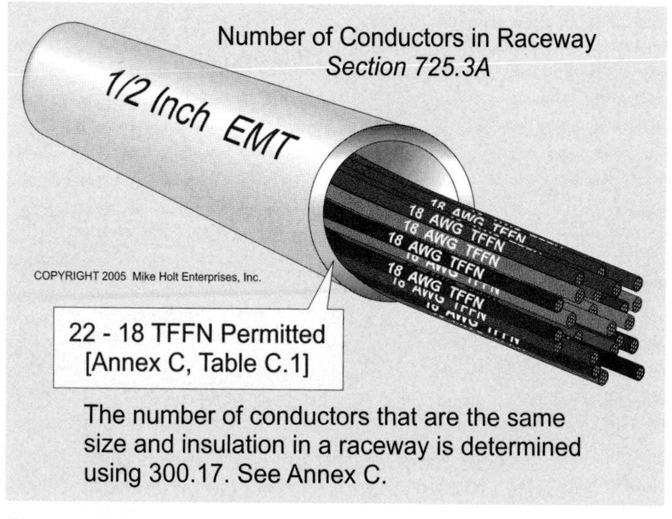
Number of Conductors in Raceway
Section 725.3A
1/2 Inch EMT
18 AWG TFFN
COPYRIGHT 2005 Mike Holt Enterprises, Inc.
22 - 18 TFFN Permitted
[Annex C, Table C.1]
The number of conductors that are the same size and insulation in a raceway is determined using 300.17. See Annex C.

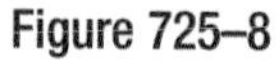
Figure 725–8

(A) Number and Size of Conductors in a Raceway. The number and size of conductors or cables in a raceway are limited in accordance with 300.17. Raceways must be large enough to permit the installation and removal of conductors without damaging the conductor's insulation.

Same Size Conductors. When all conductors in a raceway are the same size and insulation, the number of conductors permitted can be found in Annex C for the raceway type.

Question: *How many 18 TFFN fixture wires can be installed in ½ in. electrical metallic tubing?* **Figure 725-8**

(a) 40 (b) 26 (c) 30 (d) 22

Answer: *(d) 22 conductors, Annex C, Table C.1*

Different Size Conductors. When the conductors in a raceway are of different sizes, the raceway conductor fill is limited to the following percentages in accordance with Table 1 of Chapter 9. These percentages are based on conditions where the length of the conductor and number of raceway bends are within reasonable limits.

Table 1, Chapter 9

Number	Percent Fill
1 Conductor	53%
2 Conductors	31%
3 or more Conductors	40%

Table 1, Note 9. A multiconductor cable is treated as a single conductor for calculating the percentage conduit fill area. For elliptical cross section cables, the cross-sectional area calculation is based on using the major diameter of the ellipse as a circle diameter.

Step 1. The first step in raceway sizing is determining the total area of conductors. Note: The area of a cable is determined by the formula: 3.14 x (Diameter/2)²

Step 2. The second step is to select the raceway from Chapter 9, Table 4, in accordance with the percent fill listed in Chapter 9, Table 1.

Example—Raceway Sizing for Cables

Question: *What trade size electrical metallic tubing is required for the following cables:* **Figure 725-9**

- *Four–Category 5 plenum cables (⅙ in. diameter or 0.167 in.)*
- *Three–12-strand nonconductive optical fiber cables (¼ in. diameter or 0.250 in.)*
- *Two–24-strand nonconductive optical fiber cables (7/16 in. diameter or 0.438 in.)*

(a) ½ (b) ¾ (c) 1 (d) 1¼

Answer: *(d) 1¼*

Step 1. Determine the cross-sectional area of each cable: 3.14 x (D/2)².

Category 5 Cable	*= 3.14 x (0.167 in./2)²*	*= 0.0219 sq in.*
Fiber (12-strand)	*= 3.14 x (0.250 in./2)²*	*= 0.0491 sq in.*
Fiber (24-strand)	*= 3.14 x (0.438 in./2)²*	*= 0.1506 sq in.*

Step 2. Determine the total cross-sectional area for all conductors.

Category 5 Cable	*= 0.0219 sq in. x 4 cables*	*= 0.0876 sq in.*
Fiber (12-strand)	*= 0.0491 sq in. x 3 cables*	*= 0.1473 sq in.*
Fiber (24-strand)	*= 0.1506 sq in. x 2 cables*	*= 0.3012 sq in.*
Total conductor area =		*0.5361 sq in.*

Step 3. Size the raceway at 40 percent fill, Chapter 9, Table 4.
Trade size 1¼ electrical metallic tubing = 0.5980 sq in.

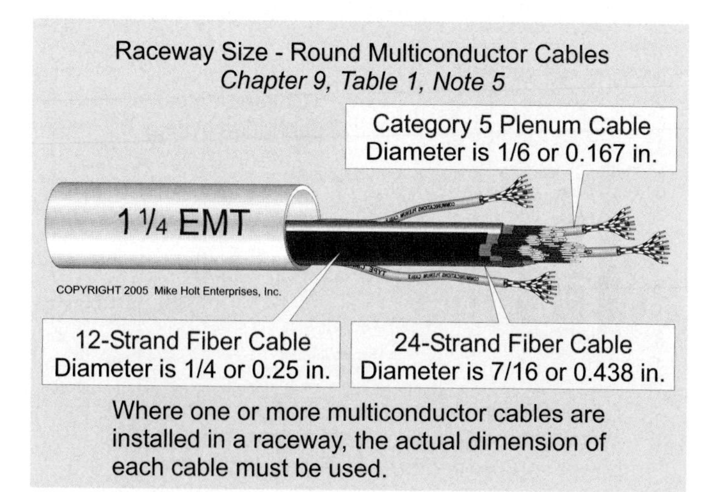
Raceway Size - Round Multiconductor Cables
Chapter 9, Table 1, Note 5
Category 5 Plenum Cable
Diameter is 1/6 or 0.167 in.
1 1/4 EMT
COPYRIGHT 2005 Mike Holt Enterprises, Inc.
12-Strand Fiber Cable
Diameter is 1/4 or 0.25 in.
24-Strand Fiber Cable
Diameter is 7/16 or 0.438 in.
Where one or more multiconductor cables are installed in a raceway, the actual dimension of each cable must be used.

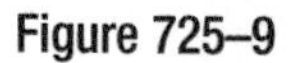
Figure 725–9

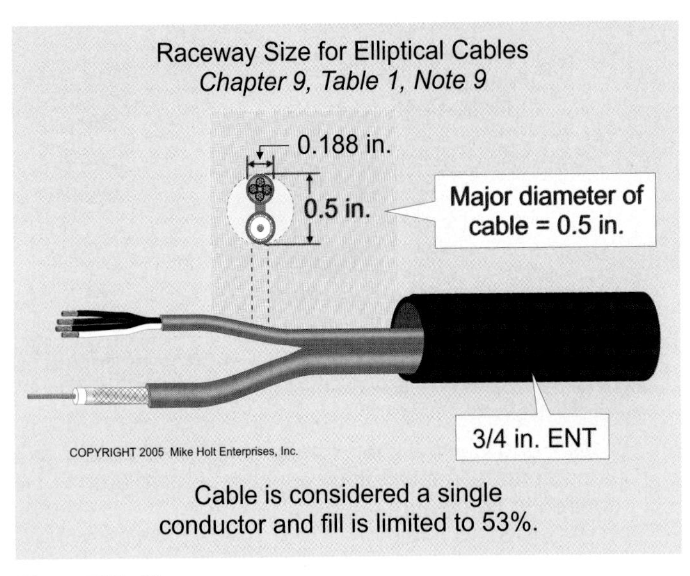

Figure 725–10

Example—Raceway Sizing for Elliptical Cables

Elliptical cables have a cross section that is taller than it is wide. The major (larger) diameter must be used for calculations. Examples of elliptical cables include two-wire (flat) NM cable, or a duplex or Siamese-type low voltage cable as shown in **Figure 725-10**.

Question: *What trade size electrical nonmetallic tubing is required for one hybrid fiber/data cable? The minor diameter of the ellipse is 3⁄16 in. (0.188 in.) and the major diameter of the ellipse as a circle diameter is ½ in. (0.50 in.).* **Figure 725-10**

(a) ½ (b) ¾ (c) 1 (d) 1¼

Answer: *(b) ¾*

The cross-sectional area of an elliptical cable is based on the major diameter of the ellipse as a circle diameter. [Note 9 of Table 1, Chapter 9].

Step 1. Determine the total cross-sectional area of the cable.
Hybrid cable = 3.14 x (0.50 in./2)^2 = 0.1963 sq in.

Step 2. Size the raceway at 53 percent fill (one conductor fill), Chapter 9, Table 4.

Trade size ½ electrical nonmetallic tubing = 0.131 sq in., too small

Trade size ¾ electrical nonmetallic tubing = 0.240 sq in., just right

Trade size 1 electrical nonmetallic tubing = 0.416 sq in., larger than required

(B) Spread of Fire or Products of Combustion. Class 1, 2, and 3 circuits installed through fire-resistant rated walls, partitions,

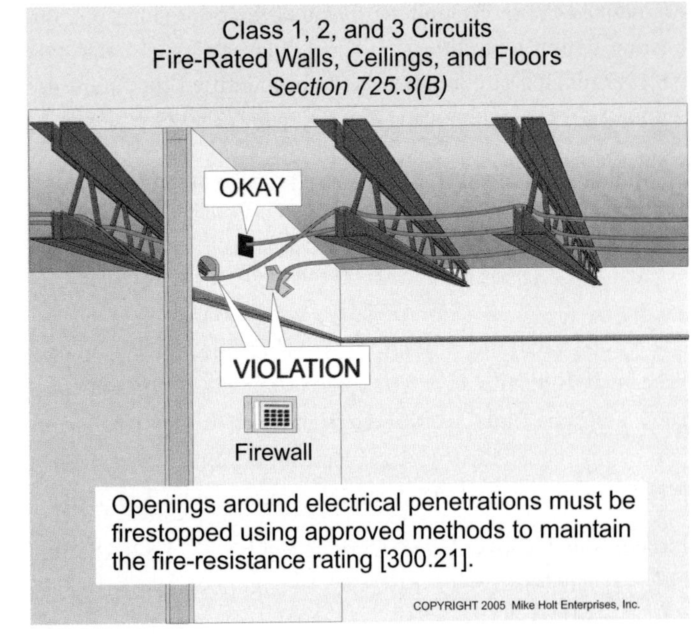

Class 1, 2, and 3 Circuits
Fire-Rated Walls, Ceilings, and Floors
Section 725.3(B)
OKAY
VIOLATION
Firewall
Openings around electrical penetrations must be firestopped using approved methods to maintain the fire-resistance rating [300.21].
COPYRIGHT 2005 Mike Holt Enterprises, Inc.

Figure 725–11

floors, or ceilings must be firestopped in accordance with the specific instructions supplied by the manufacturer for the specific type of cable and construction material (drywall, brick, etc.), as required by 300.21.

Author's Comment: Openings in fire-resistant walls, floors, and ceilings must be sealed so that the possible spread of fire or products of combustion will not be substantially increased [300.21]. **Figure 725-11**

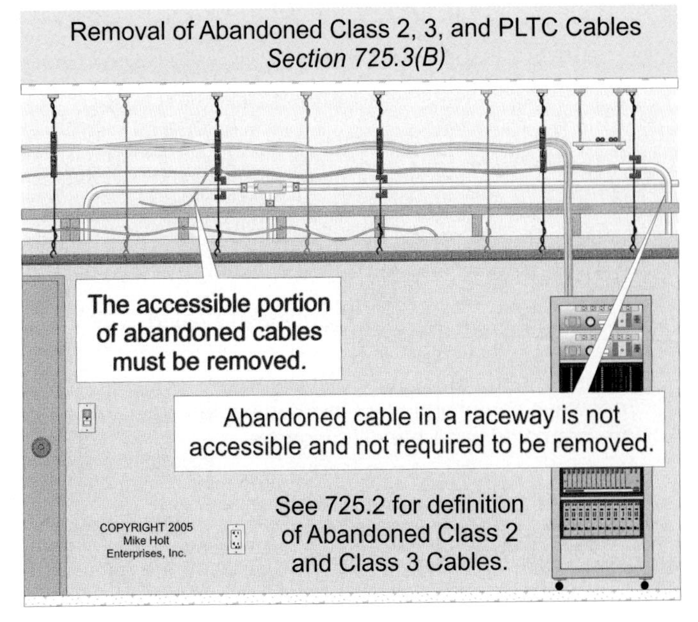

Figure 725–12

Abandoned Cable. To limit the spread of fire or products of combustion within a building, the accessible portion of cable that isn't terminated at equipment and not identified for future use with a tag must be removed [725.2]. **Figure 725-12**

> **Author's Comment:** This rule doesn't require the removal of concealed cables that are abandoned in place, which includes cables in raceways. According to the definition of "Concealed" in Article 100, cables in raceways are considered to be concealed.

(C) Ducts, Plenums, and Other Air-Handling Spaces. Where necessary for the direct action upon, or sensing of the contained air, Class 2 and Class 3 cables can be installed in ducts or plenums if they are installed in electrical metallic tubing, intermediate metal conduit, or rigid metal conduit as required by 300.22(B).

Plenum-rated Class 2 and Class 3 cables [725.82(A)], and plenum signaling raceways [725.82(I)] with plenum-rated cables can be installed above a suspended ceiling or below a raised floor used for environmental air movement [725.61(A)]. **Figure 725-13**

Nonplenum-rated Class 2 and Class 3 cables can be installed above a suspended ceiling or below a raised floor that is used for environmental air, but only if the nonplenum-rated cable is installed within electrical metallic tubing [300.22(C)(1)].

> **Author's Comment:** Type CL2 and Type CL3 cables installed beneath a raised floor in an information technology equipment room (computer room) aren't required to be plenum rated [300.22(D) and 645.5(D)(5)(c)]. **Figure 725-14**

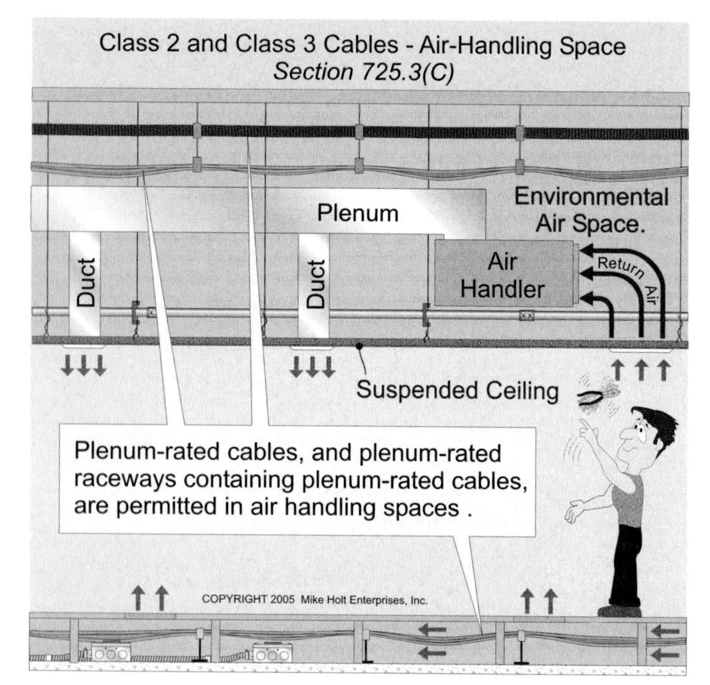
Class 2 and Class 3 Cables - Air-Handling Space
Section 725.3(C)
Plenum
Environmental Air Space.
Duct
Duct
Air Handler
Return Air
Suspended Ceiling
Plenum-rated cables, and plenum-rated raceways containing plenum-rated cables, are permitted in air handling spaces .
COPYRIGHT 2005 Mike Holt Enterprises, Inc.

Figure 725–13

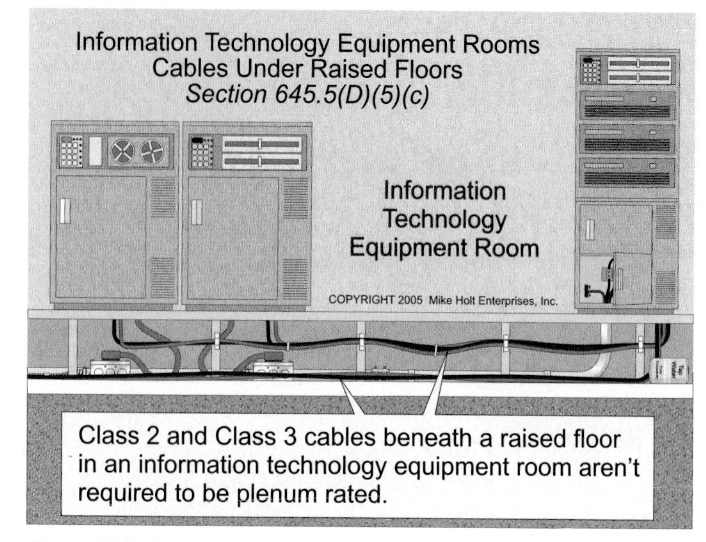
Information Technology Equipment Rooms
Cables Under Raised Floors
Section 645.5(D)(5)(c)
Information Technology Equipment Room
COPYRIGHT 2005 Mike Holt Enterprises, Inc.
Class 2 and Class 3 cables beneath a raised floor in an information technology equipment room aren't required to be plenum rated.

Figure 725–14

(D) Hazardous (Classified) Locations. Class 1, 2, and 3 circuits installed in any hazardous (classified) location must be installed in accordance with Articles 500 through 516, specifically 501.150, and 502.150. **Figure 725-15**

(E) Cable Trays. Class 1, 2, and 3 circuits installed in a cable tray must be installed in accordance with Article 392.

> **Author's Comment:** Class 2 and 3 cables in cable trays must be separated from electric light and power conductors by a solid fixed barrier, or the Class 2 or Class 3 circuits must be installed in Type MC cable [725.55(H)]. **Figure 725-16**

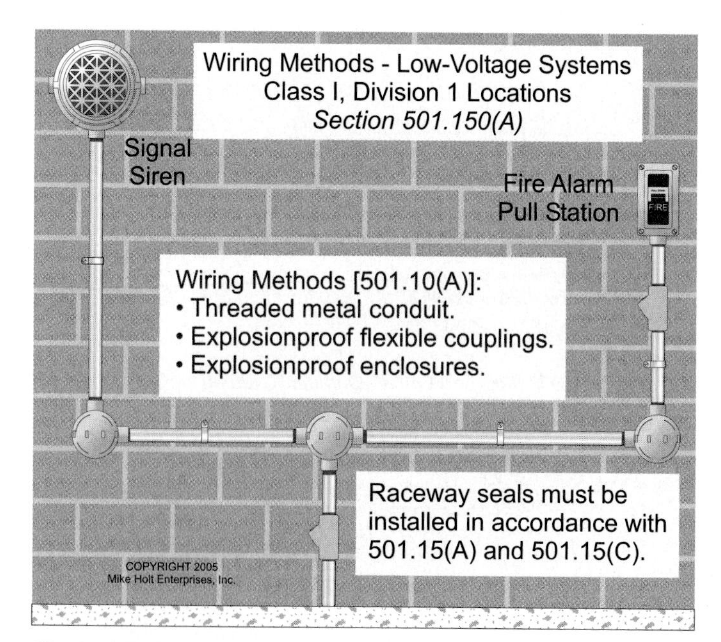
Wiring Methods - Low-Voltage Systems
Class I, Division 1 Locations
Section 501.150(A)
Signal Siren
Fire Alarm Pull Station
Wiring Methods [501.10(A)]:
• Threaded metal conduit.
• Explosionproof flexible couplings.
• Explosionproof enclosures.
Raceway seals must be installed in accordance with 501.15(A) and 501.15(C).
COPYRIGHT 2005 Mike Holt Enterprises, Inc.

Figure 725–15

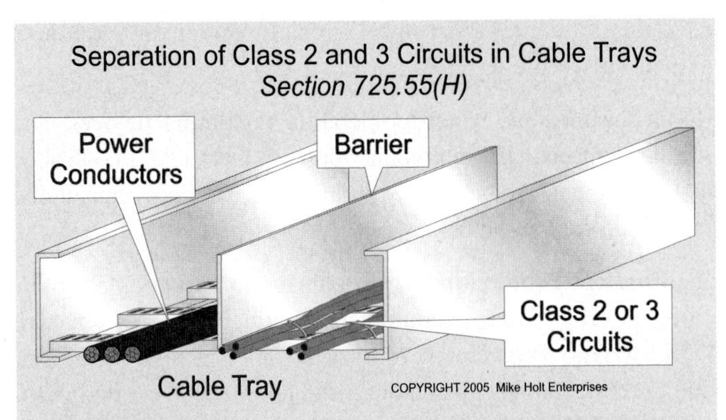

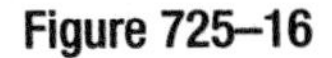
Figure 725–16

(F) Motor Control Circuits. Motor control circuit conductors tapped from the motor branch-circuit conductors must have overcurrent protection in accordance with Part VI of Article 430. **Figure 725-17**

725.7 Access to Electrical Equipment Behind Panels Designed to Allow Access. Access to equipment cannot be prohibited by an accumulation of cables that prevent the removal of suspended-ceiling panels. Cables must be located so that the suspended-ceiling panels can be moved to provide access to electrical equipment. **Figure 725-18**

725.8 Mechanical Execution of Work. Equipment and cabling must be installed in a neat and workmanlike manner.

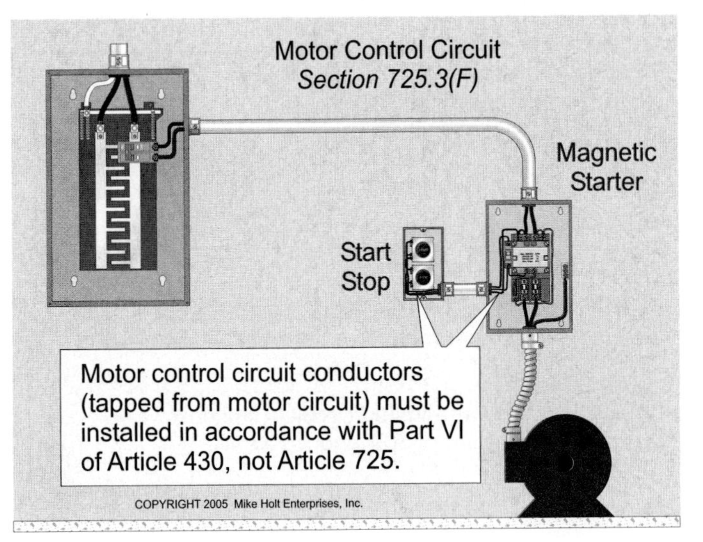

Motor Control Circuit
Section 725.3(F)
Magnetic Starter
Start
Stop
Motor control circuit conductors (tapped from motor circuit) must be installed in accordance with Part VI of Article 430, not Article 725.
COPYRIGHT 2005 Mike Holt Enterprises, Inc.

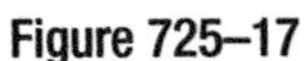
Figure 725–17

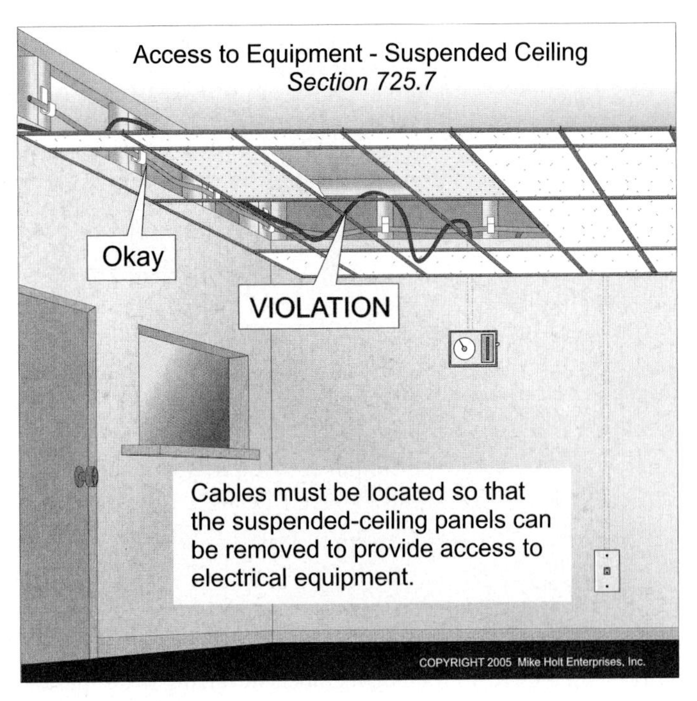

Figure 725–18

Exposed cables must be supported by the structural components of the building so that the cable will not be damaged by normal building use. Such cables must be secured by straps, staples, hangers, or similar fittings designed and installed so as not to damage the cable. **Figure 725-19**

> **Author's Comment:** Raceways and cables can be supported by independent support wires attached to the suspended ceiling in accordance with 300.11(A). **Figure 725-20**

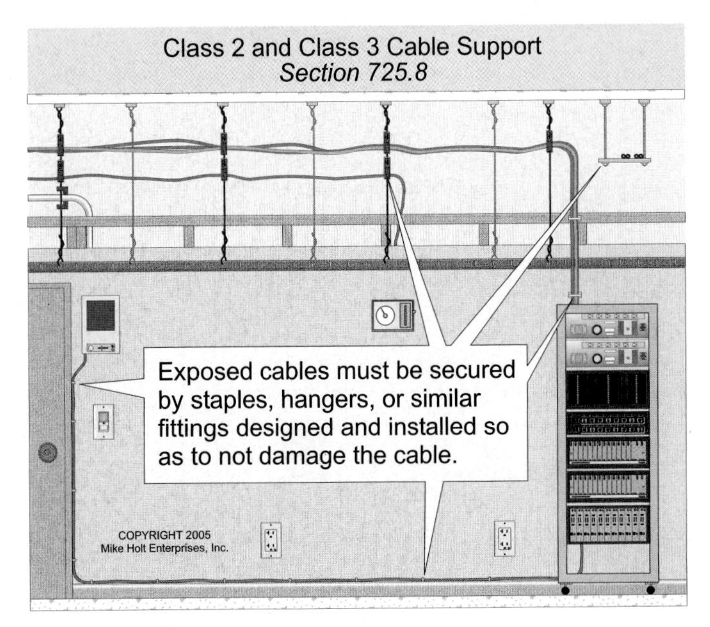

Figure 725–19

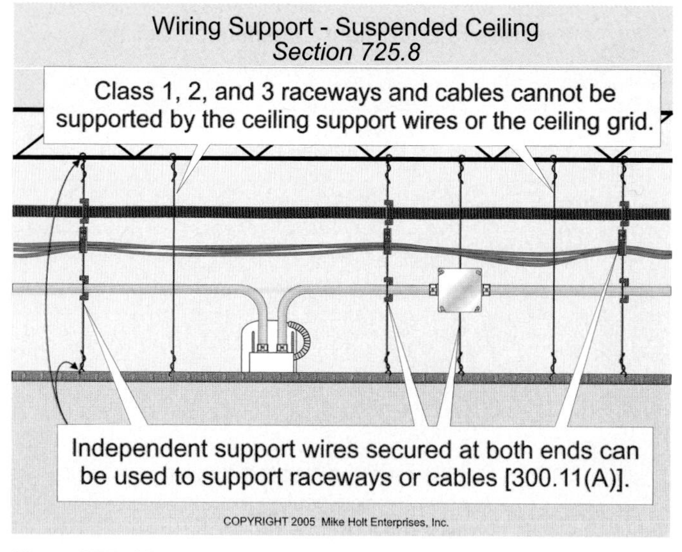
Wiring Support - Suspended Ceiling
Section 725.8
Class 1, 2, and 3 raceways and cables cannot be supported by the ceiling support wires or the ceiling grid.
Independent support wires secured at both ends can be used to support raceways or cables [300.11(A)].
COPYRIGHT 2005 Mike Holt Enterprises, Inc.

Figure 725–20

Cables run parallel to framing members or furring strips must be protected, where they are likely to be penetrated by nails or screws, by installing the wiring method so it isn't less than 1¼ in. from the nearest edge of the framing member or furring strips, or by protecting them with a 1/16 in. thick steel plate or equivalent [300.4(D)]. **Figure 725-21**

725.11 Safety-Control Equipment.

(A) Remote-Control Circuits. Class 2 and Class 3 circuits used for safety-control equipment must be reclassified as Class 1 if the failure of the remote-control circuit or equipment introduces a direct fire or life hazard.

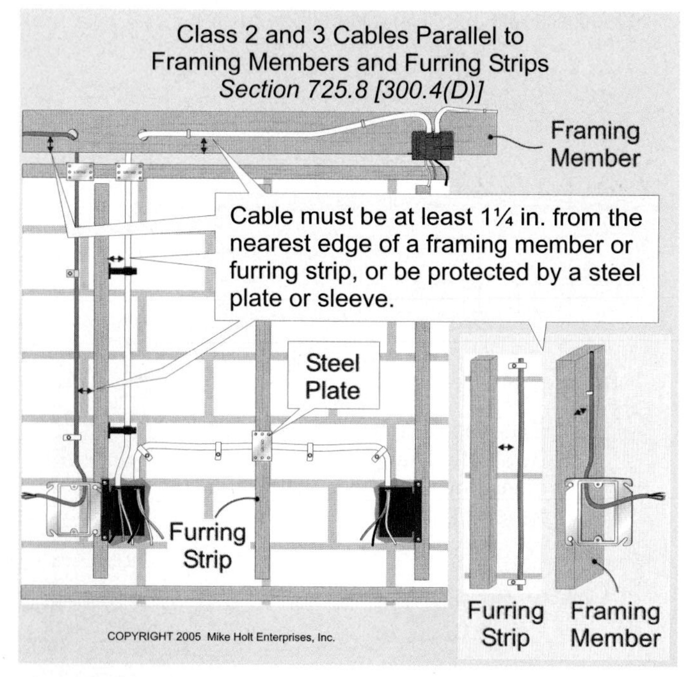
Class 2 and 3 Cables Parallel to Framing Members and Furring Strips
Section 725.8 [300.4(D)]
Framing Member
Cable must be at least 1¼ in. from the nearest edge of a framing member or furring strip, or be protected by a steel plate or sleeve.
Steel Plate
Furring Strip
Furring Strip
Framing Member
COPYRIGHT 2005 Mike Holt Enterprises, Inc.

Figure 725–21

Room thermostats, water-temperature regulating devices, and similar controls used in conjunction with electrically controlled household heating and air conditioning, are not considered safety-control equipment.

(B) Physical Protection. Where damage to remote-control circuits of safety-control equipment would introduce a hazard [725.11(A)], all conductors must be installed in rigid metal conduit, intermediate metal conduit, rigid nonmetallic conduit, electrical metallic tubing, or be suitably protected from physical damage [725.25].

725.15 Circuit Requirements.

(1) Class 1 circuits must comply with the requirements contained in Parts I and II of Article 725.

(2) Class 2 and Class 3 circuits must comply with the requirements contained in Parts I and III of Article 725.

PART II. CLASS 1 CIRCUIT REQUIREMENTS

725.21 Class 1 Circuit Classifications and Power Source Requirements.

(A) Class 1 Power-Limited Circuits. Class 1 power-limited circuits must be supplied from a power source that limits the output to 30V with no more than 1,000 VA. **Figure 725-22**

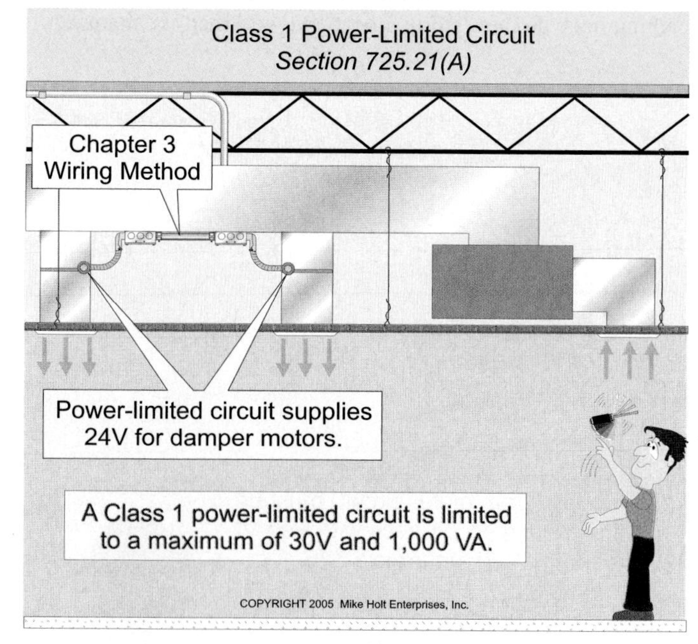
Class 1 Power-Limited Circuit
Section 725.21(A)
Chapter 3 Wiring Method
Power-limited circuit supplies 24V for damper motors.
A Class 1 power-limited circuit is limited to a maximum of 30V and 1,000 VA.
COPYRIGHT 2005 Mike Holt Enterprises, Inc.

Figure 725–22

Author's Comment: Class 1 power-limited circuits aren't that common. They are used when the voltage must be less than 30 (safe from electric shock in dry locations), but where the power demands exceed the 100 VA energy limitations of Class 2 or Class 3 circuits, such as for motorized loads like remote-controlled window blinds [Chapter 9 Table 11(A)].

(B) Class 1 Remote-Control and Signaling Circuits. Class 1 remote-control and signaling circuits must not operate at more than 600V, and the power output of the power supply isn't required to be limited.

Author's Comment: One of the most common examples of a Class 1 remote-control circuit is an electromagnetic contactor or motor starter. **Figure 725-23**

725.23 Class 1 Circuit Overcurrent Protection. Overcurrent protection for conductors 14 AWG and larger must be limited to the conductor ampacity in accordance with 240.4, Table 310.16, and 240.6(A). Overcurrent protection is not to exceed 7A for 18 AWG conductors and 10A for 16 AWG conductors. **Figure 725-24**

725.25 Class 1 Circuit Wiring Methods. Class 1 circuits must be installed in accordance with Part I of Article 300, and all Class 1 wiring must be installed as a Chapter 3 wiring method.

Author's Comment: This means that Class 1 circuits must be installed in a suitable Chapter 3 wiring method and all splices must be contained in enclosures [300.15].

725.26 Conductors of Different Circuits in Same Cable, Enclosure, or Raceway. Class 1 circuits can be installed with other circuits as specified in 725.26(A) and (B).

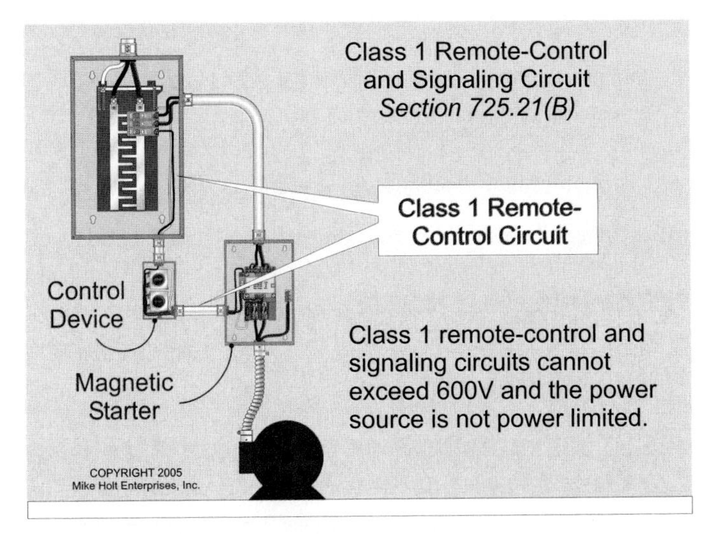

Figure 725–23

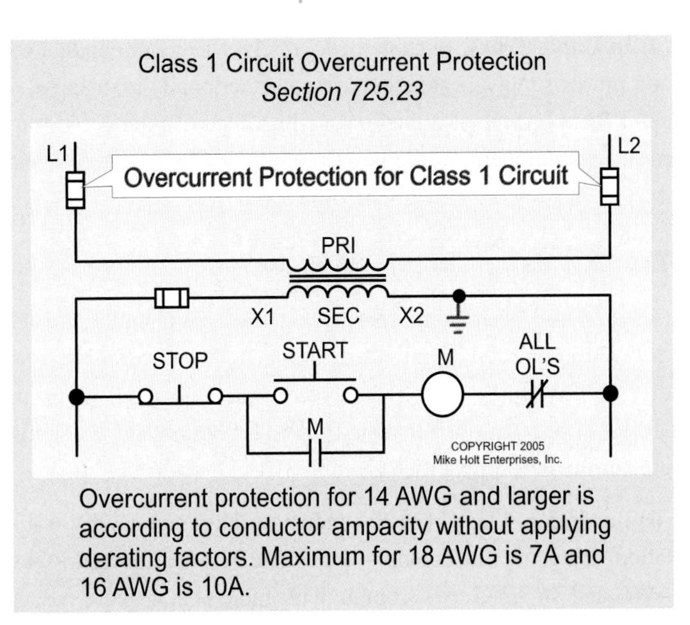

Class 1 Circuit Overcurrent Protection
Section 725.23
L1
L2
Overcurrent Protection for Class 1 Circuit
PRI
X1
SEC
X2
STOP
START
M
ALL OL'S
M
COPYRIGHT 2005
Mike Holt Enterprises, Inc.
Overcurrent protection for 14 AWG and larger is according to conductor ampacity without applying derating factors. Maximum for 18 AWG is 7A and 16 AWG is 10A.

Figure 725–24

(A) Class 1 Circuits with Other Class 1 Circuits. Two or more Class 1 circuits can be installed in the same cable, enclosure, or raceway.

(B) Class 1 Circuits with Power-Supply Circuits. Class 1 circuits are permitted with electric light and power conductors in accordance with (1) through (4).

(1) In a Cable, Enclosure, or Raceway. Class 1 circuits can be in the same cable, enclosure, or raceway with power-supply circuits, if the equipment powered is functionally associated with the Class 1 circuit. **Figure 752-25**

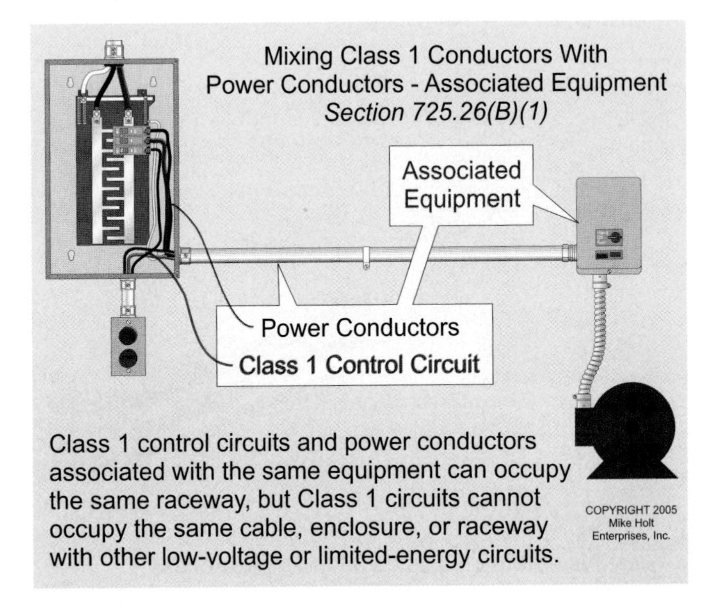

Mixing Class 1 Conductors With
Power Conductors - Associated Equipment
Section 725.26(B)(1)
Associated Equipment
Power Conductors
Class 1 Control Circuit
Class 1 control circuits and power conductors associated with the same equipment can occupy the same raceway, but Class 1 circuits cannot occupy the same cable, enclosure, or raceway with other low-voltage or limited-energy circuits.
COPYRIGHT 2005
Mike Holt
Enterprises, Inc.

Figure 725–25

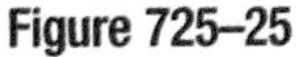

(4) In Cable Trays. In cable trays, if Class 1 circuit conductors and power-supply conductors are not functionally associated, a solid fixed barrier of a material compatible with the cable tray must separate them, or the power-supply or Class 1 circuit conductors must be in a metal-enclosed cable.

725.27 Class 1 Circuit Conductors.

(A) Size and Use. Conductors of sizes 18 AWG and 16 AWG installed in a raceway, enclosure, or listed cable are permitted if they do not supply a load that exceeds the ampacities given in 402.5. Conductors larger than 16 AWG must not supply loads greater than the ampacities given in 310.15.

(B) Insulation. Class 1 circuit conductors must have a 600V insulation rating and must comply with Table 310.13. Conductors 18 AWG and 16 AWG must comply with Table 402.3.

725.28 Number of Conductors in a Raceway.

(A) Class 1 Circuit Conductors. The number and size of conductors or cables in a raceway are limited in accordance with 300.17. Raceways must be large enough to permit the installation and removal of conductors without damaging the conductor's insulation.

Same Size Conductors. When all conductors in a raceway are the same size and insulation, the number of conductors permitted in a raceway can be found in Annex C for the raceway type.

> ***Question:*** *How many 18 TFFN fixture wires can be installed in trade size ½ electrical metallic tubing?* **Figure 725-26**
>
> *(a) 40* *(b) 26* *(c) 30* *(d) 22*
>
> ***Answer:*** *(d) 22, Annex C, Table C.1*

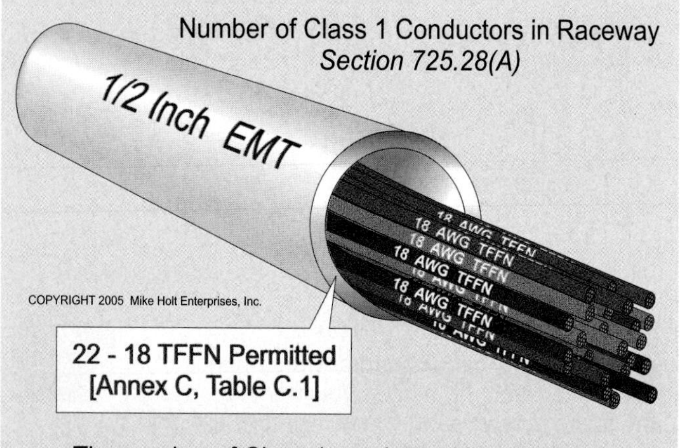

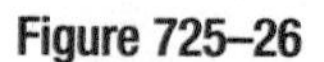
Figure 725–26

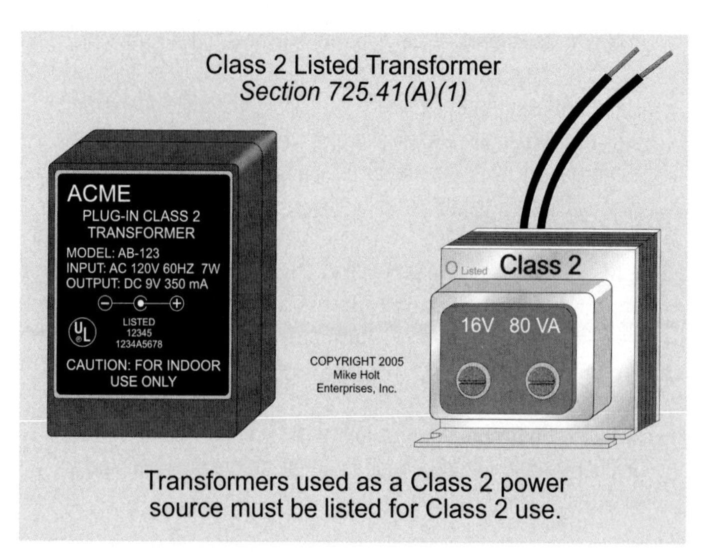

Figure 725–27

PART III. CLASS 2 AND CLASS 3 CIRCUIT REQUIREMENTS

725.41 Power Sources for Class 2 and Class 3 Circuits.

(A) Power Source. The power source for a Class 2 or a Class 3 circuit must be:

(1) A listed Class 2 or Class 3 transformer. **Figure 725-27**

(2) A listed Class 2 or Class 3 power supply.

(3) Equipment listed as a Class 2 or Class 3 power source.

Exception 2: Listed equipment, where each circuit has an energy level at or below the limits established in Chapter 9, Table 11(A) and Table 11(B), isn't required to be listed as a Class 2 or Class 3 power transformer, power supply, or power source.

> **Author's Comment:** Exception No. 2 is intended to apply to programmable controller I/O circuits.

(4) Listed information technology equipment. **Figure 725-28**

(5) A dry cell battery rated 30V or less for a Class 2 circuit.

725.42 Equipment Marking. Equipment supplying Class 2 or Class 3 circuits must be marked to indicate each circuit that is a Class 2 or Class 3 circuit.

725.51 Wiring Methods on Supply Side of the Class 2 or Class 3 Power Source. Conductors and equipment on the supply side of the Class 2 or Class 3 power source must be installed in accordance with Chapters 1 through 4. **Figure 725-29**

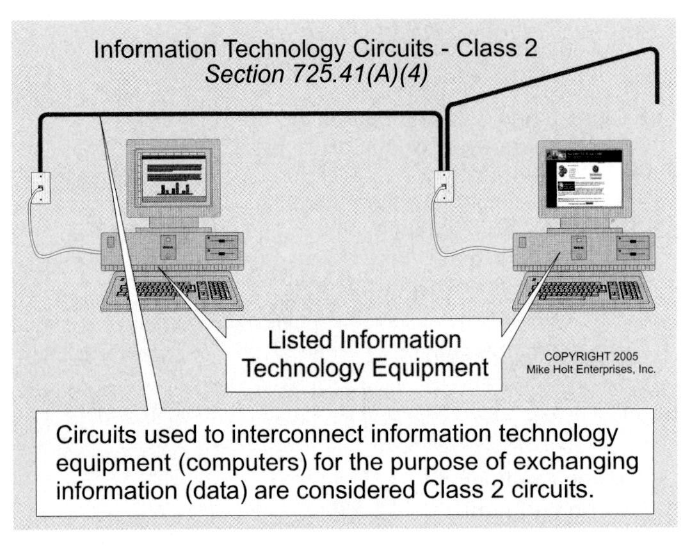
Information Technology Circuits - Class 2
Section 725.41(A)(4)
Listed Information
Technology Equipment
COPYRIGHT 2005
Mike Holt Enterprises, Inc.
Circuits used to interconnect information technology equipment (computers) for the purpose of exchanging information (data) are considered Class 2 circuits.

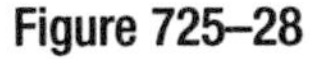
Figure 725–28

725.52 Wiring Methods on Load Side of the Class 2 or Class 3 Power Source.

(A) Class 1 Wiring Methods. Class 2 or Class 3 circuits are permitted to use a Chapter 3 wiring method [725.25].

Exception 2: Class 2 and Class 3 circuits can be reclassified as a Class 1 circuit if the Class 2 and Class 3 equipment markings required by 725.42 are eliminated and the entire circuit is installed in a Chapter 3 wiring method [725.55(D)(2)(b)].

FPN: Class 2 and Class 3 circuits reclassified as Class 1 and installed as Class 1, are no longer Class 2 or Class 3, regardless of the continued connection to a Class 2 or Class 3 power source.

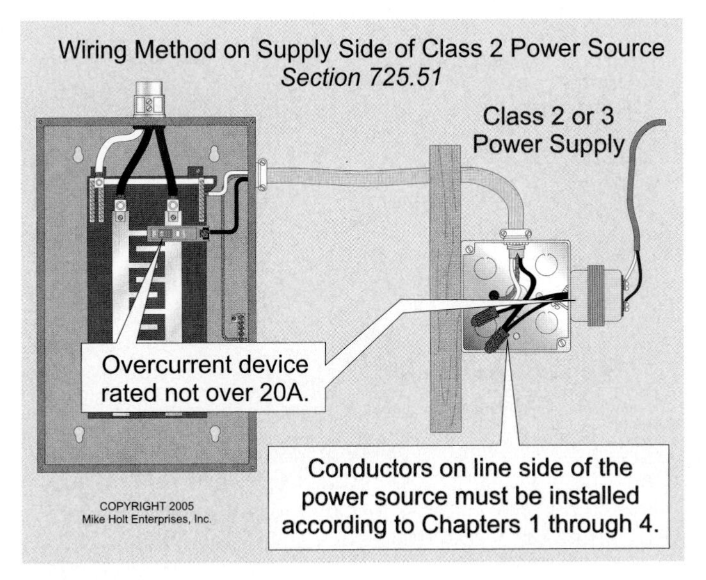
Wiring Method on Supply Side of Class 2 Power Source
Section 725.51
Class 2 or 3
Power Supply
Overcurrent device
rated not over 20A.
Conductors on line side of the power source must be installed according to Chapters 1 through 4.
COPYRIGHT 2005
Mike Holt Enterprises, Inc.

Figure 725–29

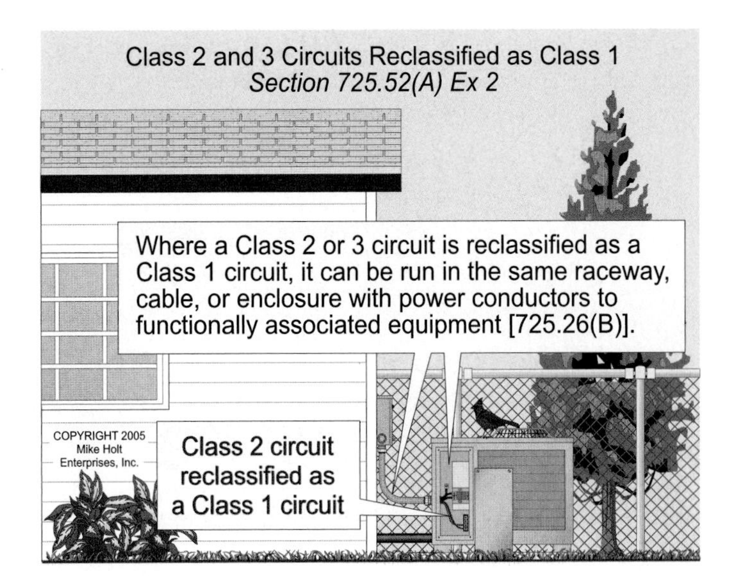
Class 2 and 3 Circuits Reclassified as Class 1
Section 725.52(A) Ex 2
Where a Class 2 or 3 circuit is reclassified as a Class 1 circuit, it can be run in the same raceway, cable, or enclosure with power conductors to functionally associated equipment [725.26(B)].
COPYRIGHT 2005
Mike Holt
Enterprises, Inc.
Class 2 circuit
reclassified as
a Class 1 circuit

Figure 725–30

Author's Comment: Class 2 or Class 3 circuits reclassified as Class 1 circuits can no longer be installed with other Class 2 or Class 3 circuits that have not been reclassified as Class 1 [725.55]. Reclassifying the circuit allows the Class 1 circuit to be installed with functionally associated power circuits in accordance with 725.26(B)(1). **Figure 725-30**

(B) Class 2 and Class 3 Wiring Methods. Class 2 and Class 3 circuit conductors must be of the type listed and marked in accordance with 725.82, and they must be installed in accordance with 725.54 and 725.61.

725.55 Separation from Other Systems.

(A) Enclosures, Raceways, or Cables. Class 2 and Class 3 circuit conductors must not be placed in any enclosure, raceway, or cable with conductors of electric light, power, Class 1, and nonpower-limited fire alarm circuits, except as permitted in (B) through (J). **Figure 725-31**

(B) Separated by Barriers. Where separated by a barrier, Class 2 and Class 3 circuits are permitted with conductors of electric light, power, Class 1, and nonpower-limited fire alarm circuits. **Figure 725-32**

Author's Comment: Separation is required to prevent a fire or shock hazard that could occur from a short between the Class 2 or Class 3 circuit and the higher-voltage circuits. Class 2 and Class 3 cables can be installed in the same raceway or enclosure with communications cables [800.133(A)], power-limited fire alarm cables [760.56], optical fiber cables [770.133(B)], CATV cables [820.133(A)], and low-power network broadband. See 725.56(E). **Figure 725-33**

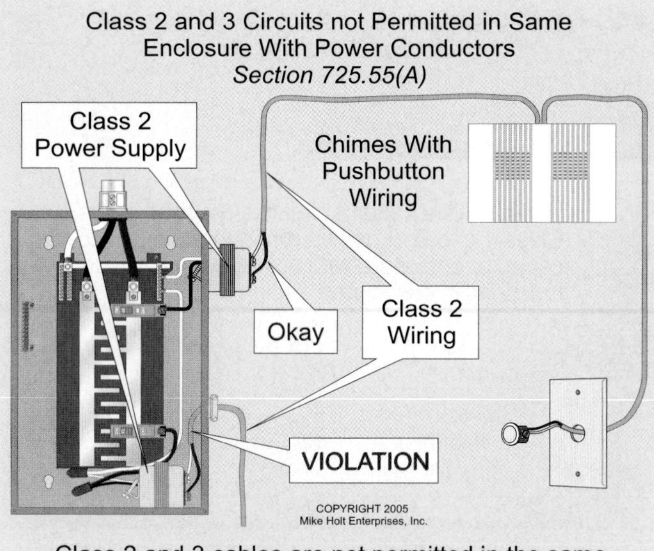

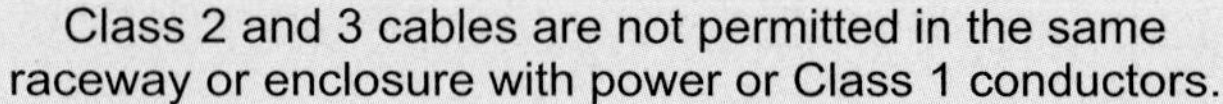

Class 2 and 3 cables are not permitted in the same raceway or enclosure with power or Class 1 conductors.

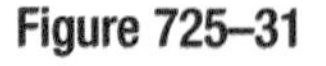
Figure 725–31

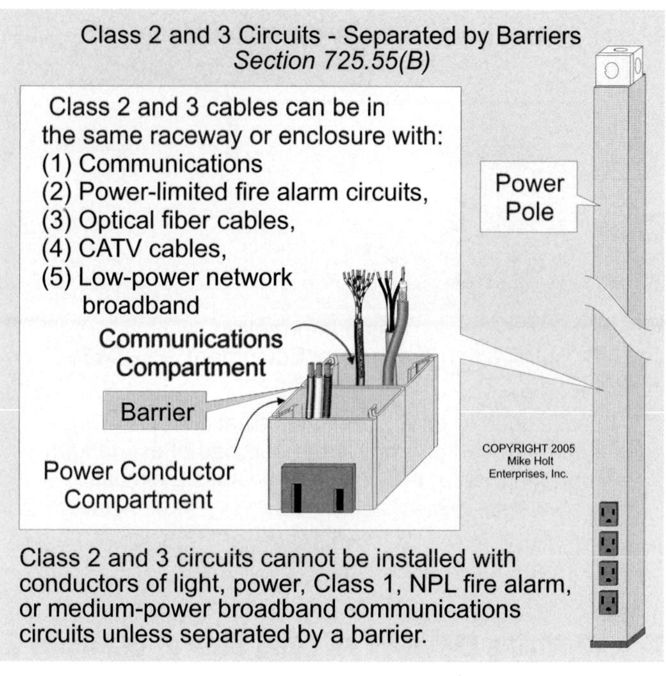

Class 2 and 3 Circuits - Separated by Barriers
Section 725.55(B)
Class 2 and 3 cables can be in the same raceway or enclosure with:
(1) Communications
(2) Power-limited fire alarm circuits,
(3) Optical fiber cables,
(4) CATV cables,
(5) Low-power network broadband
Power Pole
Communications Compartment
Barrier
Power Conductor Compartment
COPYRIGHT 2005 Mike Holt Enterprises, Inc.
Class 2 and 3 circuits cannot be installed with conductors of light, power, Class 1, NPL fire alarm, or medium-power broadband communications circuits unless separated by a barrier.

Figure 725–33

(D) Within Enclosures. Class 2 and Class 3 conductors can be mixed with electric light, power, Class 1, and nonpower-limited fire alarm conductors in enclosures, if these other conductors are introduced solely for connection to the same equipment as the Class 2 or Class 3 circuits, and:

(1) A minimum ¼ in. separation is maintained from the Class 2 or Class 3 conductors, or

(2) The conductors operate at 150V or less and comply with one of the following:

(a) The Class 2 and Class 3 circuit conductors are installed in Class 3 cable, or permitted substitute cables, and the conductors extending beyond the jacket maintain a minimum of ¼ in. separation from other conductors.

(b) The Class 2 and Class 3 circuit conductors are reclassified as a Class 1 nonpower-limited circuit in accordance with 725.52 Ex 2, and they are installed in a Chapter 3 wiring method in accordance with 725.21.

(H) Cable Trays. Where separated by a barrier, Class 2 and Class 3 circuits are permitted with conductors of electric light, power, Class 1, and nonpower-limited fire alarm circuits, or where the Class 2 or Class 3 circuits are installed in Type MC cable. **Figure 725-34**

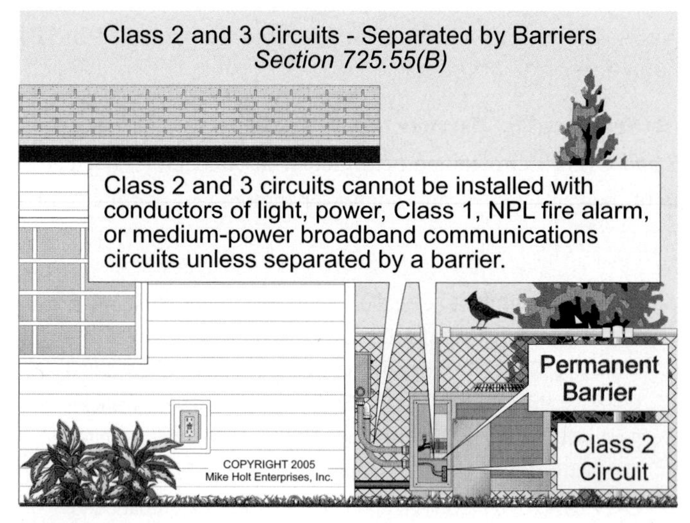

Class 2 and 3 Circuits - Separated by Barriers
Section 725.55(B)
Class 2 and 3 circuits cannot be installed with conductors of light, power, Class 1, NPL fire alarm, or medium-power broadband communications circuits unless separated by a barrier.
Permanent Barrier
Class 2 Circuit
COPYRIGHT 2005 Mike Holt Enterprises, Inc.

Figure 725–32

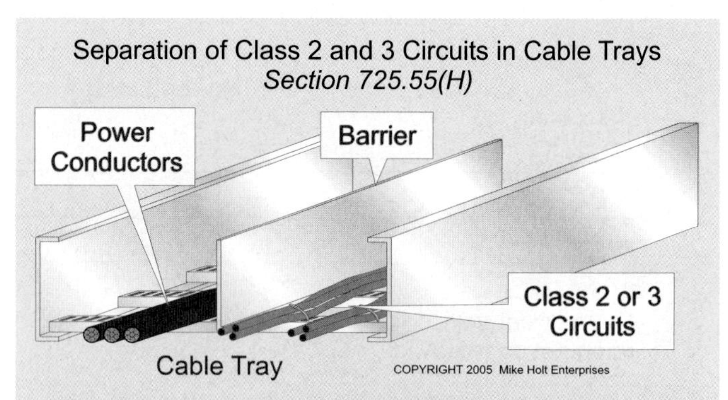

Figure 725–34

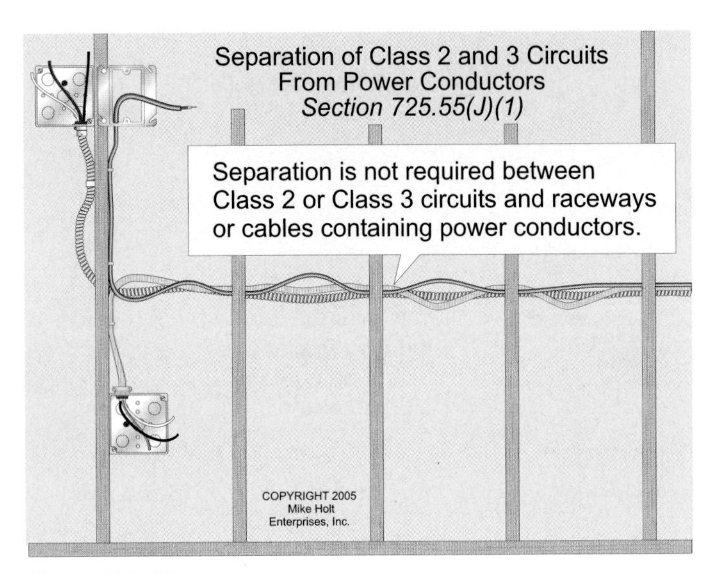

Separation of Class 2 and 3 Circuits
From Power Conductors
Section 725.55(J)(1)
Separation is not required between
Class 2 or Class 3 circuits and raceways
or cables containing power conductors.
COPYRIGHT 2005
Mike Holt
Enterprises, Inc.

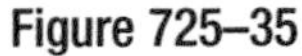
Figure 725–35

> **Author's Comment:** When Class 2 and Class 3 cables are installed in a cable tray, no separation is required from other wiring methods.

(J) Other Applications. Class 2 and Class 3 circuit conductors must be separated by not less than 2 in. from insulated conductors of electric light, power, Class 1, and nonpower-limited fire alarm circuits, unless:

(1) All of the electric light, power, Class 1, and nonpower-limited fire alarm circuit conductors or all of the Class 2 and Class 3 circuit conductors are in a raceway or in metal-sheathed, metal-clad, nonmetallic-sheathed, or underground feeder cables. **Figure 725-35**

725.56 Conductors of Different Circuits in Same Cable, Enclosure, or Raceway.

(A) Class 2 Conductors. Class 2 circuit conductors can be in the same cable, enclosure, or raceway with other Class 2 circuit conductors.

(B) Class 3 Conductors. Class 3 circuit conductors can be in the same cable, enclosure, or raceway with other Class 3 circuit conductors.

(C) Class 2 Conductors with Class 3 Conductors. Class 2 conductors are permitted within the same cable, enclosure, or raceway with Class 3 circuit conductors, provided the Class 2 circuit conductors use a Class 3 wiring method in accordance with 725.82.

> **Author's Comment:** Listed Class 2 cables have 150V insulation, whereas listed Class 3 cables are rated 300V [725.82(G)].

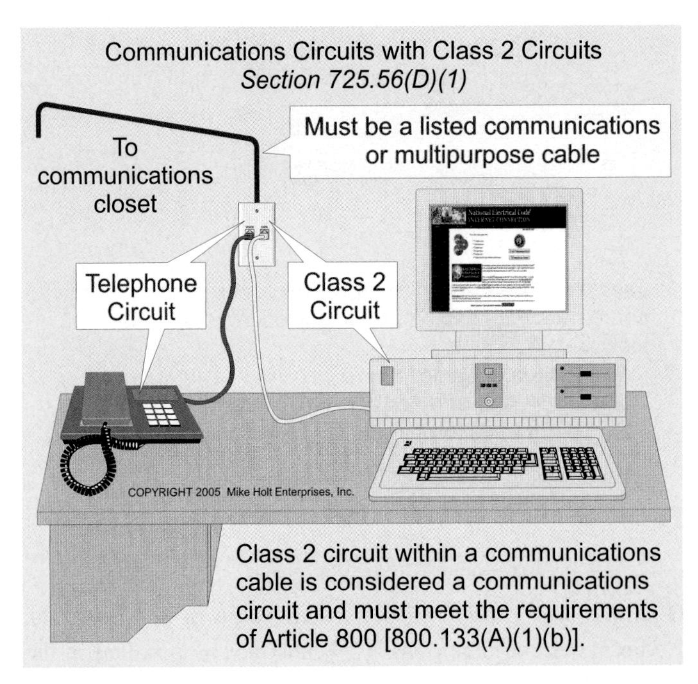

Figure 725–36

(D) Class 2 and Class 3 Circuits with Communications Circuits.

(1) Classified as Communications Circuits. Class 2 and Class 3 conductors can be within the same cables with communications conductors provided the cables are listed as communications cables or multipurpose cables and they are installed in accordance with Article 800 [800.113 and 800.133(A)(1)(b)].

> **Author's Comments:**
>
> - A common application of this requirement is when a single communications cable is used for both voice communications and data [Class 2—725.41(A)(4)]. **Figure 725-36**
> - Listed Class 2 cables have 150V insulation [725.82(G)], whereas listed communications cables have a voltage rating of 300V [800.179].

(E) Class 2 or Class 3 Cables with Other Cables. Class 2 or Class 3 cables can be in the same enclosure or raceway with jacketed cables of: **Figure 725-37**

(1) Power-limited fire alarm circuits in compliance with Article 760.

(2) Nonconductive and conductive optical fiber cables in compliance with Article 770.

(3) Communications circuits in compliance with Article 800.

(4) Coaxial cables in compliance with Article 820.

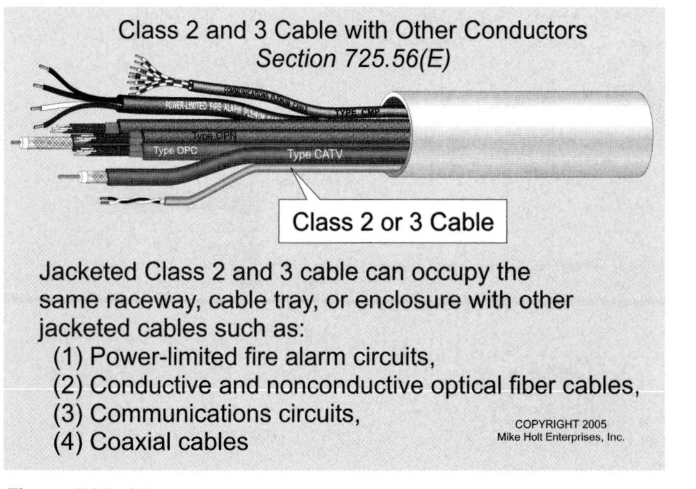
Class 2 and 3 Cable with Other Conductors
Section 725.56(E)
Type OPC
Type CATV
Type CPN
Class 2 or 3 Cable
Jacketed Class 2 and 3 cable can occupy the same raceway, cable tray, or enclosure with other jacketed cables such as:
(1) Power-limited fire alarm circuits,
(2) Conductive and nonconductive optical fiber cables,
(3) Communications circuits,
(4) Coaxial cables
COPYRIGHT 2005 Mike Holt Enterprises, Inc.

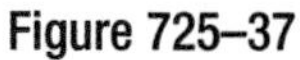

Figure 725–37

(F) Class 2 or Class 3 Circuits with Audio System Circuits. Audio system circuits [640.9(C)] must not be installed in the same cable or raceway with Class 2 or Class 3 conductors.

> **Author's Comment:** The concern is that a fault from audio amplifier circuits, to Class 2 and Class 3 circuits, has the potential of creating a hazard by disrupting the operation of alarm systems and remote-control circuits for safety control equipment.

725.57 Class 2 or Class 3 Cables Exposed to Lightning. If Class 2 or Class 3 circuit conductors extend beyond a building and are exposed to lightning or to accidental contact with conductors over 300 volts-to-ground, the following apply:

(1) Twisted Pair. Class 2 or Class 3 circuit conductors must maintain not less than 6 ft separation from lightning conductors [800.53]; the metallic sheath must be grounded or interrupted as close as practicable to the point of entrance [800.93]; and, where grounded, the cable must be grounded in accordance with 800.100.

(2) Coaxial Cable. Class 2 or Class 3 cables must maintain not less than 6 ft separation from lightning conductors [820.44(F)(3)], the metallic sheath must be grounded as close as practicable to the point of entrance [820.93], and the cable must be grounded in accordance with 820.100.

> **Author's Comment:** Most Class 2 or Class 3 circuits run overhead between buildings are susceptible to lightning, but underground installations aren't considered exposed to lightning [800.90(A) FPN 2]. To reduce electromagnetic interference and to protect from damage from voltage surges, the practice is to ground one end of underground cables to a grounding electrode [800.93 and 820.93]. **Figure 725-38**

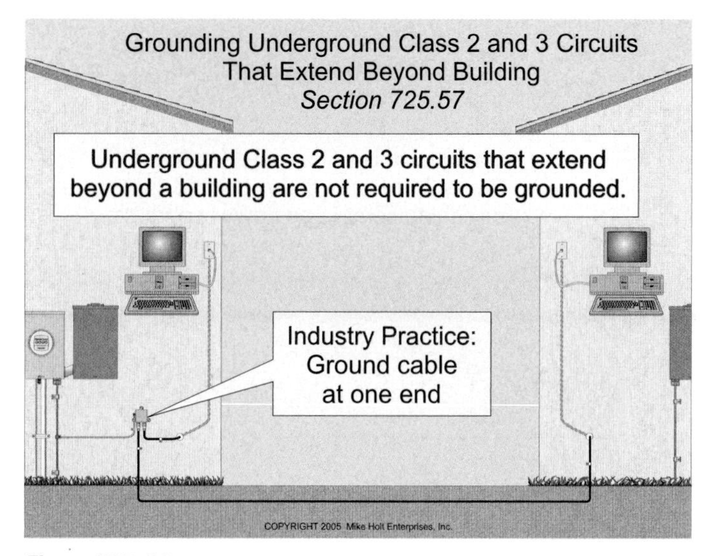
Grounding Underground Class 2 and 3 Circuits That Extend Beyond Building
Section 725.57
Underground Class 2 and 3 circuits that extend beyond a building are not required to be grounded.
Industry Practice: Ground cable at one end
COPYRIGHT 2005 Mike Holt Enterprises, Inc.

Figure 725–38

725.58 Support. Class 2 or Class 3 cables must not be supported by a raceway. **Figure 725-39**

> **Author's Comment:** Exposed cables must be supported by the structural components of the building so that the cable will not be damaged by normal building use, and cables must be secured by straps, staples, hangers, or similar fittings designed and installed so as not to damage the cable [725.8]. **Figure 725-40**

Class 2 control cables can be supported by the raceway that supplies power to the equipment controlled by the Class 2 cable [300.11(B)(2)]. **Figure 725-41**

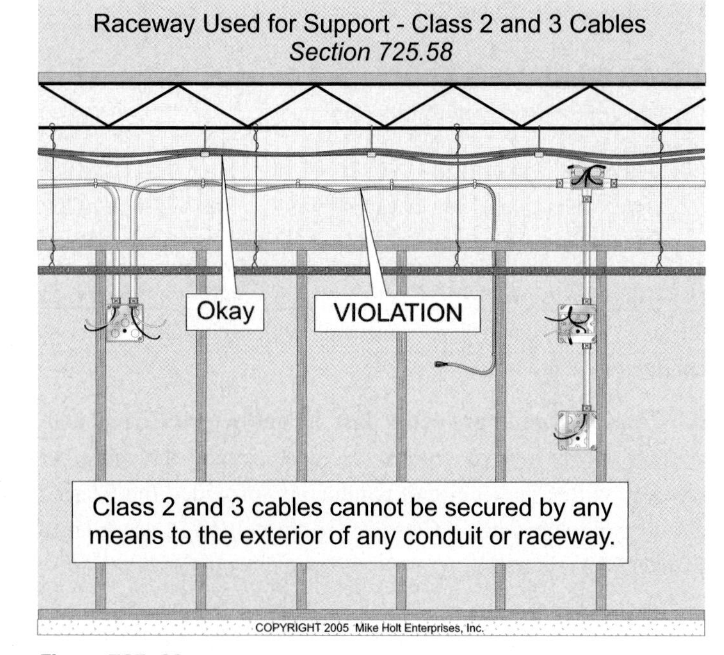
Raceway Used for Support - Class 2 and 3 Cables
Section 725.58
Okay
VIOLATION
Class 2 and 3 cables cannot be secured by any means to the exterior of any conduit or raceway.
COPYRIGHT 2005 Mike Holt Enterprises, Inc.

Figure 725–39

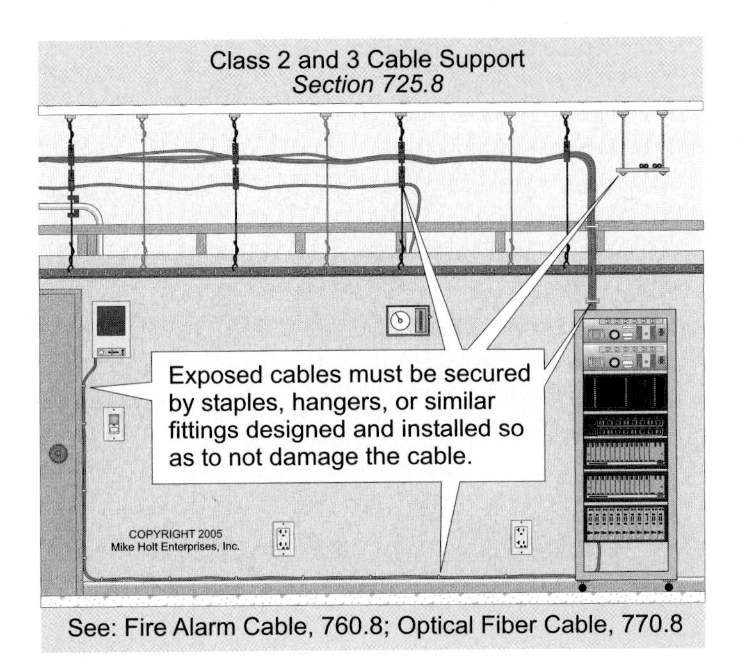
Class 2 and 3 Cable Support
Section 725.8
Exposed cables must be secured by staples, hangers, or similar fittings designed and installed so as to not damage the cable.
COPYRIGHT 2005 Mike Holt Enterprises, Inc.
See: Fire Alarm Cable, 760.8; Optical Fiber Cable, 770.8

Figure 725–40

725.61 Applications of Class 2 and Class 3 Cables.

Class 2, Class 3, and PLTC cables must comply with the requirements of (A) through (H).

(A) Ducts, Plenums, and Other Space Used for Environmental Air. Class 2 or Class 3 cables must not be run in ducts or plenums, even if plenum rated [725.3(C)], unless they act directly on the contained air, are installed using a wiring method described in 300.22(B), and are plenum rated. However, plenum-rated Class 2 or Class 3 cables can be installed above a suspended ceiling or below a raised floor that is used for environmental air. **Figure 725-42**

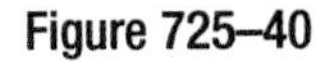

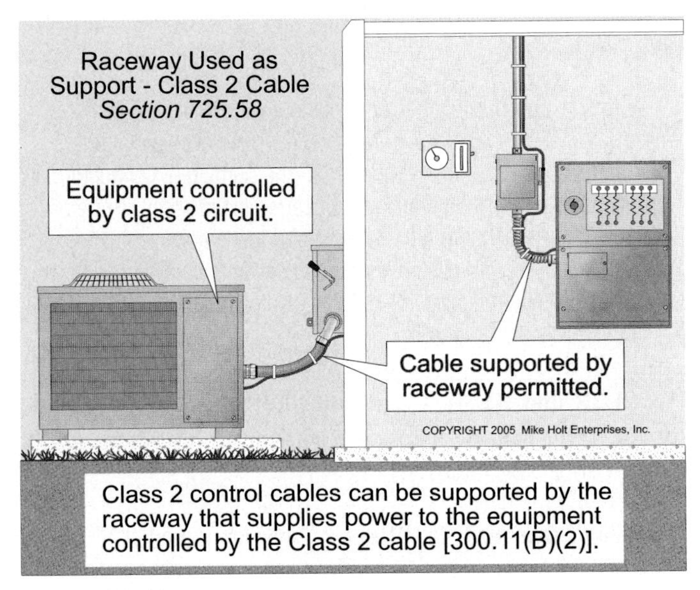
Raceway Used as Support - Class 2 Cable
Section 725.58
Equipment controlled by class 2 circuit.
Cable supported by raceway permitted.
COPYRIGHT 2005 Mike Holt Enterprises, Inc.
Class 2 control cables can be supported by the raceway that supplies power to the equipment controlled by the Class 2 cable [300.11(B)(2)].

Figure 725–41

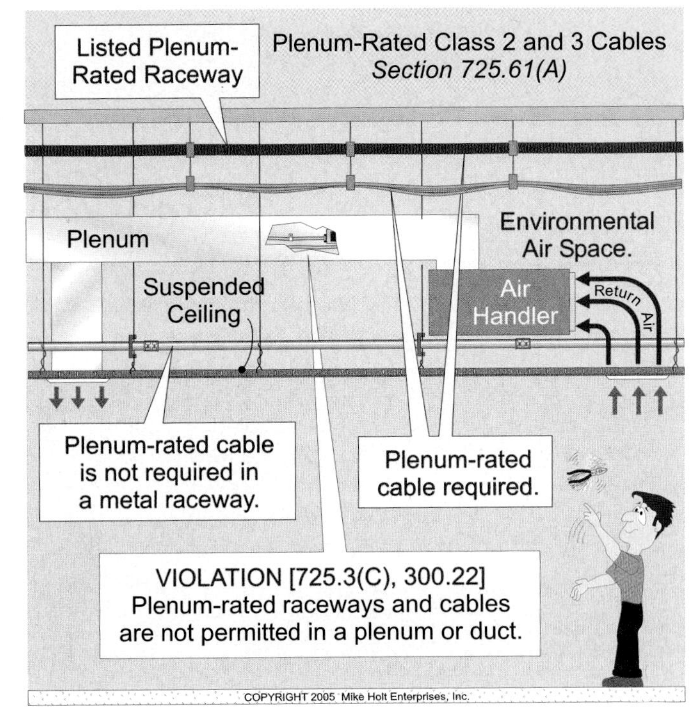
Listed Plenum-Rated Raceway
Plenum-Rated Class 2 and 3 Cables
Section 725.61(A)
Plenum
Environmental Air Space.
Suspended Ceiling
Air Handler
Return Air
Plenum-rated cable is not required in a metal raceway.
Plenum-rated cable required.
VIOLATION [725.3(C), 300.22]
Plenum-rated raceways and cables are not permitted in a plenum or duct.
COPYRIGHT 2005 Mike Holt Enterprises, Inc.

Figure 725–42

Listed plenum raceways can be installed above a suspended ceiling or below a raised floor used for environmental air [300.22(C)(2)], but only if the cables contained in these raceways are plenum-rated Types CL2P or CL3P. **Figure 725-42**

Nonplenum-rated Class 2 and Class 3 cables can be installed above a suspended ceiling or below a raised floor that is used for environmental air, but only if the cable is installed within electrical metallic tubing [300.22(C)(1)]. **Figure 725-43**

(B) Riser Rated Cables. Cables installed in risers must comply with (1), (2), or (3):

(1) Class 2 or Class 3 cables installed in vertical runs penetrating more than one floor must be riser-rated Types CL2R or CL3R. Listed riser raceways can be installed in vertical riser runs, but only if the cables for the Class 2 and Class 3 circuits contained in these raceways are riser- or plenum-rated Types CL2R, CL3R, CL2P, or CL3P.

(2) Metal Raceways. Class 2 or Class 3 cables are permitted within a metal raceway.

Author's Comments:

- When Class 2 or Class 3 cables are installed in a metal raceway, they aren't required to be riser or plenum rated.
- Metal raceways containing circuit conductors from a power-supply system that operates at 50V or less aren't required to be grounded (bonded) to an effective ground-fault current path [250.86 and 250.112(I)].

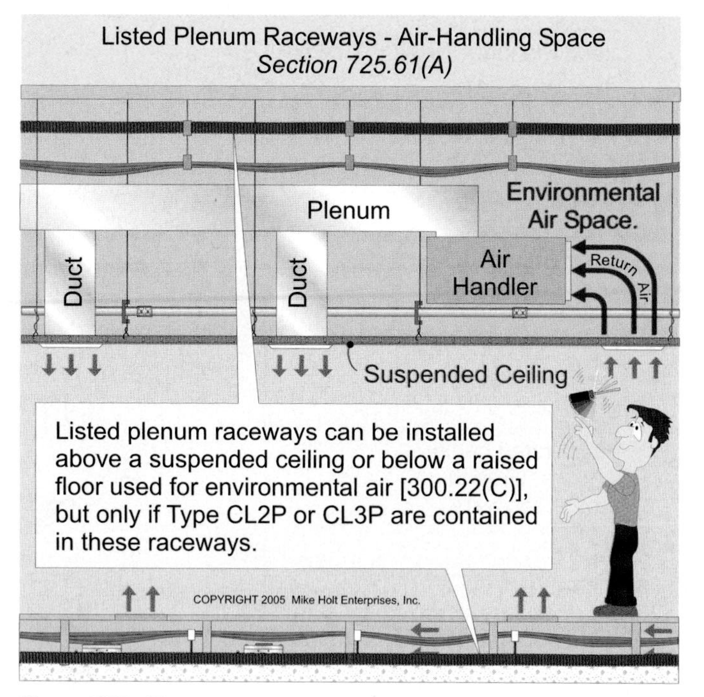
Listed Plenum Raceways - Air-Handling Space
Section 725.61(A)
Plenum
Environmental Air Space.
Duct
Duct
Air Handler
Return Air
Suspended Ceiling
Listed plenum raceways can be installed above a suspended ceiling or below a raised floor used for environmental air [300.22(C)], but only if Type CL2P or CL3P are contained in these raceways.
COPYRIGHT 2005 Mike Holt Enterprises, Inc.

Figure 725–43

(3) Types CL2, CL3, CL2X, and CL3X cables can be installed in one- and two-family dwellings without raceways. Listed general-purpose signaling raceways are permitted for use with Types CL2, CL3, CL2X, and CL3X cables.

(C) Cable Trays. Cables installed in outdoor cable trays must be Type PLTC. Cables installed in cable trays indoors must be Types PLTC, CL3P, CL3R, CL3, CL2P, CL2R, or CL2.

Listed signaling raceways can be installed in a cable tray.

(D) Hazardous (Classified) Locations. Cables installed in hazardous (classified) locations must be as described in 725.61(D)(1) through (D)(4).

(E) General-Purpose Locations. Types CL2, CL2X, CL3, or CL3X cables can be installed in locations other than ducts, plenums, or other environmental air spaces in any occupancy.

(1) Types CL2 or CL3 are permitted.

(2) Types CL2X or CL3X must be installed in a raceway or in accordance with other wiring methods covered in Chapter 3.

(3) Cables must be installed in nonconcealed spaces where the exposed length of cable does not exceed 10 ft.

(4) Listed Types CL2X and CL3X cables less than ¼ in. in diameter can be installed in one- and two-family dwellings.

(5) Listed Types CL2X and CL3X cables less than ¼ in. in diameter can be installed in nonconcealed spaces in multifamily dwellings.

Author's Comment: See Article 100 for the definition of "Multifamily Dwelling."

(G) Cable Substitutions. Class 2 and Class 3 cables can be substituted in accordance with Table 725.61.

Author's Comment: Here is a list of the abbreviations used in Table 725.61:

CM—Communications
CL—Class
TC—Tray Cable
PL—Power-Limited
P—Plenum
R—Riser
G—General
X—Dwellings in Raceway

PART VI. LISTING REQUIREMENTS

725.82 Listing and Marking of Class 2 and Class 3 Cables. Class 2 and Class 3 cables and nonmetallic signaling raceways installed within buildings must be listed in accordance with (A) through (K), and cables must be marked in accordance with 725.82(L).

Author's Comment: The *NEC* doesn't require outside or underground cable to be listed, but the cable must be approved by the authority having jurisdiction as suitable for the application in accordance with 90.4, 90.7, and 110.2. **Figure 725-44**

(A) CL2P and CL3P. CL2P and CL3P plenum cables must be listed as being suitable for use in ducts, plenums, and other space used for environmental air. See 725.61(A) for details. **Figure 725-45**

Author's Comment: Special consideration must be given to cables in areas that move or transport environmental air in order to reduce the hazards that arise from the burning of conductor insulation and of cable jackets. Because listed plenum-rated cables have adequate fire-resistant and low smoke-producing characteristics, they can be installed in environmental air space, but they cannot be installed in ducts or plenums! See 725.3(C) in this textbook for details.

(B) CL2R and CL3R. CL2R and CL3R riser cables must be listed as being suitable for use in a vertical run in a shaft, or from floor to floor. See 725.61(B).

Author's Comment: The "R" suffix stands for riser rated. Riser cables are for use in vertical shafts and have fire-resistant characteristics to prevent carrying fire from floor to floor.

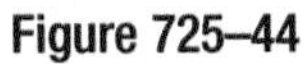
Figure 725–44

(C) CL2 and CL3. CL2 and CL3 cables must be listed as being suitable for general-purpose use. See 725.61(E).

(D) Types CL2X and CL3X. Type CL2X and CL3X limited-use cables must be listed as being suitable for use in dwellings or for use in a raceway. See 725.61(E).

(G) Class 2 and Class 3 Cable Voltage Rating. Class 2 cables must have a voltage rating not less than 150V, and Class 3 cables must have a voltage rating not less than 300V.

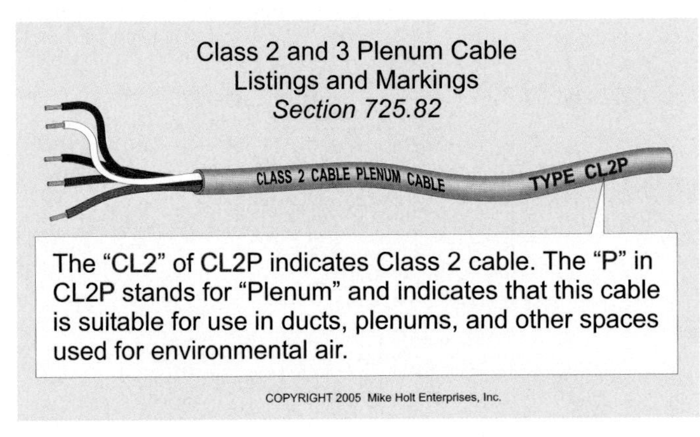

Class 2 and 3 Plenum Cable
Listings and Markings
Section 725.82
CLASS 2 CABLE PLENUM CABLE
TYPE CL2P
The "CL2" of CL2P indicates Class 2 cable. The "P" in CL2P stands for "Plenum" and indicates that this cable is suitable for use in ducts, plenums, and other spaces used for environmental air.
COPYRIGHT 2005 Mike Holt Enterprises, Inc.

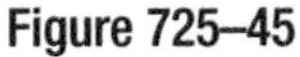
Figure 725–45

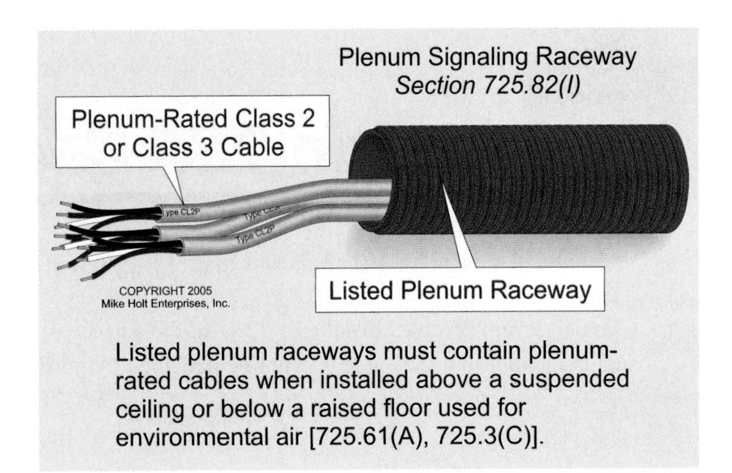

Figure 725–46

(I) Plenum Signaling Raceways. Listed plenum raceways can be installed in other environmental air spaces in accordance with 725.61(A). **Figure 725-46**

> **Author's Comment:** Listed plenum signaling raceways must contain plenum-rated cables when installed above a suspended ceiling or below a raised floor used for environmental air [725.61(A)].

(J) Riser Signaling Raceway. Riser signaling raceways must be listed as preventing the carrying of fire from floor to floor.

> **Author's Comment:** Listed riser signaling raceways must contain riser- or plenum-rated cables when installed in risers [725.61(B)]

(K) General-Purpose Signaling Raceway. General-purpose signaling raceways must be listed to resist the spread of fire.

(L) Marking. The voltage ratings must not be marked on the cables [725.82(G)].

> **FPN:** Voltage markings on cables may be misinterpreted to suggest that the cables may be suitable for Class 1, electric light, and power applications, which they're not.

Article 725 Questions

(• Indicates that 75% or fewer exam takers get the question correct.)

1. Class 1, 2, and 3 cables installed _____ on the surface of ceilings and sidewalls must be supported by the building structure in such a manner that the cable will not be damaged by normal building use.

   (a) exposed (b) concealed (c) hidden (d) a and b

2. Overcurrent protection devices for Class 1 circuit protection must be located at the point where the conductor to be protected _____.

   (a) terminates to the load
   (b) is spliced to any other conductor
   (c) receives its supply
   (d) none of these

3. •Cables and conductors of Class 2 and Class 3 circuits _____ be placed in any cable, cable tray, compartment, enclosure, manhole, outlet box, device box, raceway, or similar fitting with conductors of electric light, power, Class 1, nonpower-limited fire alarm circuits, and medium power network-powered broadband communications circuits.

   (a) may (b) must not (c) a and b (d) none of these

4. The accessible portion of abandoned Class 2 and Class 3 cables installed in duct, plenums, or other spaces used for environmental air must be _____.

   (a) removed (b) identified as abandoned (c) a or b (d) none of these

5. Class 2 and Class 3 plenum cables listed as suitable for use in ducts, plenums, and other spaces used for environmental air are _____.

   (a) CL2P and CL3P (b) CL2R and CL3R (c) CL2 and CL3 (d) PLCT

6. Class 2 and Class 3 cables listed as suitable for general-purpose use with the exception of risers, ducts, plenums, and other spaces used for environmental air, are _____.

   (a) CL2P and CL3P (b) CL2R and CL3R (c) CL2 and CL3 (d) PLCT

ARTICLE 760 Fire Alarm Systems

Introduction

Article 760 provides the requirements for the installation of wiring and equipment for fire alarm systems, including all circuits the fire alarm system controls and powers. Fire alarm systems include fire detection and alarm notification, voice communications, guard's tour, sprinkler waterflow, and sprinkler supervisory systems. NFPA 72, *National Fire Alarm Code* provides other fire alarm system requirements.

As you study this material, pay close attention to the illustrations, which highlight important requirements. Many people have difficulty understanding Article 760 from the text alone, and the figures also show you common Article 760 violations.

PART I. GENERAL

760.1 Scope. Article 760 covers the installation of wiring and equipment for fire alarm systems, including all circuits controlled and powered by the fire alarm system. **Figure 760-1**

Author's Comment: Residential smoke alarm systems, including interconnecting wiring, are not covered by Article 760 because they are not powered by a fire alarm system as defined in NFPA 72.

FPN No. 1: Fire alarm systems include fire detection and alarm notification, voice communications, guard's tour, sprinkler waterflow, and sprinkler supervisory systems. Other circuits that might be controlled or powered by the fire alarm system include elevator capture, elevator shutdown, door release, smoke door and damper control, fire door and damper control, and fan shutdown.

NFPA 72, *National Fire Alarm Code* provides the requirements for the selection, installation, performance, use, testing, and maintenance of the fire alarm system.

Author's Comments:

- Building control circuits associated with the fire alarm system, such as elevator capture and fan shutdown, must comply with Article 725 [760.3(E)]. Article 760 applies if these components are powered and directly controlled by the fire alarm system.

Fire Alarm Systems
Section 760.1
PULL DOWN
FIRE
ALARM
FIRE
FIRE
COPYRIGHT 2005 Mike Holt Enterprises, Inc.
Article 760 covers the installation of wiring and equipment for fire alarm systems.

Figure 760–1

- NFPA 101—*Life Safety Code* specifies where a fire alarm system is required.

760.2 Definitions.

Abandoned Cable. Cable that isn't terminated at equipment, and not identified for future use with a tag, will be considered as abandoned for the purposes of this article.

Author's Comment: 760.3(A) requires accessible abandoned cables to be removed.

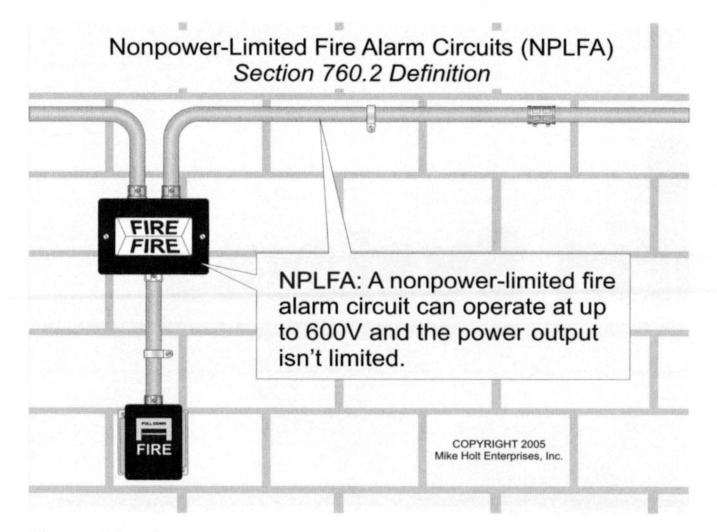

Figure 760–2

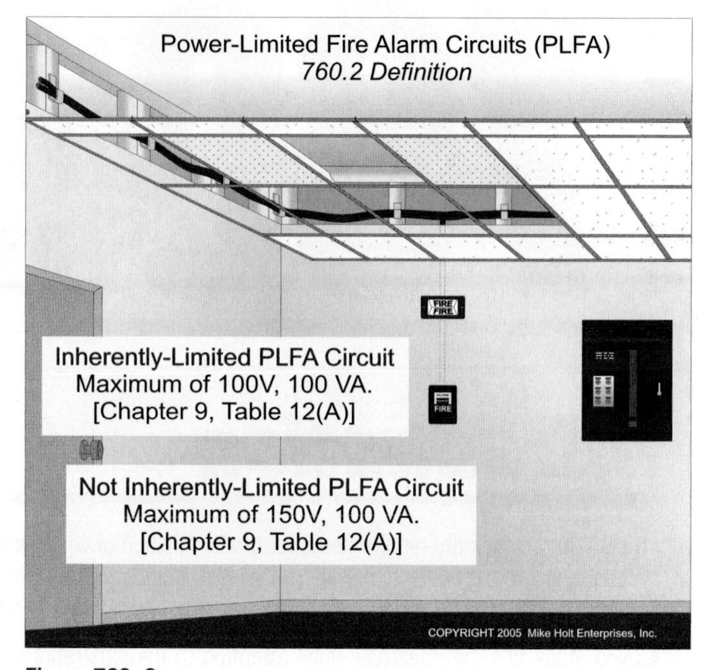

Power-Limited Fire Alarm Circuits (PLFA)
760.2 Definition
Inherently-Limited PLFA Circuit
Maximum of 100V, 100 VA.
[Chapter 9, Table 12(A)]
Not Inherently-Limited PLFA Circuit
Maximum of 150V, 100 VA.
[Chapter 9, Table 12(A)]
COPYRIGHT 2005 Mike Holt Enterprises, Inc.

Figure 760–3

Fire Alarm Circuit. The portion of the wiring system and connected equipment powered and controlled by the fire alarm system. Fire alarm circuits are classified as either nonpower-limited or power-limited.

Nonpower-Limited Fire Alarm Circuit (NPLFA). A nonpower-limited fire alarm circuit can operate at up to 600V, and the power output isn't limited [760.21]. **Figure 760-2**

Power-Limited Fire Alarm Circuit (PLFA). A power-limited fire alarm circuit must have the voltage and power limited by a listed power source that complies with 760.41 as follows: **Figure 760-3**

Inherently Limited (ac) [Chapter 9, Table 12(A)]

Voltage	Power
0 to 20V	5.0 x V
21V to 100V	100 VA

Not Inherently Limited [Chapter 9, Table 12(A)]

Voltage	Power	Overcurrent Protection
0 to 20V	5.0 x V	5A
21 to 100V	100 VA	100/V
101 to 150V	100 VA	1A

Author's Comment: Inherently limited power supplies are designed in a manner such that overcurrent on the load side of the power supply will cause interruption of the power on the load side.

760.3 Other Articles. Only those sections of Article 300 specifically referenced in Article 760 apply to fire alarm circuits and equipment.

(A) Spread of Fire or Products of Combustion. Fire alarm circuits installed through fire-resistant rated walls, partitions, floors, or ceilings must be firestopped in accordance with the instructions supplied by the manufacturer for the specific type of cable and construction material (drywall, brick, etc.), as required by 300.21.

Author's Comment: Openings in fire-resistant walls, floors, and ceilings must be sealed so that the possible spread of fire or products of combustion will not be substantially increased [300.21]. **Figure 760-4**

Abandoned Cable. To limit the spread of fire or products of combustion within a building, the accessible portion of fire alarm cable that isn't terminated at equipment, and not identified for future use with a tag, must be removed [760.2]. **Figure 760-5**

Author's Comment: This rule doesn't require the removal of concealed cables that are abandoned in place, which includes cables in raceways. According to the definition of "Concealed" in Article 100, cables in raceways are considered to be concealed.

(B) Ducts, Plenums, and Other Air-Handling Spaces. Where necessary for the direct action upon, or sensing of the contained air, fire alarm cables can be installed in ducts or plenums if they are installed in electrical metallic tubing, intermediate metal conduit, or rigid metal conduit as required by 300.22(B). **Figure 760-6**

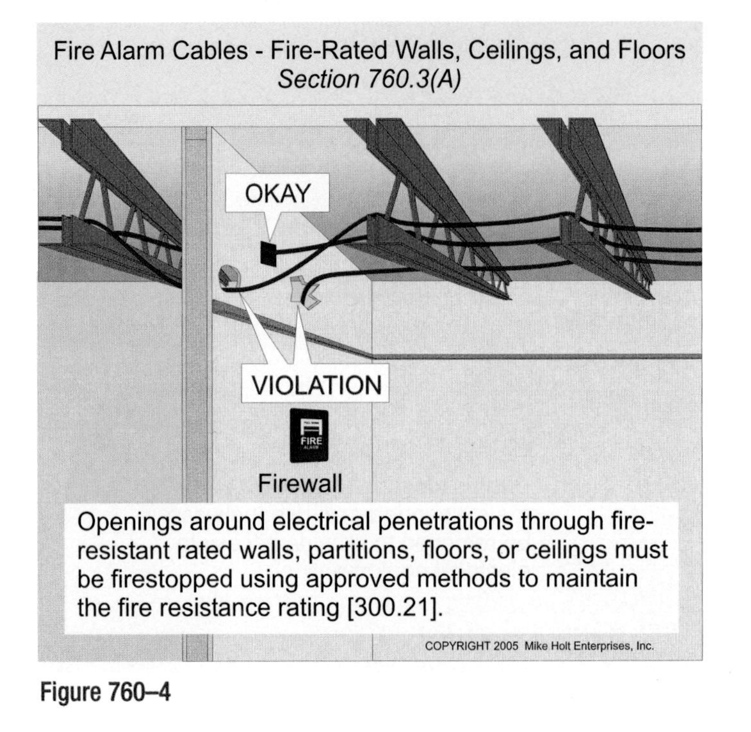
Fire Alarm Cables - Fire-Rated Walls, Ceilings, and Floors
Section 760.3(A)
OKAY
VIOLATION
Firewall
Openings around electrical penetrations through fire-resistant rated walls, partitions, floors, or ceilings must be firestopped using approved methods to maintain the fire resistance rating [300.21].
COPYRIGHT 2005 Mike Holt Enterprises, Inc.

Figure 760–4

Exception: Plenum-rated fire alarm cables are permitted above a suspended ceiling or below a raised floor used for environmental air movement [760.30(B)(1) and (2), 760.61(A), and 760.82(A)].

Nonplenum-rated fire alarm cables can be installed above a suspended ceiling or below a raised floor that is used for environmental air, but only if the cable is installed within electrical metallic tubing [300.22(C)(1)].

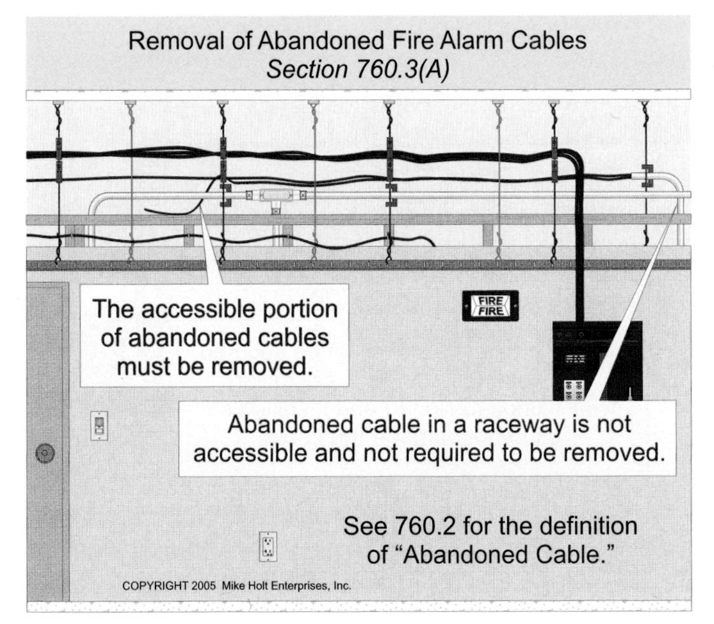
Removal of Abandoned Fire Alarm Cables
Section 760.3(A)
The accessible portion of abandoned cables must be removed.
Abandoned cable in a raceway is not accessible and not required to be removed.
See 760.2 for the definition of "Abandoned Cable."
COPYRIGHT 2005 Mike Holt Enterprises, Inc.

Figure 760–5

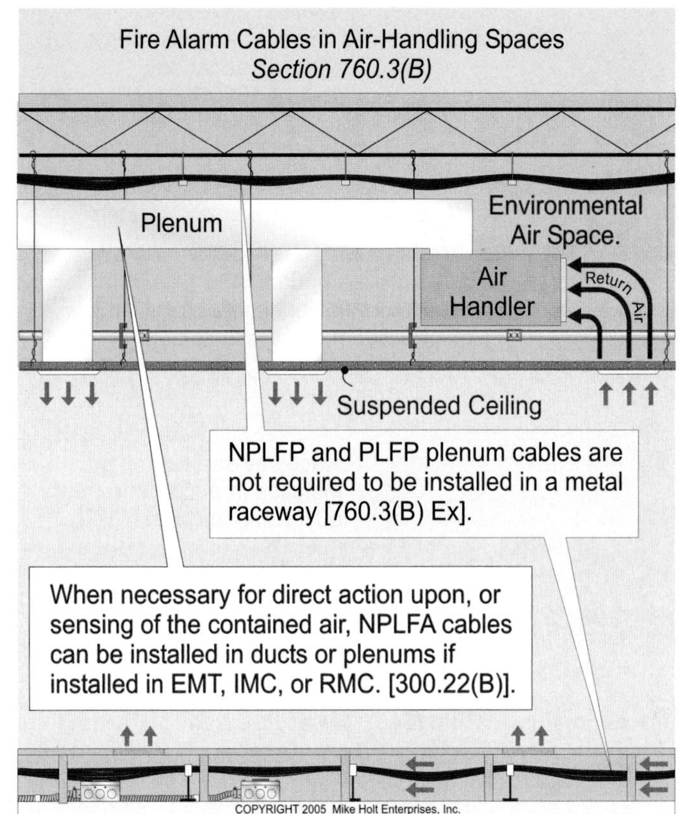
Fire Alarm Cables in Air-Handling Spaces
Section 760.3(B)
Plenum
Environmental Air Space.
Air Handler
Return Air
Suspended Ceiling
NPLFP and PLFP plenum cables are not required to be installed in a metal raceway [760.3(B) Ex].
When necessary for direct action upon, or sensing of the contained air, NPLFA cables can be installed in ducts or plenums if installed in EMT, IMC, or RMC. [300.22(B)].
COPYRIGHT 2005 Mike Holt Enterprises, Inc.

Figure 760–6

Author's Comment: Fire alarm cables installed beneath a raised floor in an information technology equipment room (computer room) aren't required to be plenum rated [300.22(D) and 645.5(D)(5)(c)]. **Figure 760-7**

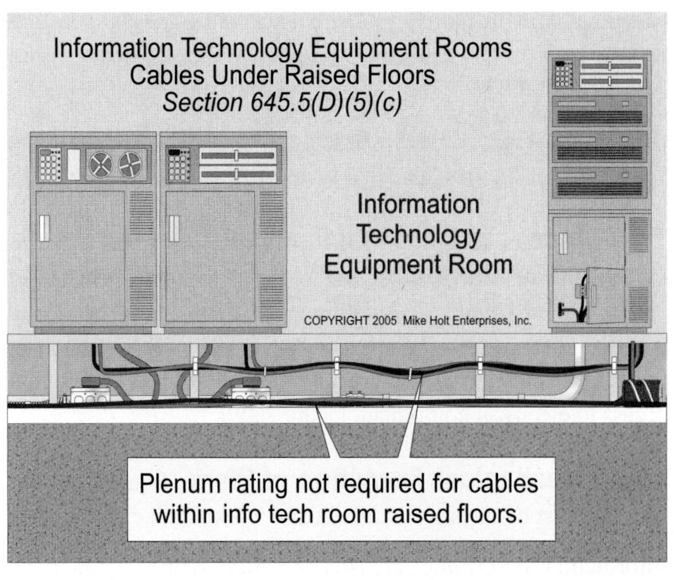
Information Technology Equipment Rooms
Cables Under Raised Floors
Section 645.5(D)(5)(c)
Information Technology Equipment Room
COPYRIGHT 2005 Mike Holt Enterprises, Inc.
Plenum rating not required for cables within info tech room raised floors.

Figure 760–7

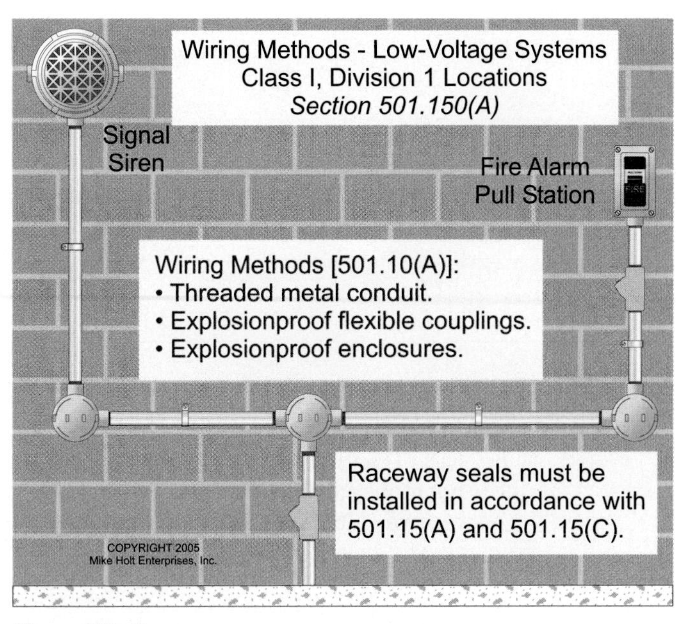
Wiring Methods - Low-Voltage Systems
Class I, Division 1 Locations
Section 501.150(A)
Signal
Siren
Fire Alarm
Pull Station
Wiring Methods [501.10(A)]:
• Threaded metal conduit.
• Explosionproof flexible couplings.
• Explosionproof enclosures.
Raceway seals must be installed in accordance with 501.15(A) and 501.15(C).
COPYRIGHT 2005
Mike Holt Enterprises, Inc.

Figure 760–8

(C) Hazardous (Classified) Locations. Fire alarm circuits installed in any hazardous (classified) location must be installed in accordance with Articles 500 through 516, specifically 501.150 and 502.150. **Figure 760-8**

(D) Corrosive, Damp, or Wet Locations. Fire alarm circuits installed in corrosive, damp, or wet locations must be identified for use in the operating environment [110.11]; must be of materials suitable for the environment in which they are to be installed [300.6]; and must be of a type suitable for the application [310.9].

(E) Building Control Circuits. Class 1, 2, and 3 circuits used for building controls (elevator capture, fan shutdown, etc.), associated with the fire alarm system, but not controlled and powered by the fire alarm system, must be installed in accordance with Article 725 [760.1].

(F) Optical Fiber Cables. Optical fiber cables utilized for fire alarm circuits must be installed in accordance with Article 770.

760.7 Access to Electrical Equipment Behind Panels Designed to Allow Access.

Access to equipment must not be prohibited by an accumulation of cables that prevent the removal of suspended-ceiling panels. Cables must be located so that the suspended-ceiling panels can be moved to provide access to electrical equipment. **Figure 760-9**

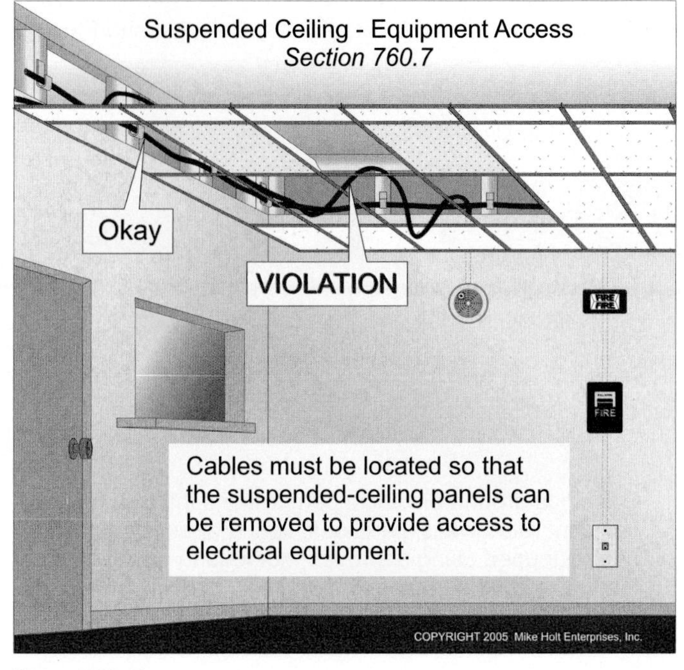
Suspended Ceiling - Equipment Access
Section 760.7
Okay
VIOLATION
Cables must be located so that the suspended-ceiling panels can be removed to provide access to electrical equipment.
COPYRIGHT 2005 Mike Holt Enterprises, Inc.

Figure 760–9

760.8 Mechanical Execution of Work.

Equipment and cabling must be installed in a neat and workmanlike manner.

Information describing industry practices can be found in ANSI/NECA 305, *Standard for Fire Alarm System Job Practices.*

Author's Comment: For more information about this standard, visit http://www.necaneis.org/.

Exposed cables must be supported by the structural components of the building so that the cable will not be damaged by normal building use. Cables must be secured by straps, staples, hangers, or similar fittings designed and installed so as not to damage the cable. **Figure 760-10**

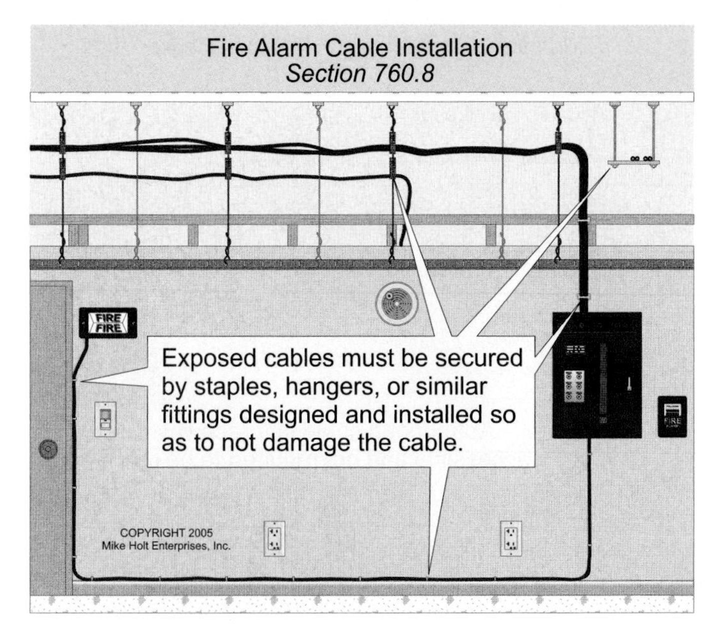
Fire Alarm Cable Installation
Section 760.8
Exposed cables must be secured by staples, hangers, or similar fittings designed and installed so as to not damage the cable.
COPYRIGHT 2005
Mike Holt Enterprises, Inc.

Figure 760–10

Author's Comment: Raceways and cables can be supported by independent support wires attached to the suspended ceiling [300.11(A)].

Cables run parallel to framing members or furring strips must be protected where they are likely to be penetrated by nails or screws, by installing the wiring method so it isn't less than 1¼ in. from the nearest edge of the framing member or furring strips, or by protecting them with a 1⁄16 in. thick steel plate or equivalent [300.4(D)].

760.9 Fire Alarm Circuit and Equipment Grounding (Bonding). Grounding (bonding) isn't required for fire alarm circuits that operate at 50V or less [250.20(A)].

760.10 Fire Alarm Circuit Identification. Fire alarm circuits must be identified at all terminal and junction locations. The identification must be in a manner that will prevent unintentional interference with the fire alarm circuits during testing and servicing. **Figure 760-11**

Author's Comment: Marking of conduits and cables is not required, contrary to some manufacturer's marketing efforts [760.42].

760.11 Fire Alarm Circuit Cables Exposed to Lightning. Where twisted-pair power-limited fire alarm circuit conductors extend beyond a building, they must be installed in accordance with Article 800, or they must be installed in accordance with Part I of Article 300.

760.15 Fire Alarm Circuit Requirements.

(A) Nonpower-Limited Fire Alarm (NPLFA) Circuits. Nonpower-limited fire alarm (NPLFA) circuits must comply with the requirements contained in Parts I and II of this article.

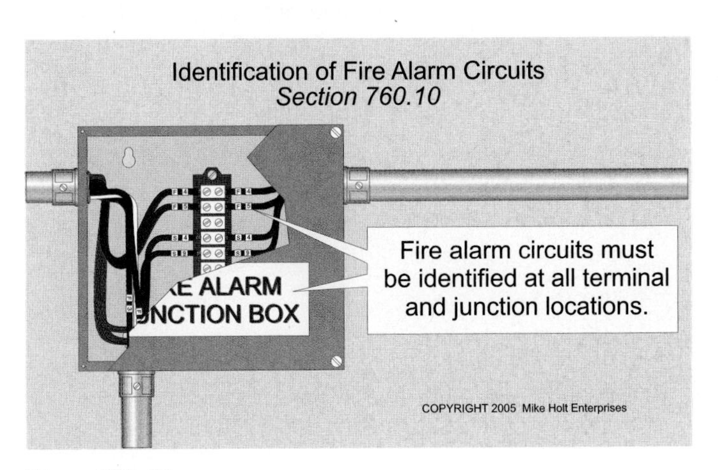
Identification of Fire Alarm Circuits
Section 760.10
Fire alarm circuits must be identified at all terminal and junction locations.
COPYRIGHT 2005 Mike Holt Enterprises

Figure 760–11

(B) Power-Limited Fire Alarm (PLFA) Circuits. Power-limited fire alarm (PLFA) circuits must comply with the requirements contained in Parts I and III of this article.

PART II. NONPOWER-LIMITED FIRE ALARM (NPLFA) CIRCUITS

760.21 GFCI and AFCI Protection. The power source for a nonpower-limited fire alarm circuit must not be supplied through a ground-fault circuit interrupter (GFCI) or arc-fault circuit interrupter (AFCI). **Figure 760-12**

Author's Comment: This GFCI/AFCI limitation only applies to the circuit that supplies a nonpower-limited fire alarm system. Smoke alarms connected to a 15 or 20A, 120V alarm circuit must be AFCI protected if located in the bedroom of a dwelling unit [210.12(B)], because it isn't supplied by a fire alarm panel in accordance with NFPA 72, *National Fire Alarm Code.*

760.23 NPLFA Circuit Overcurrent Protection. Overcurrent protection for conductors 14 AWG and larger must be limited to the conductor ampacity in accordance with Table 310.16 and 240.6(A). Overcurrent protection is not to exceed 7A for 18 AWG conductors and 10A for 16 AWG conductors.

760.25 NPLFA Circuit Wiring Methods. Nonpower-limited fire alarm circuits must be installed in accordance with Part I of Article 300, and all wiring must be in a Chapter 3 wiring method.

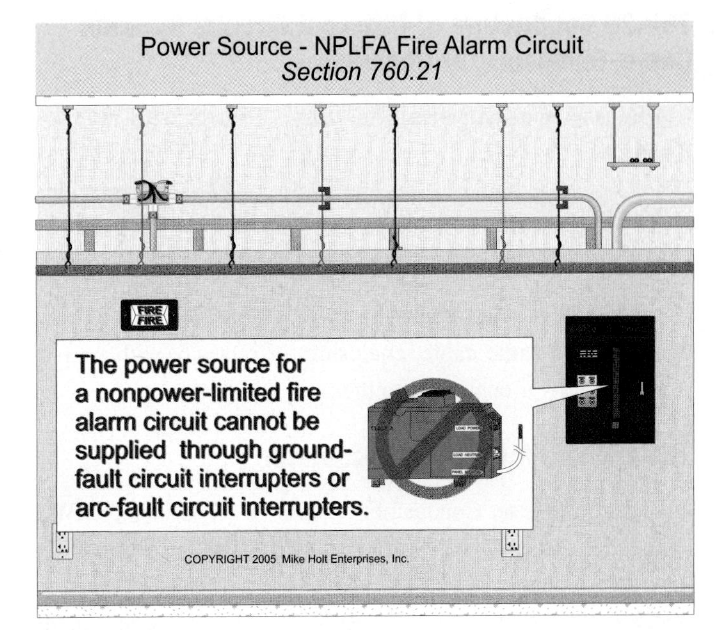
Power Source - NPLFA Fire Alarm Circuit
Section 760.21
FIRE
The power source for a nonpower-limited fire alarm circuit cannot be supplied through ground-fault circuit interrupters or arc-fault circuit interrupters.
COPYRIGHT 2005 Mike Holt Enterprises, Inc.

Figure 760–12

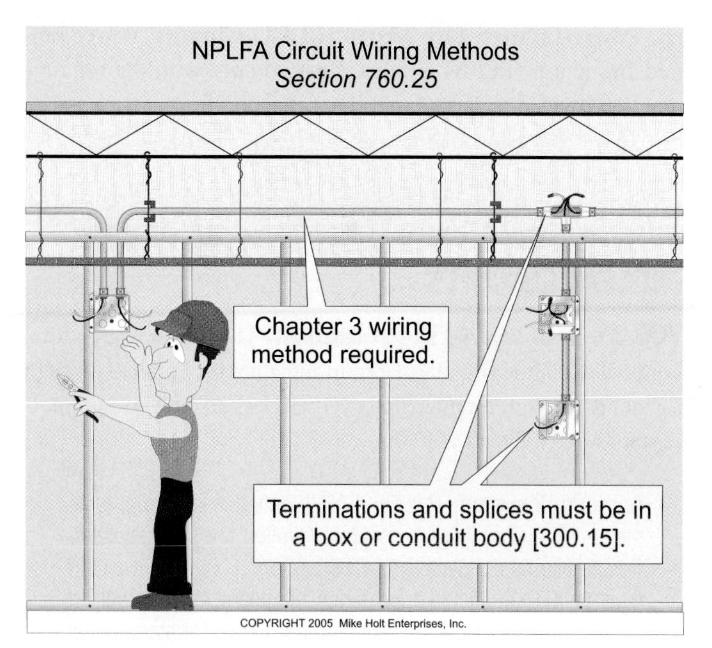
NPLFA Circuit Wiring Methods
Section 760.25
Chapter 3 wiring method required.
Terminations and splices must be in a box or conduit body [300.15].
COPYRIGHT 2005 Mike Holt Enterprises, Inc.

Figure 760–13

Author's Comment: This means that NPLFA circuits must be installed as a suitable Chapter 3 wiring method and all splices must be contained in enclosures [300.15], except for splices and terminations in fire alarm devices and utilization equipment such as detectors, pull stations, and indicating devices when nonpower-limited multiconductor cable is used [760.30(A)(1)]. **Figure 760-13**

Exception 1: Multiconductor NPLFA cables for fire alarm circuits operating at 150V or less [760.30].

760.26 Conductors of Different Circuits in Same Cable, Enclosure, or Raceway.

(A) Class 1 Control and Signaling Circuits with NPLFA Circuits. Class 1 and nonpower-limited fire alarm circuits can occupy the same cable, enclosure, or raceway, provided all conductors are insulated for the maximum voltage of any conductor.

(B) Fire Alarm with Power-Supply Circuits. NPLFA circuits can be in the same cable, enclosure, or raceway with power-supply circuits if connected to the same equipment.

760.27 NPLFA Circuit Conductors.

(A) Sizes and Use. Conductors of sizes 18 AWG and 16 AWG installed in a raceway, enclosure, or listed cable are permitted if they do not supply a load that exceeds the ampacities given in 402.5. Conductors larger than 16 AWG must not supply loads greater than the ampacities given in 310.15.

(B) Insulation. Nonpower-limited fire alarm circuit conductors must have a 600V insulation rating and must comply with Table 310.13. Conductors 18 AWG and 16 AWG must comply with Table 402.3.

760.28 Number of Conductors in a Raceway.

The number of NPLFA conductors in a raceway must comply with the fill requirements contained in 300.17.

Same Size Conductors. When all conductors in a raceway are the same size and insulation, the number of conductors permitted can be found in Annex C for the raceway type.

Question: *How many 18 TFFN fixture wires can be installed in ½ in. electrical metallic tubing?* **Figure 725-14**

(a) 40 *(b) 26* *(c) 30* *(d) 22*

Answer: *(d) 22, Annex C, Table C.1*

PART III. POWER-LIMITED FIRE ALARM (PLFA) CIRCUITS

760.41 Power Sources for PLFA Circuits.

The power source for a power-limited fire alarm circuit must not be supplied through ground-fault circuit interrupters (GFCI) or arc-fault circuit interrupters (AFCI).

Author's Comment: This GFCI/AFCI limitation only applies to the circuit that supplies a power-limited fire alarm system. Smoke detectors connected to a 15 or 20A, 120V circuit must be AFCI protected if located in the bedroom of a dwelling unit [210.12(B)], because according to NFPA 72, *National Fire Alarm Code*, the circuit for the smoke detectors isn't the power source of a power-limited fire alarm circuit.

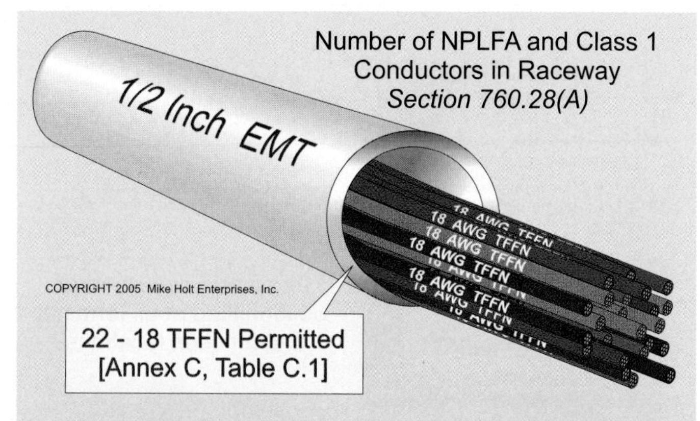
Number of NPLFA and Class 1 Conductors in Raceway
Section 760.28(A)
1/2 Inch EMT
18 AWG TFFN
COPYRIGHT 2005 Mike Holt Enterprises, Inc.
22 - 18 TFFN Permitted [Annex C, Table C.1]
The number of NPLFA and Class 1 conductors in a raceway is determined using 300.17. See Annex C.

Figure 760–14

760.42 Equipment Marking. Fire alarm equipment supplying PLFA circuits must be durably marked to indicate each circuit that is a power-limited fire alarm circuit.

Author's Comment: Fire alarm circuits must be marked at all terminal and junction locations [760.10].

FPN: Power-limited circuits can be reclassified and installed as nonpower-limited circuits if the power-limited fire alarm circuit markings at the fire alarm panel terminals are eliminated [760.52(A) Ex 3].

760.52 Wiring Methods and Materials on Load Side of the PLFA Power Source. Nonpower-limited fire alarm circuits are permitted to use wiring methods and materials in accordance with 760.52(A), (B), or a combination of (A) and (B).

(A) Using NPLFA Wiring Methods. Power-limited fire alarm circuits are permitted to use nonpower-limited fire alarm circuit wiring methods as identified in 760.25.

Exception 3: Power-limited fire alarm circuits can be reclassified and installed as nonpower-limited circuits if the power-limited fire alarm circuit markings required by 760.42 are eliminated, and the entire circuit is installed in accordance with Part II of this article, specifically 760.25 through 760.28.

FPN: Power-limited circuits reclassified and installed as nonpower-limited are no longer classified as power-limited, regardless of their continued connection to a power-limited fire alarm power source.

Author's Comment: PLFA circuits reclassified as NPLFA circuits can no longer be installed with other PLFA circuits that have not been reclassified as NPLFA [760.54].

(B) PLFA Wiring Methods and Materials. Power-limited fire alarm conductors and cables described in 760.82 must be installed as detailed in (1), (2), or (3).

(1) Exposed or Fished in Concealed Spaces. Power-limited fire alarm circuits installed exposed must be adequately supported and protected against physical damage or be installed in a raceway.

Author's Comment: Exposed cables must be supported by the structural components of the building so that the cable will not be damaged by normal building use. Cables must be secured by straps, staples, hangers, or similar fittings designed and installed so as not to damage the cable. Cables run parallel to framing members or furring strips must be protected where they are likely to be penetrated by nails or screws, by installing the wiring method so it isn't less than 1¼ in. from the nearest edge of the framing member or furring strips, or it must be protected by a 1⁄16 in. thick steel plate or the equivalent [760.8].

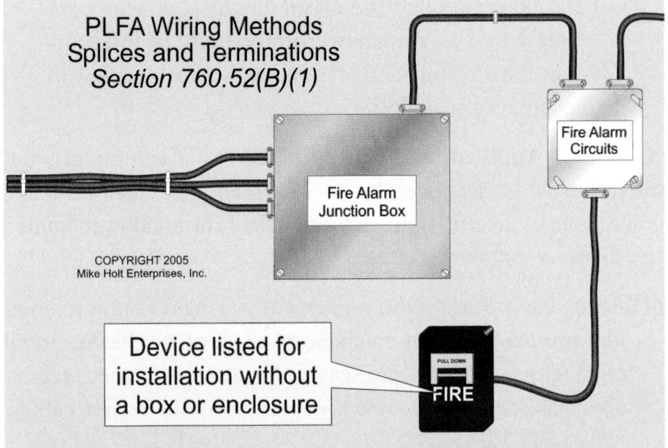
PLFA Wiring Methods
Splices and Terminations
Section 760.52(B)(1)
Fire Alarm Circuits
Fire Alarm Junction Box
COPYRIGHT 2005
Mike Holt Enterprises, Inc.
Device listed for installation without a box or enclosure
FIRE
Splices and terminations of power-limited fire alarm circuits must be made in listed fittings, boxes, enclosures, fire alarm devices, or utilization equipment.

Figure 760–15

Cable splices or terminations must be made in listed fittings, boxes, enclosures, fire alarm devices, or utilization equipment. **Figure 760-15**

760.55 Separation from Other Circuit Conductors.

(A) General. Power-limited fire alarm conductors must not be placed in any enclosure, raceway, or cable with conductors of electric light, power, Class 1, or nonpower-limited fire alarm circuits.

(B) Separated by Barriers. Where separated by a barrier, power-limited fire alarm circuits are permitted with Class 1 and nonpower-limited fire alarm circuits.

Author's Comment: Separation is required to prevent a fire or shock hazard that could occur from a short between the fire alarm circuit and the higher-voltage circuits.

(D) Associated Systems Within Enclosures. PLFA conductors can be mixed with electric light, power, Class 1, and nonpower-limited fire alarm circuit conductors in enclosures where these other conductors are introduced solely for connection to the same equipment as the nonpower-limited circuits, and:

(1) A minimum of ¼ in. separation is maintained from the PLFA conductors, or

(2) The conductors operate at 150 volts or less and comply with one of the following:

(a) The power-limited fire alarm circuit conductors are installed in Types FPL, FPLR, FPLP, or permitted substitute cables and these conductors (extending beyond the jacket) are separated by a minimum of ¼ in. from other conductors.

(b) The power-limited fire alarm circuit conductors are reclassified as a nonpower-limited circuit in accordance with 760.52 Ex 3, and they are installed in accordance with 760.25.

(G) Other Applications. Power-limited fire alarm circuit conductors must be separated by not less than 2 in. from insulated conductors of electric light, power, Class 1, or nonpower-limited fire alarm circuits unless:

(1) All of the electric light, power, Class 1, and nonpower-limited fire alarm circuit conductors, or all of the PLFA circuit conductors, are in a raceway or in metal-sheathed, metal-clad, nonmetallic-sheathed, or underground feeder cables. **Figure 760-16**

760.56 PLFA Circuits, Class 2, Class 3, and Communications Circuits.

(A) PLFA, Communications, and Class 3 Circuits. Power-limited fire alarm circuits, communications circuits, or Class 3 circuits can be in the same cable, enclosure, or raceway.

(B) PLFA and Class 2 Circuits. Class 2 circuits can be within the same cable, enclosure, or raceway as conductors of power-limited fire alarm circuits provided the Class 2 circuit conductor insulation isn't less than that required for the power-limited fire alarm circuit.

> **Author's Comment:** Listed Class 2 cables have a voltage rating of 150V [725.82(G)], whereas listed power-limited fire alarm cables have a voltage rating of 300V [760.82(C)].

(D) Audio System Circuits and PLFA Circuits. Audio system circuits [640.9(C)] using Class 2 or Class 3 wiring methods must not be installed in the same cable or raceway with power-limited fire alarm conductors or cables.

> **Author's Comment:** The concern is that a fault from audio amplifier circuits to fire alarm circuits has the potential to create a hazard by disrupting the operation of fire alarm systems. However, this new restriction does not apply to the voice annunciation audio circuits that are supplied and controlled from a fire alarm panel and commonly required in high-rise and similar applications.

760.58 Support. PLFA cables must not be supported by a raceway. **Figure 760-17**

760.61 Applications of Listed PLFA Cables.

(A) Ducts, Plenums, and Other Space Used for Environmental Air. Power-limited fire alarm cables must not be installed exposed in ducts or plenums, unless they act directly on the contained air, are installed using a wiring method described in 300.22(B), and are plenum rated. However, plenum-rated power-limited fire alarm cables can be installed above a suspended ceiling or below a raised floor that is used for environmental air. **Figure 760-18**

> **Author's Comments:**
>
> - Nonplenum-rated cables can be installed above a suspended ceiling or below a raised floor that is used for environmental air, but only if the nonplenum-rated cable is installed within electrical metallic tubing [300.22(C)(1)].
> - Fire alarm cables installed beneath a raised floor in an information technology equipment room (computer room) aren't required to be plenum rated [300.22(D) and 645.5(D)(5)(c)].

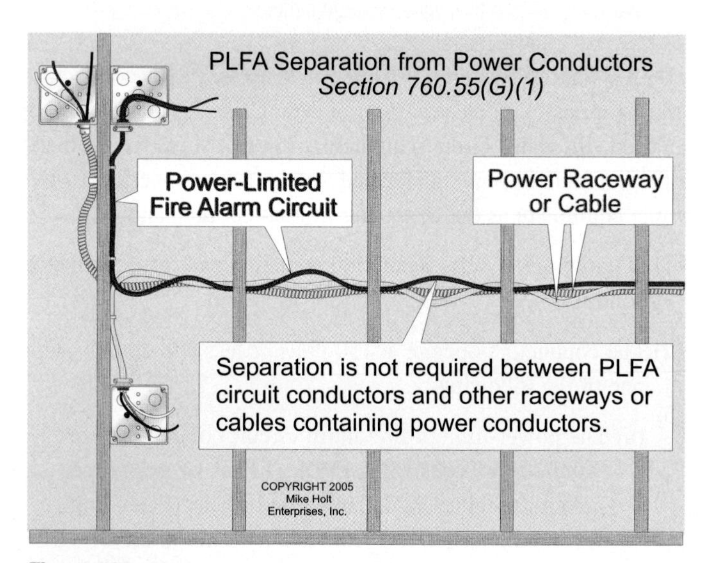
PLFA Separation from Power Conductors
Section 760.55(G)(1)
Power-Limited Fire Alarm Circuit
Power Raceway or Cable
Separation is not required between PLFA circuit conductors and other raceways or cables containing power conductors.
COPYRIGHT 2005 Mike Holt Enterprises, Inc.

Figure 760–16

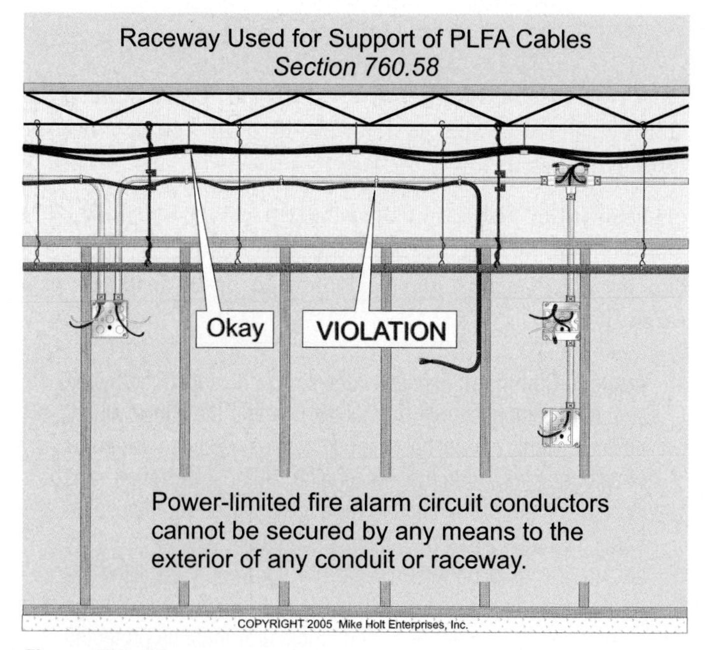
Raceway Used for Support of PLFA Cables
Section 760.58
Okay
VIOLATION
Power-limited fire alarm circuit conductors cannot be secured by any means to the exterior of any conduit or raceway.
COPYRIGHT 2005 Mike Holt Enterprises, Inc.

Figure 760–17

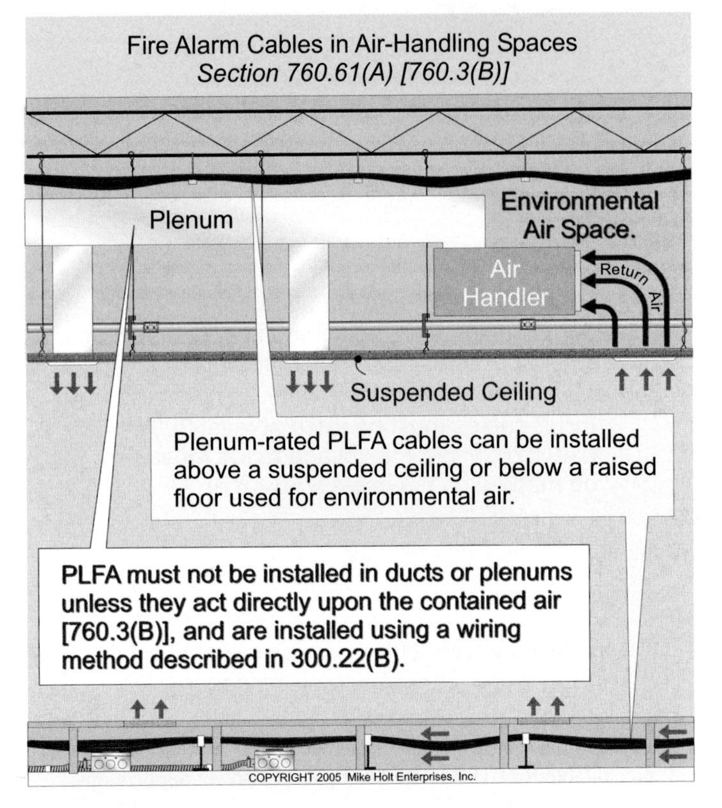
Fire Alarm Cables in Air-Handling Spaces
Section 760.61(A) [760.3(B)]
Plenum
Environmental Air Space.
Air Handler
Return Air
Suspended Ceiling
Plenum-rated PLFA cables can be installed above a suspended ceiling or below a raised floor used for environmental air.
PLFA must not be installed in ducts or plenums unless they act directly upon the contained air [760.3(B)], and are installed using a wiring method described in 300.22(B).
COPYRIGHT 2005 Mike Holt Enterprises, Inc.

Figure 760–18

(B) Riser.

(1) Power-limited fire alarm cables installed in vertical runs penetrating more than one floor must be riser rated Types FPLR or FPLP [760.82(E)].

(2) Power-limited fire alarm cables installed in vertical runs penetrating more than one floor aren't required to be riser rated if they are installed within a metal raceway.

> **Author's Comment:** Metal raceways containing circuit conductors from a power-supply system that operates at 50V or less generally aren't required to be grounded (bonded) to an effective ground-fault current path [250.86 and 250.112(I)]. However, if they are run outside and overhead, they may be required to be grounded (bonded) by 250.20(A) or "protected" as required by 800.90.

(3) Type FPL cable is permitted in one- and two-family dwellings.

(C) Other Wiring in Buildings. Power-limited fire alarm cables installed in locations other than ducts, plenums, or other environmental air spaces must be of Type FPL, or they must be installed in a raceway [760.82(F)]. However, nonpower-limited fire alarm wiring methods may be used—some of which do not require raceways [760.30].

(D) Cable Substitutions. Power-limited fire alarm circuit cables can be substituted in accordance with Table 760.61.

PART IV. LISTING REQUIREMENTS

760.81 Listing and Marking of NPLFA Cables.

Nonpower-limited fire alarm cables within buildings must be constructed in accordance with (A) and (B), be listed and resistant to the spread of fire in accordance with (C) through (F), and be marked in accordance with (G).

(A) NPLFA Conductor Materials. Conductors must be 18 AWG or larger, solid or stranded copper.

(B) Insulation. Insulated circuit conductors 14 AWG and larger must have a 600V insulation rating and must comply with Table 310.13. Conductors 18 AWG and 16 AWG must comply with 760.27.

(C) Type NPLFP. Type NPLFP plenum-rated nonpower-limited fire alarm cable must be listed for use in environmental air spaces and have adequate fire-resistant and low smoke-producing characteristics.

(D) Type NPLFR. Type NPLFR nonpower-limited fire alarm riser cable must be listed as being suitable for use in a vertical run in a shaft, or from floor to floor, and have fire-resistant characteristics capable of preventing the carrying of fire from floor to floor.

(E) Type NPLF. Type NPLF general-purpose nonpower-limited fire alarm cable must be listed for general-purpose fire alarm use and be listed as being resistant to the spread of fire.

(G) NPLFA Cable Markings. Nonpower-limited fire alarm circuit cable can be marked with a voltage rating of 150 volts.

760.82 Listing and Marking of PLFA Cables.

Power-limited fire alarm cable installed within buildings must be listed in accordance with (A) through (H) and must be marked in accordance with (I).

> **Author's Comment:** The *NEC* doesn't require outside or underground cable to be listed, but the cable must be approved by the authority having jurisdiction as suitable for the application in accordance with 90.4, 90.7, and 110.2.

(A) Conductor Materials. Conductors must be solid or stranded copper.

(B) Conductor Size. The size of conductors in a multiconductor cable must not be smaller than 26 AWG. Single conductors must not be smaller than 18 AWG.

(C) Ratings. Power-limited fire alarm cable must have a voltage rating not less than 300V and this rating must not be marked on the cable [760.82(I)].

(D) Type FPLP. Type FPLP power-limited fire alarm plenum cables must be listed for use in ducts, plenums, and other environmental air spaces [760.61(A)]. **Figure 760-19**

Author's Comment: Special consideration must be given to cables in areas that move or transport environmental air in order to reduce the hazards that arise from the burning of conductor insulation and of cable jackets. Because listed plenum-rated cables have adequate fire-resistant and low smoke-producing characteristics, they can be installed in environmental air space, but they cannot be installed in ducts or plenums! See 760.3(B) in the textbook for details.

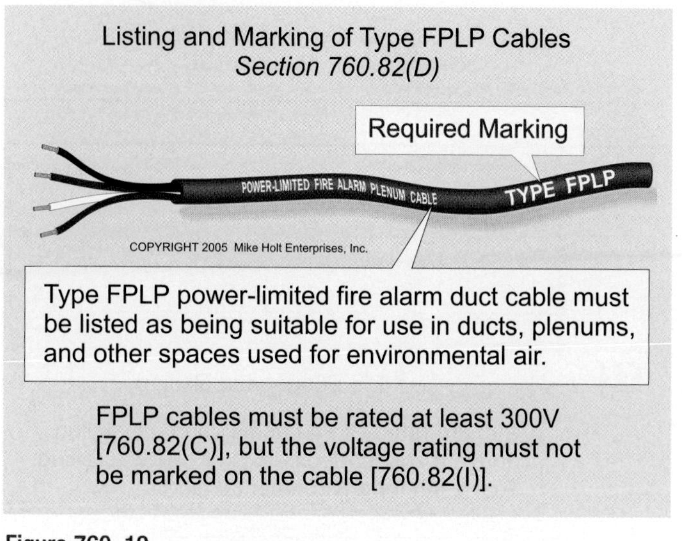

Listing and Marking of Type FPLP Cables
Section 760.82(D)
Required Marking
POWER-LIMITED FIRE ALARM PLENUM CABLE
TYPE FPLP
COPYRIGHT 2005 Mike Holt Enterprises, Inc.
Type FPLP power-limited fire alarm duct cable must be listed as being suitable for use in ducts, plenums, and other spaces used for environmental air.
FPLP cables must be rated at least 300V [760.82(C)], but the voltage rating must not be marked on the cable [760.82(I)].

Figure 760–19

(E) Type FPLR. Type FPLR power-limited fire alarm riser cables must be listed as being suitable for use in a vertical run in a shaft, or from floor to floor [760.61(B)].

(F) Type FPL. Type FPL power-limited fire alarm cables must be listed as being suitable for general-purpose use [760.61(C)].

(H) Coaxial Cables. Coaxial cables must be listed as Types FPLP, FPLR, or FPL.

(I) Cable Marking. The 300V rating of power-limited fire alarm cable must not be marked on the cable [760.81(C)].

FPN: Voltage markings on cables may suggest that the cables are suitable for Class 1, electric light, and power applications, which they're not.

Article 760 Questions

1. Fire alarm circuits installed in ducts, plenums, and other air-handling spaces must be installed in accordance with the requirements contained in 300.22.

   (a) True (b) False

2. Fire alarm cables installed _____ to framing members must be protected against physical damage from penetration by screws or nails by 1¼ in. separation from the framing member or by a suitable metal plate in accordance with 300.4(D).

   (a) exposed (b) concealed (c) parallel (d) all of these

3. Class 1 and nonpower-limited fire alarm circuits can occupy the same cable, enclosure, or raceway, provided all conductors are insulated for the maximum voltage of any conductor.

   (a) True (b) False

4. Splices and terminations of nonpower-limited fire alarm circuits must be made in _____ fittings, boxes, enclosures, fire alarm devices, or utilization equipment.

   (a) identified (b) listed (c) approved (d) none of these

5. Power-limited fire alarm circuits are permitted to be reclassified and installed as _____ if the power-limited fire alarm circuit markings required by 760.42 are eliminated, and the entire circuit is installed in a Chapter 3 wiring method in accordance with Part II of Article 760.

   (a) intrinsically safe circuits (b) ground-fault protected circuits
   (c) nonpower-limited circuits (d) Class I Division 2

6. Power-limited fire alarm (PLFA) cables can be supported by strapping, taping, or attaching to the exterior of a conduit or raceway.

   (a) True (b) False

7. Coaxial cables used in power-limited fire alarm systems must have an overall insulation rating of not less than _____.

   (a) 100V (b) 300V (c) 600V (d) 1,000V

Notes

ARTICLE

770 Optical Fiber Cables and Raceways

Introduction

Article 770 provides the requirements for installing optical fiber cables and special raceways for optical fiber cables. It also contains the requirements for composite cables, often called "hybrid," that combine optical fibers with metallic current-carrying conductors.

While we normally think of Article 300 in connection with wiring methods, you need to use only Article 770 methods for fiber-optic cables, except where Article 770 makes specific references to Article 300. The first such reference, in 770.3, is to 300.21. This addresses requirements for stopping the spread of combustion. Another Chapter 3 reference is to 300.22, which applies when installing optical fiber cables and optical fiber raceways in ducts, plenums, or other air-handling spaces.

Article 770 doesn't refer to 300.15, so boxes aren't required for splices or terminations of optical fiber cable. The FPN in 770.113 states that splice cases and terminal boxes are typically used as enclosures for splicing, or terminating and splicing, optical fiber cables. While this is a good practice, it's not required—no FPN is an enforceable *Code* requirement [90.5(C)].

Article 90 states that the *NEC* isn't a design guide or installation manual. Article 770 doesn't tell how to ensure your system will meet, and test out to, the performance requirements you need or your contract specifies. For example, it doesn't mention bend radii. It doesn't tell how to install and test cable safely either, but that doesn't mean you should look into a cable, even if you cannot see any light coming through it. Light used in fiber-optic circuits usually isn't visible, but it can still damage your eyes.

PART I. GENERAL

770.1 Scope. Article 770 covers the installation of optical fiber cables, which contain optical fibers used to transmit light for control, signaling, and communication. This article also contains the installation requirements for optical fiber raceways, which contain and support the optical fiber cables, as well as the requirements for composite cables that combine optical fibers with current-carrying conductors.

> **Author's Comment:** The growth of high-tech applications and significant technological development of optical fibers and the equipment used to send and receive light pulses has increased the use of fiber optics. Since optical fiber cable isn't affected by electromagnetic interference, there has been a large growth in its uses in communications for voice, data transfer, data processing, and computer control of machines and processes.

770.2 Definitions.

Abandoned Cable. Cable that isn't terminated at equipment, and not identified for future use with a tag, is considered as abandoned for the purpose of this article.

> **Author's Comment:** 770.3(A) requires abandoned cables to be removed.

Optical Fiber Raceway. A raceway designed for enclosing and routing nonconductive optical fiber cables. **Figure 770-1**

> **Author's Comment:** According to the *UL General Information for Electrical Equipment Directory* (White Book), only nonconductive optical fiber cable (for example, without electrical conductors) can be installed in an optical fiber raceway.

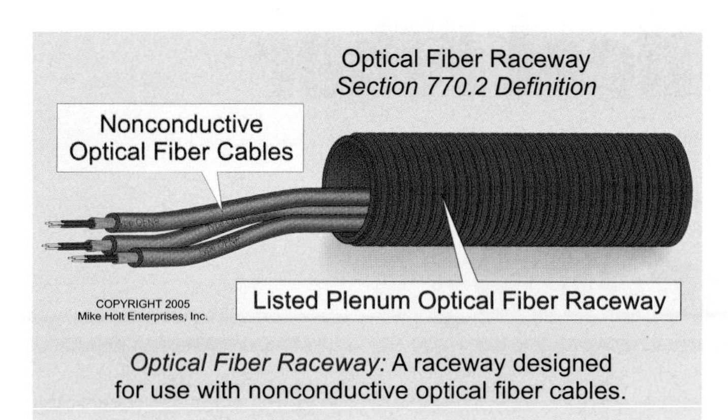
Optical Fiber Raceway
Section 770.2 Definition
Nonconductive
Optical Fiber Cables
Listed Plenum Optical Fiber Raceway
COPYRIGHT 2005
Mike Holt Enterprises, Inc.
Optical Fiber Raceway: A raceway designed
for use with nonconductive optical fiber cables.

Figure 770–1

Point of Entrance. The point at which the cable emerges from an external wall, from a concrete floor slab, or from a rigid metal conduit or an intermediate metal conduit grounded to an electrode in accordance with 800.100.

Author's Comment: Understanding the definition of this term is important for determining the maximum length of unlisted underground plastic innerduct [770.12(D)] or optical fiber cable inside a building [770.113 Ex 1].

770.3 Locations and Other Articles.

Only those sections of Article 300 specifically referenced in Article 770 apply to optical fiber cables and raceways.

Author's Comment: Article 770 doesn't reference 300.15, so boxes aren't required for splices or terminations of optical fiber cable. On the other hand, composite (hybrid) optical fiber cables [770.9(C)] are considered electrical cables and must comply with the appropriate requirements of *NEC* Chapters 1 through 4. **Figure 770-2**

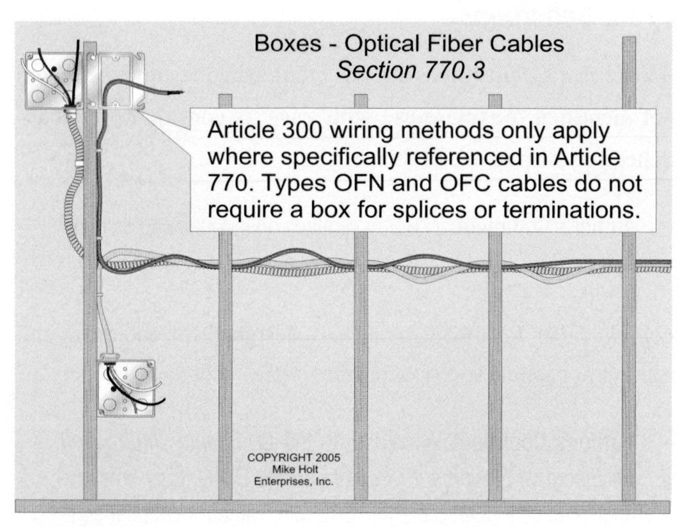
Boxes - Optical Fiber Cables
Section 770.3
Article 300 wiring methods only apply
where specifically referenced in Article
770. Types OFN and OFC cables do not
require a box for splices or terminations.
COPYRIGHT 2005
Mike Holt
Enterprises, Inc.

Figure 770–2

(A) Spread of Fire or Products of Combustion. Optical fiber cables and optical fiber raceways installed through fire-resistant-rated walls, partitions, floors, or ceilings must be firestopped in accordance with the instructions supplied by the manufacturer for the specific type of cable and construction material (drywall, brick, etc.), as required by 300.21.

Author's Comment: Openings in fire-resistant walls, floors, and ceilings must be sealed so that the possible spread of fire or products of combustion will not be substantially increased [300.21]. **Figure 770-3**

Abandoned Cable. To limit the spread of fire or products of combustion within a building, the accessible portion of optical fiber cable that isn't terminated at equipment, and not identified for future use with a tag, must be removed [770.2]. **Figure 770-4**

Author's Comment: This rule doesn't require the removal of concealed cables that are abandoned in place, which includes cables in raceways. According to the definition of "Concealed" in Article 100, cables in raceways are considered to be concealed.

(B) Ducts, Plenums, and Other Air-Handling Spaces. Where necessary for the direct action upon, or sensing of the contained air, optical fiber cables can be installed in ducts or plenums if they are installed in electrical metallic tubing, intermediate metal conduit, or rigid metal conduit as required by 300.22(B). This is highly unlikely with optical fiber cables.

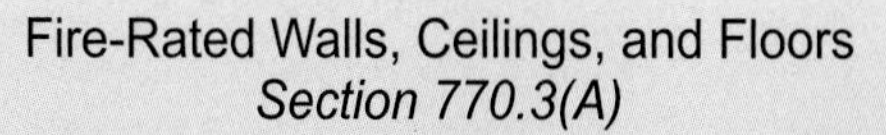
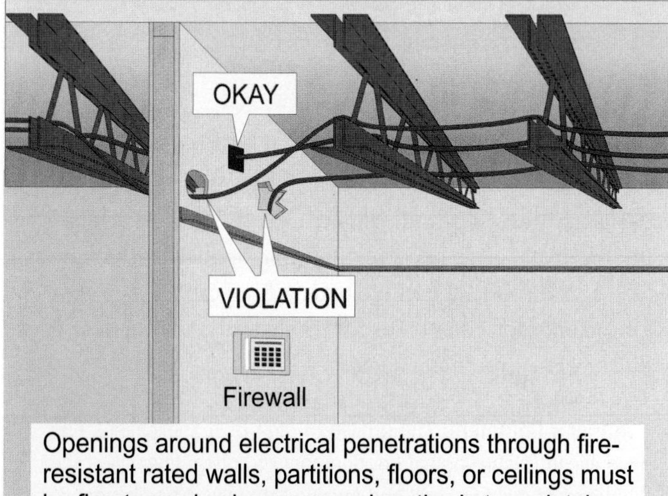
Fire-Rated Walls, Ceilings, and Floors
Section 770.3(A)
OKAY
VIOLATION
Firewall
Openings around electrical penetrations through fire-resistant rated walls, partitions, floors, or ceilings must be firestopped using approved methods to maintain the fire-resistance rating [300.21].
COPYRIGHT 2005 Mike Holt Enterprises, Inc.

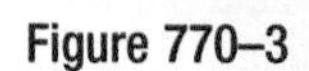
Figure 770–3

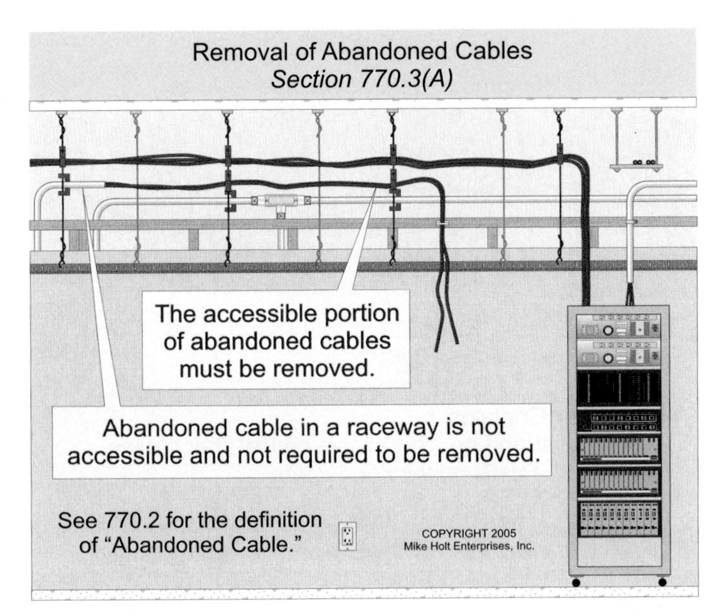
Removal of Abandoned Cables
Section 770.3(A)
The accessible portion of abandoned cables must be removed.
Abandoned cable in a raceway is not accessible and not required to be removed.
See 770.2 for the definition of "Abandoned Cable."
COPYRIGHT 2005 Mike Holt Enterprises, Inc.

Figure 770–4

Exception: Plenum-rated optical fiber cables [770.154(A)] and plenum-rated optical fiber raceways (with plenum-rated fire optical fiber cables) [770.182(A)] are permitted above a suspended ceiling or below a raised floor used for environmental air movement. **Figure 770-5**

> **Author's Comment:** Optical fiber cables installed beneath a raised floor in an information technology equipment room (computer room) aren't required to be plenum rated [300.22(D) and 645.5(D)(5)(c)]. **Figure 770-6**

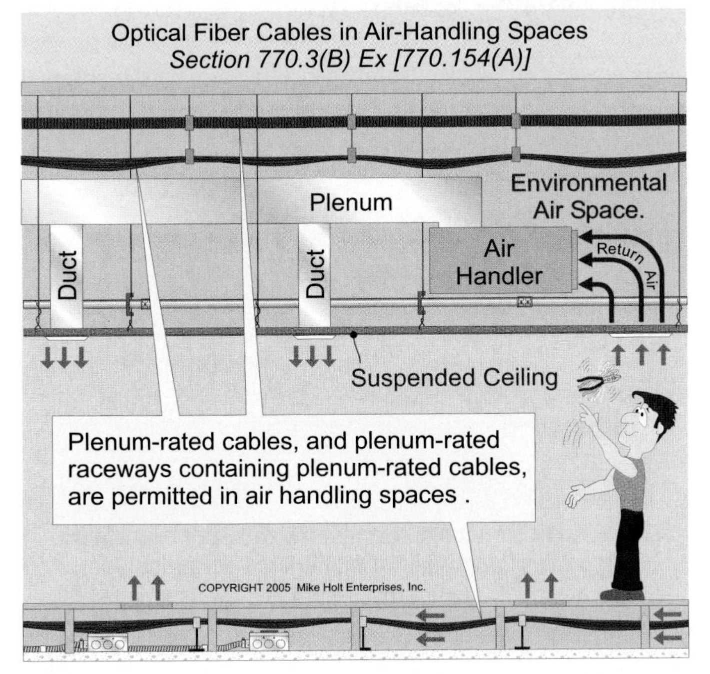
Optical Fiber Cables in Air-Handling Spaces
Section 770.3(B) Ex [770.154(A)]
Plenum
Environmental Air Space.
Duct
Duct
Air Handler
Return Air
Suspended Ceiling
Plenum-rated cables, and plenum-rated raceways containing plenum-rated cables, are permitted in air handling spaces .
COPYRIGHT 2005 Mike Holt Enterprises, Inc.

Figure 770–5

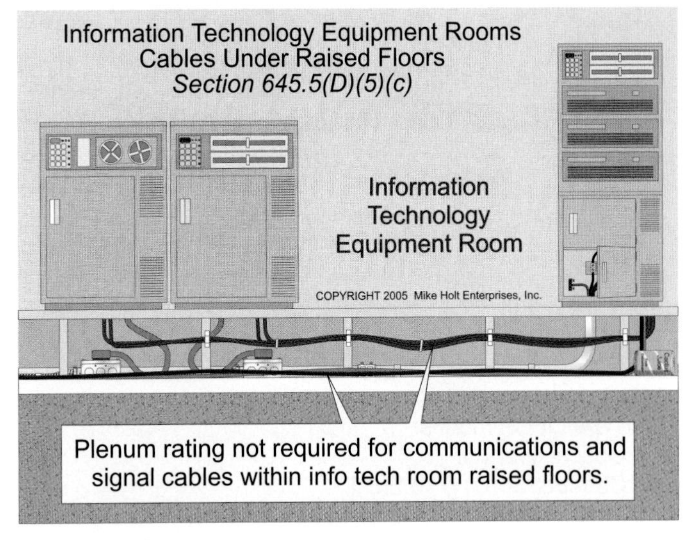
Information Technology Equipment Rooms
Cables Under Raised Floors
Section 645.5(D)(5)(c)
Information Technology Equipment Room
COPYRIGHT 2005 Mike Holt Enterprises, Inc.
Plenum rating not required for communications and signal cables within info tech room raised floors.

Figure 770–6

770.6 Optical Fiber Cables. Optical fiber cables transmit light for control, signaling, and communications through an optical fiber. **Figure 770-7**

770.9 Types.

(A) Nonconductive Cable. Nonconductive optical fiber cable contains nothing that can conduct electricity, so it cannot energize or be energized even when closely associated with electrical conductors. **Figure 770-8A**

(B) Conductive Cable. Conductive optical fiber cables contain conductive members such as metallic strength members, metallic vapor barriers, and metallic armor or sheath. **Figure 770-8B**

(C) Composite. Composite optical fiber cables contain optical fibers and current-carrying electrical conductors.

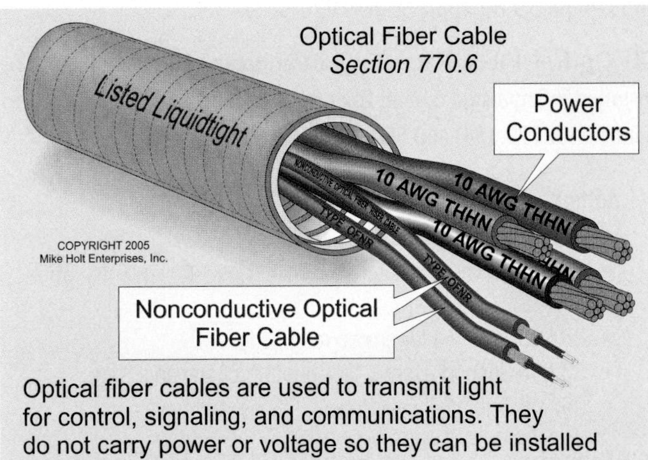
Optical Fiber Cable
Section 770.6
Listed Liquidtight
Power Conductors
10 AWG THHN
10 AWG THHN
10 AWG THHN
TYPE OFNR
TYPE OFNR
COPYRIGHT 2005 Mike Holt Enterprises, Inc.
Nonconductive Optical Fiber Cable
Optical fiber cables are used to transmit light for control, signaling, and communications. They do not carry power or voltage so they can be installed with power conductors [770.133].

Figure 770–7

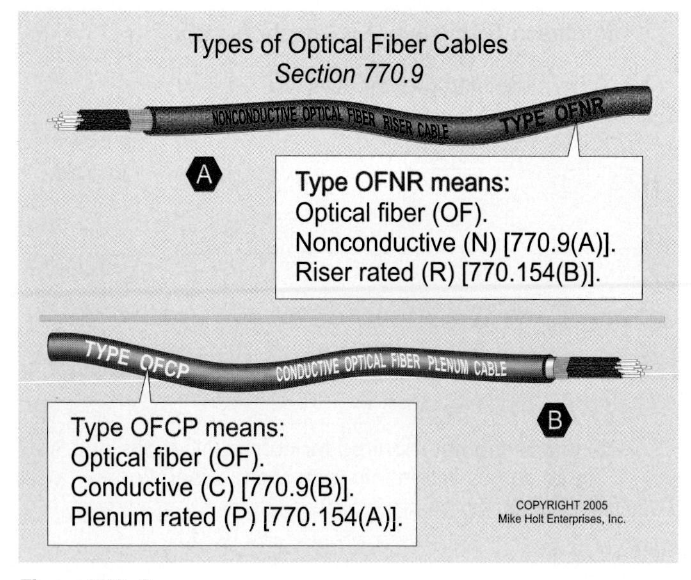

Figure 770–8

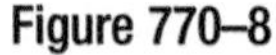

Author's Comment: Article 770 permits the use of composite cables only where the optical fibers and current-carrying electrical conductors are functionally associated [770.133(A)].

770.12 Raceways for Optical Fiber Cables.

Installations of raceways for optical fiber cables must comply with (A) through (D):

(A) Chapter 3 Raceways. Listed optical fiber cable can be installed in any type of Chapter 3 wiring method, where installed in accordance with Chapter 3. If optical fiber cables are installed within the raceway without current-carrying conductors, the raceway fill tables of Chapter 3 and Chapter 9 do not apply. Where nonconductive optical fiber cables are installed with electric conductors in a raceway, the raceway fill tables of Chapter 3 and Chapter 9 do apply. **Figure 770-9**

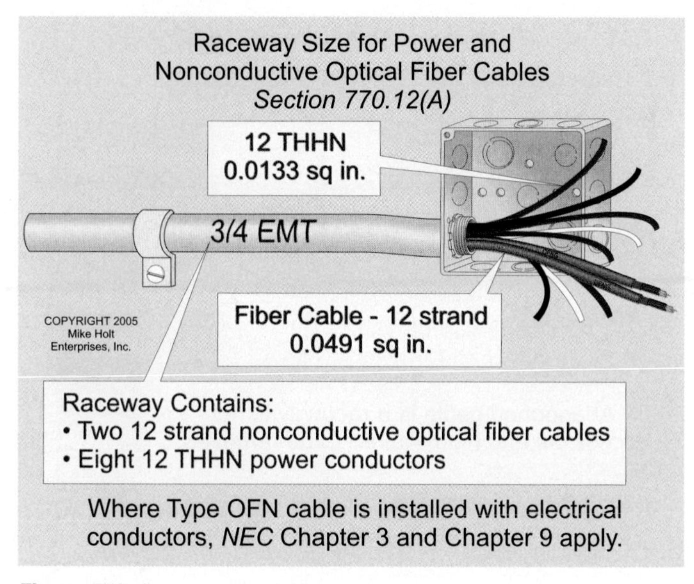

Figure 770–9

(B) Optical Fiber Raceway. Listed optical fiber cable can be installed in any listed optical fiber raceway that is installed in accordance with 770.154, and 362.24 through 362.56. **Figure 770-10**

Author's Comment:

- 362.24 Bending radius.
- 362.26 Maximum total bends between pull points, 360 degrees.
- 362.28 Trimmed to remove rough edges.
- 362.30 Support every 3 ft, within 3 ft of any enclosure.
- 362.48 Joints between tubing, fittings, and boxes.

(C) Innerduct. Listed optical fiber raceway installed in accordance with 770.154 can be installed as innerduct in any Chapter 3 raceway.

(D) Entering Buildings. Unlisted underground or outside plant construction plastic innerduct entering the building from the outside must be terminated and firestopped at the point of entrance.

Author's Comment: According to 770.2, the point of entrance is defined as the point where the cable emerges from an external wall, from a concrete floor slab, or from a rigid metal conduit or an intermediate metal conduit grounded (bonded) to an electrode in accordance with 800.100.

770.21 Access to Electrical Equipment Behind Panels Designed to Allow Access.

Access to equipment must not be prohibited by an accumulation of cables that prevent the removal of suspended-ceiling panels. Cables must be located so that the suspended-ceiling panels can be moved to provide access to electrical equipment.

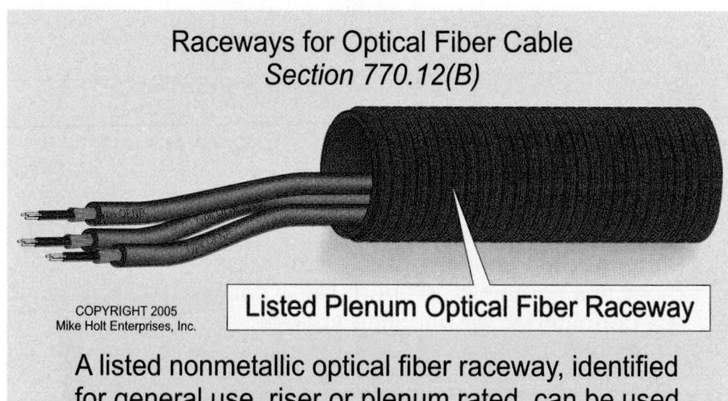

Figure 770–10

770.24 Mechanical Execution of Work. Equipment and cabling must be installed in a neat and workmanlike manner.

FPN: Information describing industry practices can be found in ANSI/NECA/BICSI 568, *Standard for Installing Commercial Building Telecommunications Cabling*.

Author's Comment: For more information about this standard, visit http://www.necaneis.org/.

Exposed cables must be supported by the structural components of the building so that the cable will not be damaged by normal building use. Such cables must be secured by straps, staples, hangers, or similar fittings designed and installed so as not to damage the cable. **Figure 770-11**

Cables run parallel to framing members or furring strips must be protected where they are likely to be penetrated by nails or screws by installing the wiring method so it isn't less than 1¼ in. from the nearest edge of the framing member or furring strips, or is protected by a 1⁄16 in. thick steel plate or equivalent [300.4(D)].

Fiber-optic raceways and cable assemblies must be securely fastened in place and the ceiling-support wires or ceiling grid cannot be used to support optical fiber raceways or cables [300.11]. **Figure 770-12**

Author's Comment: Raceways and cables can be supported by independent support wires attached to the suspended ceiling in accordance with 300.11(A).

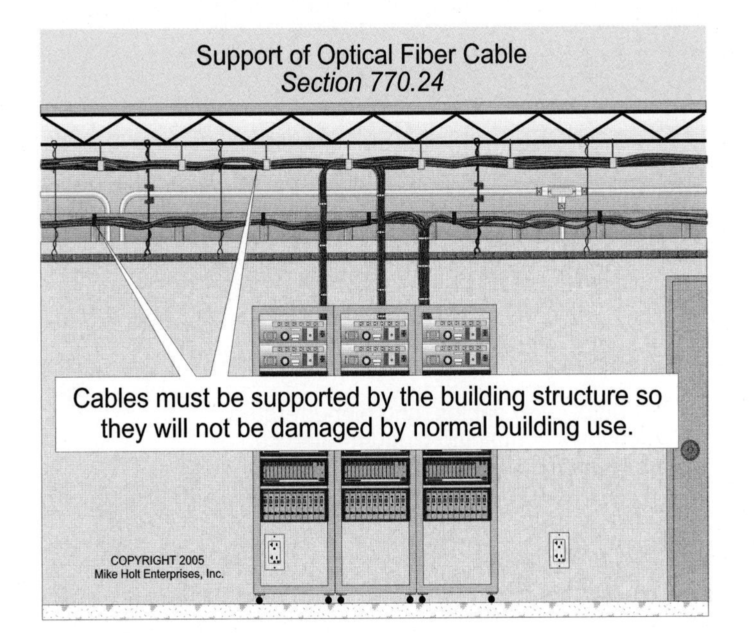

Figure 770–11

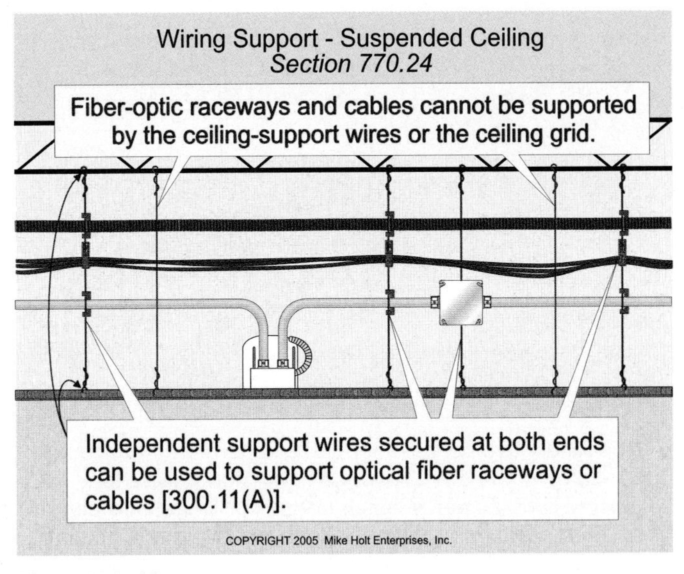

Wiring Support - Suspended Ceiling
Section 770.24
Fiber-optic raceways and cables cannot be supported by the ceiling-support wires or the ceiling grid.
Independent support wires secured at both ends can be used to support optical fiber raceways or cables [300.11(A)].
COPYRIGHT 2005 Mike Holt Enterprises, Inc.

Figure 770–12

PART III. CABLES WITHIN BUILDINGS

770.113 Listing of Optical Fiber Cables. Optical fiber cables installed within buildings must be listed in accordance with 770.154 and 770.179, and marked in accordance with Table 770.113.

Author's Comment: The *NEC* doesn't require outside or underground cable to be listed, but the cable must be approved by the authority having jurisdiction as suitable for the application in accordance with 90.4, 90.7, and 110.2.

Exception 1: Optical fiber cable is not required to be listed and marked where the length of the cable within the building, from its point of entrance, doesn't exceed 50 ft and the cable terminates in an enclosure.

Author's Comment: The point of entrance is defined as the point where the cable emerges from an external wall, from a concrete floor slab, or from a rigid metal conduit or an intermediate metal conduit grounded to an electrode in accordance with 800.100(B) [770.2].

Exception 2: Optical fiber cable is not required to be listed and marked if the cable entering the building is installed in intermediate metal conduit, rigid metal conduit, rigid nonmetallic conduit, or electrical metallic tubing.

770.133 Installation of Optical Fibers and Electrical Conductors.

(A) With Conductors for Electric Light, Power, and Class 1. Optical fibers are permitted within the same composite cable

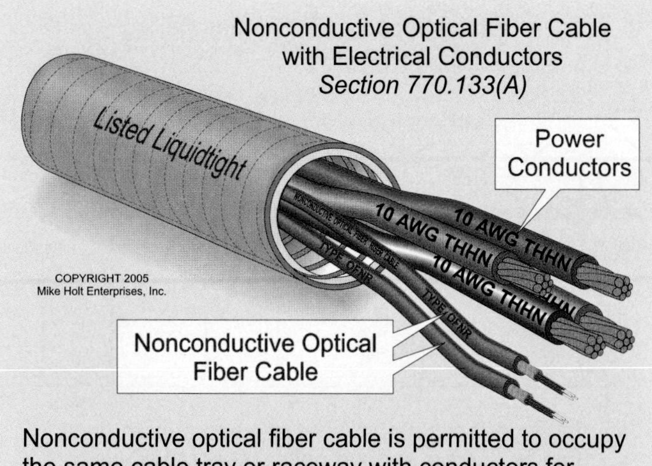

Figure 770–13

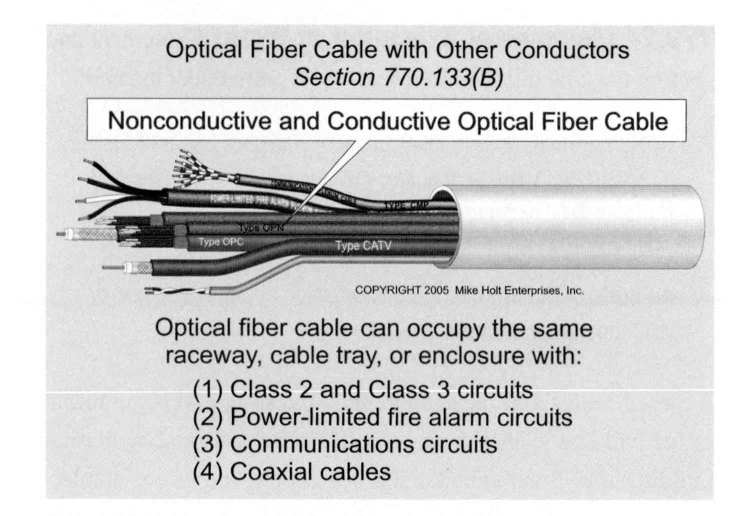

Optical Fiber Cable with Other Conductors
Section 770.133(B)
Nonconductive and Conductive Optical Fiber Cable
Type CMP
Type OPN
Type OPC
Type CATV
COPYRIGHT 2005 Mike Holt Enterprises, Inc.
Optical fiber cable can occupy the same
raceway, cable tray, or enclosure with:
(1) Class 2 and Class 3 circuits
(2) Power-limited fire alarm circuits
(3) Communications circuits
(4) Coaxial cables

Figure 770–14

with electric light, power, Class 1, or nonpower-limited fire alarm circuits, if the functions of the optical fibers and the electrical conductors are associated.

Nonconductive optical fiber cables are permitted to occupy the same cable tray or raceway with conductors for electric light, power, Class 1, or nonpower-limited fire alarm circuits. **Figure 770-13**

Conductive optical fiber cables must not occupy the same cable tray or raceway with conductors for electric light, power, Class 1, or nonpower-limited fire alarm circuits.

Nonconductive optical fiber cables cannot occupy a cabinet, outlet box, panel, or similar enclosure housing the electrical terminations of an electric light, power, and Class 1 circuit.

Exception 1: Nonconductive optical fiber cables are permitted to occupy the same cabinet, outlet box, panel, or similar enclosure housing the electrical terminations of an electric light, power, or Class 1 circuit, if the nonconductive optical fiber cable is functionally associated with the electric light, power, Class 1, or nonpower-limited fire alarm circuit.

(B) With Other Conductors. Optical fiber cables can be installed in the same cable tray, enclosure, or raceway with any of the following jacketed cables: **Figure 770-14**

(1) Class 2 and Class 3 circuits in compliance with Article 725.

(2) Power-limited fire alarm circuits in compliance with Article 760.

(3) Communications circuits in compliance with Article 800.

(4) Coaxial cables in compliance with Article 820.

770.154 Applications of Listed Optical Fiber Cables and Raceways.

Nonconductive and conductive optical fiber cables must comply with any of the requirements given in (A) through (E) or the applicable requirements of the cables substituted as permitted in (F).

(A) Ducts or Plenums Used for Environmental Air. Optical fiber cables must not be installed in ducts or plenums, unless they act directly on the contained air, are installed using a wiring method described in 300.22(B), and are plenum rated. However, plenum-rated Types OFNP or OFCP cables [770.179(A)] can be installed above a suspended ceiling or below a raised floor that is used for environmental air. **Figure 770-15**

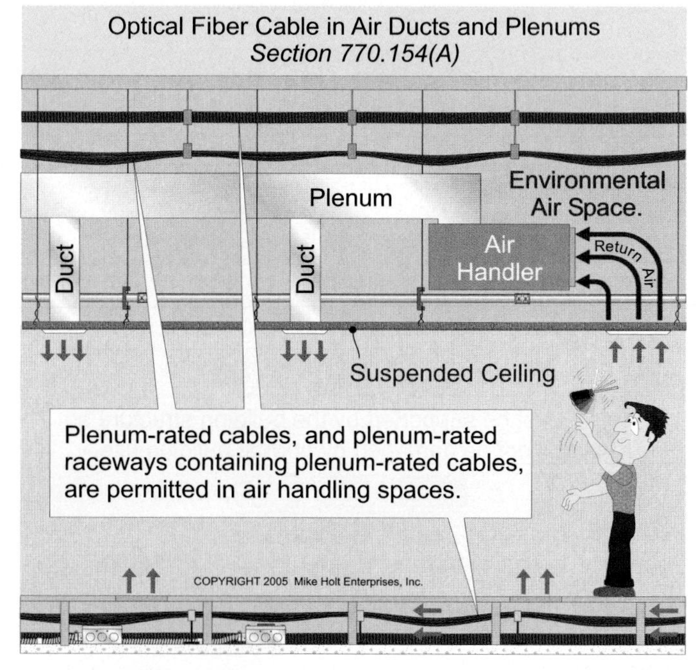

Figure 770–15

Plenum-rated optical fiber raceways [770.170(E)] are permitted above a suspended ceiling or below a raised floor that is used for environmental air, if the raceway contains Types OFNP or OFCP plenum-rated cables. **See Figure 770-15.**

Author's Comments:

- Nonplenum-rated cables can be installed above a suspended ceiling or below a raised floor that is used for environmental air, but only if the nonplenum-rated cable is installed within electrical metallic tubing [300.22(C)(1)].
- Optical fiber cables installed beneath a raised floor in an information technology equipment room (computer room) aren't required to be plenum rated [300.22(D) and 645.5(D)(5)(c)]. **Figure 770-16**

(B) Riser. Cables installed in risers must comply with (1), (2), or (3):

(1) Cables installed in vertical runs penetrating more than one floor must be riser rated Types OFNR or OFCR. Floor penetrations requiring Types OFNR or OFCR must contain only cables suitable for riser or plenum use. If installed in risers, optical fiber raceways with Types OFNP, OFCP, OFNR, and OFCR cables are permitted.

(2) Metal Raceways. Types OFNG, OFN, and OFCG cables must be encased in a metal raceway or located in a fireproof shaft having a firestop at each floor.

Author's Comment: Metal raceways containing circuit conductors from a power-supply system that operates at 50V or less aren't required to be grounded (bonded) to an effective ground-fault current path [250.86 and 250.112(I)].

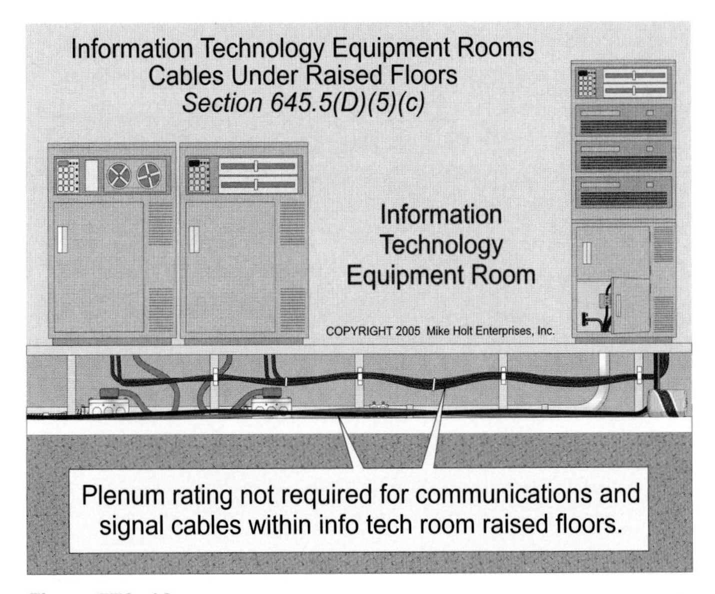

Figure 770–16

(3) One- and Two-Family Dwellings. Types OFNG, OFN, OFCG, or OFC cables can be installed in one- and two-family dwellings.

(C) Other Wiring Within Buildings. Cables installed in building locations other than ducts, plenums, or other environmental air spaces must be Types OFNG, OFN, OFCG, or OFC. Such cables can be installed in listed general-purpose optical fiber raceways.

(E) Cable Trays. Optical fiber cables can be installed in cable trays.

FPN: Optical fiber cables aren't required to be listed for use in cable trays.

(F) Cable Substitutions. Permitted substitutions for optical fiber cables are listed in Table 770.154.

PART IV. LISTING REQUIREMENTS

770.179 Listing Requirements for Optical Fiber Cables. Optical fiber cables must be listed in accordance with (A) through (D), and optical fiber raceways must be listed in accordance with (A) through (C).

(A) Types OFNP and OFCP. Type OFNP nonconductive and Type OFCP conductive optical fiber plenum cables must be listed as being suitable for use in ducts, plenums, and other space used for environmental air, and must also be listed as having adequate fire-resistant and low smoke-producing characteristics. **Figure 770-17**

Author's Comment: Special consideration must be given to cables in areas that move or transport environmental air in order to reduce the hazards that arise from the burning of conductor insulation and of cable jackets. Because listed plenum-rated

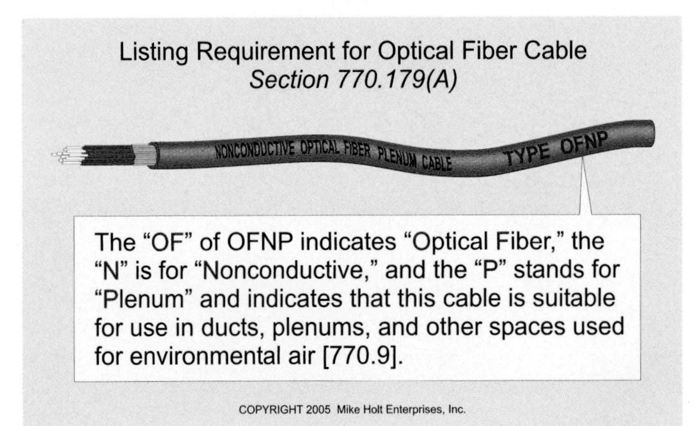

Listing Requirement for Optical Fiber Cable
Section 770.179(A)
NONCONDUCTIVE OPTICAL FIBER PLENUM CABLE TYPE OFNP
The "OF" of OFNP indicates "Optical Fiber," the "N" is for "Nonconductive," and the "P" stands for "Plenum" and indicates that this cable is suitable for use in ducts, plenums, and other spaces used for environmental air [770.9].
COPYRIGHT 2005 Mike Holt Enterprises, Inc.

Figure 770–17

cables have adequate fire-resistant and low smoke-producing characteristics, they can be installed in environmental air space, but they cannot be installed in ducts or plenums! See 770.3(B) in this textbook for details.

(B) Types OFNR and OFCR. Type OFNR nonconductive and Type OFCR conductive optical fiber riser cables must be listed as being suitable for use in a vertical run in a shaft, or from floor to floor, and have fire-resistant characteristics capable of preventing the carrying of fire from floor to floor.

(C) Types OFNG and OFCG. Types OFNG and OFCG nonconductive and conductive general-purpose optical fiber cables must be listed as being suitable for general-purpose use, with the exception of risers and plenums, and must also be listed as being resistant to the spread of fire.

(D) Types OFN and OFC. Types OFN and OFC nonconductive and conductive optical fiber cables must be listed as being suitable for general-purpose use, with the exception of risers, plenums, and other environmental air spaces.

770.182 Listing Requirements for Optical Fiber Raceways.

(A) Plenum Optical Fiber Raceway. Plenum optical fiber raceways must be listed as having adequate fire-resistant and low smoke-producing characteristics. **Figure 770-18**

Author's Comment: Where plenum-rated optical fiber raceways are used in other air-handling spaces, only plenum-rated optical fiber cable is permitted [770.154(A)].

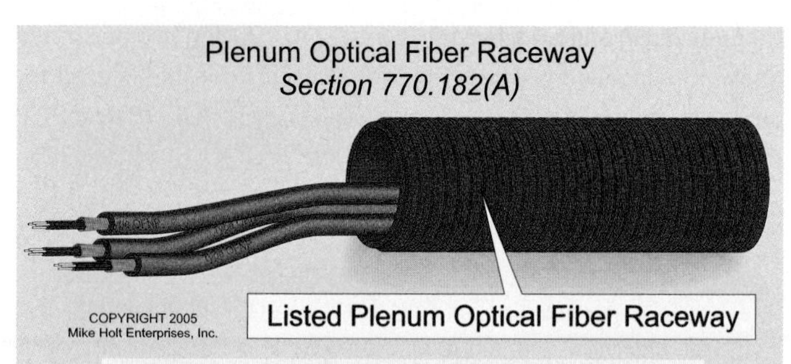

Figure 770–18

(B) Riser Optical Fiber Raceway. Riser optical fiber raceways must be listed as having fire-resistant characteristics capable of preventing the carrying of fire from floor to floor.

Author's Comment: Where plenum-rated optical fiber raceways are used in a riser, only nonconductive plenum-rated optical fiber cable is permitted [770.154(B)(1)].

(C) General-Purpose Optical Fiber Cable Raceway. General-purpose optical fiber cable raceways must be listed as being resistant to the spread of fire.

Article 770 Questions

1. Optical fiber cable not terminated at equipment and not identified for future use with a tag is considered abandoned.

   (a) True (b) False

2. When optical fiber cable is installed in a raceway, the raceway must be of a type permitted in Chapter 3 and the raceway must be installed in accordance with Chapter 3 requirements.

   (a) True (b) False

3. Where exposed to contact with electric light or power conductors, the noncurrent-carrying metallic members of optical fiber cables entering buildings must be _____.

   (a) grounded as close as possible to emergence through an exterior wall
   (b) grounded as close as possible to emergence through a concrete floor slab
   (c) interrupted as close to the point of entrance as practicable by an insulating joint
   (d) any of these

4. Optical fibers are permitted in the same cable, and conductive and nonconductive optical fiber cables are permitted in the same cable tray, enclosure, or raceway with conductors of Class 2 and Class 3 circuits in compliance with Article 725.

   (a) True (b) False

5. Optical fiber cables listed as suitable for general-purpose use, with the exception of risers, plenums, and other space used for environmental air are Types _____.

   (a) OFNP and OFCP (b) OFNR and OFCR (c) OFNG and OFCG (d) OFN and OFC

Notes

CHAPTER 8

Communications Systems

Introduction

Chapter 8 of the *National Electrical Code* covers the wiring requirements for communications systems such as telephones, radio and TV antennas, satellite dishes, closed-circuit television (CCTV), and cable TV (CATV) systems, as well as network-powered broadband communications systems not under the exclusive control of the communications utility. **Figure 1**

Communications systems aren't subject to the general requirements contained in Chapters 1 through 4 or the special requirements of Chapters 5 through 7, except where there's a specific reference in Chapter 8 to a rule in one of those chapters [90.3]. **Figure 2**

Also, installations of communications equipment under the exclusive control of communications utilities located outdoors, or in building spaces used exclusively for such installations, are exempt from the *NEC* [90.2(B)(4)]. **Figure 3**

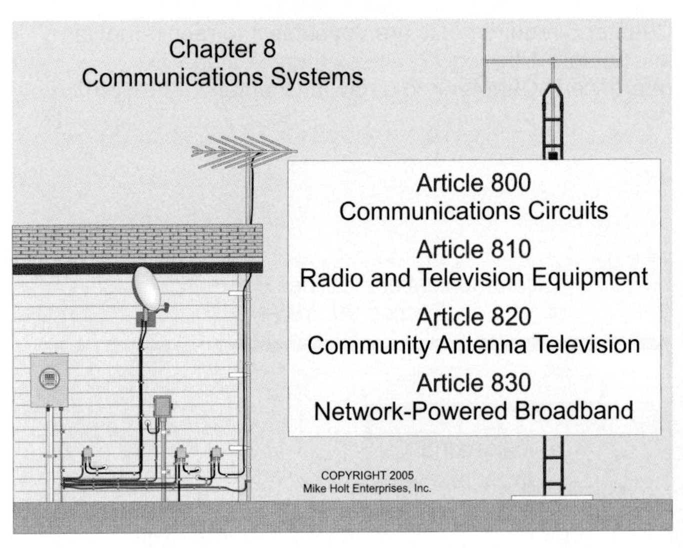

Figure 1

Article 800. Communications Circuits. Article 800 covers the installation requirements for telephone wiring and for other related telecommunications purposes such as computer Local Area Networks (LANs), and outside wiring for fire and burglar alarm systems connected to central stations.

Article 810. Radio and Television Equipment. This article covers antenna systems for radio and television receiving equipment, amateur radio transmitting and receiving equipment, and certain features of transmitter safety. It also includes antennas such as multi-element, vertical rod and dish, and the wiring and cabling that connect them to the equipment.

Article 820. Community Antenna Television (CATV) and Radio Distribution Systems. Article 820 covers the installation of coaxial cables to distribute limited-energy high-frequency signals for television, cable TV, and closed-circuit television (CCTV), which is often used for security purposes. This article also covers premises wiring of satellite TV systems where the dish antenna is outside and covered by Article 810.

Article 830. Network-Powered Broadband Communications Systems. This article contains the installation requirements for network-powered broadband communications systems where powered from the communications utility network for voice, audio, video, data, and interactive services through a network interface unit (NIU). An example of a network-powered broadband communications system is hybrid fiber-coaxial (HFC) cable used for video/audio conferencing or interactive multimedia entertainment systems.

Code Arrangement
Section 90.3

General Requirements

- Chapter 1 - General
- Chapter 2 - Wiring and Protection
- Chapter 3 - Wiring Methods and Materials
- Chapter 4 - Equipment for General Use

Chapters 1 through 4 apply to all applications.

Special Requirements

- Chapter 5 - Special Occupancies
- Chapter 6 - Special Equipment
- Chapter 7 - Special Conditions

Chapters 5 through 7 can supplement or modify the general requirements of Chapters 1 through 4.

- Chapter 8 - Communications Systems

Chapter 8 requirements are not subject to requirements in Chapters 1 through 7, unless there is a specific reference in Chapter 8 to a rule in Chapters 1 through 7.

COPYRIGHT 2005 Mike Holt Enterprises, Inc.

Figure 2

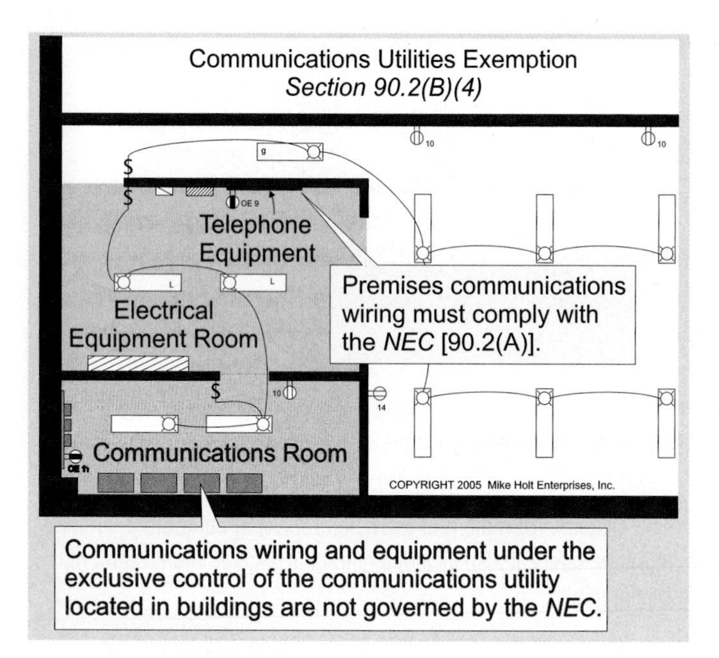
Communications Utilities Exemption
Section 90.2(B)(4)
Telephone Equipment
Electrical Equipment Room
Premises communications wiring must comply with the NEC [90.2(A)].
Communications Room
COPYRIGHT 2005 Mike Holt Enterprises, Inc.
Communications wiring and equipment under the exclusive control of the communications utility located in buildings are not governed by the NEC.

Figure 3

Article 830 actually covers only one specific broadband technology, and does not apply to most cable modems, or fiber-to-the-home (FTTH) technologies.

Author's Comment: The output wiring (voice, audio, video, data, and interactive services signals) from a network interface unit must be installed in accordance with the requirements contained in the following articles [830.3(D)]:

- Article 725 – Class 2 and Class 3 Circuits
- Article 760 – Fire Alarm Circuits
- Article 770 – Optical Fiber Cables and Raceways
- Article 800 – Communications Circuits
- Article 820 – Community Antenna Television and Radio Distribution Systems (CATV)

ARTICLE

Communications Circuits

Introduction

This article has its roots in telephone technology. Consequently, it addresses telephone and related systems that use twisted-pair wiring. Here are a few key points to remember from Article 800:

- Don't attach incoming cables to the service-entrance power mast.
- It is critical to determine the "point of entrance" for these circuits.
- Ground the primary protector as close as practicable to the point of entrance.
- Keep the grounding (earthing) conductor for the primary protector as straight and as short as possible.
- If you locate cables above a suspended ceiling, route and support them to allow access via ceiling panel removal.
- Keep these cables separated from lightning protection circuits.
- If you install cables in a Chapter 3 raceway, you must do so in conformance with the *NEC* requirements for the raceway system.
- Special labeling and marking provisions apply—follow them carefully.

PART I. GENERAL

800.1 Scope. This article covers wiring and equipment for telephone systems connected to a central station system, and other types of telephone systems using similar equipment and methods of installation.

Author's Comment: The telephone utility typically provides the twisted-pair cable to a terminal board at a building or structure (**Figure 800-1**). Sometimes this is called a Network Interface Device (NID). Twisted-pair wiring from the NID to the premises falls within the scope of Article 800. **Figure 800-2**

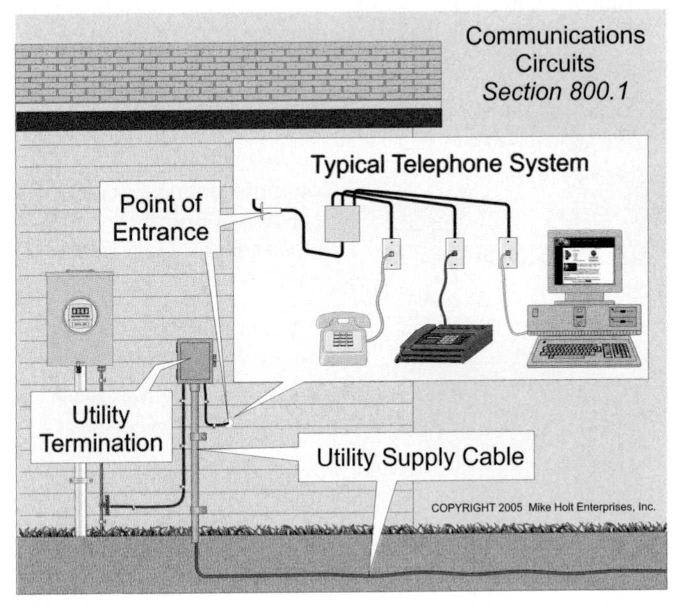

Figure 800–1

800.2 Definitions.

Abandoned Cable. Cable that isn't terminated at equipment, and not identified for future use with a tag, is considered as abandoned for the purpose of this article.

Author's Comment: 800.3(C) requires abandoned cables to be removed.

Cable. A factory assembly of two or more conductors having an overall covering.

Point of Entrance. The point at which the cable emerges from an external wall, a concrete floor, or rigid metal conduit or intermediate metal conduit that is grounded (bonded) to an electrode in accordance with 800.100. **Figure 800-3**

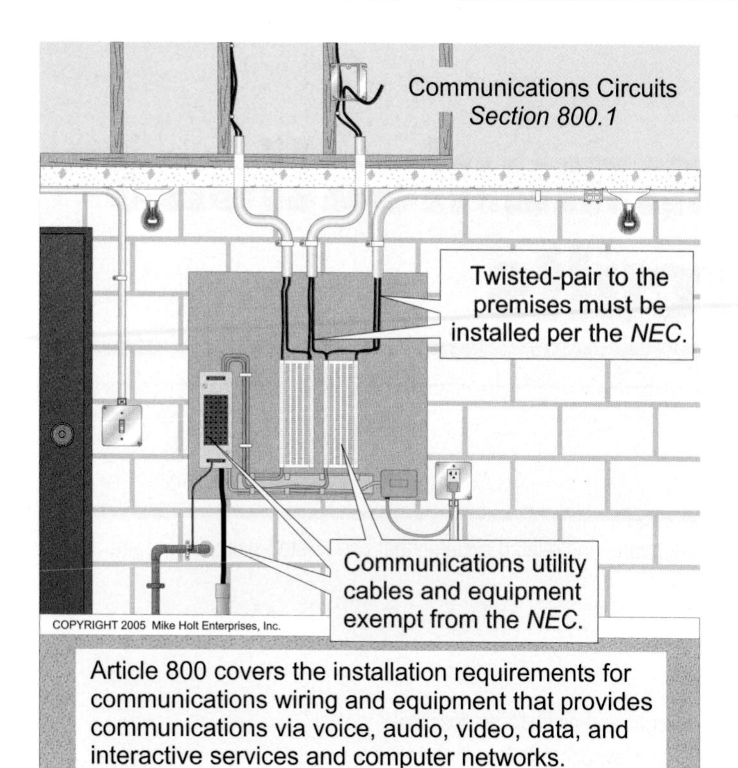
Communications Circuits
Section 800.1
Twisted-pair to the premises must be installed per the NEC.
Communications utility cables and equipment exempt from the NEC.
COPYRIGHT 2005 Mike Holt Enterprises, Inc.
Article 800 covers the installation requirements for communications wiring and equipment that provides communications via voice, audio, video, data, and interactive services and computer networks.

Figure 800–2

Author's Comment: The term "Point of Entrance" is important for properly locating the primary protector [800.90(B)] and the application of grounding (bonding) the cable [800.93].

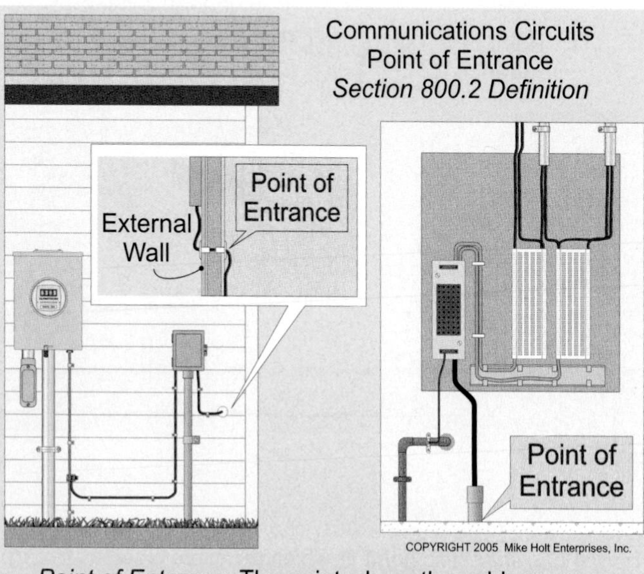
Communications Circuits
Point of Entrance
Section 800.2 Definition
Point of Entrance
External Wall
Point of Entrance
COPYRIGHT 2005 Mike Holt Enterprises, Inc.
Point of Entrance: The point where the cable emerges from an external wall or concrete floor slab.

Figure 800–3

800.3 Other Articles.

(B) Hazardous (Classified) Locations. Communications circuits installed in any hazardous (classified) location must be installed in accordance with Articles 500 through 516, specifically 501.150 and 502.150.

(C) Spread of Fire or Products of Combustion. Communications circuits installed through fire-resistant rated walls, partitions, floors, or ceilings must be firestopped in accordance with the instructions supplied by the manufacturer for the specific type of cable and construction material (drywall, brick, etc.), as required by 300.21.

Author's Comment: Openings in fire-resistant walls, floors, and ceilings must be sealed so that the possible spread of fire or products of combustion will not be substantially increased [300.21]. **Figure 800-4**

Abandoned Cable. To limit the spread of fire or products of combustion within a building, the accessible portion of communications cable that isn't terminated at equipment, and not identified for future use with a tag, must be removed [800.2]. **Figure 800-5**

Author's Comment: This rule doesn't require the removal of concealed cables that are abandoned in place, which includes cables in raceways. According to the definition of "Concealed" in Article 100, cables in raceways are considered to be concealed.

(D) Other Space Used for Environmental Air. Communications cables must not be installed exposed in ducts or plenums [800.133(D)].

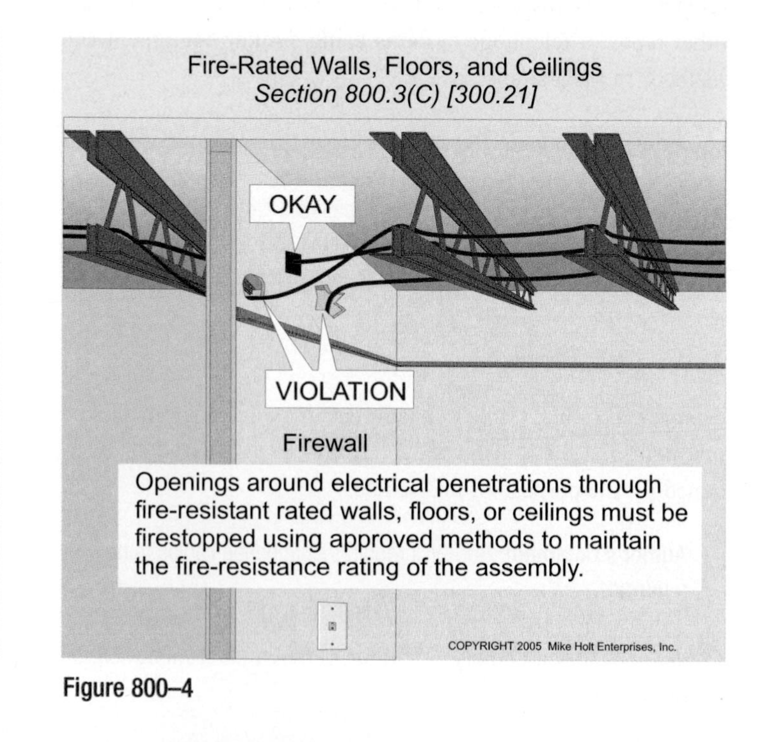
Fire-Rated Walls, Floors, and Ceilings
Section 800.3(C) [300.21]
OKAY
VIOLATION
Firewall
Openings around electrical penetrations through fire-resistant rated walls, floors, or ceilings must be firestopped using approved methods to maintain the fire-resistance rating of the assembly.
COPYRIGHT 2005 Mike Holt Enterprises, Inc.

Figure 800–4

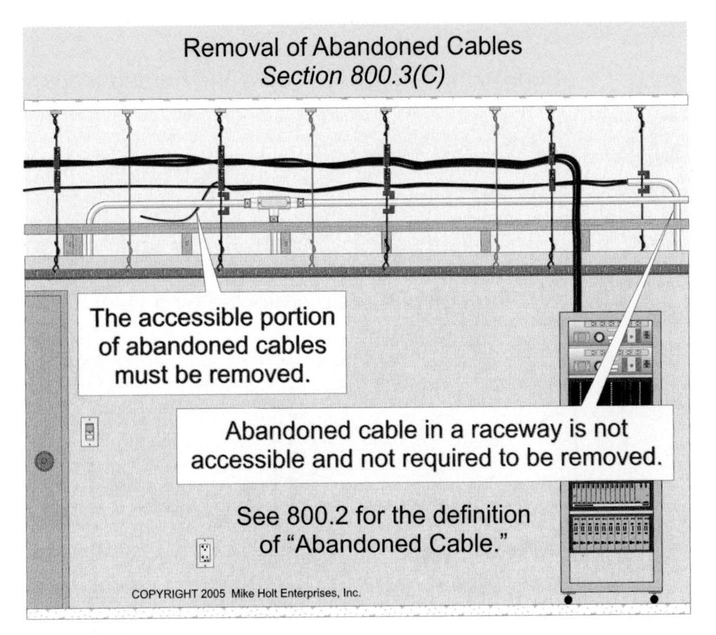
Removal of Abandoned Cables
Section 800.3(C)
The accessible portion of abandoned cables must be removed.
Abandoned cable in a raceway is not accessible and not required to be removed.
See 800.2 for the definition of "Abandoned Cable."
COPYRIGHT 2005 Mike Holt Enterprises, Inc.

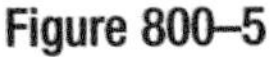
Figure 800–5

Plenum-rated communications cables (CMP) can be installed above a suspended ceiling or below a raised floor that is used for environmental air [800.154(A)]. **Figure 800-6**

> **Author's Comment:** Communications cables installed beneath a raised floor in an information technology equipment room (computer room) aren't required to be plenum rated [300.22(D) and 645.5(D)(5)(c)]. **Figure 800-7**

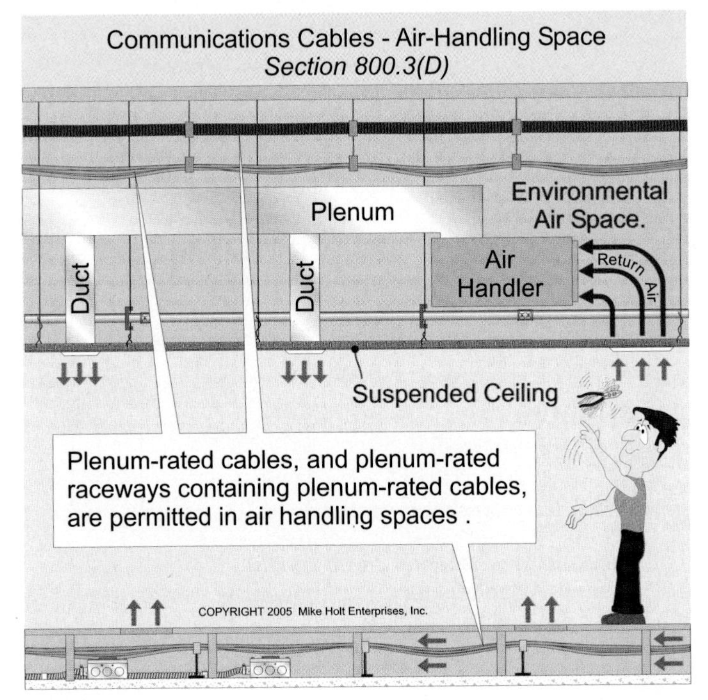
Communications Cables - Air-Handling Space
Section 800.3(D)
Plenum
Environmental Air Space.
Duct
Duct
Air Handler
Return Air
Suspended Ceiling
Plenum-rated cables, and plenum-rated raceways containing plenum-rated cables, are permitted in air handling spaces .
COPYRIGHT 2005 Mike Holt Enterprises, Inc.

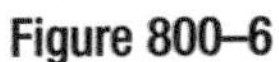
Figure 800–6

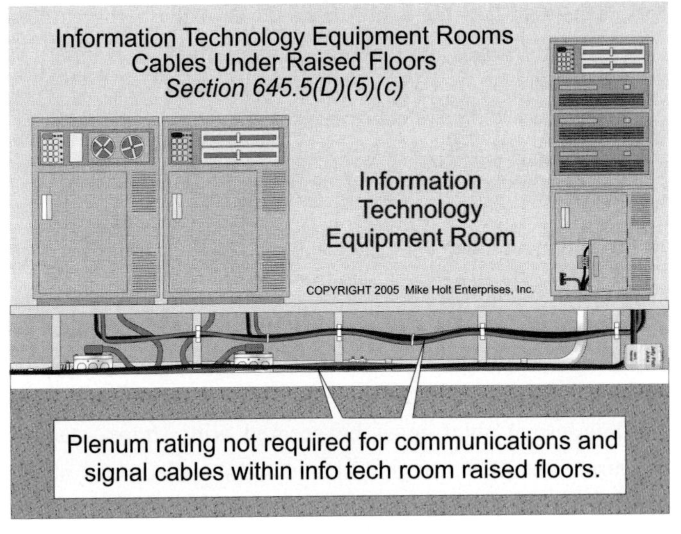
Information Technology Equipment Rooms
Cables Under Raised Floors
Section 645.5(D)(5)(c)
Information Technology Equipment Room
COPYRIGHT 2005 Mike Holt Enterprises, Inc.
Plenum rating not required for communications and signal cables within info tech room raised floors.

Figure 800–7

800.18 Installation of Equipment. Communications equipment must be listed [800.170] and must be installed in accordance with manufacturer's instructions [110.3(B)].

800.21 Access to Electrical Equipment Behind Panels Designed to Allow Access. Access to equipment must not be prohibited by an accumulation of cables that prevent the removal of suspended-ceiling panels. Communications cables must be located so that the suspended-ceiling panels can be moved to provide access to electrical equipment. **Figure 800-8**

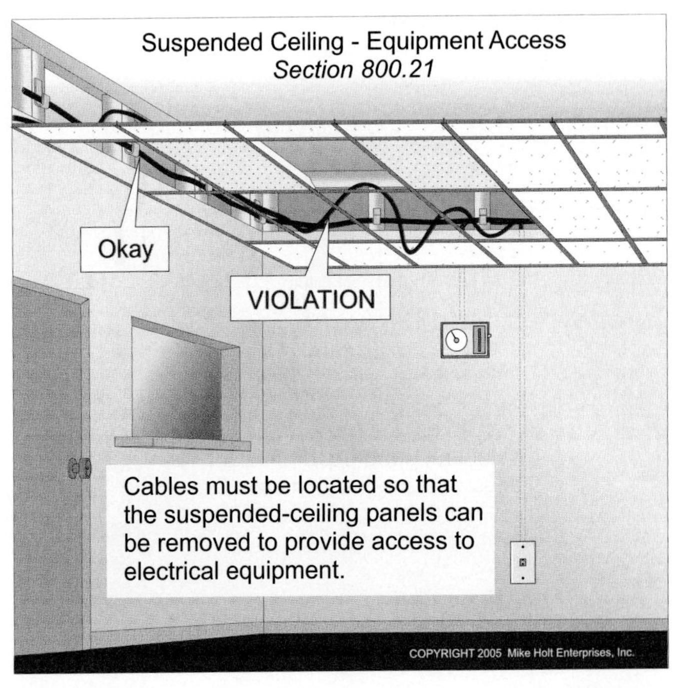
Suspended Ceiling - Equipment Access
Section 800.21
Okay
VIOLATION
Cables must be located so that the suspended-ceiling panels can be removed to provide access to electrical equipment.
COPYRIGHT 2005 Mike Holt Enterprises, Inc.

Figure 800–8

800.24 Mechanical Execution of Work. Equipment and cabling must be installed in a neat and workmanlike manner.

FPN: Information describing industry practices can be found in ANSI/NECA/BICSI 568, *Standard for Installing Commercial Building Telecommunications Cabling.*

Author's Comment: For more information about this standard, visit http://www.necaneis.org/.

Exposed cables must be supported by the structural components of the building so that the cable will not be damaged by normal building use. Cables must be secured with straps, staples, hangers, or similar fittings designed and installed so as not to damage the cable. **Figure 800-9**

Cables run parallel to framing members or furring strips must be protected where they are likely to be penetrated by nails or screws, by installing the wiring method so it isn't less than 1¼ in. from the nearest edge of the framing member or furring strips, or is protected by a 1/16 in. thick steel plate or equivalent [300.4(D)]. **Figure 800-10**

Communications raceways and cable assemblies must be securely fastened in place and ceiling-support wires or the ceiling grid cannot be used to support communications raceways or cables [300.11]. **Figure 800-11**

Author's Comment: Raceways and cables can be supported by independent support wires attached to the suspended ceiling in accordance with 300.11(A).

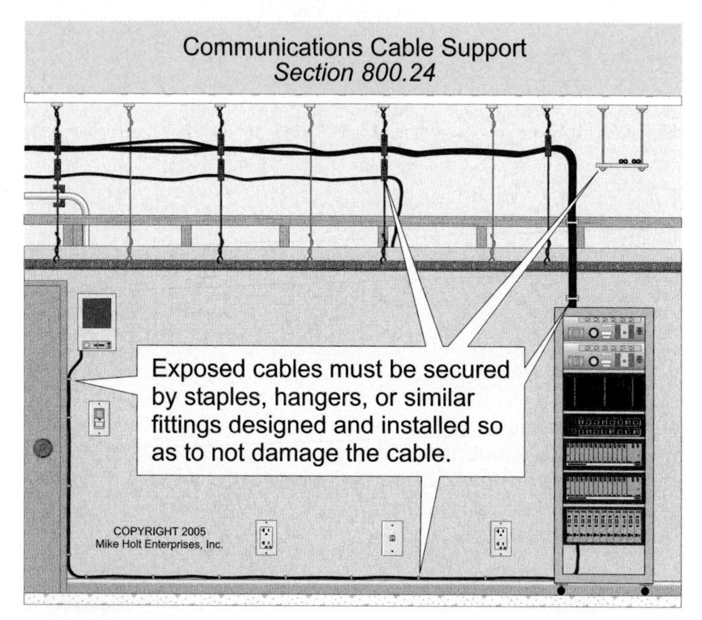

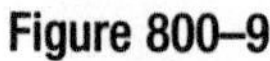
Figure 800–9

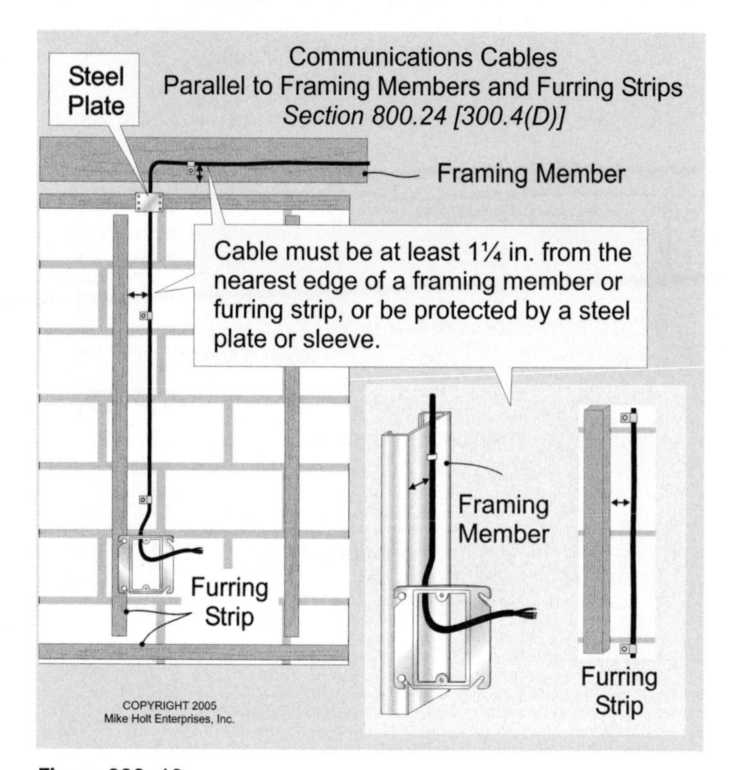

Figure 800–10

800.44 Overhead Communications Wires and Cables. Communications cables must have a height not less than 10 ft above swimming and wading pools, diving structures and observation stands, and towers or platforms [680.8(B)].

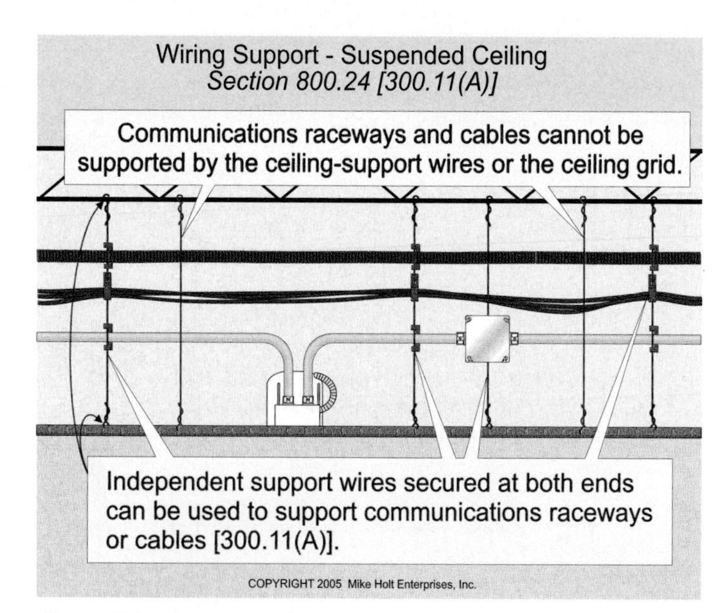

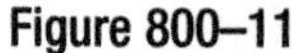
Figure 800–11

PART II. CONDUCTORS OUTSIDE AND ENTERING BUILDINGS

800.47 Underground Circuits Entering Buildings.

Underground communications wires and cables entering buildings must comply with (A) and (B).

(A) With Electric Light and Power Conductors. Underground communications wires and cables in a handhole or manhole must be separated from exposed electric light, power, Class 1, or non-power-limited fire alarm circuit conductors by a suitable barrier.

Author's Comment: See Article 100 for the definition of "Handhole."

800.53 Lightning Conductors. Where practicable, a separation not less than 6 ft must be maintained between communications wiring and lightning protection conductors. **Figure 800-12**

PART III. PROTECTION

800.90 Primary Protection.

(A) Application. A listed primary protector is required for each communications circuit.

(B) Location. The primary protector must be located as close as practicable to the point of entrance.

FPN: See 800.2 for the definition of *point of entrance*.

Author's Comment: Selecting a primary protector location to achieve the shortest practicable primary protector grounding (earthing) conductor helps reduce differences in potential between communications circuits and other metallic systems during lightning events.

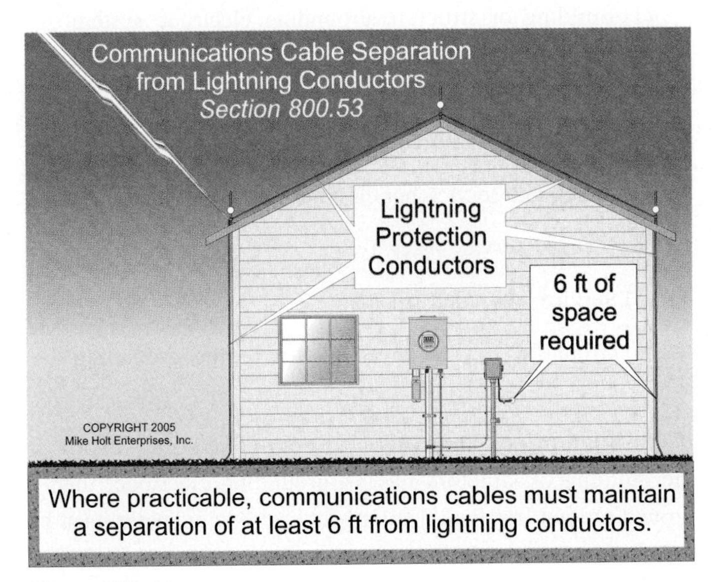

Communications Cable Separation
from Lightning Conductors
Section 800.53
Lightning
Protection
Conductors
6 ft of
space
required
COPYRIGHT 2005
Mike Holt Enterprises, Inc.
Where practicable, communications cables must maintain a separation of at least 6 ft from lightning conductors.

Figure 800–12

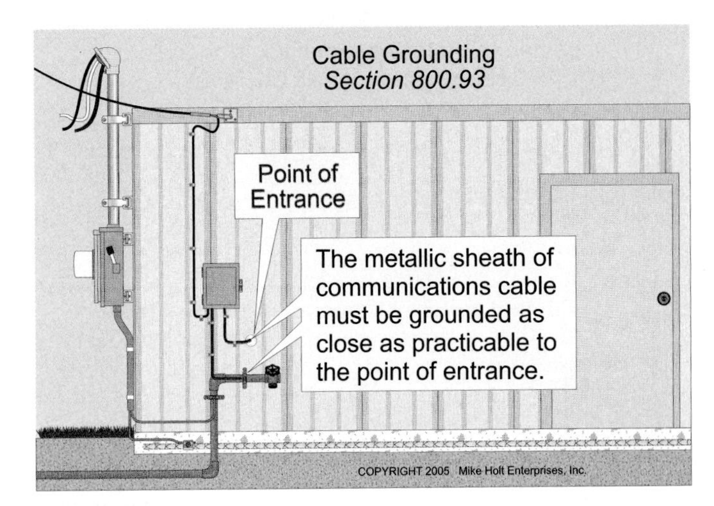

Figure 800–13

800.93 Cable Grounding. The metallic sheath of communications cable must be grounded or interrupted by an insulating joint as close as practicable to the point of entrance of the phone cable to the building or structure in accordance with 800.100. **Figure 800-13**

PART IV. GROUNDING METHODS

800.100 Cable and Primary Protector Grounding. The metallic sheath of communications cables, where required to be grounded [800.93], and primary protectors must be grounded in accordance with (A) through (D).

(A) Grounding Conductor. The grounding conductor must be:

(1) Insulation. The grounding conductor must be insulated and must be listed as suitable for the purpose.

(2) Material. The grounding conductor must be copper or other corrosion-resistant conductive material, stranded or solid.

(3) Size. The grounding conductor must not be smaller than 14 AWG.

(4) Length. The grounding conductor must be as short as practicable. In one- and two-family dwellings, the grounding conductor must not exceed 20 ft. **Figure 800-14**

FPN: Limiting the length of the grounding conductor helps limit potential (voltage) differences between the building's power and communications systems during lightning events.

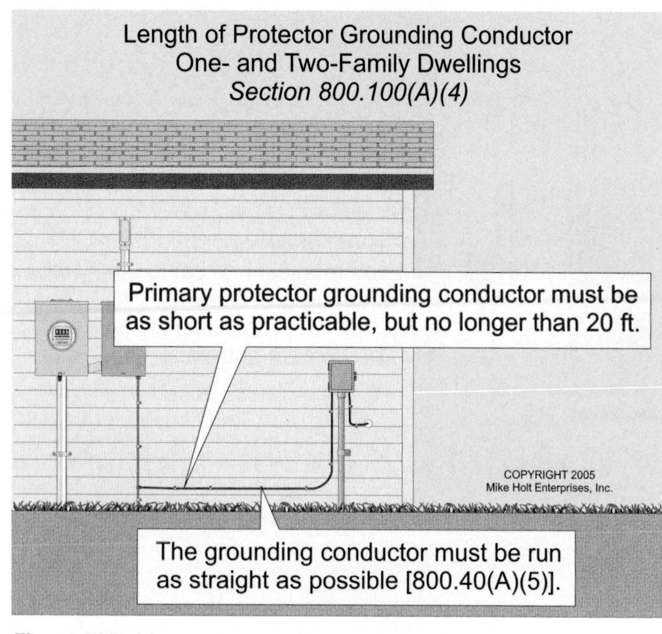

Length of Protector Grounding Conductor
One- and Two-Family Dwellings
Section 800.100(A)(4)
Primary protector grounding conductor must be as short as practicable, but no longer than 20 ft.
COPYRIGHT 2005
Mike Holt Enterprises, Inc.
The grounding conductor must be run as straight as possible [800.40(A)(5)].

Figure 800–14

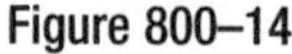

Author's Comment: The bonding of all external systems (communications and power) that enter a building to a single point minimizes the possibility of equipment damage due to the potential (voltage) differences between the systems. **Figure 800-15**

Exception: Where it isn't practicable to limit the grounding conductor to 20 ft for one- and two-family dwellings, a separate ground rod not less than 5 ft long [800.100(B)(2)(2)] with fittings suitable for the application [800.100(C)] must be installed. The additional ground rod must be bonded to the power grounding electrode system with a minimum 6 AWG conductor [800.100(D)].

(5) Run in Straight Line. The grounding conductor to the electrode must be run in as straight a line as practicable.

Author's Comment: Lightning doesn't like to travel around corners or through loops, which is why the grounding conductor should be run as straight as practicable.

(6) Mechanical Protection. The grounding conductor must be guarded from physical damage. Where run in a metal raceway, both ends of the raceway must be bonded to the grounding conductor.

Author's Comment: Installing the grounding conductor in a nonmetallic raceway, when the authority having jurisdiction judges that physical protection is required, is a better practice than using a metal raceway.

(B) Electrode. The grounding conductor must be connected in accordance with (1) or (2).

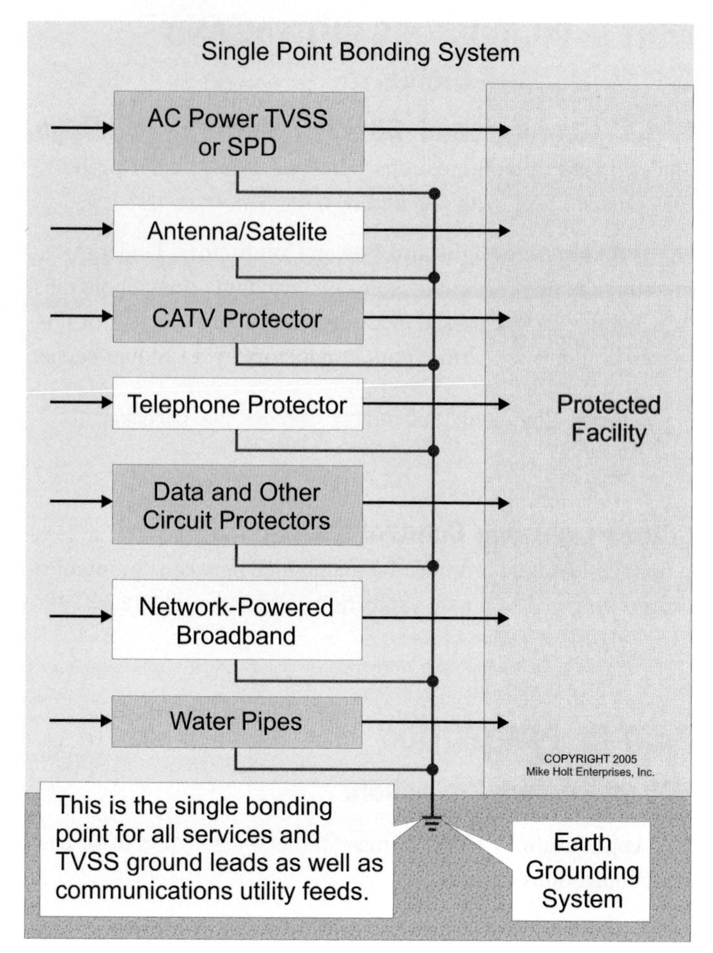

Figure 800–15

(1) In Buildings or Structures With Grounding Means. The grounding conductor must terminate to the nearest accessible: **Figure 800-16**

(1) Building or structure grounding electrode system [250.50].

(2) Interior metal water piping system, within 5 ft from its point of entrance [250.52(A)(1)].

(3) Accessible service bonding means [250.94].

(4) Metallic service raceway.

(5) Service equipment enclosure.

(6) Grounding electrode conductor or the grounding electrode conductor metal enclosure.

(2) In Buildings or Structures Without Grounding Means. If the building or structure has no grounding electrode, then a ground rod not less than 5 ft long and ½ in. in diameter must be installed for the communications system [800.100(B)(2)(2)]. **Figure 800-17**

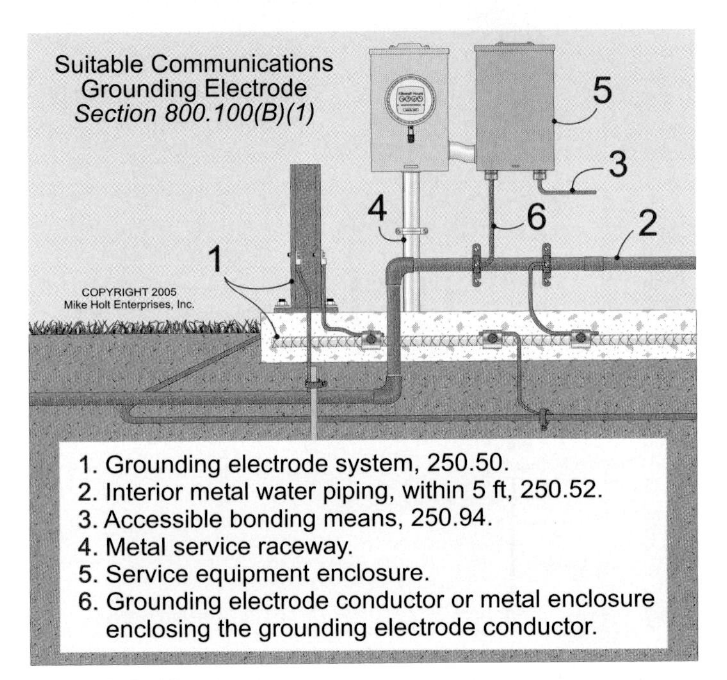
Suitable Communications
Grounding Electrode
Section 800.100(B)(1)
COPYRIGHT 2005
Mike Holt Enterprises, Inc.
1. Grounding electrode system, 250.50.
2. Interior metal water piping, within 5 ft, 250.52.
3. Accessible bonding means, 250.94.
4. Metal service raceway.
5. Service equipment enclosure.
6. Grounding electrode conductor or metal enclosure enclosing the grounding electrode conductor.

Figure 800–16

Author's Comment: The reason communications ground rods only need to be 5 ft long is because that's the length the phone company used before the *NEC* contained requirements for communications systems. Phone company ground rods were only 5 ft long because that's the length that would fit in their equipment trailers.

(C) Electrode Connection. Terminations at the electrode must be by exothermic welding, listed lug, listed pressure connector, or listed clamp. Grounding fittings that are concrete-encased or buried in the earth must be listed for direct burial and marked "DB" [250.70]. **Figure 800-18**

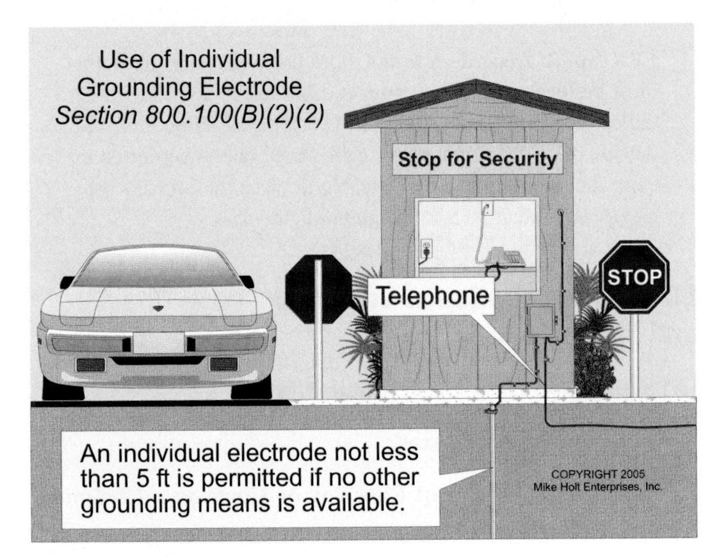
Use of Individual
Grounding Electrode
Section 800.100(B)(2)(2)
Stop for Security
Telephone
STOP
An individual electrode not less than 5 ft is permitted if no other grounding means is available.
COPYRIGHT 2005
Mike Holt Enterprises, Inc.

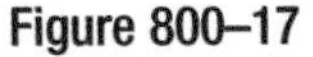
Figure 800–17

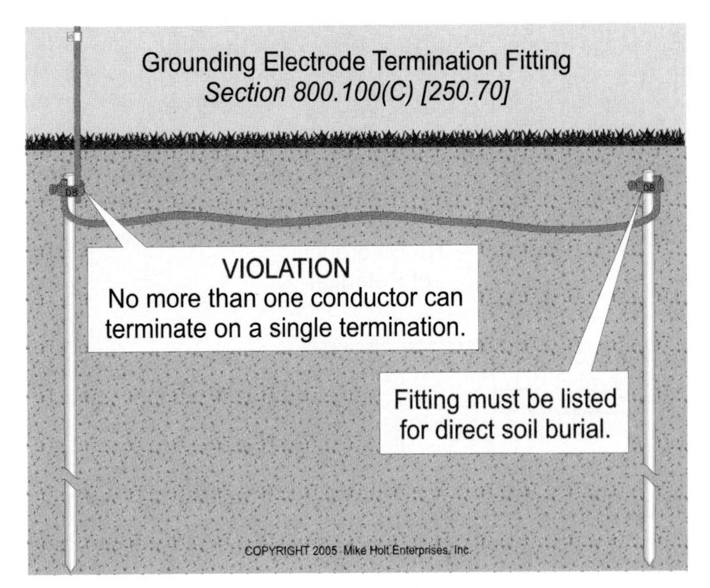
Grounding Electrode Termination Fitting
Section 800.100(C) [250.70]
VIOLATION
No more than one conductor can terminate on a single termination.
Fitting must be listed for direct soil burial.
COPYRIGHT 2005 Mike Holt Enterprises, Inc.

Figure 800–18

(D) Bonding of Electrodes. If a separate grounding electrode, such as a ground rod, is installed for a communications system, it must be bonded to the building's power grounding electrode system with a minimum 6 AWG conductor. **Figure 800-19**

FPN No. 2: Bonding all systems to the same single point (intersystem bonding) helps reduce differences in potential between the power and communications systems during lightning events. **Figure 800-20**

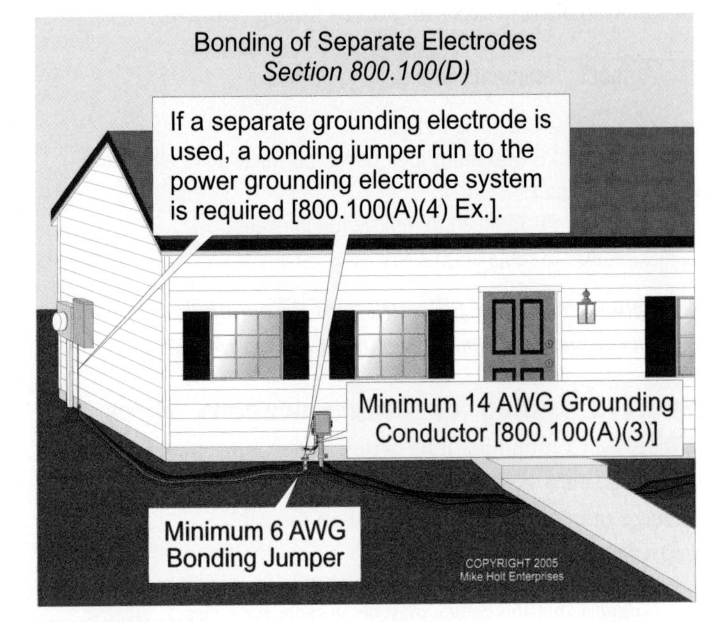
Bonding of Separate Electrodes
Section 800.100(D)
If a separate grounding electrode is used, a bonding jumper run to the power grounding electrode system is required [800.100(A)(4) Ex.].
Minimum 14 AWG Grounding Conductor [800.100(A)(3)]
Minimum 6 AWG Bonding Jumper
COPYRIGHT 2005
Mike Holt Enterprises

Figure 800–19

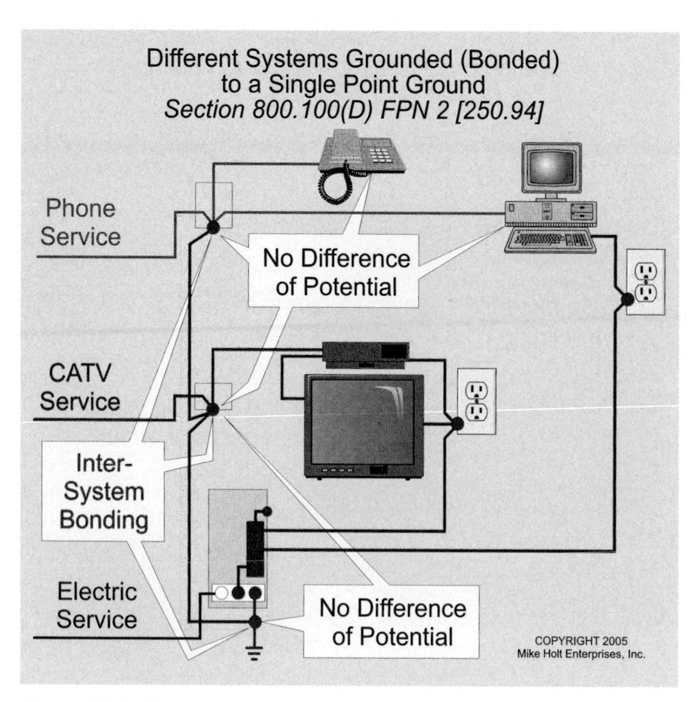

Figure 800–20

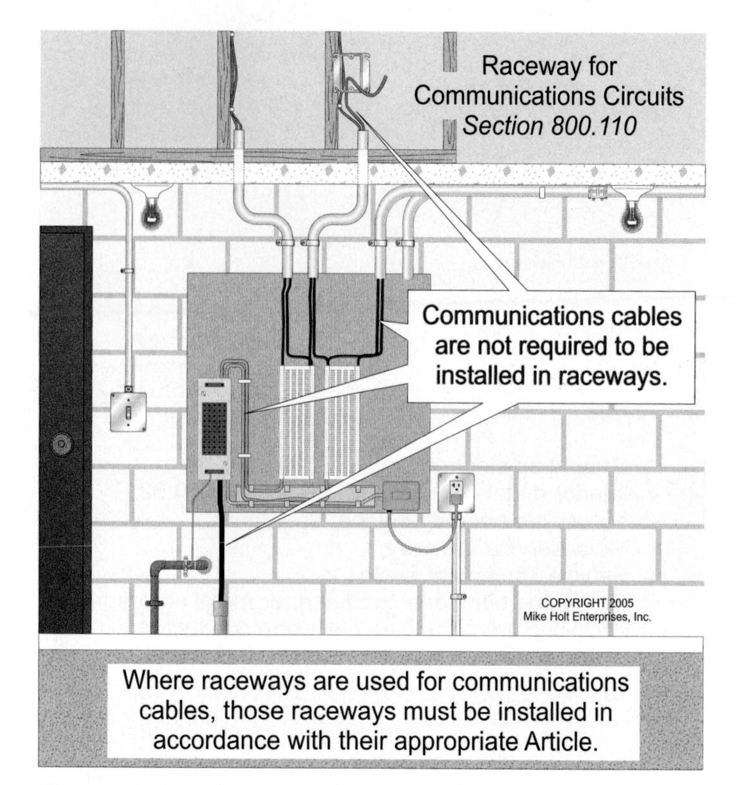

Figure 800–21

PART V. COMMUNICATIONS WIRES AND CABLES WITHIN BUILDINGS

800.110 Raceways for Communications Circuits.

Where communications cables are installed in a Chapter 3 raceway, the raceway must comply with the Chapter 3 wiring method installation requirements. **Figure 800-21**

Where communications cables are installed in a nonmetallic communications raceway [800.182], the raceway must be installed in accordance with 362.24 through 362.56.

Author's Comment:

- 362.24 Bending radius.
- 362.26 Maximum total bends between pull points, 360 degrees.
- 362.28 Trimmed to remove rough edges.
- 362.30 Supported every 3 ft.

Exception: Maximum fill restrictions do not apply to communications cables and conductors installed in a raceway.

800.113 Listing of Communications Wires and Cables. Communications wires and cables installed within buildings must be listed in accordance with 800.133. The cable voltage rating is not to be marked on the cable.

FPN: Voltage markings on cables may be misinterpreted to suggest that the cables may be suitable for Class 1, electric light, and power applications.

Exception 2: Communications cable is not required to be listed if the length of the cable within the building, from its point of entrance, doesn't exceed 50 ft and the cable terminates in an enclosure.

Author's Comment: The point of entrance is defined as the point where the cable emerges from an external wall, from a concrete floor slab, or from a rigid metal conduit or an intermediate metal conduit grounded (bonded) to an electrode in accordance with 800.100 [800.2].

FPN No. 2: According to 800.90(B), the primary protector must be located as close as practicable to the point at which the cable enters the building [800.90(B)]. Therefore, unlisted outside plant communications cables may not be permitted to enter the building if it's practicable to place the primary protector closer than 50 ft to the point of entrance.

800.133 Installation of Communications Circuits and Equipment.

(A) Separation from Other Conductors.

(1) In Raceways and Boxes.

(a) With Other Cables in Raceway or Enclosure. Communications cables can be in the same raceway or enclosure with cables of any of the following: **Figure 800-22**

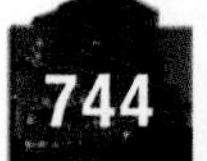

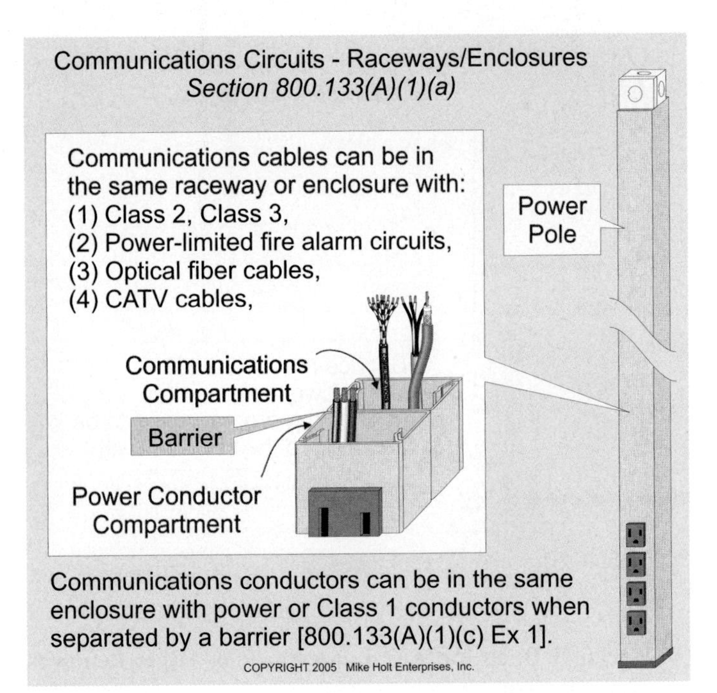

Communications Circuits - Raceways/Enclosures
Section 800.133(A)(1)(a)
Communications cables can be in the same raceway or enclosure with:
(1) Class 2, Class 3,
(2) Power-limited fire alarm circuits,
(3) Optical fiber cables,
(4) CATV cables,
Power Pole
Communications Compartment
Barrier
Power Conductor Compartment
Communications conductors can be in the same enclosure with power or Class 1 conductors when separated by a barrier [800.133(A)(1)(c) Ex 1].
COPYRIGHT 2005 Mike Holt Enterprises, Inc.

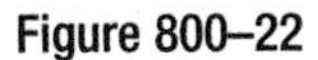
Figure 800–22

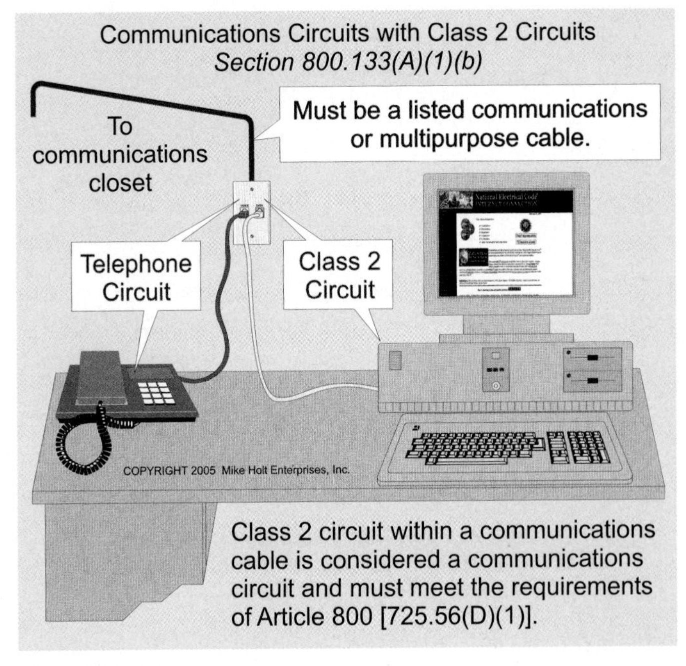

Communications Circuits with Class 2 Circuits
Section 800.133(A)(1)(b)
To communications closet
Must be a listed communications or multipurpose cable.
Telephone Circuit
Class 2 Circuit
COPYRIGHT 2005 Mike Holt Enterprises, Inc.
Class 2 circuit within a communications cable is considered a communications circuit and must meet the requirements of Article 800 [725.56(D)(1)].

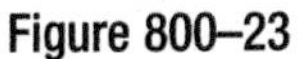
Figure 800–23

(1) Class 2 and Class 3 circuits in compliance with Article 725.

(2) Power-limited fire alarm circuits in compliance with Article 760.

(3) Optical fiber cables in compliance with Article 770.

(4) Coaxial circuits in compliance with Article 820.

(b) Class 1, Class 2, and Class 3 Circuits. Class 1 circuits must not be run in the same cable with communications circuits.

Class 2 and Class 3 conductors can be within the same cable with communications conductors, provided the Class 2 or Class 3 circuits use communications cables in accordance with Article 800 [725.56(D)(1)].

Author's Comments:

- A common application of this requirement is when a single cable is used for both voice communications and data [Class 2 – 725.41(A)(4)]. **Figure 800-23**
- Listed Class 2 cables have a voltage rating of 150V [725.82(G)], whereas communications cables have a voltage rating of 300V [800.179].

(c) With Power Conductors in Same Raceway or Enclosure. Communications conductors must not be placed in any raceway, compartment, outlet box, junction box, or similar fitting with conductors of electric light, power, Class 1, or nonpower-limited fire alarm circuits.

Exception 1: Communications circuits can be within the same enclosure with conductors of electric light, power, Class 1, and nonpower-limited fire alarm circuits, where separated by a permanent barrier or listed divider.

Author's Comment: Separation is required to prevent a fire or shock hazard that could occur from a short between the communications circuit and the higher-voltage circuits.

Exception 2: Communications conductors can be mixed with power conductors if the power circuit conductors are only introduced to supply power to communications equipment. The power circuit conductors must maintain a minimum ½ in. separation from the communications circuit conductors.

(2) Other Applications. Communications circuits must maintain 2 in. of separation from electric light, power, Class 1, or nonpower-limited fire alarm circuit conductors.

Exception 1: Separation by distance isn't required if all electric light, power, Class 1, or nonpower-limited fire alarm circuit conductors are in a raceway or in metal-sheathed, metal-clad, nonmetallic-sheathed, or underground feeder cables, or all communications cables are in a raceway. **Figure 800-24**

(B) Cable Trays. Types CMP, CMR, CMG, and CM communications cables [800.179], and listed communications raceways [800.182], can be installed in cable trays [392.3(A)].

(C) Support of Conductors. Communications cables must not be supported by a raceway. **Figure 800-25**

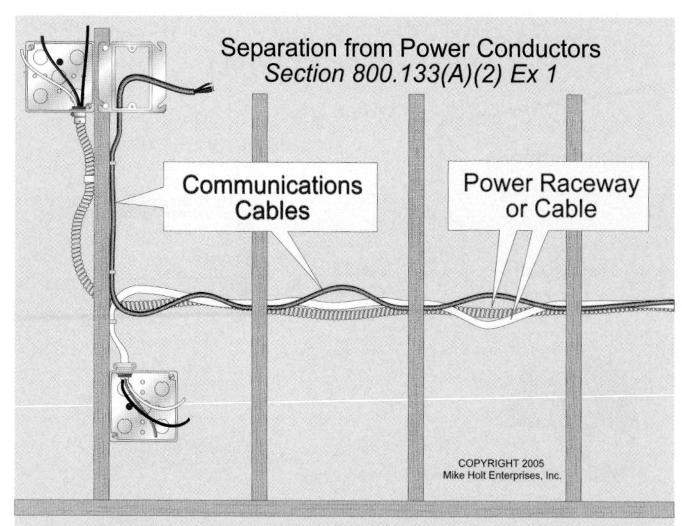

Figure 800–24

> **Author's Comment:** Exposed cables must be supported by the structural components of the building so that the cable will not be damaged by normal building use. The cables must be secured by straps, staples, hangers, or similar fittings designed and installed so as not to damage the cable [800.24].

Exception: Overhead (aerial) spans of communications wiring can be attached to a raceway-type mast intended for the attachment and support of such conductors.

> **Author's Comment:** Communications cables must not be supported by, or attached to, the power service mast [230.28]. **Figure 800-26**

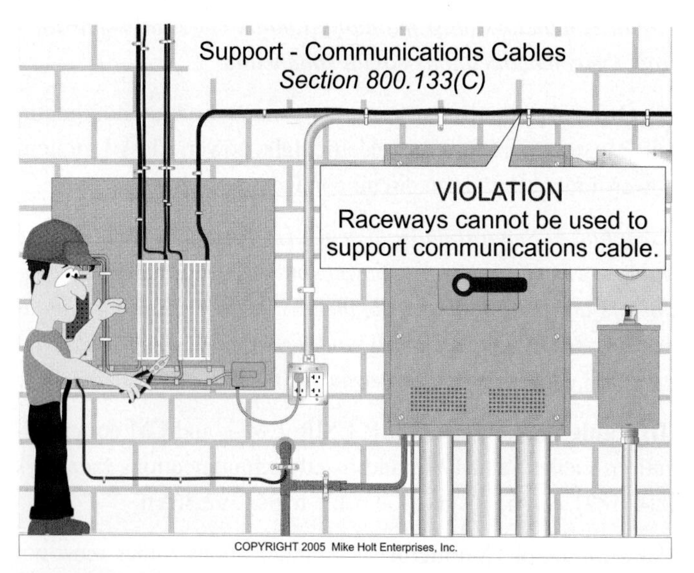

Support - Communications Cables
Section 800.133(C)
VIOLATION
Raceways cannot be used to support communications cable.
COPYRIGHT 2005 Mike Holt Enterprises, Inc.

Figure 800–25

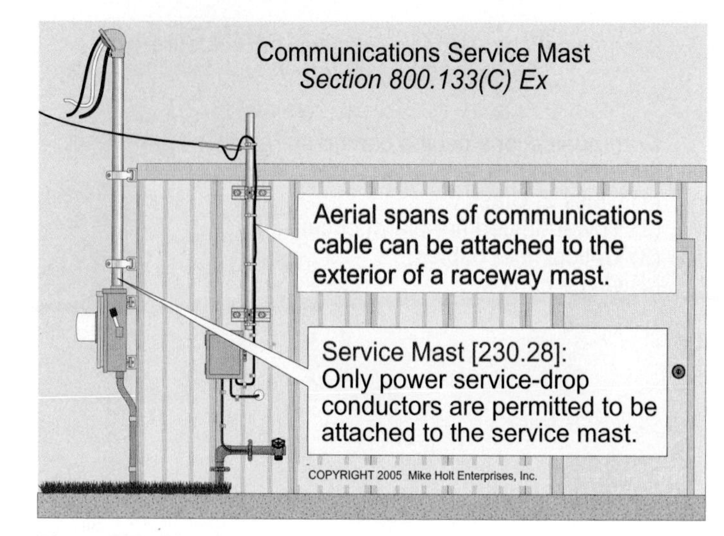

Figure 800–26

(D) Wiring in Ducts for Dust, Loose Stock, or Vapor Removal. Communications cables must not be installed in manufactured ducts that transport dust, loose stock, or vapors [300.22(A)].

800.154 Applications of Listed Communications Wires, Cables, and Raceways.

Communications wires and cables must comply with (A) through (F) or, where cable substitutions are made, in accordance with (G).

(A) Ducts or Plenums Used for Environmental Air. Communications cables must not be installed in ducts for handling dust, loose stock, or vapors [300.22(A), 800.3(B), and 800.133(D)].

Where necessary for the direct action upon, or sensing of the contained air, communications cables can be installed in ducts or plenums if they are installed in electrical metallic tubing, intermediate metal conduit, or rigid metal conduit as required by 300.22(B).

Plenum-rated Type CMP cables [800.179(A)] can be installed above a suspended ceiling or below a raised floor that is used for environmental air [300.22(C)].

Plenum-rated communications raceways [800.182] are permitted above a suspended ceiling or below a raised floor that is used for environmental air, if the raceway contains Type CMP plenum-rated cables. **Figure 800-27**

> **Author's Comment:** Nonplenum-rated cables can be installed above a suspended ceiling or below a raised floor that is used for environmental air, but only if the nonplenum-rated cable is installed within electrical metallic tubing [300.22(C)(1)].

(B) Riser. Cables installed in risers must comply with (1), (2), or (3).

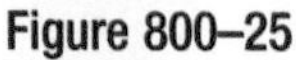

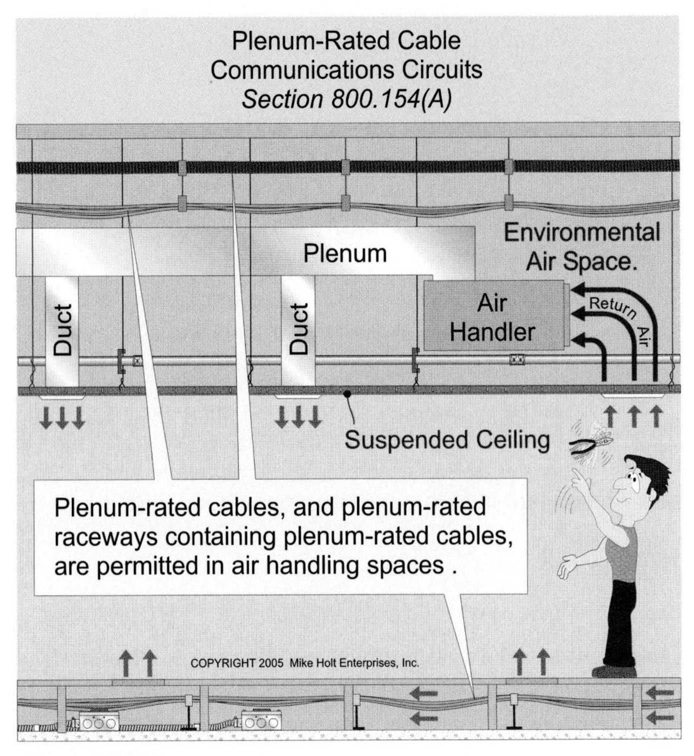

Figure 800–27

(1) Exposed Cables. Cables installed in vertical runs penetrating more than one floor must be riser rated Type CMR [800.179(B)]. Floor penetrations requiring Type CMR must contain only cables suitable for riser or plenum use. Where installed in risers, communications raceways with Types CMR and CMP cables are permitted.

(2) Cables in Metal Raceways. Type CM cables must be encased in a metal raceway or located in a fireproof shaft having a fire-stop at each floor.

Author's Comments:

- When communications cables are installed in a metal raceway, they aren't required to be riser or plenum rated.
- Metal raceways containing circuit conductors from power-supply systems that operate at 50V or less aren't required to be grounded (bonded) to an effective ground-fault current path [250.86 and 250.162(A)].

(3) One- and Two-Family Dwellings. Types CM and CMX cables can be installed in one- and two-family dwellings.

(D) Cable Trays. Types CMP, CMR, CMG, and CM communications cables can be installed in cable trays.

(E) Other Wiring Within Buildings. Cables installed in building locations other than covered in (1) through (5) must be:

(1) Types CMG [800.179(C)] or CM [800.179(D)]. Listed communications general-purpose raceways must be permitted.

(2) In Raceways. Listed communications cables of any type can be installed in a Chapter 3 raceway.

(3) Nonconcealed Spaces. Type CMX communications cable [800.182(A)], if in nonconcealed spaces where the exposed length of cable doesn't exceed 10 ft.

(4) One- and Two-Family Dwellings. Type CMX communications cables [800.182(A)] that are less than ¼ in. in diameter if in one- or two-family dwellings.

(5) Multifamily Dwellings. Type CMX communications cables [800.182(A)] that are less than ¼ in. in diameter if in nonconcealed spaces in multifamily dwellings.

(G) Cable Substitutions. The substitutions for communications cables listed in Table 800.154 are permitted.

PART VI. LISTING REQUIREMENTS

800.179 Communications Wires and Cables. Communications wires and cables must have a voltage rating not less than 300V and must be listed in accordance with (A) through (E). **Figure 800-28**

Author's Comment: Voltage markings on cables may be misinterpreted to suggest that the cables may be suitable for Calss 1, electric light, and power applications, which they're not [800.113 FPN].

(A) Type CMP. Type CMP plenum-rated communications cables must be listed as being suitable for use in ducts, plenums, and other environmental air spaces and must also be listed as having adequate fire-resistant and low smoke-producing characteristics [800.154(A)]. **Figure 800-29**

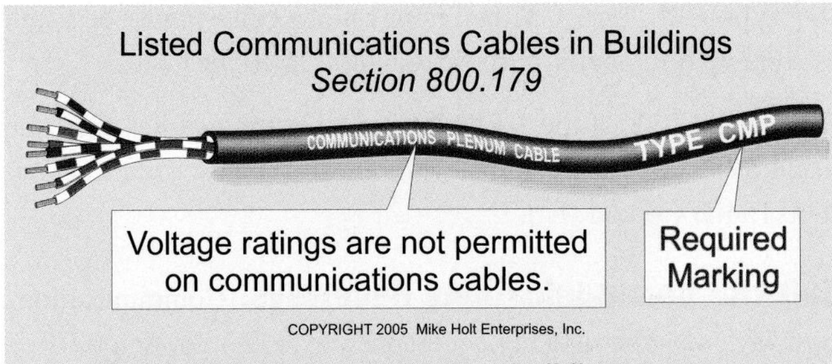

Communications cables in a building must be listed as being suitable for the purpose in accordance with 800.179. The cable must be rated at least 300V but the voltage rating cannot be marked on the cable [800.113].

Figure 800–28

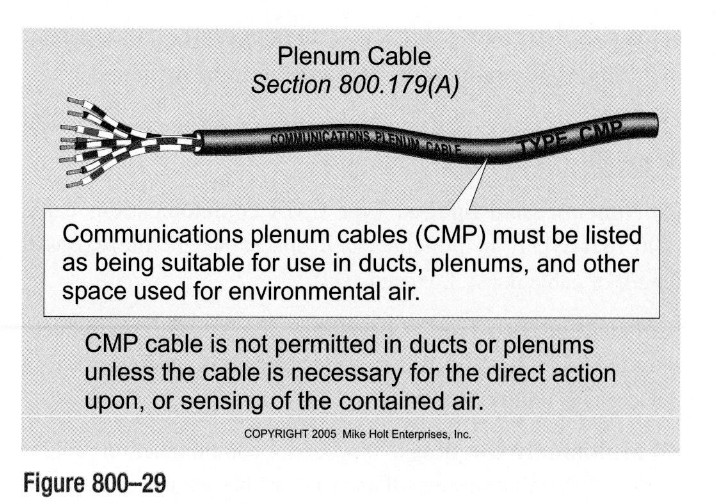

Figure 800–29

Author's Comments:

- Plenum-rated cables are not permitted within a duct or plenum, unless the cable is necessary for the direct action upon, or sensing of, the contained air [300.22(B) and 800.154(A)].
- Special consideration must be given to cables in areas that move or transport environmental air in order to reduce the hazards that arise from the burning of conductor insulation and the jackets of cables. Because listed plenum-rated cables have adequate fire-resistant and low smoke-producing characteristics, they can be installed in other environmental air spaces, but they cannot be installed in ducts or plenums! See 800.3(B) in this textbook for details.

(B) Type CMR. Type CMR communications riser cables must be listed as suitable for use in a vertical run in a shaft or from floor to floor [800.154(B)].

(C) Type CMG. Type CMG general-purpose communications cable must be listed as being suitable for general-purpose communications use, with the exception of risers and plenums, and must also be listed as being resistant to the spread of fire.

(D) Type CM. Type CM communications cables must be listed as suitable for general-purpose use [800.154(E)].

(E) Type CMX. Type CMX communications cables must be listed as suitable for use in dwellings and for use in a raceway [800.154(B)(3) and (E)].

800.182 Communications Raceways. Communications raceways must be listed in accordance with (A) through (C).

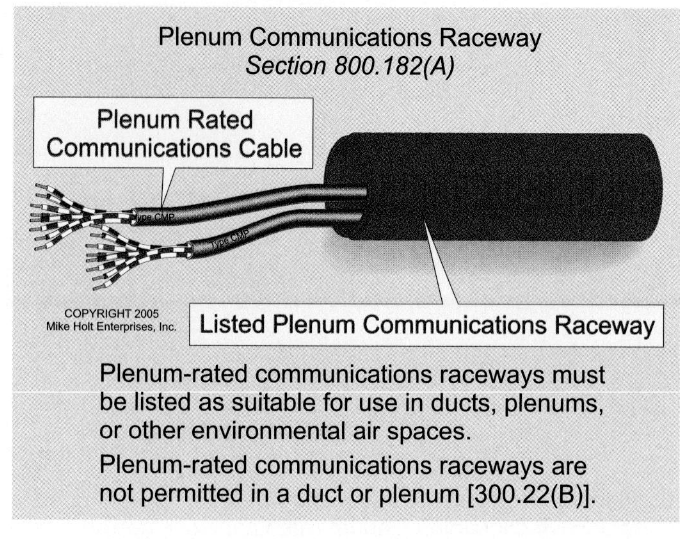

Figure 800–30

(A) Plenum-Rated Communications Raceways. Plenum-rated communications raceways must be listed as suitable for use in ducts, plenums, or other environmental air spaces. **Figure 800-30**

Author's Comments:

- Plenum-rated communications raceways are not permitted in a duct or plenum space [300.22(B)].
- Where used in environmental air spaces, a listed plenum-rated communications raceway must contain only Type CMP cables [800.154(A)].
- Special consideration must be given to raceways in areas that move or transport environmental air in order to reduce the hazards that arise from the burning conductor insulation and cable jackets. Because listed plenum-rated cables have adequate fire-resistant and low smoke-producing characteristics, they can be installed in other environmental air spaces.

(B) Riser Communications Raceway. Riser-rated communications raceways must be listed as suitable for running vertically through more than one floor.

Author's Comment: Listed riser rated communications raceways that run vertically and penetrate more than one floor must only contain Types CMR and CMP cables [800.154(B)(1)].

(C) General-Purpose Communications Raceway. General-purpose communications raceways installed in general-purpose areas must only contain Types CM, CMR, or CMP cables [800.154(E)(1)].

Article 800 Questions

1. Communications cables installed _____ on the surface of ceilings and sidewalls must be supported by the building structure in such a manner that the cable will not be damaged by normal building use.

   (a) exposed (b) concealed (c) hidden (d) a and b

2. In one- and two-family dwellings, the primary protector grounding conductor for communications systems must be as short as practicable, not to exceed _____ in length.

   (a) 5 ft (b) 8 ft (c) 10 ft (d) 20 ft

3. Communications wires and cables are not required to be listed and marked where the length of the cable within the building, measured from its point of entrance, does not exceed_____ and the cable enters the building from the outside and is terminated in an enclosure or on a listed primary protector.

   (a) 25 ft (b) 30 ft (c) 50 ft (d) 100 ft

4. Communications cable risers penetrating more than one floor, or cables installed in vertical runs in a shaft, must be Type CMR. Listed communications cables are also allowed when _____.

   (a) encased in metal raceways
   (b) located in a fireproof shaft with firestops at each floor
   (c) a or b
   (d) none of these

5. Communications _____ cable must be listed as being suitable for use in a vertical run in a shaft, or from floor to floor, and must also be listed as having fire-resistant characteristics capable of preventing the carrying of fire from floor to floor.

   (a) plenum (b) riser (c) general-purpose (d) none of these

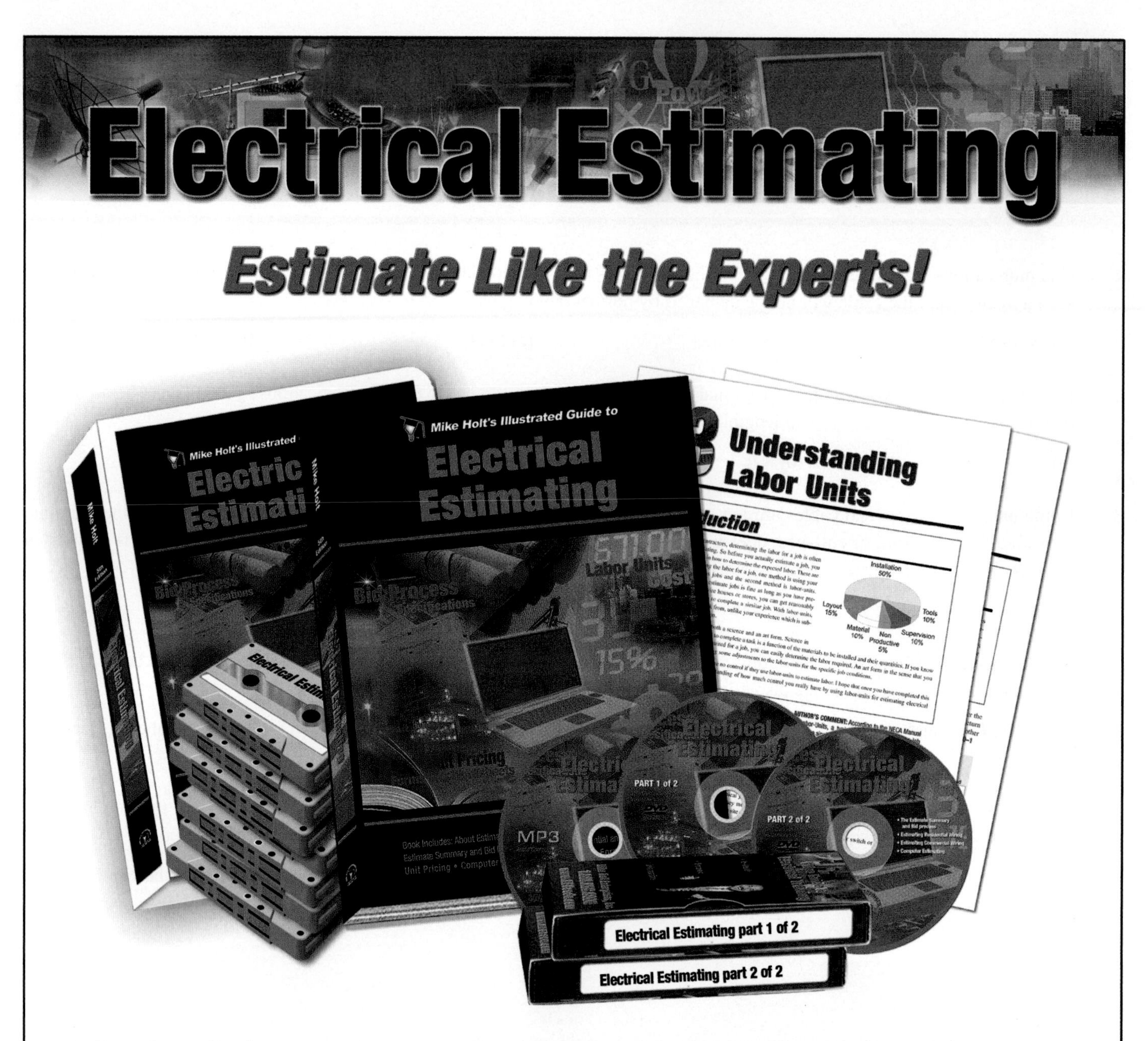

Running a business requires an understanding of how to estimate a job. This program will show you how to take-off a job, how to determine the material cost, how to accurately estimate labor cost, and how to determine your overhead, profit, and break-even point. This course also gives tips on selling and marketing your business. You'll learn how to estimate and project manage residential, commercial, and industrial projects; and how to make more money on every job by proper estimating and project management. Watch the videos or DVDs at home or office and listen to the cassettes or MP3s while driving. You'll quickly learn how to make more money on every job.

Call us today at 1.888.NEC.Code, or visit us online at www.NECcode.com, for the latest information and pricing.

ARTICLE 810 Radio and Television Equipment

Introduction

This article covers transmitter and receiver equipment—and the wiring and cabling associated with that equipment. Here are a few key points to remember from Article 810:

- Avoid contact with conductors of other systems.
- Don't attach antennas or other equipment to the service-entrance power mast.
- If the mast isn't grounded properly, voltage surges caused by nearby lightning strikes can destroy it.
- Keep the grounding conductor straight, and protect it from physical damage.
- If the mast isn't bonded properly, you risk flashovers and possible electrocution.
- Keep in mind that the purpose of bonding is to prevent a difference of potential between metallic objects and other conductive items, such as swimming pools. Thus, Article 810 provides several different bonding requirements.
- Clearances are critical. Article 810 provides extensive clearance requirements. For example, it provides separate clearance requirements for indoor and outdoor locations.

PART I. GENERAL

810.1 Scope. Article 810 contains the installation requirements for the wiring of television and radio receiving equipment, such as digital satellite receiving equipment for television signals and amateur radio equipment antennas. **Figure 810-1**

Author's Comment: Article 810 covers:

- VHF/UHF antennas, which receive local television signals.
- Satellite antennas, which are often referred to as satellite dishes. Large satellite dish antennas (often about 6 ft in diameter) usually have a motor that moves the dish to focus on different satellites. The smaller satellite dish antennas (18 in. in diameter) are usually aimed at a single satellite.
- Roof-mounted antennas for AM/FM/XM radio reception.
- Amateur radio transmitting and receiving equipment, including HAM radio equipment (a noncommercial (amateur) communications system).

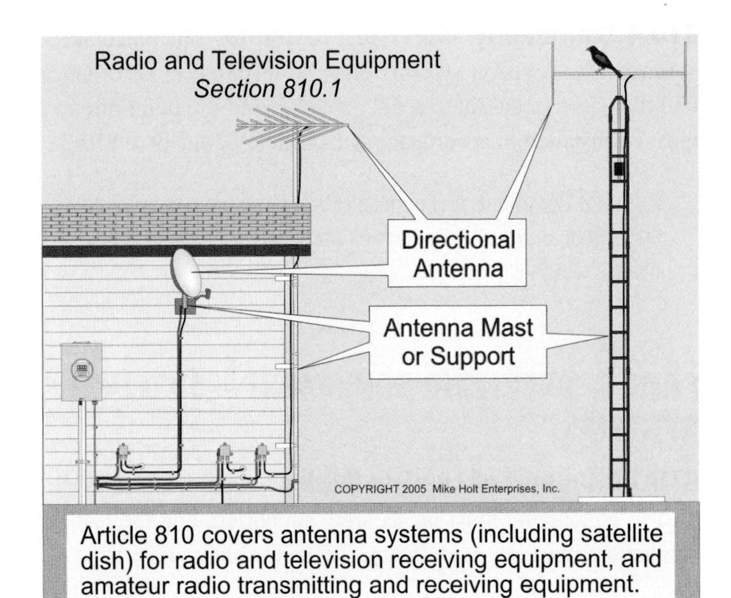

Radio and Television Equipment
Section 810.1
Directional Antenna
Antenna Mast or Support
COPYRIGHT 2005 Mike Holt Enterprises, Inc.
Article 810 covers antenna systems (including satellite dish) for radio and television receiving equipment, and amateur radio transmitting and receiving equipment.

Figure 810–1

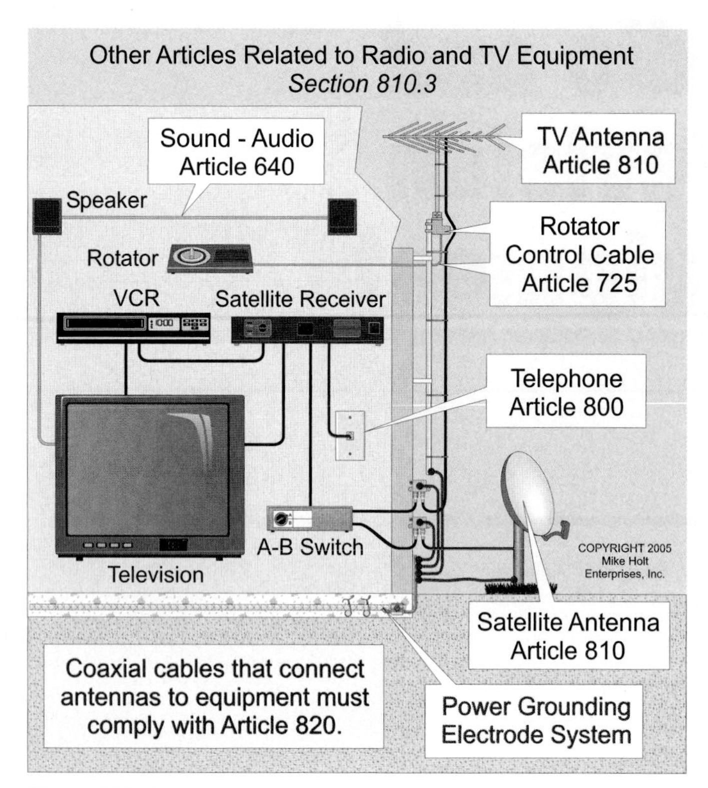

Figure 810–2

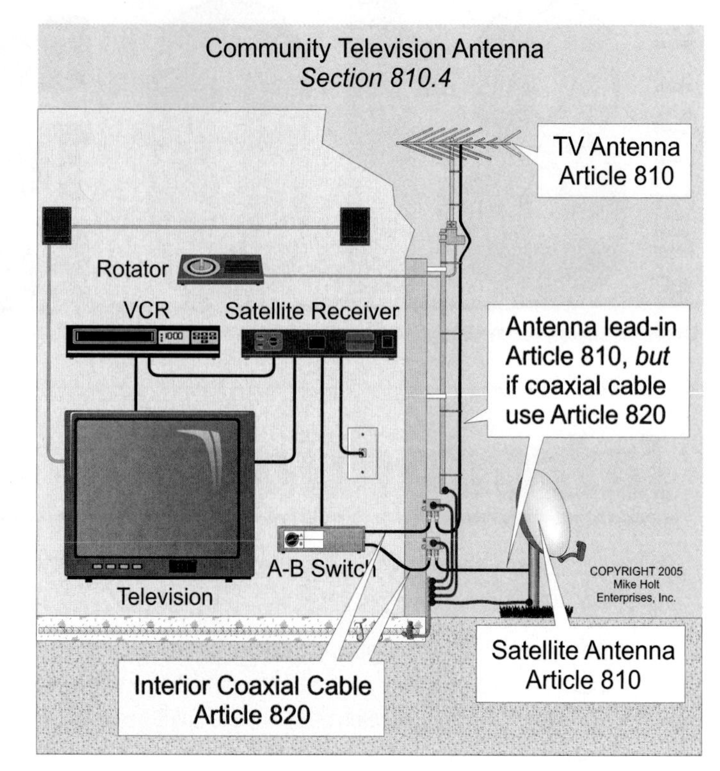

Figure 810–3

810.3 Other Articles. Wiring from the power source to Article 810 equipment must be installed in accordance with Chapters 1 through 4. Wiring for audio equipment must comply with Article 640 and coaxial cables that connect antennas to equipment must be installed in accordance with Article 820. **Figure 810-2**

810.4 Community Television Antenna. The antenna for community television systems must be installed in accordance with this article, but the coaxial cable beyond the point of entry must be installed in accordance with Article 820. **Figure 810-3**

> **Author's Comment:** A community TV antenna is used for multiple-occupancy facilities, such as apartments, condominiums, motels, and hotels.

PART II. RECEIVING EQUIPMENT—ANTENNA SYSTEMS

810.12 Support of Lead-In Cables. Outdoor antennas and lead-in conductors must be securely supported, and the lead-in conductors must be securely attached to the antenna.

> **Author's Comment:** Outdoor antennas aren't permitted to be attached to the electric service mast [230.28]. **Figure 810-4**

810.13 Avoid Contact with Conductors of Other Systems. Outdoor antennas and lead-in conductors must be kept at least 2 ft from exposed electric power conductors to avoid the possibility of accidental contact.

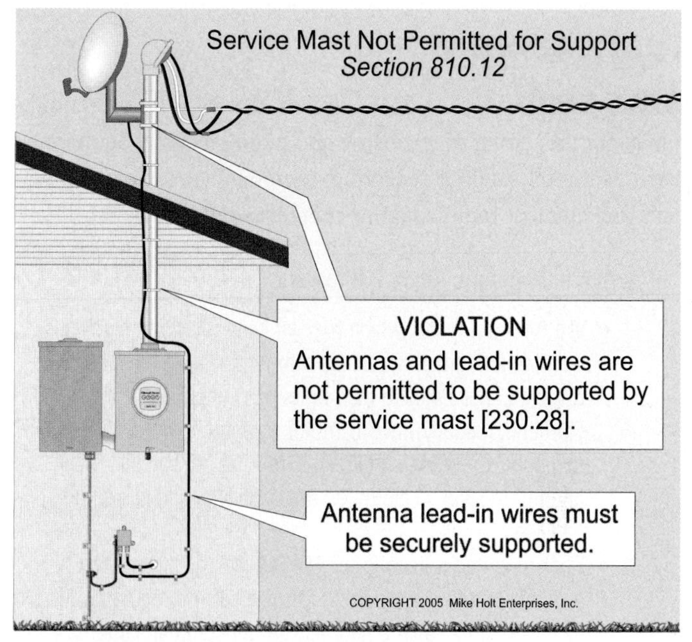

Figure 810–4

Author's Comment: According to the NFPA *NEC Handbook*, "One of the leading causes of electrical shock and electrocution is the accidental contact of radio, television, and amateur radio transmitting and receiving antennas, and equipment with light or power conductors. Extreme caution should therefore be exercised during this type of installation, and periodic visual inspections should be conducted thereafter."

810.15 Grounding. Outdoor masts and metal structures that support antennas must be grounded in accordance with 810.21.

Author's Comment: An antenna mast located within 5 ft of an outdoor swimming pool must be bonded to the pool's common bonding grid [680.26(B)(5)]. **Figure 810-5**

810.18 Clearances.

(A) Outside of Buildings. Lead-in conductors attached to buildings must be installed so that they cannot swing closer than 2 ft to the conductors of circuits of 250V or less, or closer than 10 ft to the conductors of circuits of over 250V.

Lead-in conductors must be kept at least 6 ft from lightning protection conductors.

Underground antenna lead-in conductors must maintain a separation not less than 12 in. from electric power conductors. **Figure 810-6**

Exception: Separation is not required where the underground antenna lead-in conductors or the electric power conductors are installed in a raceway or cable armor.

Author's Comment: The *NEC* does not specify a burial depth for antenna lead-in wires.

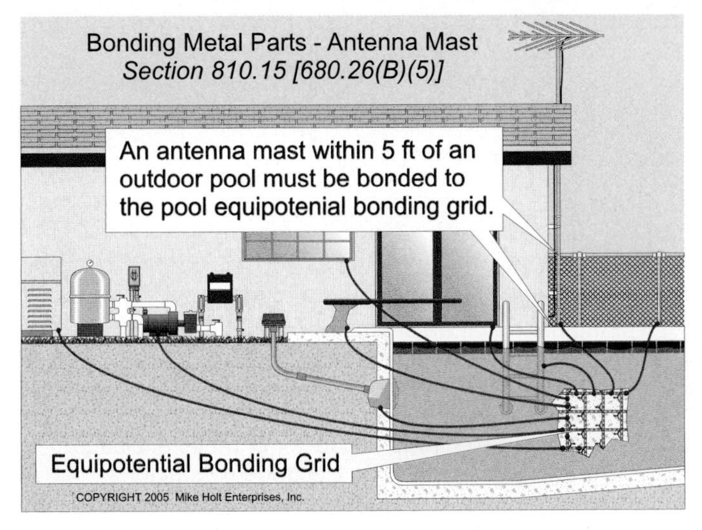

Bonding Metal Parts - Antenna Mast
Section 810.15 [680.26(B)(5)]
An antenna mast within 5 ft of an outdoor pool must be bonded to the pool equipotenial bonding grid.
Equipotential Bonding Grid
COPYRIGHT 2005 Mike Holt Enterprises, Inc.

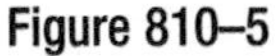
Figure 810–5

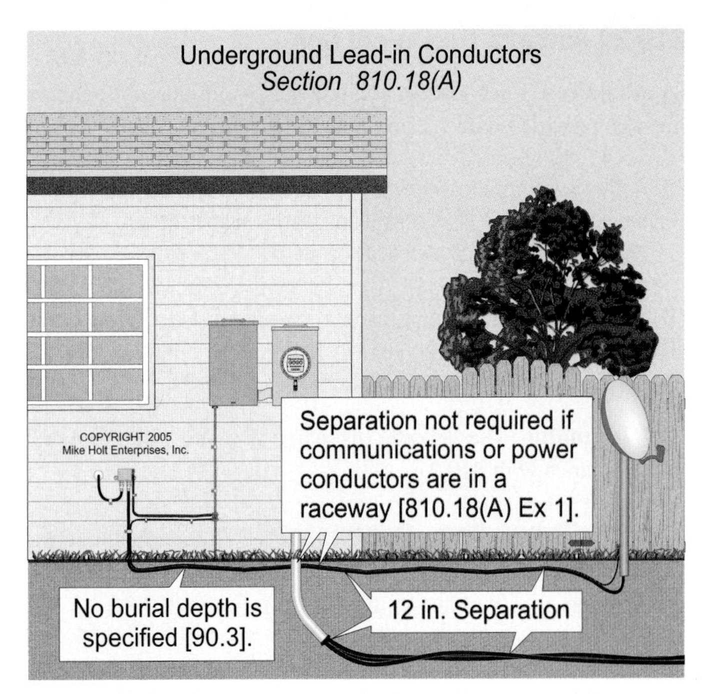

Figure 810–6

(B) Indoors. Indoor antenna and lead-in conductors must not be less than 2 in. from electric light and power conductors.

Exception 1: Separation by distance isn't required if the antenna lead-in conductors or the electric light and power conductors are installed in a raceway or cable armor.

(C) Enclosures. Indoor antenna lead-in conductors can be in the same enclosure with electric power conductors where separated by an effective, permanently installed barrier. **Figure 810-7**

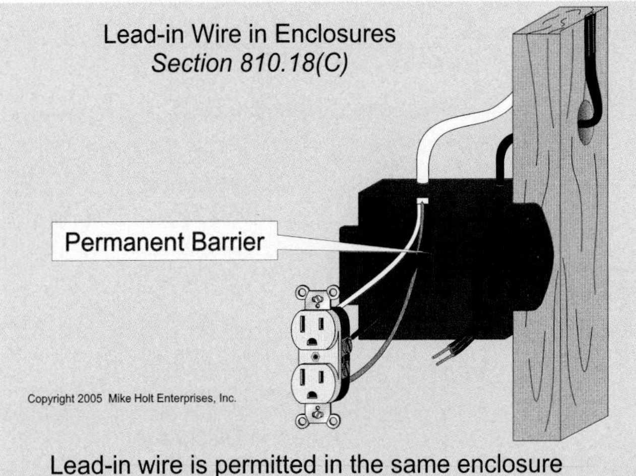

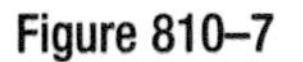
Figure 810–7

810.20 Antenna Discharge Unit.

(A) Required. Each lead-in conductor from an outdoor antenna must be provided with a listed antenna discharge unit.

> **Author's Comment:** Antennas located indoors (like in an attic) are not required to have an antenna discharge unit, nor are they required to be grounded in accordance with 810.20(C).

(B) Location. The antenna discharge unit must be located outside or inside the building, nearest the point of entrance, but not near combustible material.

(C) Grounding. The antenna discharge unit must be grounded in accordance with 810.21.

810.21 Grounding Conductors.

The antenna mast [810.15] and antenna discharge unit [810.20(C)] must be grounded as specified in (A) through (K). **Figure 810-8**

> **Author's Comment:** Grounding the lead-in antenna cables and the mast helps prevent voltage surges caused by static discharge or nearby lightning strikes from reaching the center conductor of the lead-in coaxial cable. Because the satellite sits outdoors, wind creates a static charge on the antenna as well as on the cable attached to it. This charge can build up on both antenna and cable until it jumps across an air space, often passing through the electronics inside the low noise block downconverter feedhorn (LNBF) or receiver. Grounding the coaxial cable and dish to the building grounding electrode system helps to dissipate this static charge.
>
> Nothing can prevent damage from a direct lightning strike. But grounding with proper surge protection can help reduce damage to satellite and other equipment from nearby lightning strikes.

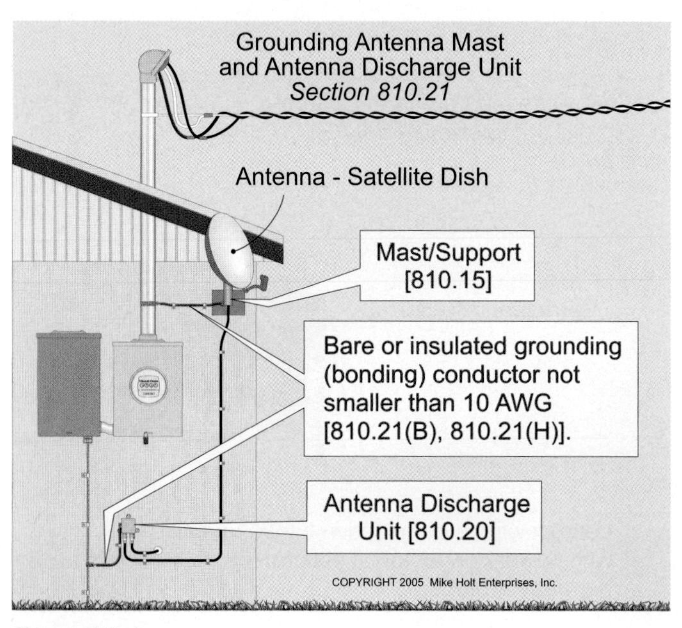

Figure 810–8

(A) Material. The grounding conductor to the electrode [810.21(F)] must be copper or other corrosion-resistant conductive material, stranded or solid.

(B) Insulation. The grounding conductor isn't required to be insulated.

(C) Supports. The grounding conductor must be securely fastened in place.

(D) Protection Against Physical Damage. The grounding conductor must be guarded from physical damage. If the grounding conductor is run in a metal raceway, both ends of the raceway must be bonded to the grounding conductor.

> **Author's Comment:** Installing the grounding conductor in a nonmetallic raceway, when the authority having jurisdiction judges that physical protection is required, is a better practice than using a metal raceway.

(E) Run in Straight Line. The grounding conductor must be run in as straight a line as practicable.

> **Author's Comment:** Lightning doesn't like to travel around corners or through loops, which is why the grounding conductor should be run as straight as practicable.

(F) Electrode.

(1) The grounding conductor must terminate to the nearest accessible: **Figure 810-9**

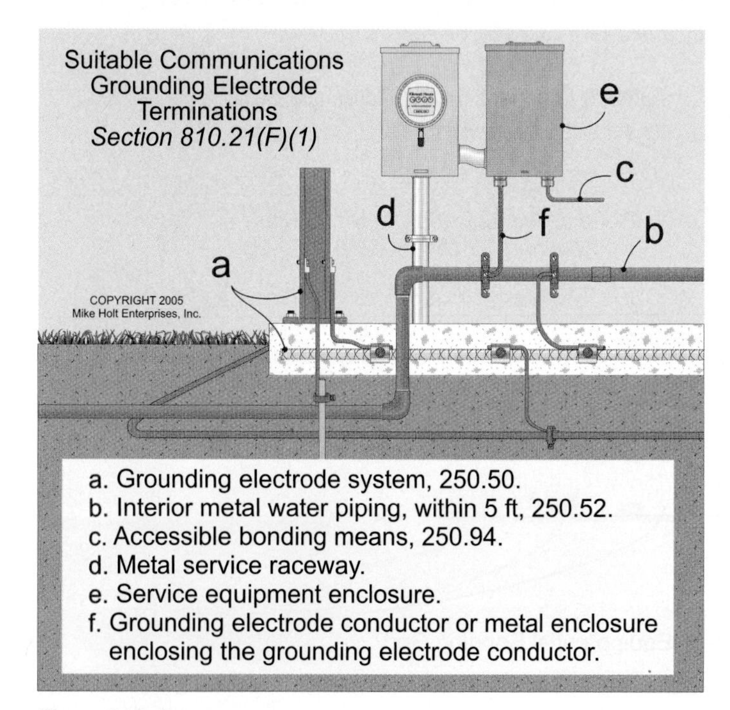

Figure 810–9

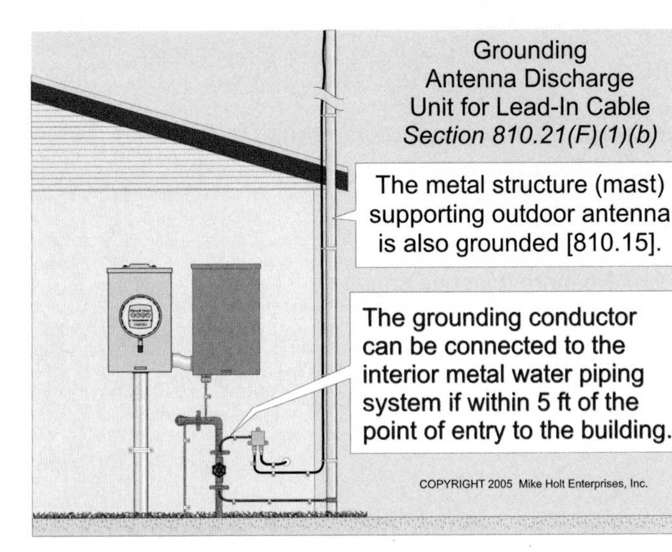

Grounding
Antenna Discharge
Unit for Lead-In Cable
Section 810.21(F)(1)(b)
The metal structure (mast) supporting outdoor antenna is also grounded [810.15].
The grounding conductor can be connected to the interior metal water piping system if within 5 ft of the point of entry to the building.
COPYRIGHT 2005 Mike Holt Enterprises, Inc.

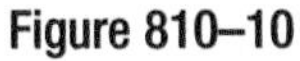
Figure 810–10

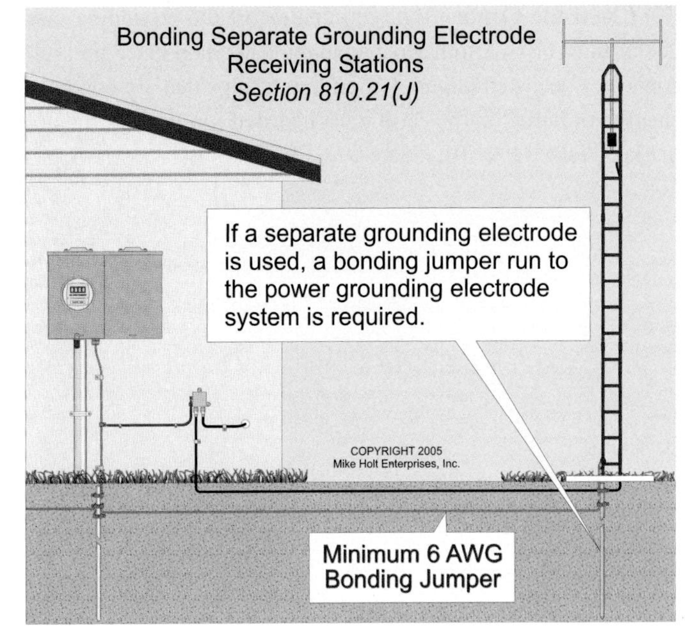

Bonding Separate Grounding Electrode
Receiving Stations
Section 810.21(J)
If a separate grounding electrode is used, a bonding jumper run to the power grounding electrode system is required.
COPYRIGHT 2005 Mike Holt Enterprises, Inc.
Minimum 6 AWG
Bonding Jumper

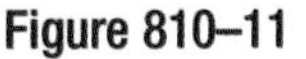
Figure 810–11

a. Building or structure grounding electrode system [250.50].

b. Interior metal water piping system, within 5 ft from its point of entrance [250.52(A)(1)]. **Figure 810-10**

c. Accessible service bonding means [250.94].

d. Metallic service raceway.

e. Service equipment enclosure.

f. Grounding electrode conductor or the grounding electrode conductor metal enclosure.

(G) Inside or Outside Building. The grounding conductor can be run either inside or outside the building.

(H) Size. The grounding conductor must not be smaller than 10 AWG copper or 17 AWG copper-clad steel or bronze.

Author's Comment: Copper-clad steel or bronze wire (17 AWG) is often molded into the jacket of the coaxial cable to simplify the grounding of the satellite dish by eliminating the need to run a separate ground wire to the dish [810.21(F)(1)].

(J) Bonding of Electrodes. If a ground rod is installed to serve as the ground for the radio and television equipment, it must be connected to the building's power grounding electrode system with a minimum 6 AWG conductor. **Figure 810-11**

Author's Comment: The bonding of separate system electrodes (building and radio and television equipment electrode) reduces voltages that may develop between the building's power and the radio and television equipment grounding electrode system during lightning events. **Figure 810-12**

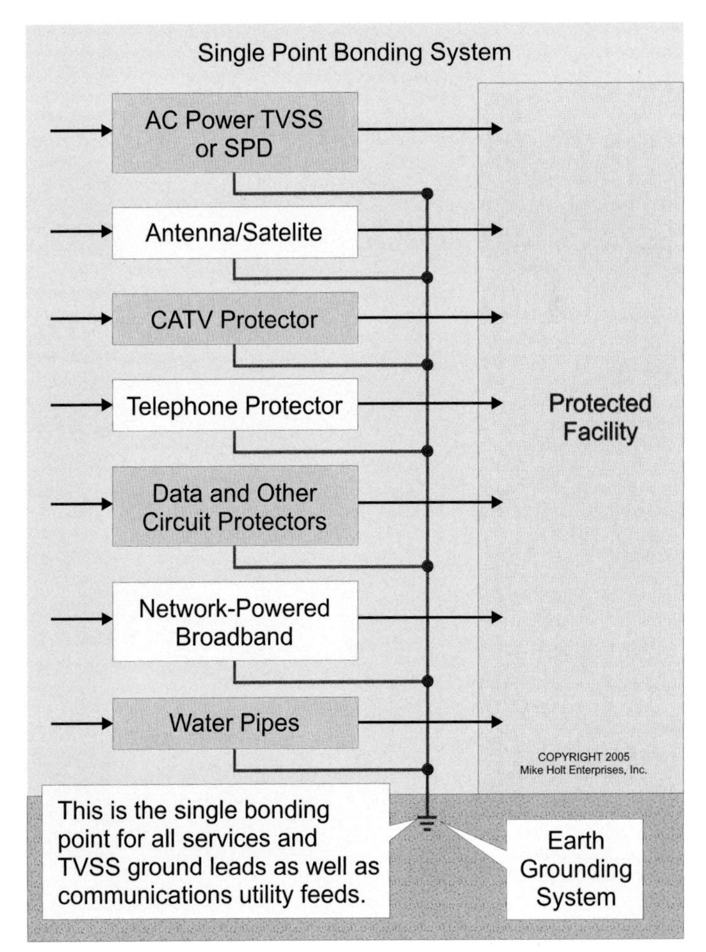

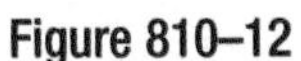
Figure 810–12

(K) Electrode Connection. Termination of the grounding conductor must be by exothermic welding, listed lug, listed pressure connector, or listed clamp. Grounding fittings that are concrete-encased or buried in the earth must be listed for direct burial and marked "DB" [250.70]. **Figure 810-13**

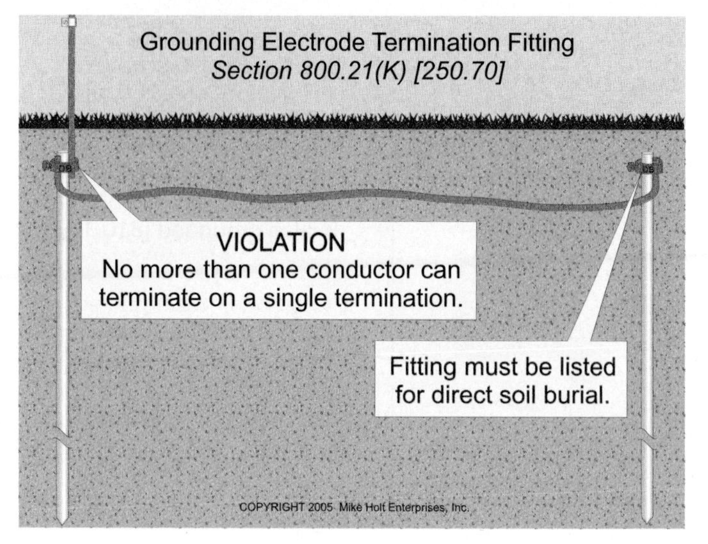
Grounding Electrode Termination Fitting
Section 800.21(K) [250.70]
VIOLATION
No more than one conductor can terminate on a single termination.
Fitting must be listed for direct soil burial.
COPYRIGHT 2005 Mike Holt Enterprises, Inc.

Figure 810–13

Article 810 Questions

1. Coaxial cables that connect antennas to equipment must be installed in accordance with Article 820.

   (a) True (b) False

2. Outdoor antennas and lead-in conductors must be securely supported and the lead-in conductors must be securely attached to the antenna, but they must not be attached to the electric service mast.

   (a) True (b) False

3. Underground antenna conductors for radio and television receiving equipment must be separated at least _____ from any light, power, or Class 1 circuit conductors.

   (a) 6 ft (b) 5 ft (c) 12 in. (d) 18 in.

4. Each conductor of a lead-in from an outdoor antenna must be provided with a listed antenna discharge unit.

   (a) True (b) False

5. The grounding conductor for an antenna mast or antenna discharge unit must be run to the grounding electrode in as straight a line as practicable.

   (a) True (b) False

Notes

Community Antenna Television (CATV) and Radio Distribution Systems

Introduction

This article focuses on the distribution of television and radio signals within a facility or on a property via cable, rather than their transmission or reception via antenna. These signals are limited-energy, but they are high frequency.

- As with Article 800, you must determine the "point of entrance" for these circuits.
- Ground the incoming cable as close as practicable to the point of entrance.
- If cables are located above a suspended ceiling, route and support them to allow access via ceiling panel removal.
- Clearances are critical, and Article 820 provides extensive clearance requirements. For example, Article 820 requires at least 6 ft of clearance between coaxial cable and lightning conductors.
- If you use a separate grounding electrode, you must run a bonding jumper to the power grounding system.

PART I. GENERAL

820.1 Scope. Article 820 covers the installation of coaxial cables for distributing high-frequency signals typically employed in community antenna television (CATV) systems. **Figure 820-1**

Author's Comment: Coaxial cables that connect antennas to television and radio receiving equipment [810.3] and community television systems [810.4] must be installed in accordance with this article. **Figure 820-2**

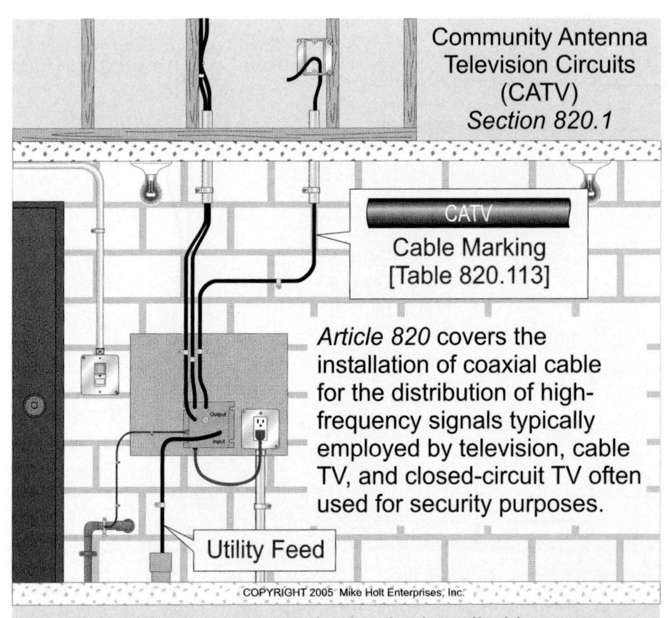

Community Antenna Television Circuits (CATV) Section 820.1
CATV
Cable Marking [Table 820.113]
Article 820 covers the installation of coaxial cable for the distribution of high-frequency signals typically employed by television, cable TV, and closed-circuit TV often used for security purposes.
Utility Feed
COPYRIGHT 2005 Mike Holt Enterprises, Inc.

Coaxial cables are not required to be installed in raceways.

Figure 820–1

820.2 Definitions.

Abandoned Cable. Cable that isn't terminated at equipment, and not identified for future use with a tag, is considered as abandoned for the purpose of this article.

Author's Comment: 820.3(A) requires the accessible portions of abandoned cables to be removed.

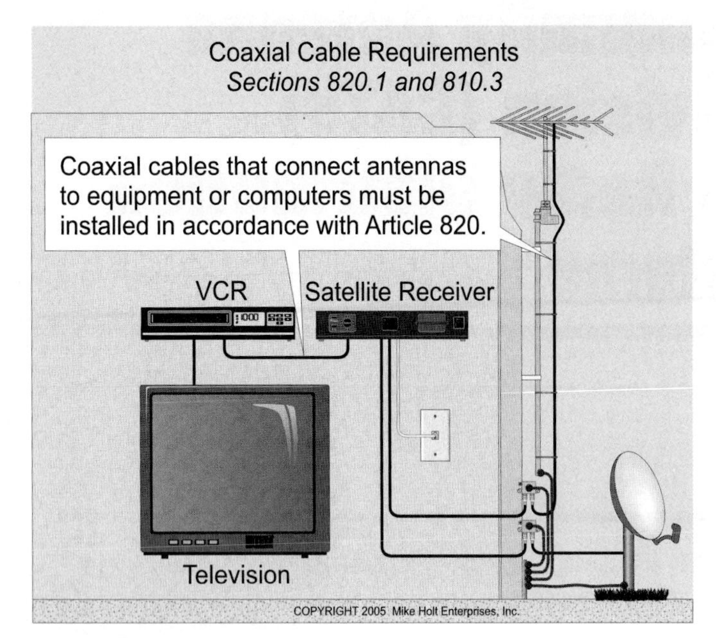
Coaxial Cable Requirements
Sections 820.1 and 810.3
Coaxial cables that connect antennas to equipment or computers must be installed in accordance with Article 820.
VCR
Satellite Receiver
Television
COPYRIGHT 2005 Mike Holt Enterprises, Inc.

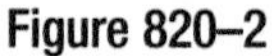
Figure 820–2

Point of Entrance. The point at which the cable emerges from an external wall, from a concrete floor slab, or from a rigid metal conduit or an intermediate metal conduit grounded (bonded) to an electrode in accordance with 820.100. **Figure 820-3**

> **Author's Comment:** Understanding the definition of this term is important both for the application of cable grounding [820.93], and for determining the maximum length of unlisted cable that can enter a building from the outside [820.113, Ex 3].

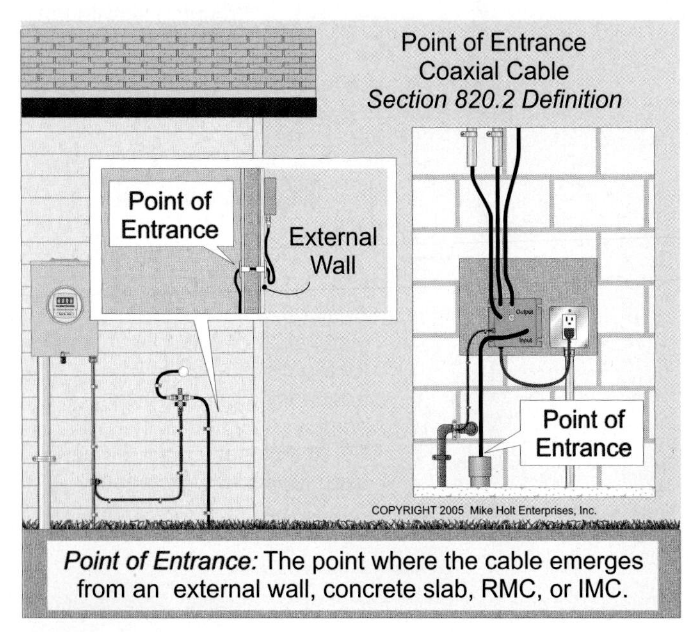
Point of Entrance
Coaxial Cable
Section 820.2 Definition
Point of Entrance
External Wall
Point of Entrance
COPYRIGHT 2005 Mike Holt Enterprises, Inc.
Point of Entrance: The point where the cable emerges from an external wall, concrete slab, RMC, or IMC.

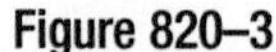
Figure 820–3

820.3 Locations and Other Articles.

Circuits and equipment must comply with (A) through (G).

(A) Spread of Fire or Products of Combustion. Coaxial cables installed through fire-resistant rated walls, partitions, floors, or ceilings must be firestopped in accordance with the instructions supplied by the manufacturer for the specific type of cable and construction material (drywall, brick, etc.), as required by 300.21.

> **Author's Comment:** Openings in fire-resistant walls, floors, and ceilings must be sealed so that the possible spread of fire or products of combustion will not be substantially increased [300.21].

Abandoned Cable. To limit the spread of fire or products of combustion within a building, the accessible portion of cable that isn't terminated at equipment, and not identified for future use with a tag, must be removed. **Figure 820-4**

> **Author's Comment:** This rule doesn't require the removal of concealed cables that are abandoned in place, which includes cables in raceways. According to the definition of "Concealed" in Article 100, cables in raceways are considered to be concealed.

(B) Ducts, Plenums, and Other Air-Handling Spaces. Where necessary for the direct action upon, or sensing of the contained air, coaxial cables can be installed in ducts or plenums if they are installed in electrical metallic tubing, intermediate metal conduit, or rigid metal conduit as required by 300.22(B).

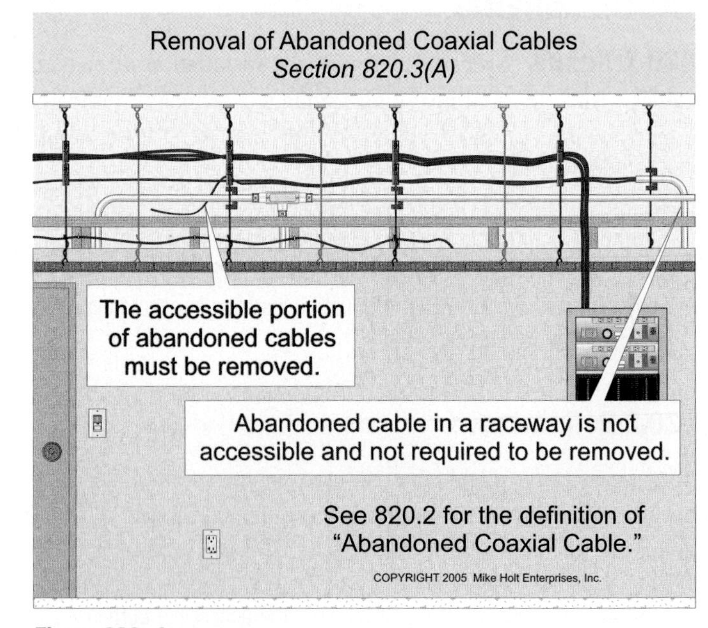
Removal of Abandoned Coaxial Cables
Section 820.3(A)
The accessible portion of abandoned cables must be removed.
Abandoned cable in a raceway is not accessible and not required to be removed.
See 820.2 for the definition of "Abandoned Coaxial Cable."
COPYRIGHT 2005 Mike Holt Enterprises, Inc.

Figure 820–4

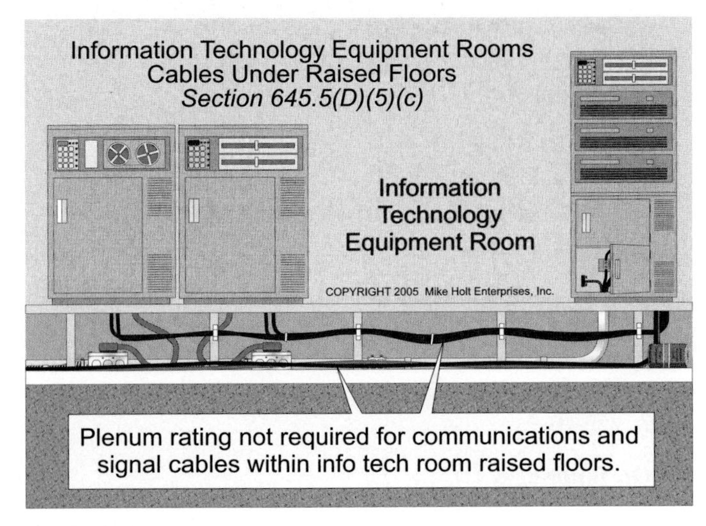

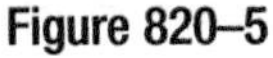

Figure 820–5

Exception: Plenum-rated coaxial cables [820.179(A)] and plenum-rated CATV raceways [820.182(A)] with Type CATVP coaxial cables [820.154(A)] are permitted above a suspended ceiling or below a raised floor used for environmental air movement.

> **Author's Comment:** Coaxial cables installed beneath a raised floor in an information technology equipment room (computer room) aren't required to be plenum rated [300.22(D) and 645.5(D)(5)(c)]. **Figure 820-5**

(C) Installation and Use. Equipment must be installed in accordance with manufacturer's instructions [110.3(B)].

(D) Optical Fiber Cables. Optical fiber cable must be installed in accordance with Article 770.

(E) Communications Circuits. Twisted-pair conductor cable used for communications circuits must comply with Article 800.

820.15 Energy Limitations. CATV coaxial cable is permitted to deliver low-energy power from a power-limited power supply to equipment associated with the radio frequency distribution system at a maximum of 60V from a power supply that has energy-limiting characteristics, such as from boosters, some splitters, amplifiers, etc.

820.21 Access to Electrical Equipment Behind Panels Designed to Allow Access. Access to equipment must not be prohibited by an accumulation of cables that prevent the removal of suspended-ceiling panels. Coaxial cables must be located so that the suspended-ceiling panels can be moved to provide access to electrical equipment. **Figure 820-6**

820.24 Mechanical Execution of Work. Equipment and cabling must be installed in a neat and workmanlike manner.

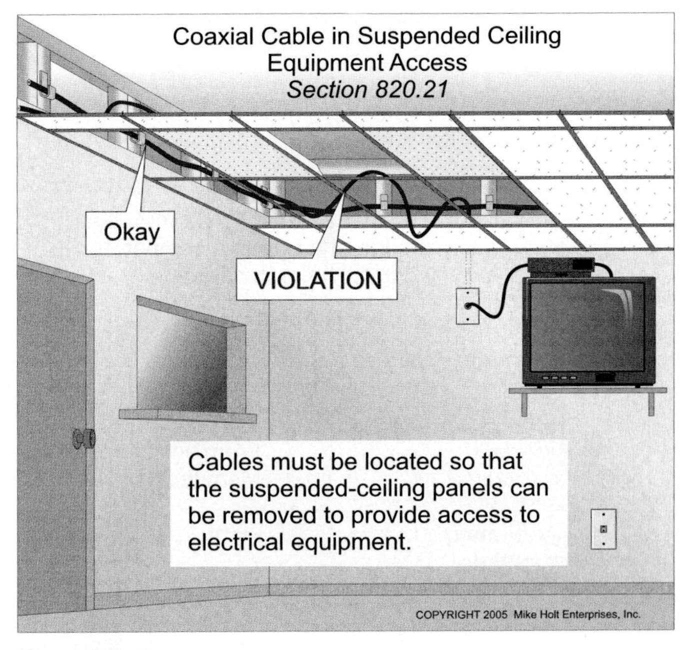

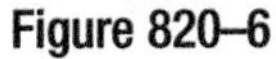

Figure 820–6

> **FPN:** Information that describes industry practices can be found in ANSI/NECA/BICSI 568, *Standard for Installing Commercial Building Telecommunications Cabling.*

> **Author's Comment:** For more information about this standard, visit http://www.necaneis.org/.

Exposed cables must be supported by the structural components of the building so that the cable will not be damaged by normal building use. Cables must be secured by straps, staples, hangers, or similar fittings designed and installed so as not to damage the cable. **Figure 820-7**

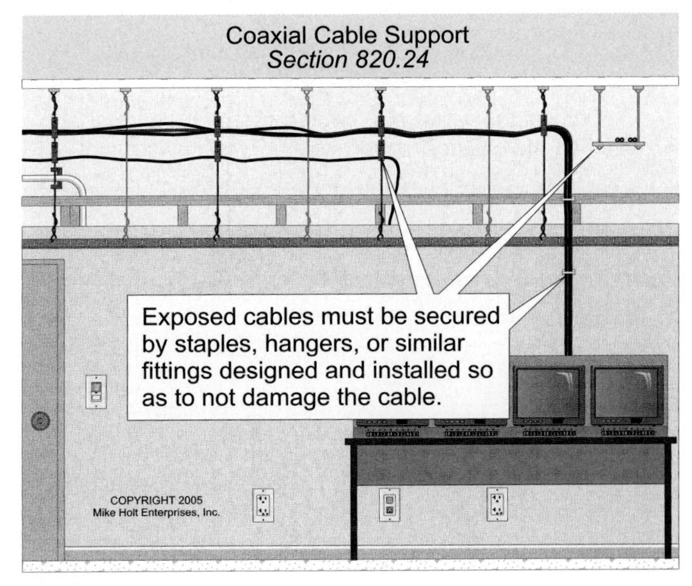

Figure 820–7

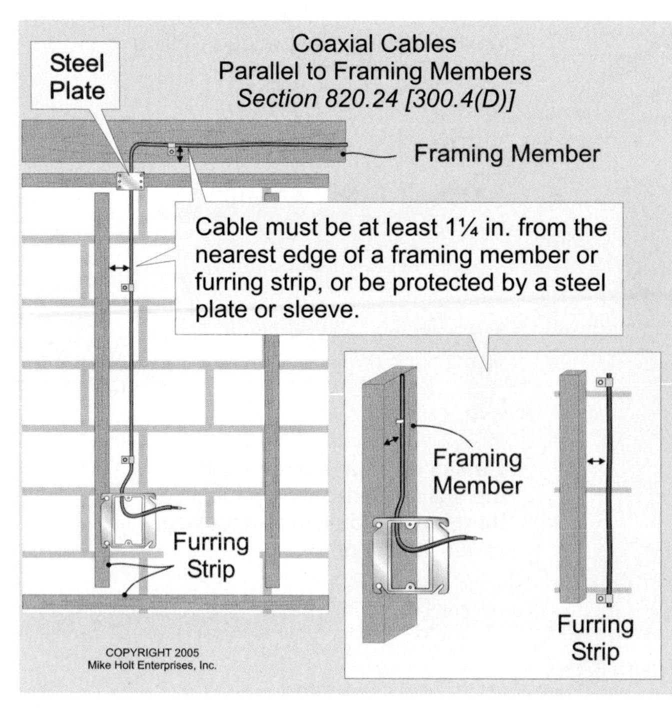
Steel Plate
Coaxial Cables
Parallel to Framing Members
Section 820.24 [300.4(D)]
Framing Member
Cable must be at least 1¼ in. from the nearest edge of a framing member or furring strip, or be protected by a steel plate or sleeve.
Framing Member
Furring Strip
Furring Strip
COPYRIGHT 2005
Mike Holt Enterprises, Inc.

Figure 820–8

Cables run parallel to framing members or furring strips must be protected where they are likely to be penetrated by nails or screws, by installing the cables so they aren't less than 1¼ in. from the nearest edge of the framing member or furring strips, or by protecting the cable with a 1⁄16 in. thick steel plate [300.4(D)]. **Figure 820-8**

Raceways that contain coaxial cables must be securely fastened in place. Ceiling-support wires or the ceiling grid must not be used to support raceways or cables [300.11]. **Figure 820-9**

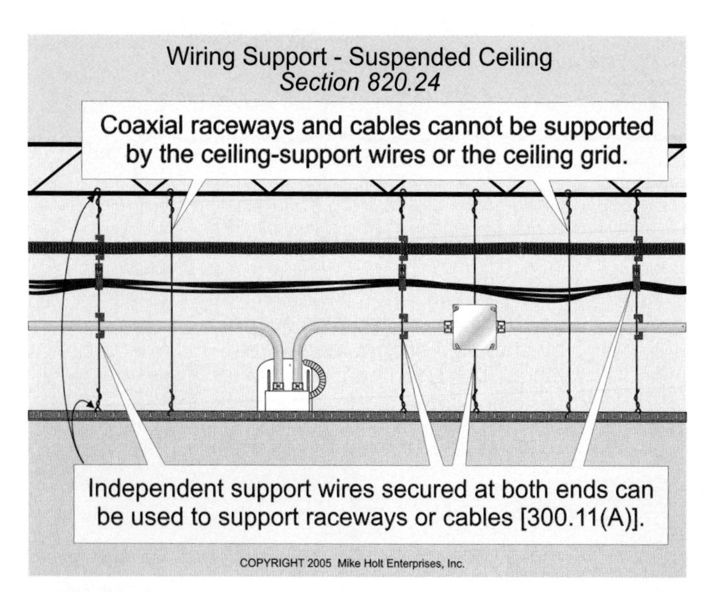
Wiring Support - Suspended Ceiling
Section 820.24
Coaxial raceways and cables cannot be supported by the ceiling-support wires or the ceiling grid.
Independent support wires secured at both ends can be used to support raceways or cables [300.11(A)].
COPYRIGHT 2005 Mike Holt Enterprises, Inc.

Figure 820–9

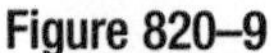

Author's Comment: Raceways and cables can be supported by independent support wires attached to the suspended ceiling in accordance with 300.11(A).

PART II. CABLES OUTSIDE AND ENTERING BUILDINGS

820.44 Overhead Cables.

(C) On Masts. Aerial coaxial cables must not be attached to a mast that encloses or supports power and lighting conductors. **Figure 820-10**

(F) On Buildings.

(1) Electric Light and Power. The coaxial cable must have a separation not less than 4 in. from electric light and power conductors.

Author's Comment: CATV coaxial cables must have a height not less than 10 ft above swimming and wading pools, diving structures and observation stands, or towers or platforms [680.8(B)]. **Figure 820-11**

(3) Lightning Conductors. Where practicable, a separation not less than 6 ft must be maintained between coaxial cables and lightning protection conductors.

820.47 Underground Circuits Entering Buildings.

(A) Underground Systems. Underground coaxial cables in a pedestal or handhole enclosure must be in a section permanently separated from exposed electric light, power, or Class 1 circuit conductors by a suitable barrier.

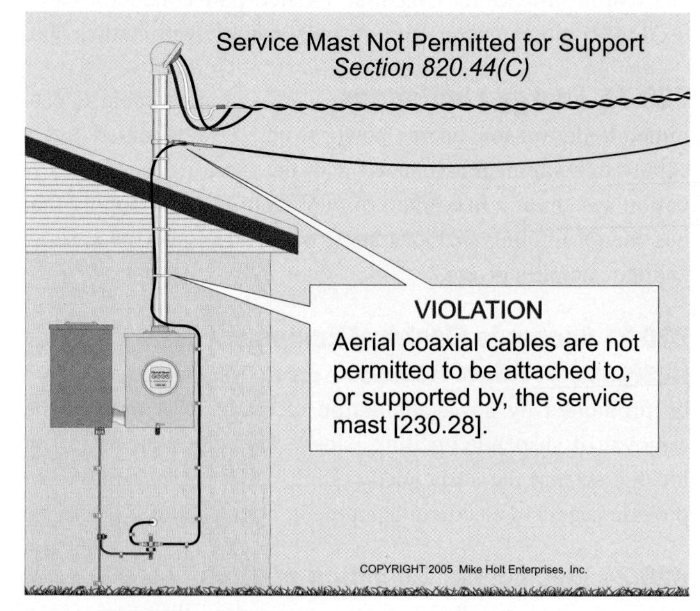
Service Mast Not Permitted for Support
Section 820.44(C)
VIOLATION
Aerial coaxial cables are not permitted to be attached to, or supported by, the service mast [230.28].
COPYRIGHT 2005 Mike Holt Enterprises, Inc.

Figure 820–10

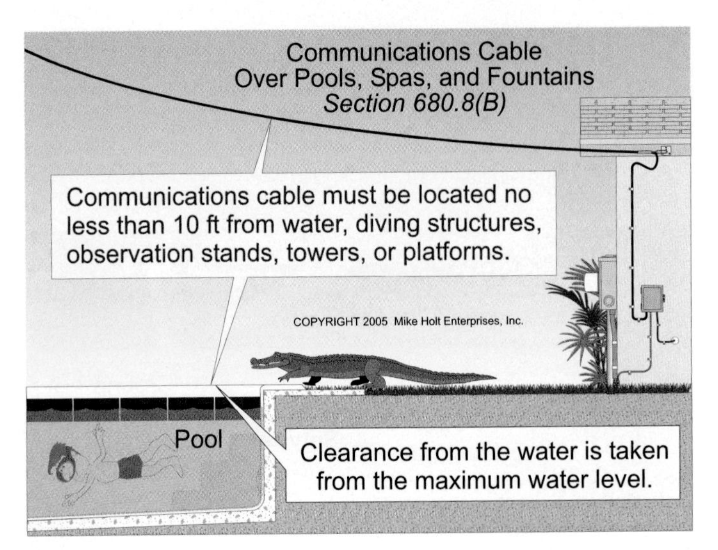

Communications Cable
Over Pools, Spas, and Fountains
Section 680.8(B)
Communications cable must be located no less than 10 ft from water, diving structures, observation stands, towers, or platforms.
COPYRIGHT 2005 Mike Holt Enterprises, Inc.
Pool
Clearance from the water is taken from the maximum water level.

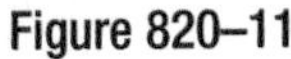

Figure 820–11

(B) **Direct-Buried Cables and Raceways.** Direct-buried coaxial cable must be separated by not less than 12 in. from underground light, power, or Class 1 circuit conductors. **Figure 820-12**

Exception 1: Underground coaxial cables need not be separated from service conductors if the service conductors or coaxial cables are installed in a raceway or cable armor.

Exception 2: Underground coaxial cables need not be separated from feeder or branch-circuit power conductors, if the power conductors are installed in a raceway or in metal-sheathed, metal-clad, UF, or USE cables, or the coaxial cables have metal cable armor or are installed in a raceway.

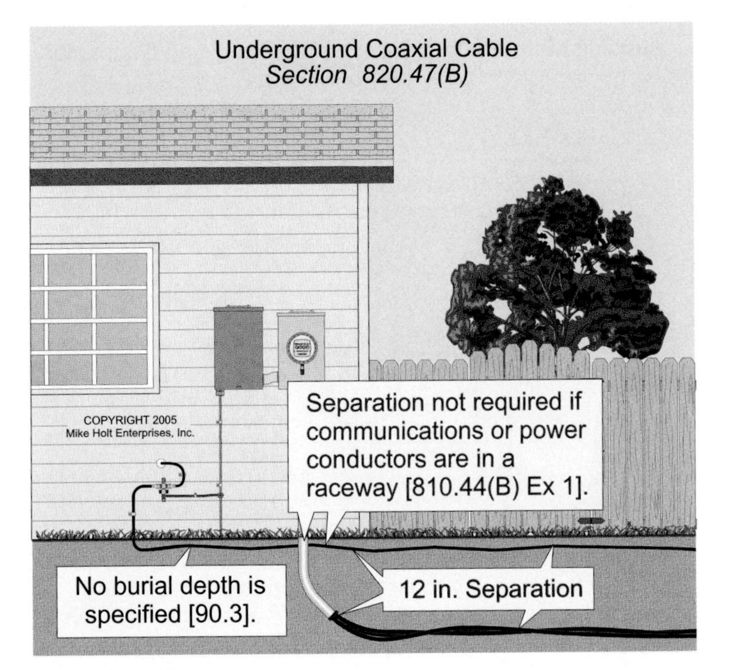

Figure 820–12

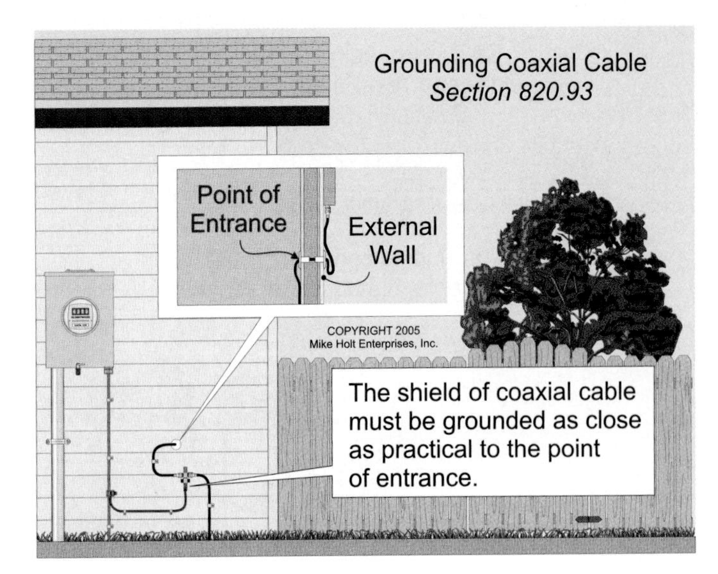

Figure 820–13

PART III. PROTECTION

820.93 Grounding Cable. The outer metallic sheath of coaxial cable must be grounded as close as practicable to the point of entrance to the building or structure in accordance with 820.100. **Figure 820-13**

Mobile Home. The outer metallic sheath of coaxial cable must be grounded at the mobile home disconnect in accordance with 820.100.

> **FPN:** Limiting the length of the grounding conductor helps limit damage to equipment because of a potential (voltage) difference between CATV and other systems during lightning events.

PART IV. GROUNDING METHODS

820.100 Cable Grounding. The shield of the coaxial cable must be grounded in accordance with (A) through (D) where required to be grounded by 820.93.

(A) Grounding Conductor.

(1) Insulation. The grounding conductor must be insulated and must be listed as suitable for the purpose.

(2) Material. The grounding conductor must be copper or other corrosion-resistant conductive material, stranded or solid.

(3) Size. The grounding conductor must not be smaller than 14 AWG, and is not required to be larger than 6 AWG. It must have a current-carrying capacity equal to the outer conductor of the coaxial cable.

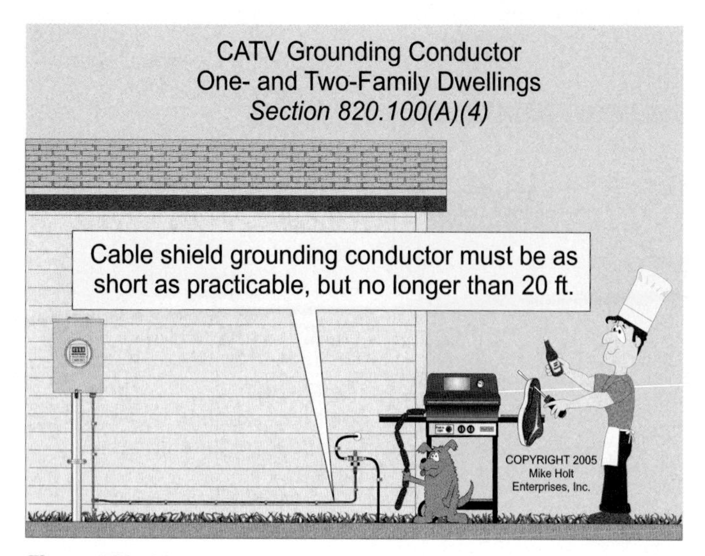
CATV Grounding Conductor
One- and Two-Family Dwellings
Section 820.100(A)(4)
Cable shield grounding conductor must be as short as practicable, but no longer than 20 ft.
COPYRIGHT 2005
Mike Holt
Enterprises, Inc.

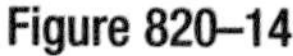

Figure 820–14

(4) Length. The grounding conductor must be as short as practicable.

In one- and two-family dwellings, the grounding conductor must not exceed 20 ft. **Figure 820-14**

> **FPN:** Limiting the length of the grounding conductor will help to reduce voltages between the building's power and CATV systems during lightning events.

Exception: Where it isn't practicable to limit the coaxial grounding conductor to 20 ft for one- and two-family dwellings, a separate ground rod not less than 8 ft long [820.100(B)(2)], with fittings suitable for the application [820.100(C)], must be installed. The additional ground rod must be bonded to the power grounding electrode system with a minimum 6 AWG conductor [820.100(D)]. **Figure 820-15**

(5) Run in Straight Line. The grounding conductor to the electrode must be run in as straight a line as practicable.

> **Author's Comment:** Lightning doesn't like to travel around corners or through loops, which is why the grounding conductor should be run as straight as practicable.

(6) Protection Against Physical Damage. The grounding conductor must be guarded from physical damage and where run in a metal raceway, both ends of the raceway must be bonded to the grounding conductor. **Figure 820-16**

> **Author's Comment:** Installing the grounding conductor in a nonmetallic raceway, when the authority having jurisdiction judges that physical protection is not required, is a better practice.

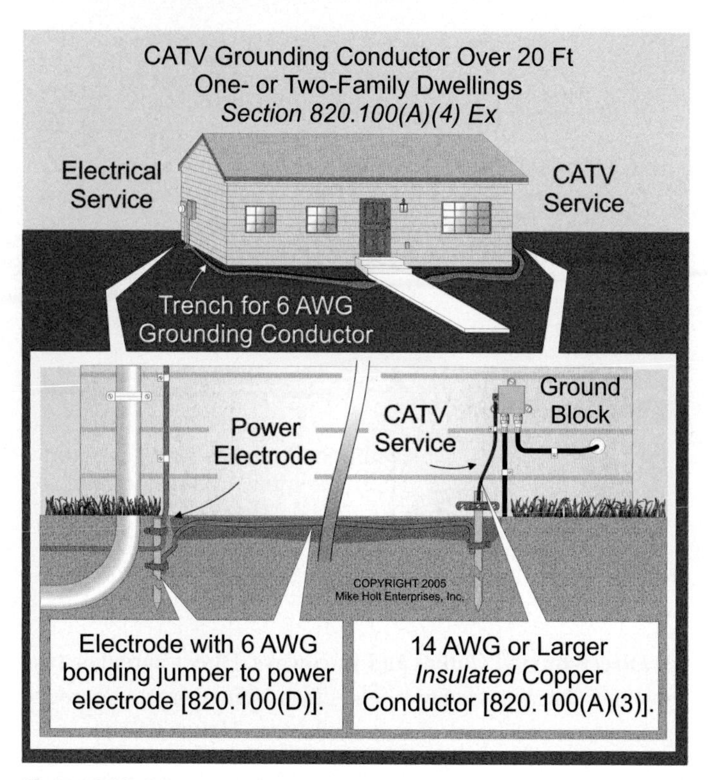
CATV Grounding Conductor Over 20 Ft
One- or Two-Family Dwellings
Section 820.100(A)(4) Ex
Electrical Service
CATV Service
Trench for 6 AWG Grounding Conductor
Power Electrode
CATV Service
Ground Block
COPYRIGHT 2005
Mike Holt Enterprises, Inc.
Electrode with 6 AWG bonding jumper to power electrode [820.100(D)].
14 AWG or Larger *Insulated* Copper Conductor [820.100(A)(3)].

Figure 820–15

(B) Electrode. The grounding conductor must be connected in accordance with 820.100(B)(1) through (B)(2).

(1) In Buildings or Structures With Grounding Means. The grounding conductor must terminate at the nearest accessible: **Figure 820-17**

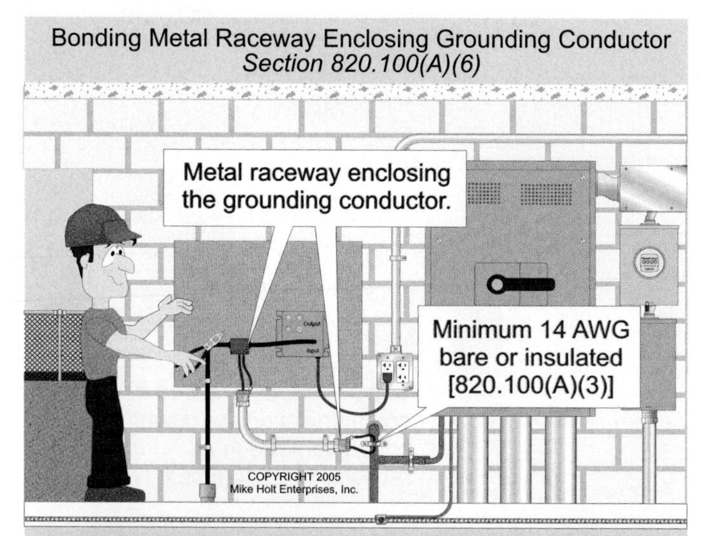
Bonding Metal Raceway Enclosing Grounding Conductor
Section 820.100(A)(6)
Metal raceway enclosing the grounding conductor.
Minimum 14 AWG bare or insulated [820.100(A)(3)]
COPYRIGHT 2005
Mike Holt Enterprises, Inc.
Where a grounding conductor is in a metal raceway, both ends of the raceway must be bonded to the grounding conductor, or to the same terminal or electrode to which the grounding conductor is connected.

Figure 820–16

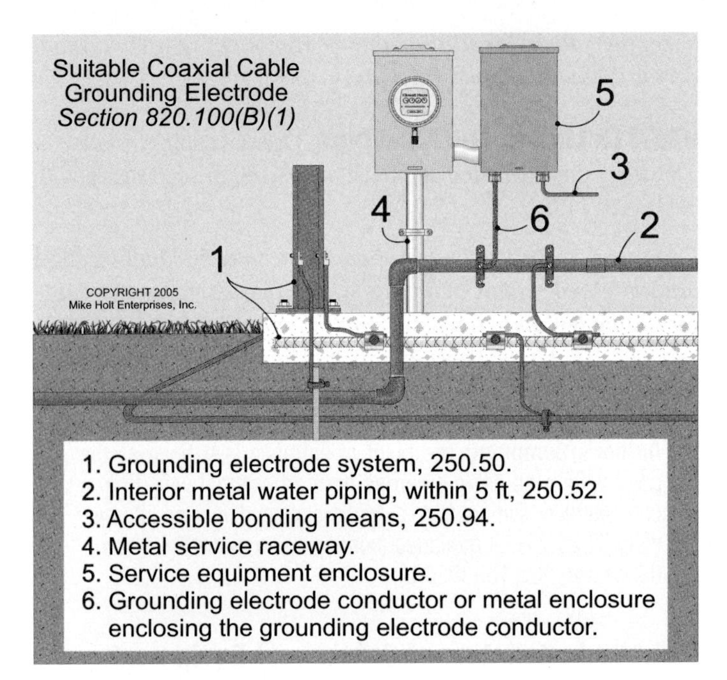

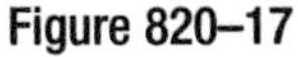
Figure 820–17

(1) Building or structure grounding electrode system [250.50].

(2) Interior metal water piping system, within 5 ft from its point of entrance [250.52(A)(1)].

(3) Accessible service bonding means [250.94].

(4) Metallic service raceway.

(5) Service equipment enclosure.

(6) Grounding electrode conductor or the grounding electrode conductor metal enclosure.

(7) The grounding electrode conductor or the grounding electrode of a building or structure disconnecting means that is grounded in accordance with 250.32.

(2) In Buildings or Structures Without Grounding Means. If the building or structure has no grounding electrode, such as a building supplied by a single branch circuit [250.32(A)(1) Ex], then a ground rod not less than 8 ft long and ½ in. in diameter must be installed to ground the CATV coaxial cable [820.100(B)(2)(2)].

(C) Electrode Connection. Terminations to the electrode must be by exothermic welding, listed lug, listed pressure connector, or clamp. Grounding fittings that are concrete-encased or buried in the earth must be listed for direct burial and marked "DB" [250.70]. **Figure 820-18**

(D) Bonding of Electrodes. If a separate grounding electrode, such as a ground rod, is installed for the CATV system, it must be bonded to the building's power grounding electrode system with a minimum 6 AWG conductor. **Figure 820-19**

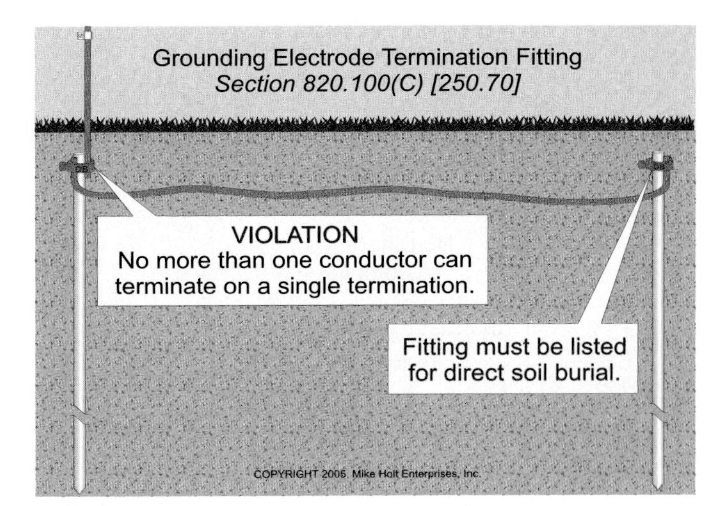

Figure 820–18

FPN No. 2: Bonding all systems to the same single point (intersystem bonding) helps reduce differences in potential between the systems during lightning events. **Figure 820-20**

PART V. CABLES WITHIN BUILDINGS

820.110 Raceways for Coaxial Cables. Where coaxial cables are installed in a Chapter 3 raceway, the raceway must comply with the Chapter 3 installation requirements for the wiring method. **Figure 820-21**

Where coaxial cables are installed in nonmetallic raceways listed for coaxial cables, they must comply with 820.182(A), (B), or (C) and they must be installed in accordance with 362.24 through 362.56.

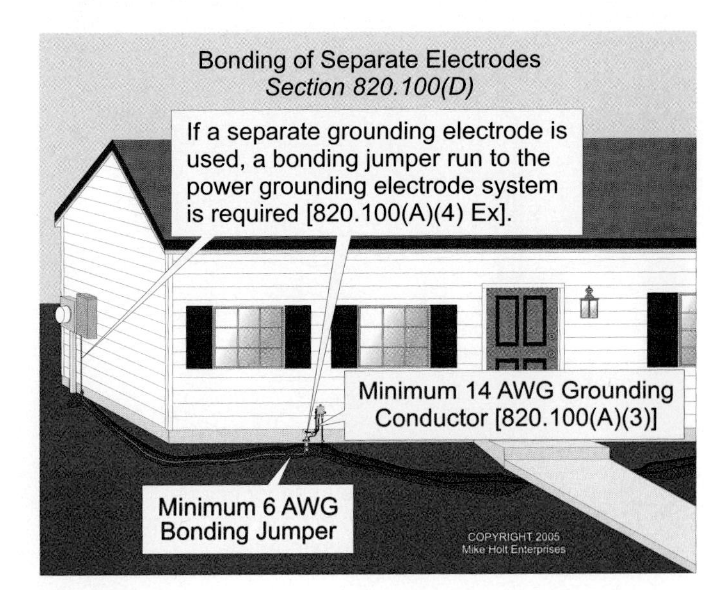

Figure 820–19

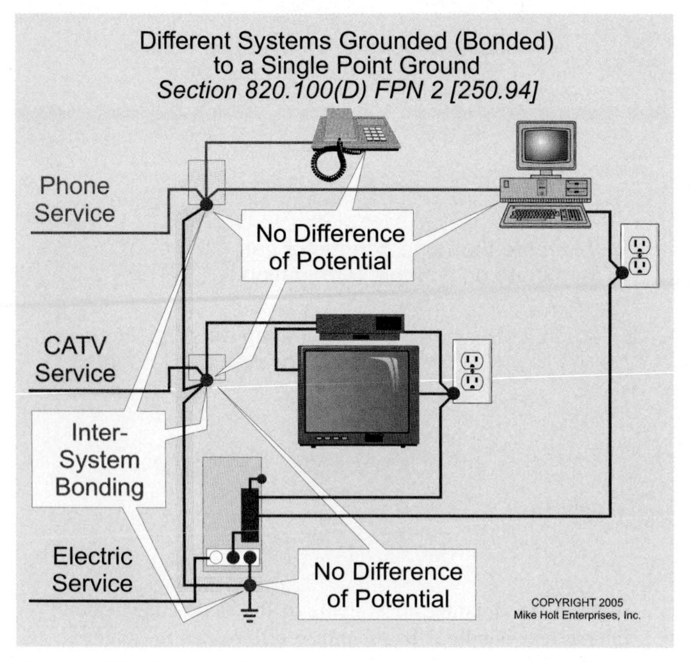

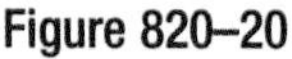
Figure 820–20

Author's Comment:

- 362.24 Bending radius.
- 362.26 Maximum total bends between pull points, 360 degrees.
- 362.28 Trimmed to remove rough edges.
- 362.30 Support every 3 ft, within 3 ft of any enclosure.
- 362.48 Joints between tubing, fittings, and boxes.

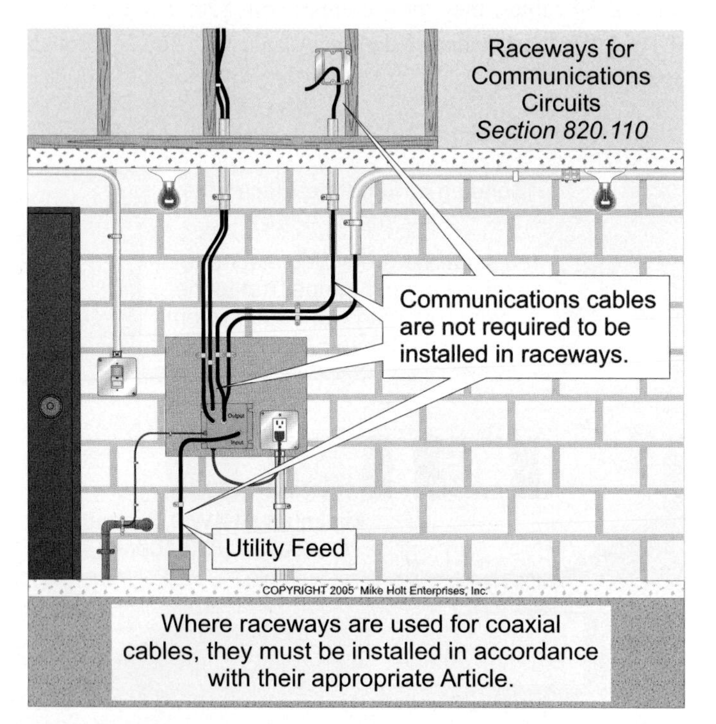

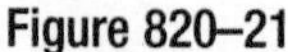
Figure 820–21

Exception: Raceway fill limitations do not apply when coaxial cables are installed in a raceway.

820.113 Listing and Markings. Coaxial cables installed in a building must be listed and must be marked in accordance with Table 820.113.

Exception 2: Unlisted outside coaxial communications cables are permitted within buildings where the length of the cable within the building, from its point of entrance [820.2], doesn't exceed 50 ft and the cable enters from outside and terminates at a grounding block.

Author's Comment: The point of entrance is defined as the point where the cable emerges from an external wall, from a concrete floor slab, or from a rigid metal conduit or an intermediate metal conduit grounded (bonded) to an electrode in accordance with 820.100 [820.2].

820.133 Installation of Cables and Equipment.

(A) Separation from Other Conductors.

(1) In Raceways and Boxes.

(1) With Other Cables in Raceway or Enclosure. Coaxial cables can be in the same raceway or enclosure as jacketed cables of any of the following: **Figure 820-22**

(a) Class 2 and Class 3 circuits in compliance with Article 725.

(b) Power-limited fire alarm circuits in compliance with Article 760.

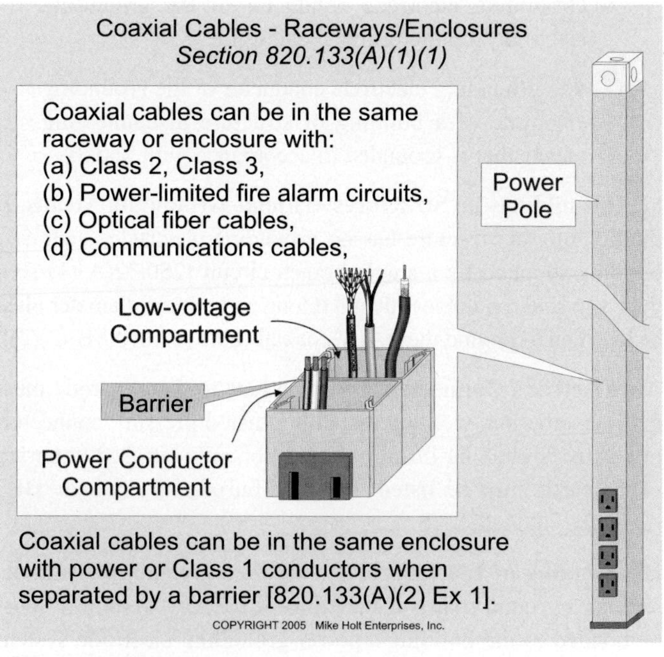

Coaxial Cables - Raceways/Enclosures
Section 820.133(A)(1)(1)
Coaxial cables can be in the same
raceway or enclosure with:
(a) Class 2, Class 3,
(b) Power-limited fire alarm circuits,
(c) Optical fiber cables,
(d) Communications cables,
Power
Pole
Low-voltage
Compartment
Barrier
Power Conductor
Compartment
Coaxial cables can be in the same enclosure
with power or Class 1 conductors when
separated by a barrier [820.133(A)(2) Ex 1].
COPYRIGHT 2005 Mike Holt Enterprises, Inc.

Figure 820–22

(c) Nonconductive and conductive optical fiber cables in compliance with Article 770

(d) Communications circuits in compliance with Article 800.

(2) With Power Conductors in Same Raceway or Enclosure. Coaxial cable must not be in any raceway or enclosure with conductors of electric light, power, Class 1, or nonpower-limited fire alarm circuits.

Exception 1: Coaxial cables are permitted with conductors of electric light, power, Class 1, and nonpower-limited fire alarm circuits, where separated by a permanent barrier or listed divider. **Figure 820-23**

Author's Comment: Separation is required to prevent a fire or shock hazard that could occur from a short between the higher-voltage circuits and the coaxial cable.

Exception 2: Coaxial cables can be mixed in enclosures other than raceways or cables with power conductors if the power circuit conductors are only introduced to supply power to coaxial cable system distribution equipment. The power circuit conductors must maintain a minimum of ¼ in. separation from the communications circuit conductors.

(2) Other Applications. Coaxial cables must maintain 2 in. of separation from electric light, power, Class 1, or nonpower-limited fire alarm circuit conductors.

Exception 1: Separation by distance isn't required if all electric light, power, Class 1, or nonpower-limited fire alarm circuit conductors are in a raceway or in metal-sheathed, metal-clad, nonmetallic-sheathed, or underground feeder cables, or all coaxial cables are in a raceway. **Figure 800-24**

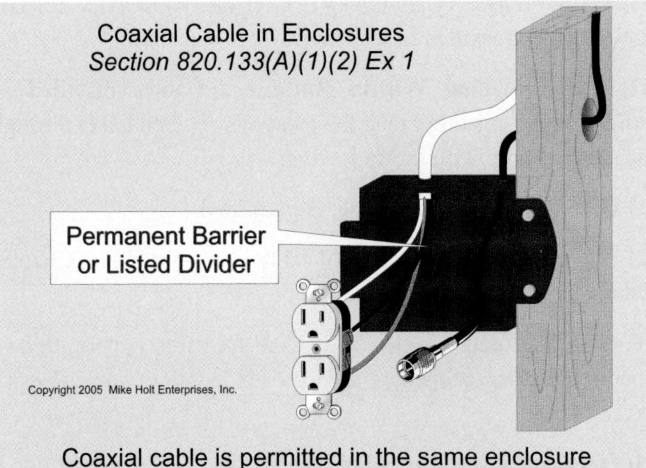

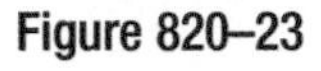
Figure 820–23

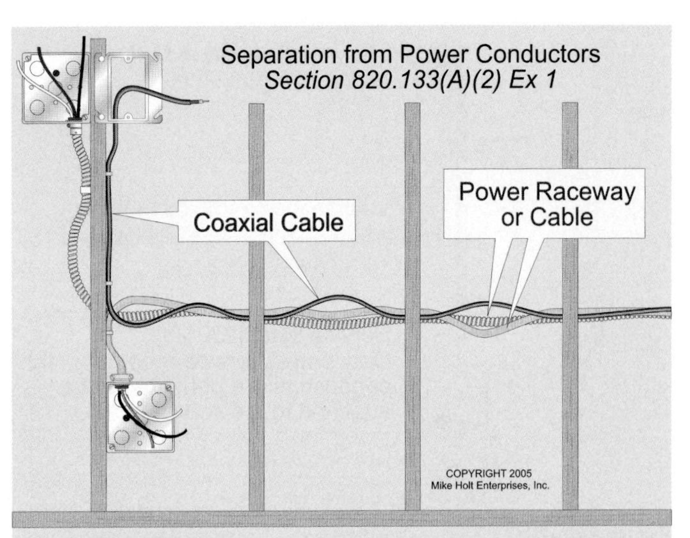

Figure 820–24

(C) Support of Cables. Coaxial cables must not be supported by a raceway. **Figure 800-25**

Author's Comment: Exposed cables must be supported by the structural components of the building so that the cable will not be damaged by normal building use. Cables must be secured by straps, staples, hangers, or similar fittings designed and installed so as not to damage the cable [820.24].

Exception: Overhead (aerial) spans of coaxial cables can be attached to a raceway-type mast intended for the attachment and support of such conductors. **Figure 800-26**

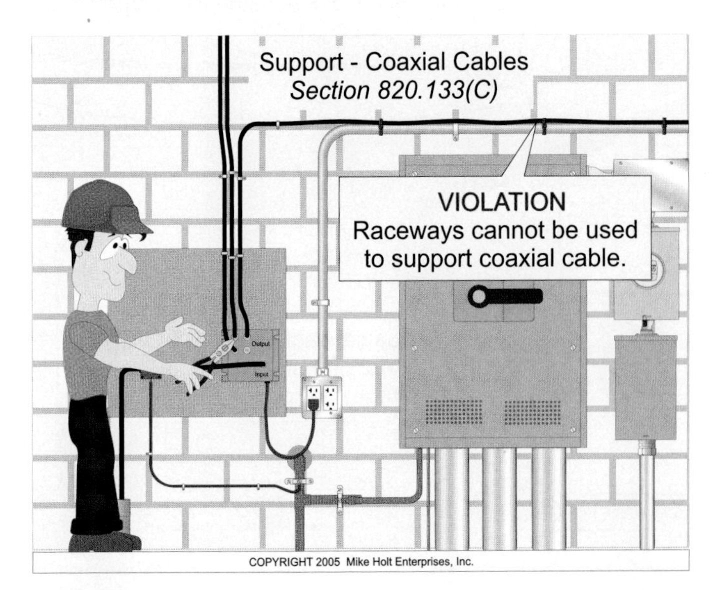

Support - Coaxial Cables
Section 820.133(C)
VIOLATION
Raceways cannot be used
to support coaxial cable.
COPYRIGHT 2005 Mike Holt Enterprises, Inc.

Figure 820–25

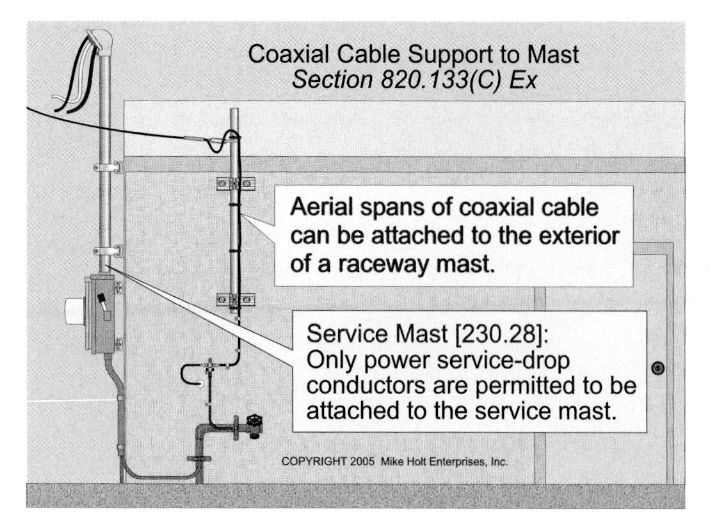

Figure 820–26

> **Author's Comment:** Coaxial cables must not be supported by, or attached to, the power service mast [230.28 and 820.44(C)]. **Figure 800-27**

820.154 Applications of Listed CATV Cables and Raceways. CATV cables must comply with (A) through (D), or where cable substitutions are made, in accordance with Table 820.154.

(A) Ducts or Plenums Used for Environmental Air. Coaxial cables must not be installed in ducts or plenums [820.3(B)].

Where necessary for the direct action upon, or sensing of the contained air, communications cables can be installed in ducts or plenums if they are installed in electrical metallic tubing, intermediate metal conduit, or rigid metal conduit as required by 300.22(B).

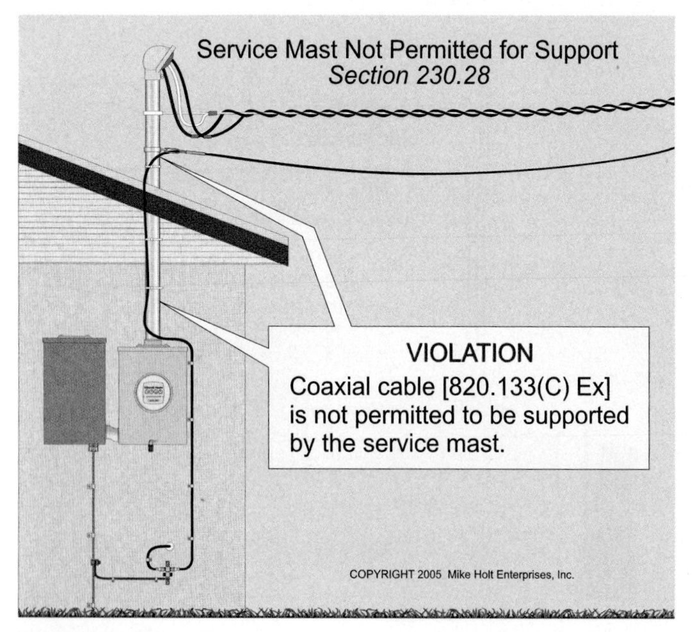

Figure 820–27

Plenum-rated Type CATVP cables [820.179(A)] can be installed above a suspended ceiling or below a raised floor that is used for environmental air [300.22(C)].

> **Author's Comment:** Nonplenum-rated cables can be installed above a suspended ceiling or below a raised floor that is used for environmental air, but only if the nonplenum-rated cable is installed within electrical metallic tubing [300.22(C)(1)].

(B) Riser Space. Coaxial cables installed in risers must comply with (1) through (3).

(1) Cables in Vertical Runs. Coaxial cables run vertically and penetrating more than one floor must be Type CATVR [820.179(B)] or Type CATVP [820.179(A)].

(2) Metal Raceways. Listed coaxial cable Types CATV and CATVX are permitted within a metal raceway.

Author's Comments:

- When coaxial cables are installed in a metal raceway, they aren't required to be riser or plenum rated.
- Metal raceways containing circuit conductors from power-supply systems that operate at 50V or less aren't required to be grounded (bonded) to an effective ground-fault current path [250.86 and 250.112(I)].

(3) One- and Two-Family Dwellings. Types CATV [820.179(C)] and CATVX [820.179(D)] cables can be installed in one- and two-family dwellings.

> **FPN:** See 820.3(A) for the fire-stop requirements for floor penetrations.

(C) Cable Trays. Types CATVP, CATVR, and CATV coaxial cables can be installed in a cable tray.

(D) Other Wiring Within Buildings. Cables installed in building locations other than the locations covered in (1) through (5) must comply with the following:

(1) General. Type CATV cable is permitted.

(2) In Raceways. Type CATVX cable is permitted in a Chapter 3 raceway.

(3) Nonconcealed Spaces. Type CATVX cable is permitted in nonconcealed spaces where the exposed length of cable doesn't exceed 10 ft.

(4) One- and Two-Family Dwellings. Type CATVX cables that are less than ⅜ in. in diameter can be installed in one- or two-family dwellings.

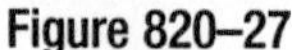

(5) Multifamily Dwellings. Type CATVX cables that are less than ⅜ in. in diameter can be installed in nonconcealed spaces in multifamily dwellings.

PART VI. LISTING REQUIREMENTS

820.179 Coaxial Cables. Coaxial cables must be listed in accordance with (A) through (D).

(A) Plenum CATV Cable. Type CATVP cable must be listed as being suitable for use in ducts, plenums, and other environmental air spaces and must also be listed as having adequate fire-resistant and low smoke-producing characteristics [820.154(A)]. **Figure 820-28**

> **Author's Comment:** Special consideration must be given to cables installed in areas that move or transport environmental air in order to reduce the hazards that arise from the burning of conductor insulation, the jackets of cables, and nonmetallic raceways. Because listed plenum-rated cables have adequate fire-resistant and low smoke-producing characteristics, they can be installed in environmental air spaces, but they cannot be installed in ducts or plenums! See 820.3(B) in this textbook for details.

(B) Type CATVR. Type CATVR cable must be listed as being suitable for use in a vertical run in a shaft, or from floor to floor, and have fire-resistant characteristics capable of preventing the carrying of fire from floor to floor [820.154(B) and 820.154(B)(1)].

(C) Type CATV. Type CATV cable must be listed as being suitable for general-purpose use, with the exception of risers, plenums, and other environmental air spaces [820.154(D)(1)].

(D) Type CATVX. Type CATVX limited-use community antenna television cable must be listed as being suitable for use in dwellings and for use in raceways [820-61(D)(3)].

820.182 CATV Raceways. Nonmetallic CATV raceways must be listed in accordance with (A) through (C).

(A) Plenum Coaxial Raceways. Plenum-rated raceways for coaxial cables must be listed as being suitable for use in ducts, plenums, or other environmental air spaces. **Figure 820-29**

> **Author's Comment:** Where used in environmental air spaces, listed plenum-rated communications raceways must contain Type CATVP cable [820.179(A)].

(B) Riser Coaxial Raceway. Riser-rated raceways for coaxial cables must be listed as being suitable to be run vertically through more than one floor.

> **Author's Comment:** Listed riser-rated coaxial raceways run vertically and penetrating more than one floor can only contain CATVR and CATVP cables [820.179(B)].

(C) General-Purpose Coaxial Raceway. General-purpose raceways for coaxial cables can contain CATV, CATVX, CATVR, or CATVP cables.

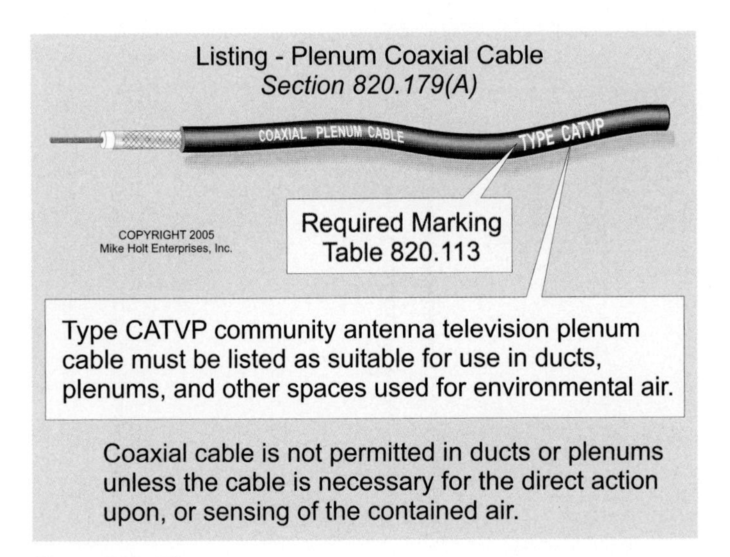
Listing - Plenum Coaxial Cable
Section 820.179(A)
COAXIAL PLENUM CABLE
TYPE CATVP
Required Marking
Table 820.113
COPYRIGHT 2005
Mike Holt Enterprises, Inc.
Type CATVP community antenna television plenum cable must be listed as suitable for use in ducts, plenums, and other spaces used for environmental air.
Coaxial cable is not permitted in ducts or plenums unless the cable is necessary for the direct action upon, or sensing of the contained air.

Figure 820–28

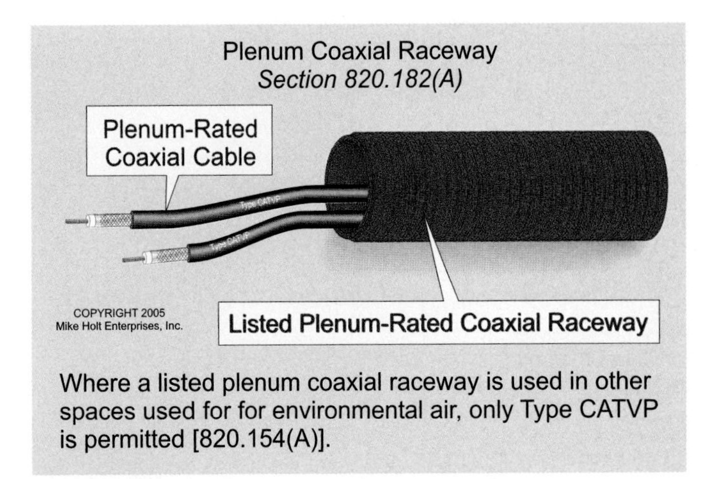
Plenum Coaxial Raceway
Section 820.182(A)
Plenum-Rated
Coaxial Cable
COPYRIGHT 2005
Mike Holt Enterprises, Inc.
Listed Plenum-Rated Coaxial Raceway
Where a listed plenum coaxial raceway is used in other spaces used for for environmental air, only Type CATVP is permitted [820.154(A)].

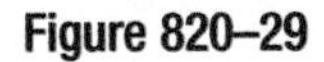

Figure 820–29

Article 820 Questions

1. CATV cable not terminated at equipment and not identified for future use with a tag is considered abandoned.

   (a) True
   (b) False

2. CATV cables installed _____ on the surface of ceilings and sidewalls must be supported by the building structure in such a manner that the cable will not be damaged by normal building use.

   (a) exposed
   (b) concealed
   (c) hidden
   (d) a and b

3. The conductor used to ground the outer cover of a coaxial cable must be _____.

   (a) insulated
   (b) 14 AWG minimum
   (c) bare
   (d) a and b

4. The grounding conductor for a CATV system must be connected to the nearest accessible location included on the list in 820.100(B)(1) when the building _____.

   (a) has a grounding means
   (b) is without a grounding means
   (c) has an emergency transfer switch
   (d) is wired using a metallic cable or raceway system

5. Coaxial cable is permitted to be placed in a raceway, compartment, outlet box, or junction box with the conductors of light or power circuits, or Class 1 circuits when _____.

   (a) installed in rigid metal conduit
   (b) separated by a permanent barrier
   (c) insulated
   (d) none of these

Network-Powered Broadband Communications Systems

Introduction

In previous Code cycles, Article 830 dealt with a very narrow part of the electrical industry. Broadband was not widely implemented. Today, however, broadband is becoming standard in hotel rooms, offices, and new homes. Thus, Article 830 is now mainstream. It's the last Article, but not the least, in the *NEC.*

A primary goal of Article 830 is to prevent the spread of combustion. Keeping this goal in mind will help you understand all the provisions and requirements of the Article.

As with the other Chapter 8 Articles, defining the point of entrance is critical, and certain requirements in the other Chapter 8 Articles also apply in Article 830. For example, you cannot leave the accessible portion of abandoned cables in place.

To get a clear handle on what constitutes a network-powered broadband system, refer to Table 830.4—it shows you which systems qualify for coverage by Article 830.

This Article consists of five parts. Part I provides the foundation for compliance with Article 830. It provides definitions, general rules, listing and marking requirements, and other basic information.

Part II specifically addresses cables outside buildings, with additional requirements if those cables enter buildings. Part V addresses wiring methods within buildings.

Part III addresses protection. To apply protection correctly, you must have clearly defined the point of entrance. Part IV provides the requirements for grounding methods. You will need to know this to correctly apply 830.33 (in Part II), but also need to know this for other purposes.

To further expand your ability to perform work related to Article 830, you may wish to take advantage of educational resources available from BICSI (www.bicsi.org). Here, you will be shown what you need to know to perform work that is *Code*-compliant.

PART I. GENERAL

830.1 Scope. Article 830 contains the installation requirements for network-powered broadband communications circuits that provide voice, audio, video, data, and interactive services through a network interface unit [820.2].

> **FPN No. 1:** A typical network-powered broadband system includes a cable supplying power and broadband signal to a network interface unit that converts the broadband signal into voice, audio, video, data, and interactive services signals. Typical cables are coaxial cable with both broadband signal and power on the center conductor, composite cable with a pair of conductors for power, and composite optical fiber cable with a pair of conductors for power. **Figure 830-1**

> **FPN No. 2:** The requirements contained in Article 830 apply to the wiring to the network interface unit at the premises when not under the exclusive control of the communications utility [90.2(B)(4)]. **Figure 830-2**

Author's Comment: Since the network-powered broadband circuit is typically under the exclusive control of the communications utility, there's no need to cover that subject in this textbook. However, the premises wiring (voice, audio, video, data, and interactive services signals) from the network interface unit must be installed in accordance with the following articles [830.3(D)]: **Figure 830-3**

- Article 725 – Class 2 and Class 3 Circuits
- Article 760 – Power-limited Fire Alarm Circuits

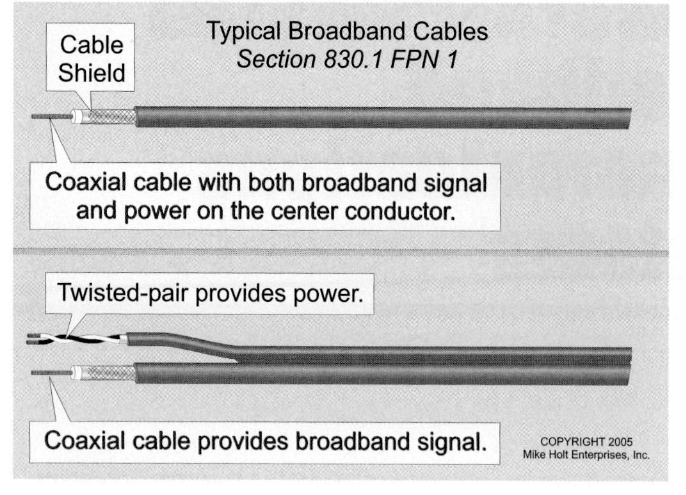

Figure 830–1

- Article 770 – Optical Fiber Cables
- Article 800 – Communications Circuits
- Article 820 – Community Antenna Television and Radio Distribution Systems

830.2 Definitions.

Network Interface Unit (NIU). A device that converts a broadband signal into component voice, audio, video, data, and interactive services signals. The NIU provides isolation between the network power and the premises signal circuits. **Figure 830-4**

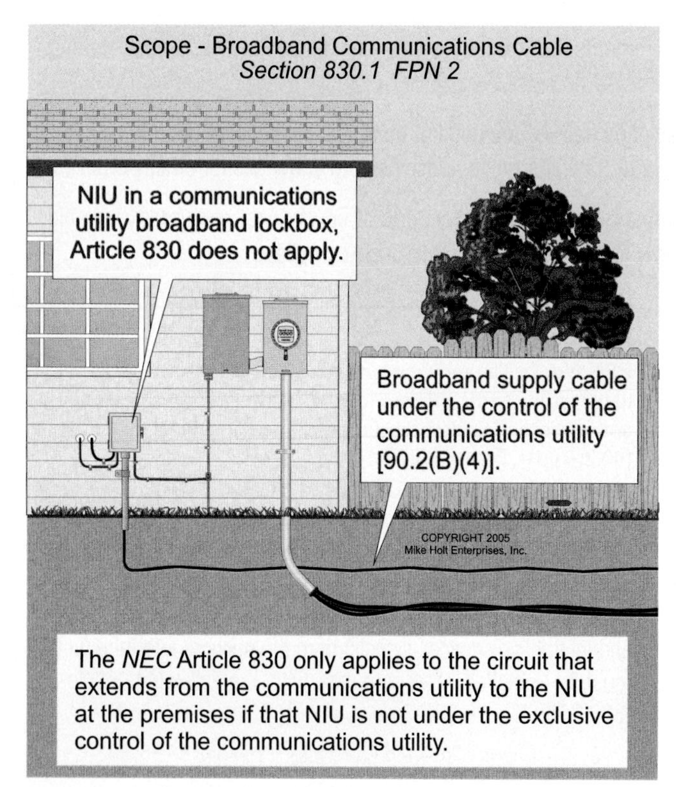

Scope - Broadband Communications Cable
Section 830.1 FPN 2
NIU in a communications utility broadband lockbox, Article 830 does not apply.
Broadband supply cable under the control of the communications utility [90.2(B)(4)].
COPYRIGHT 2005 Mike Holt Enterprises, Inc.
The NEC Article 830 only applies to the circuit that extends from the communications utility to the NIU at the premises if that NIU is not under the exclusive control of the communications utility.

Figure 830–2

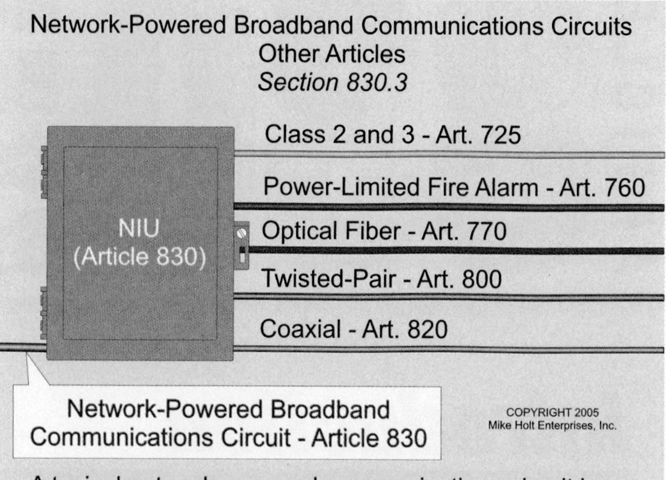

Figure 830–3

Network-Powered Broadband Communications Circuit. The circuit extending from the communications utility serving terminal or tap up to, and including, the network interface unit (NIU).

> **FPN:** A typical single-family network-powered communications circuit consists of a communications drop or communications service cable and an NIU.

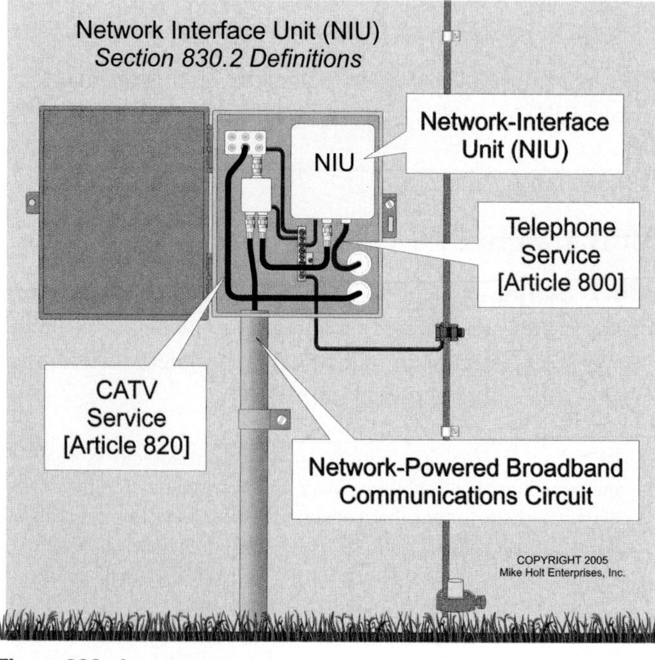

Figure 830–4

Article 830 Questions

1. Network-powered broadband cable not terminated at equipment and not identified for future use with a tag is considered abandoned.

   (a) True (b) False

Notes

CHAPTER 9

Tables—Conductor and Raceway Specifications

Introduction

As you look at how the *NEC* is organized, you can see that understanding the first four Chapters is critical. The next four Chapters apply to special occupancies, special equipment, special conditions, and communications systems—which means that understanding these may not be critical for you, unless you are working with systems they apply to.

Following those four Chapters are Chapter 9 (Tables) and the Annexes. It might be tempting to think that Chapter 9 is for use as a "just in case reference" and the Annexes are "just optional reading." Don't give in to that temptation. These resources can help you save calculation time, better understand *Code* requirements, and reduce errors in application. The key to realizing such advantages is taking the time to study Chapter 9 tables before you actually need the knowledge on the job.

Chapter 9 refers you to Annex C, which contains conduit and tubing fill tables for conductors and fixture wires of the same size. A question that commonly arises regarding this arrangement is, "Why doesn't the *NEC* include the Annex C tables in Chapter 9?" The answer is that Annex C provides tables "for information only," while Chapter 9 provides tables for meeting *Code* requirements. However, those Annex C tables can be very helpful.

If you are familiar with Chapter 9 and Annex C, you will be able to meet *NEC* requirements more efficiently, accurately, and confidently—whether on an exam or in the field.

- Table 1 contains the percent of conductor fill for sizing raceways based on common conditions where the length of the pull and the number of bends are within reasonable limits. For certain conditions, a larger size conduit or a lesser conduit fill should be considered [Table 1 FPN No. 1].
- Table 4 contains the dimensions and the percent area of raceways necessary in determining raceway fill or sizing.
- Table 5 contains the dimensions of conductors necessary in determining raceway fill or sizing.
- Table 8 contains the properties for bare conductors, such as the circular mil area, the cross-sectional area, as well as the dc resistance for dc voltage drop calculations.
- Table 9 contains the ac resistance for voltage drop calculations.

Table 1—Conductor Fill Percentage

The number and size of conductors in any raceway must not be more than will permit dissipation of the heat and ready installation or withdrawal of the conductors without damage to the conductors or to their insulation [300.17]. Where conductors are installed in conduit or tubing, the allowable conductor fill must not exceed that permitted by Table 1 of Chapter 9.

Table 1—Conductor Percent Fill

Number of Conductors	All Conductor Types
1	53%
2	31%
Over 2	40%

> **FPN No. 1:** The maximum percentage of allowable conductor fill is based on common conditions where the length of the conductor and number of raceway bends are within reasonable limits. **Figure 900-1**

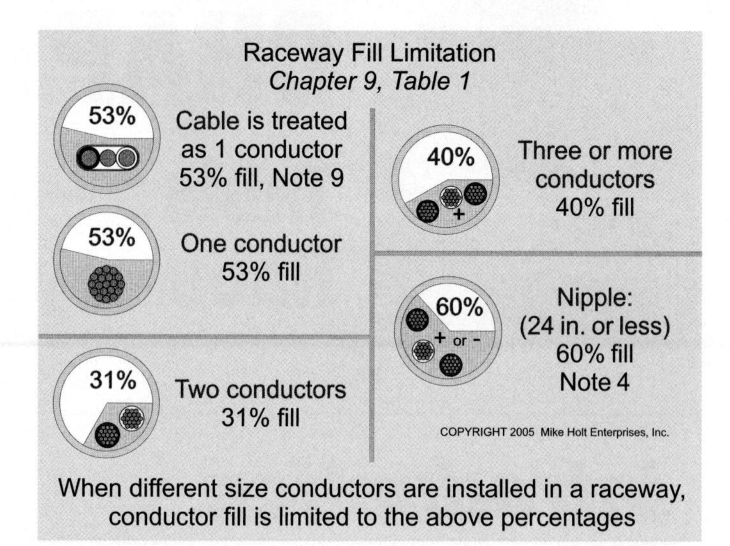

Figure 900–1

Table 1, Note 1—Conductors all the Same Size and Insulation. When all conductors are the same size and insulation, the number of conductors permitted in a raceway can be determined by simply looking at the tables located in Annex C, which is based on maximum percent fill as listed in Table 1 of Chapter 9.

Table 1, Note 3—Equipment Grounding (Bonding) Conductors. When equipment grounding (bonding) conductors are installed in a raceway, the actual area of the conductor must be used when calculating raceway fill. **Figure 900-2**

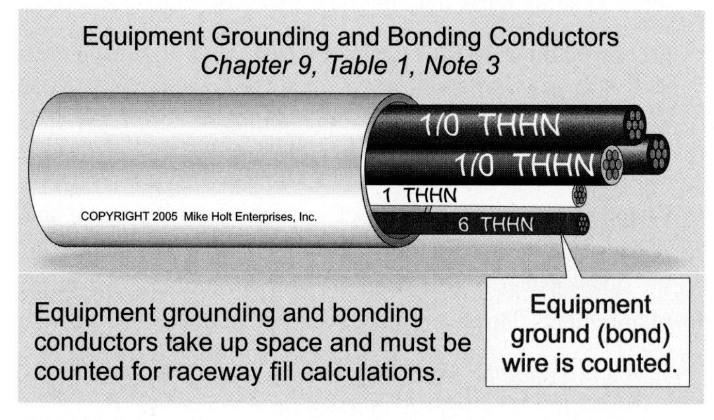

Figure 900–2

> **Author's Comment:** Chapter 9, Table 5 can be used to determine the cross-sectional area of insulated conductors and Chapter 9, Table 8 can be used to determine the cross-sectional area of bare conductors [Note 8 of Table 1, Chapter 9].

Table 1, Note 4—Nipples; Raceways not Exceeding 24 Inches. The cross-sectional areas of conduit and tubing can be found in Table 4 of Chapter 9. When a conduit or tubing raceway does not exceed 24 in. in length, it's called a nipple. Nipples are permitted to be filled to 60% of their total cross-sectional area. **Figure 900-3**

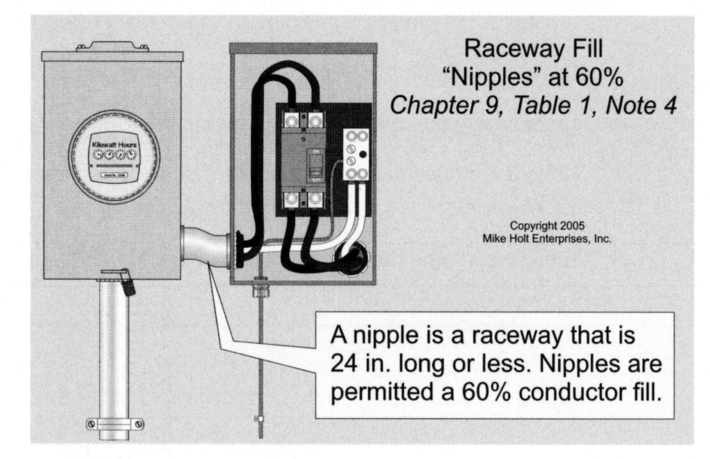

Raceway Fill
"Nipples" at 60%
Chapter 9, Table 1, Note 4
Copyright 2005
Mike Holt Enterprises, Inc.
A nipple is a raceway that is 24 in. long or less. Nipples are permitted a 60% conductor fill.

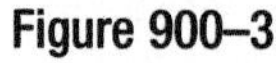

Figure 900–3

Table 1, Note 7—Rounding Up. When the calculated number of conductors (all of the same size and insulation) results in a decimal of 0.8 or larger, the next higher whole number can be used. But, be careful—this only applies when the conductors are all the same size (cross-sectional area including insulation) and for raceways over 24 in. in length.

Table 1, Note 8—Bare Conductors. The dimensions for bare conductors are listed in Table 8 of Chapter 9.

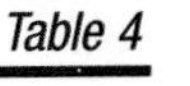

Table 4—Raceway Dimensions

Table 4 is primarily used to determine the cross-sectional area for conductor fill based on the number of conductors in accordance with Table 1 of Chapter 9.

Question: *What is the area permitted for conductor fill for trade size 2 EMT having a length of 20 inches?* **Figure 900-4**

(a) 2.067 sq in. *(b) 3.356 sq in.*
(c) 2.013 sq in. *(d) 1.342 sq in.*

Answer: (c) 2.013 sq in.

Question: *What is the area permitted for conductor fill for trade size 2 EMT having a length of 30 in. containing four conductors?*

(a) 2.067 sq in. *(b) 3.356 sq in.*
(c) 2.013 sq in. *(d) 1.342 sq in.*

Answer: (d) 1.342 sq in.

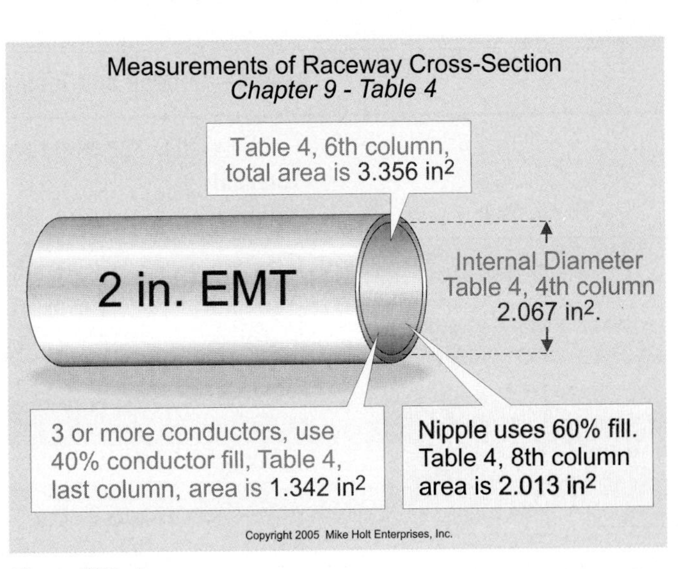

Figure 900–4

Table 4—Conduit and Tubing Cross-Sectional Area

Article 358 Electrical Metallic Tubing (EMT)

Trade Size	Internal Diameter	100% Area	60% Area Nipple	40% Area
½	0.622 in.	0.304 in.^{2}	0.182 in.^{2}	0.122 in.^{2}
¾	0.824 in.	0.533 in.^{2}	0.320 in.^{2}	0.213 in.^{2}
1	1.049 in.	0.864 in.^{2}	0.519 in.^{2}	0.346 in.^{2}
1¼	1.380 in.	1.496 in.^{2}	0.897 in.^{2}	0.598 in.^{2}
1½	1.610 in.	2.036 in.^{2}	1.221 in.^{2}	0.814 in.^{2}
2	2.067 in.	3.356 in.^{2}	2.013 in.^{2}	1.342 in.^{2}
2½	2.731 in.	5.858 in.^{2}	3.515 in.^{2}	2.343 in.^{2}
3	3.356 in.	8.846 in.^{2}	5.307 in.^{2}	3.538 in.^{2}
3½	3.834 in.	11.545 in.^{2}	6.927 in.^{2}	4.618 in.^{2}
4	4.334 in.	14.753 in.^{2}	8.852 in.^{2}	5.901 in.^{2}

Table 4 continued...

Table 4—Conduit and Tubing Cross-Sectional Area (continued)

Article 352 Rigid Nonmetallic Conduit (RNC)				
Trade Size	Internal Diameter	100% Area	60% Area Nipple	40% Area
½	0.602 in.	0.285 in.^{2}	0.171 in.^{2}	0.114 in.^{2}
¾	0.804 in.	0.508 in.^{2}	0.305 in.^{2}	0.203 in.^{2}
1	1.029 in.	0.832 in.^{2}	0.499 in.^{2}	0.333 in.^{2}
1¼	1.36 in.	1.453 in.^{2}	0.872 in.^{2}	0.581 in.^{2}
1½	1.59 in.	1.986 in.^{2}	1.191 in.^{2}	0.794 in.^{2}
2	2.047 in.	3.291 in.^{2}	1.975 in.^{2}	1.316 in.^{2}
2½	2.445 in.	4.695 in.^{2}	2.817 in.^{2}	1.878 in.^{2}
3	3.042 in.	7.268 in.^{2}	4.361 in.^{2}	2.907 in.^{2}
3½	3.521 in.	9.737 in.^{2}	5.842 in.^{2}	3.895 in.^{2}
4	3.998 in.	12.554 in.^{2}	7.532 in.^{2}	5.022 in.^{2}
5	5.016 in.	19.761 in.^{2}	11.856 in.^{2}	7.904 in.^{2}
6	6.031 in.	28.567 in.^{2}	17.14 in.^{2}	11.427 in.^{2}

Table 5—Conductor Cross-Sectional Area

Table 5 of Chapter 9 lists the cross-sectional area of insulated conductors and fixture wires.

Question: *What is the cross-sectional area for one 14 THHN conductor?* **Figure 900-5**

(a) 0.0206 sq in. *(b) 0.0172 sq in.*
(c) 0.0097 sq in. *(d) 0.0278 sq in.*

Answer: (c) 0.0097 sq in.

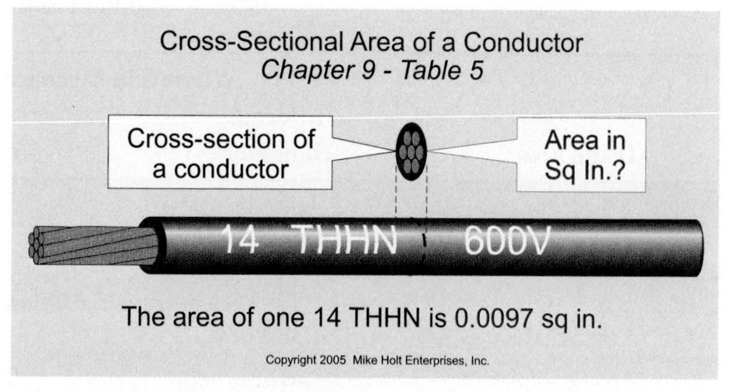

Figure 900–5

Table 5—Dimensions of Insulated Conductors and Fixture Wires

Dimensions of THHN/THWN Conductors			
Size	Area	Size	Area
14	0.0097 in.^{2}	1/0	0.1855 in.^{2}
12	0.0133 in.^{2}	2/0	0.2223 in.^{2}
10	0.0211 in.^{2}	3/0	0.2679 in.^{2}
8	0.0366 in.^{2}	4/0	0.3237 in.^{2}
6	0.0507 in.^{2}	250	0.397 in.^{2}
4	0.0824 in.^{2}	300	0.4608 in.^{2}
3	0.0973 in.^{2}	350	0.5242 in.^{2}
2	0.1158 in.^{2}	500	0.7073 in.^{2}
1	0.1562 in.^{2}	750	1.0496 in.^{2}

Table 8—Conductor Properties

Chapter 9, Table 8 contains conductor properties such as cross-sectional area in circular mils, number of strands per conductor, cross-sectional area in sq in. for bare conductors, and dc resistance at 75°C for both copper and aluminum conductors.

Question: *What is the cross-sectional area for one 10 AWG bare conductor with seven strands?* **Figure 900-6**

(a) 0.006 sq in. *(b) 0.011 sq in.*
(c) 0.038 sq in. *(d) a or b*

Answer: (b) 0.011 sq in.

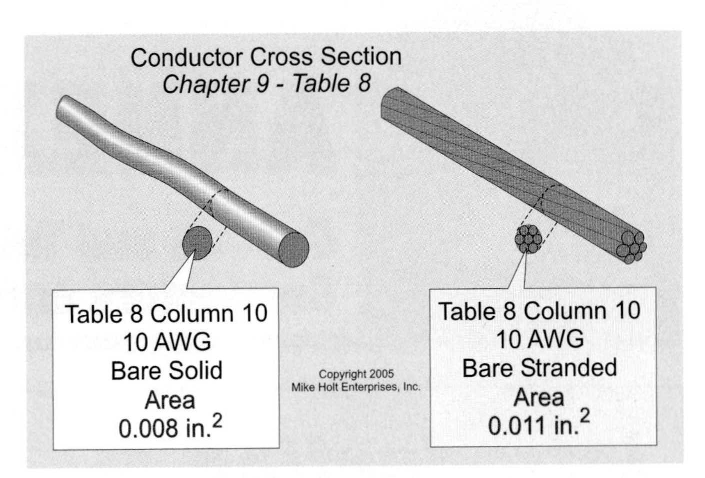

Figure 900–6

Table 9—AC Impedance for Conductors in Conduit or Tubing

Chapter 9, Table 9, contains the ac impedance for copper and aluminum conductors.

Table 8 and Table 9 Stranded Conductor Properties of Copper

Conductor Size	Circular Mils	in.^{2} Area	Conductor Resistance dc (R)	ac (Z)
14	4,110	0.004	3.14	3.10
12	6,530	0.006	1.98	2.00
10	10,380	0.011	1.24	1.20
8	16,510	0.017	0.778	0.78
6	26,240	0.027	0.491	0.49
4	41,740	0.042	0.308	0.31
3	52,620	0.053	0.245	0.25
2	66,360	0.067	0.194	0.20
1	83,690	0.087	0.154	0.16
1/0	105,600	0.109	0.122	0.12
2/0	133,100	0.137	0.0967	0.10
3/0	167,800	0.173	0.0766	0.079
4/0	211,600	0.219	0.0608	0.063
250	250,000	0.260	0.0515	0.054
300	300,000	0.312	0.0429	0.045
350	350,000	0.364	0.0367	0.039
400	400,000	0.416	0.0321	0.035
500	500,000	0.519	0.0258	0.029
600	600,000	0.626	0.0214	0.025
750	750,000	0.782	0.0171	0.021
1000	1,000,000	1.042	0.0129	0.018

ANNEX C

Conduit and Tubing Fill Tables for Conductors and Fixture Wires of the Same Size

Introduction

Annex C contains conduit and tubing fill tables for conductors and fixture wires of the same size (total cross-sectional area including insulation [Chapter 9 Table 1 Note 1]). Annex C provides tables "for information only," while Chapter 9 provides tables for meeting *Code* requirements. Remember that if the conductors are not all the same size, you will need to use the Chapter 9 tables. If the conductors are all the same size, Annex C will provide a way to more quickly find the correct raceway size.

Tables C.1 through C.12(A) are based on maximum percent fill as listed in Table 1 of Chapter 9.

- Table C.1—Conductors and fixture wires in electrical metallic tubing (EMT)
- Table C.1(A)—Compact conductors in electrical metallic tubing (EMT)
- Table C.2—Conductors and fixture wires in electrical nonmetallic tubing (ENT)
- Table C.2(A)—Compact conductors in electrical nonmetallic tubing (ENT)
- Table C.3—Conductors and fixture wires in flexible metal conduit (FMC)
- Table C.3(A)—Compact conductors in flexible metal conduit (FMC)
- Table C.4—Conductors and fixture wires in intermediate metal conduit (IMC)
- Table C.4(A)—Compact conductors in intermediate metal conduit (IMC)
- Table C.5—Conductors and fixture wires in liquidtight flexible nonmetallic conduit (gray type) (LFNC-B)
- Table C.5(A)—Compact conductors in liquidtight flexible nonmetallic conduit (gray type) (LFNC-B)
- Table C.6—Conductors and fixture wires in liquidtight flexible nonmetallic conduit (orange type) (LFNC-A)
- Table C.6(A)—Compact conductors in liquidtight flexible nonmetallic conduit (orange type) (LFNC-A)
- Table C.7—Conductors and fixture wires in liquidtight flexible metallic conduit (LFMC)
- Table C.7(A)—Compact conductors in liquidtight flexible metal conduit (LFMC)
- Table C.8—Conductors and fixture wires in rigid metal conduit (RMC)
- Table C.8(A)—Compact conductors in rigid metal conduit (RMC)
- Table C.9—Conductors and fixture wires in rigid nonmetallic conduit (RNC) Schedule 80
- Table C.9(A)—Compact conductors in rigid nonmetallic conduit (RNC) Schedule 80
- Table C.10—Conductors and fixture wires in rigid nonmetallic conduit (RNC) Schedule 40
- Table C.10(A)—Compact conductors in rigid nonmetallic conduit (RNC) Schedule 40
- Table C.11—Conductors and fixture wires in Type A, rigid PVC conduit
- Table C.11(A)—Compact conductors in Type A, rigid PVC conduit
- Table C.12—Conductors and fixture wires in Type EB, PVC conduit
- Table C.12(A)—Compact conductors in Type EB, PVC conduit

Question: *How many 3/0 THHN conductors can be installed in trade size 2 EMT?* **Figure 900-7**

(a) 3 *(b) 4* *(c) 5* *(d) 6*

Answer: (c) 5 conductors
Annex C, Table C.1

Question: *How many compact 1/0 THHN conductors can be installed in trade size 1¼ RNC?*

(a) 3 *(b) 4* *(c) 5* *(d) 6*

Answer: (b) 4 conductors
Annex C, Table C.10

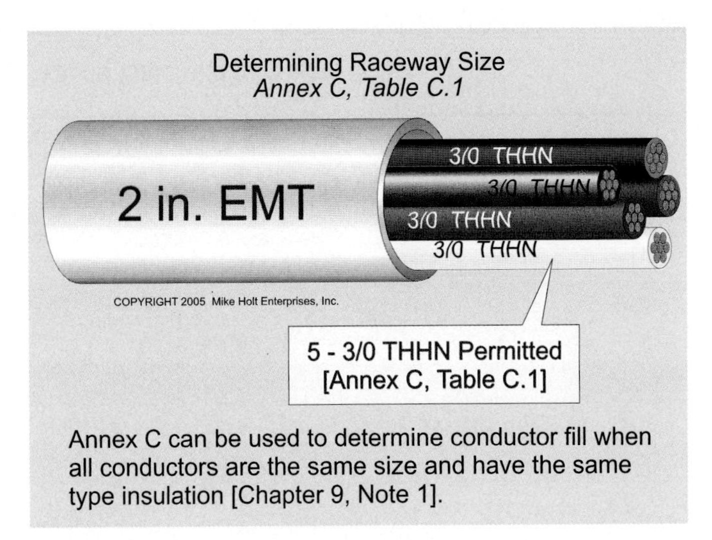

Figure 900–7

Table C.1 Maximum Number of THHN/THWN Conductors in EMT

Conductor	Raceway Trade Size									
Size	½	¾	1	1¼	1½	2	2½	3	3½	4
14	12									
12	9									
10	5	10								
8	3	6	9							
6		4	7	12						
4			4	7	10					
3			3	6	8					
2			3	5	7	11				
1				4	5	8				
1/0				3	4	7	12			
2/0					3	6	10			
3/0					3	5	8			
4/0						4	7	11		
250						3	6	9	11	
300						3	5	7	10	
350							4	6	9	11
400							4	6	8	10
500							3	5	6	8

Table C.10 Maximum Number of THHN/THWN Conductors in RNC

Conductor Size	Raceway Trade Size									
	½	¾	1	1¼	1½	2	2½	3	3½	4
14	11									
12	8									
10	5	9								
8	3	5	9							
6		4	6	11						
4			4	7	9					
3			3	6	8					
2			3	5	7	11				
1				3	5	8	12			
1/0				3	4	7	10			
2/0					3	6	8			
3/0					3	5	7	11		
4/0						4	6	9	12	
250						3	4	7	10	12
300						3	4	6	8	11
350							3	5	7	9
400							3	5	6	8
500							3	4	5	5

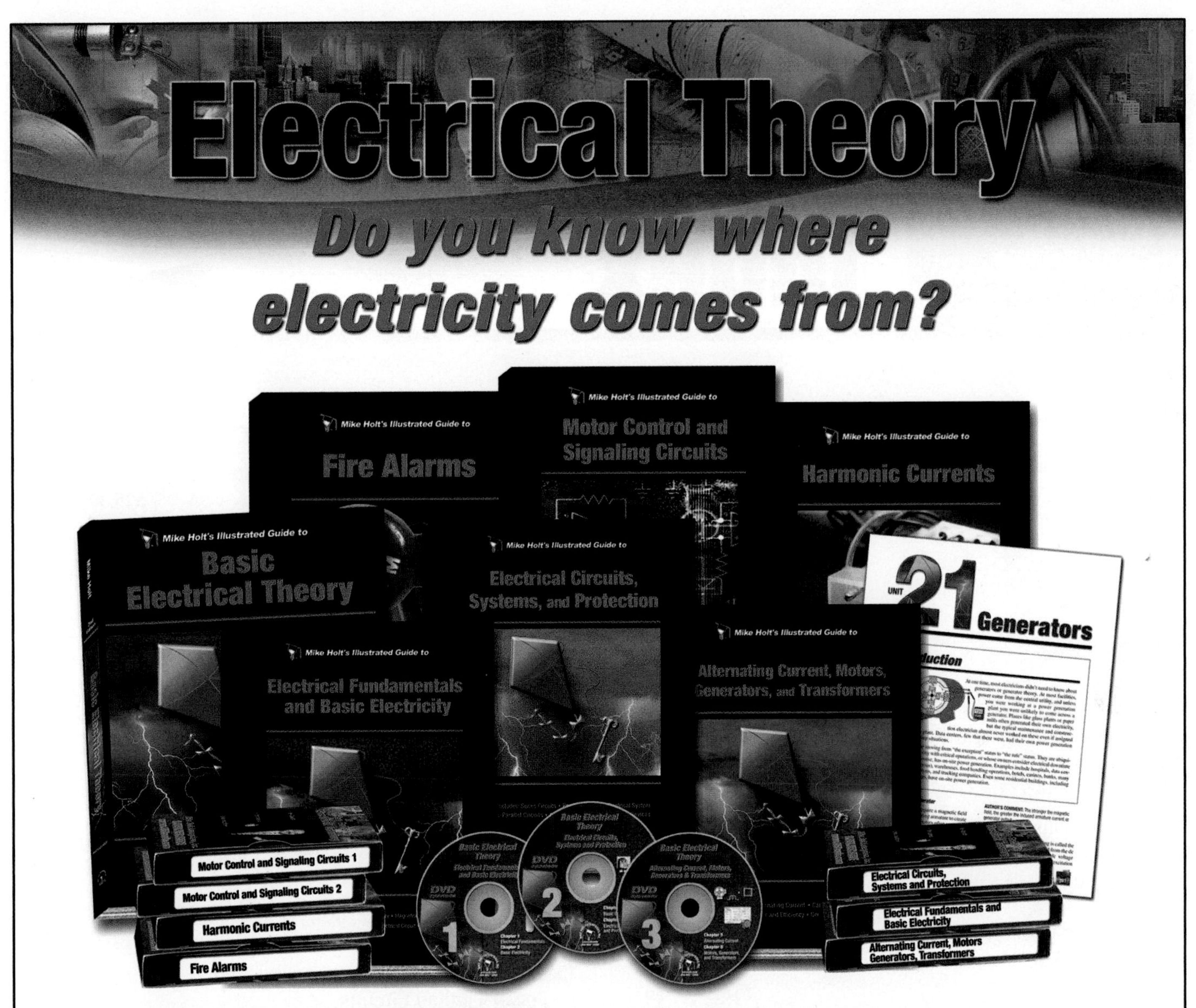

To be able to say yes to that question, you must understand a bit about the physics of matter. What value does a brief study of the nature of matter have for the student of electrical theory? The understanding that comes from that study lays the foundation for understanding electrical theory. Only when you know the theory can you truly have confidence in the practical aspects of your electrical work.

Mike's Basic Electrical Theory full-color textbook provides hundreds of illustrated graphics, detailed examples, practice questions, practice exams, summaries, and a comprehensive practice final exam. Follow along with our Videos or DVDs. Subjects include: Electrical Fundamentals and Basic Electricity; Electrical Circuits, Systems, and Protection; Alternating Current; and Motors, Generators, and Transformers. Additional Theory topics available: Fire Alarms; Motor Controls and Signaling Circuits; and Harmonic Currents.

Call us today at 1.888.NEC.Code, or visit us online at www.NECcode.com, for the latest information and pricing.

Index

E

T

W

Mike Holt's Electrical Theory Libraries

NAME COMPANY TITLE

MAILING ADDRESS CITY STATE ZIP

SHIPPING ADDRESS CITY STATE ZIP

PHONE FAX E-MAIL ADDRESS WEB SITE

❑ CHECK ❑ VISA ❑ MASTER CARD ❑ DISCOVER ❑ AMEX ❑ MONEY ORDER

CREDIT CARD # : ______________________ EXP. DATE: ______

3 or 4 digit security number on front for AmEx on back for all others: ________

	Code	Description	Price	Discount	Sale Price
❑	UTLIB	Electrical Theory Ultimate Library w/Videos	~~$495~~	25% DISCOUNT!	$371.25
❑	UTLIBD	Electrical Theory Ultimate Library w/DVDs	~~$495~~	25% DISCOUNT!	$371.25
❑	ETLIBV	Electrical Theory Standard Library w/Videos	~~$275~~	25% DISCOUNT!	$206.25
❑	ETLIBD	Electrical Theory Standard Library w/DVDs	~~$275~~	25% DISCOUNT!	$206.25

Sales Tax **FLORIDA RESIDENTS ONLY** add 6% $ ________

Shipping: 4% of Total Price (or Minimum $7.50) $ ________

TOTAL DUE $ ________

No other discounts apply. Not valid on previous orders.

Coupon Code UND205

Mike Holt Enterprises, Inc. • 10320 NW 53rd St. Sunrise, FL 33351

FAX 1.954.720.7944 • www.NECcode.com

Discount 25% off

Electrical Theory Libraries

Do you know where electricity comes from? To be able to say yes to that question, you must understand a bit about the physics of matter. What value does a brief study of the nature of matter have for the student of electrical theory? Only when you know the theory can you truly have confidence in the practical aspects of your electrical work.

Mike Holt's Calculations Libraries

NAME COMPANY TITLE

MAILING ADDRESS CITY STATE ZIP

SHIPPING ADDRESS CITY STATE ZIP

PHONE FAX E-MAIL ADDRESS WEB SITE

❑ CHECK ❑ VISA ❑ MASTER CARD ❑ DISCOVER ❑ AMEX ❑ MONEY ORDER

CREDIT CARD # : ______________________ EXP. DATE: ______

3 or 4 digit security number on front for AmEx on back for all others: ________

	Code	Description	Price	Discount	Sale Price
❑	05CAJV	Journeyman Calculation Library w/Videos	~~399~~	25% DISCOUNT!	$299.25
❑	05CAJD	Journeyman Calculation Library w/DVDs	~~399~~	25% DISCOUNT!	$299.25
❑	05CAVM	Master/Contractor Calculation Library w/Videos	~~599~~	25% DISCOUNT!	$449.25
❑	05CADM	Master/Contractor Calculation Library w/DVDs	~~599~~	25% DISCOUNT!	$449.25

Sales Tax **FLORIDA RESIDENTS ONLY** add 6% $ ________

Shipping: 4% of Total Price (or Minimum $7.50) $ ________

TOTAL DUE $ ________

No other discounts apply. Not valid on previous orders.

Coupon Code UND205

Mike Holt Enterprises, Inc. • 10320 NW 53rd St. Sunrise, FL 33351

FAX 1.954.720.7944 • www.NECcode.com

Discount 25% off

Calculations Libraries

Electrical Calculations must be mastered to become successful in the electrical trade. You must understand how to perform the important electrical calculations specified in the *National Electrical Code*. Our Calculations libraries include the Electrical *NEC* Exam Preparation textbook which covers Theory, *Code*, and Calculations in great detail. The detailed videos/DVDs are taped from live classes and take you step-by-step through the calculations.

Mike Holt's Online Training Courses

NAME COMPANY TITLE

MAILING ADDRESS CITY STATE ZIP

SHIPPING ADDRESS CITY STATE ZIP

PHONE FAX E-MAIL ADDRESS WEB SITE

❑ CHECK ❑ VISA ❑ MASTER CARD ❑ DISCOVER ❑ AMEX ❑ MONEY ORDER

CREDIT CARD # : ______________________ EXP. DATE: ______

3 or 4 digit security number on front for AmEx on back for all others: ________

	Code	Description	Price	Discount	Sale Price
❑	ETOL1	Electrical Theory Online Training part 1	~~$89~~	20% DISCOUNT!	$71.20
❑	ETOL2	Electrical Theory Online Training part 2	~~$89~~	20% DISCOUNT!	$71.20
❑	ETOL3	Electrical Theory Online Training part 3	~~$89~~	20% DISCOUNT!	$71.20
❑	05CCOLP1	2005 Code Change Online Training part 1	~~$89~~	20% DISCOUNT!	$71.20
❑	05CCOLP2	2005 Code Change Online Training part 2	~~$89~~	20% DISCOUNT!	$71.20
❑	05WAOLP	2005 Code Change Online Training part 1-2	~~$178~~	20% DISCOUNT!	$142.40
❑	05GBOLP	Grounding versus Bonding Online Training part 1-2	~~$178~~	20% DISCOUNT!	$142.40

Sales Tax **FLORIDA RESIDENTS ONLY** add 6% $ ________

Shipping: 4% of Total Price (or Minimum $7.50) $ ________

TOTAL DUE $ ________

No other discounts apply. Not valid on previous orders.

Coupon Code UND205

Mike Holt Enterprises, Inc. • 10320 NW 53rd St. Sunrise, FL 33351

FAX 1.954.720.7944 • www.NECcode.com

Discount 20% off

Any Online Training Course

Interactive Online Training is available. Many states are now accepting online testing for Continuing Education credits. Go online today and see if your state is on the list and view selected chapters for FREE. Don't get frustrated trying to find a local class or miss valuable time at work if your state accepts this convenient solution.

Mike Holt's Electrical Theory Libraries

NAME ____ COMPANY ____ TITLE ____

MAILING ADDRESS ____ CITY ____ STATE ____ ZIP ____

SHIPPING ADDRESS ____ CITY ____ STATE ____ ZIP ____

PHONE ____ FAX ____ E-MAIL ADDRESS ____ WEB SITE ____

❑ CHECK ❑ VISA ❑ MASTER CARD ❑ DISCOVER ❑ AMEX ❑ MONEY ORDER

CREDIT CARD # : ____ EXP. DATE: ____

3 or 4 digit security number on front for AmEx on back for all others: ____

	Item	Description	List	Discount	Price
❑	UTLIB	Electrical Theory Ultimate Library w/Videos	~~$495~~	25% DISCOUNT!	$371.25
❑	UTLIBD	Electrical Theory Ultimate Library w/DVDs	~~$495~~	25% DISCOUNT!	$371.25
❑	ETLIBV	Electrical Theory Standard Library w/Videos	~~$275~~	25% DISCOUNT!	$206.25
❑	ETLIBD	Electrical Theory Standard Library w/DVDs	~~$275~~	25% DISCOUNT!	$206.25

Sales Tax **FLORIDA RESIDENTS ONLY** add 6% $ ____

Shipping: 4% of Total Price (or Minimum $7.50) $ ____

TOTAL DUE $ ____

No other discounts apply. Not valid on previous orders.

Coupon Code UND205

Mike Holt Enterprises, Inc. • 10320 NW 53rd St. Sunrise, FL 33351

FAX 1.954.720.7944 • www.NECcode.com

Discount 25% off

Electrical Theory Libraries

Do you know where electricity comes from? To be able to say yes to that question, you must understand a bit about the physics of matter. What value does a brief study of the nature of matter have for the student of electrical theory? Only when you know the theory can you truly have confidence in the practical aspects of your electrical work.

Mike Holt's Calculations Libraries

NAME ____ COMPANY ____ TITLE ____

MAILING ADDRESS ____ CITY ____ STATE ____ ZIP ____

SHIPPING ADDRESS ____ CITY ____ STATE ____ ZIP ____

PHONE ____ FAX ____ E-MAIL ADDRESS ____ WEB SITE ____

❑ CHECK ❑ VISA ❑ MASTER CARD ❑ DISCOVER ❑ AMEX ❑ MONEY ORDER

CREDIT CARD # : ____ EXP. DATE: ____

3 or 4 digit security number on front for AmEx on back for all others: ____

	Item	Description	List	Discount	Price
❑	05CAJV	Journeyman Calculation Library w/Videos	~~399~~	25% DISCOUNT!	$299.25
❑	05CAJD	Journeyman Calculation Library w/DVDs	~~399~~	25% DISCOUNT!	$299.25
❑	05CAVM	Master/Contractor Calculation Library w/Videos	~~599~~	25% DISCOUNT!	$449.25
❑	05CADM	Master/Contractor Calculation Library w/DVDs	~~599~~	25% DISCOUNT!	$449.25

Sales Tax **FLORIDA RESIDENTS ONLY** add 6% $ ____

Shipping: 4% of Total Price (or Minimum $7.50) $ ____

TOTAL DUE $ ____

No other discounts apply. Not valid on previous orders.

Coupon Code UND205

Mike Holt Enterprises, Inc. • 10320 NW 53rd St. Sunrise, FL 33351

FAX 1.954.720.7944 • www.NECcode.com

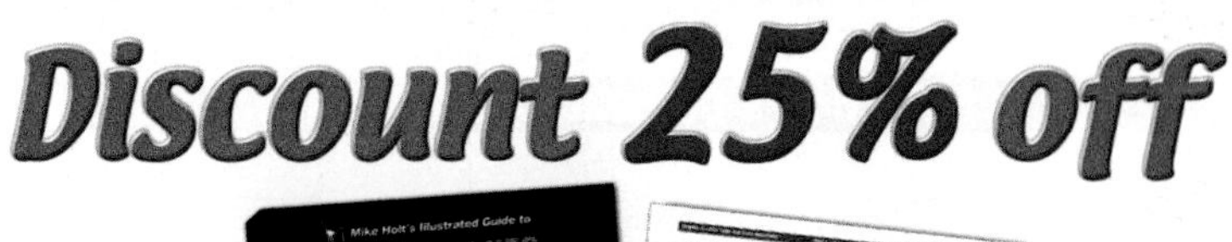

Calculations Libraries

Electrical Calculations must be mastered to become successful in the electrical trade. You must understand how to perform the important electrical calculations specified in the *National Electrical Code*. Our Calculations libraries include the Electrical *NEC* Exam Preparation textbook which covers Theory, *Code*, and Calculations in great detail. The detailed videos/DVDs are taped from live classes and take you step-by-step through the calculations.

Mike Holt's Online Training Courses

NAME ____ COMPANY ____ TITLE ____

MAILING ADDRESS ____ CITY ____ STATE ____ ZIP ____

SHIPPING ADDRESS ____ CITY ____ STATE ____ ZIP ____

PHONE ____ FAX ____ E-MAIL ADDRESS ____ WEB SITE ____

❑ CHECK ❑ VISA ❑ MASTER CARD ❑ DISCOVER ❑ AMEX ❑ MONEY ORDER

CREDIT CARD # : ____ EXP. DATE: ____

3 or 4 digit security number on front for AmEx on back for all others: ____

	Item	Description	List	Discount	Price
❑	ETOL1	Electrical Theory Online Training part 1	~~$89~~	20% DISCOUNT!	$71.20
❑	ETOL2	Electrical Theory Online Training part 2	~~$89~~	20% DISCOUNT!	$71.20
❑	ETOL3	Electrical Theory Online Training part 3	~~$89~~	20% DISCOUNT!	$71.20
❑	05CCOLP1	2005 Code Change Online Training part 1	~~$89~~	20% DISCOUNT!	$71.20
❑	05CCOLP2	2005 Code Change Online Training part 2	~~$89~~	20% DISCOUNT!	$71.20
❑	05WAOLP	2005 Code Change Online Training part 1-2	~~$178~~	20% DISCOUNT!	$142.40
❑	05GBOLP	Grounding versus Bonding Online Training part 1-2	~~$178~~	20% DISCOUNT!	$142.40

Sales Tax **FLORIDA RESIDENTS ONLY** add 6% $ ____

Shipping: 4% of Total Price (or Minimum $7.50) $ ____

TOTAL DUE $ ____

No other discounts apply. Not valid on previous orders.

Coupon Code UND205

Mike Holt Enterprises, Inc. • 10320 NW 53rd St. Sunrise, FL 33351

FAX 1.954.720.7944 • www.NECcode.com

Discount 20% off

Any Online Training Course

Interactive Online Training is available. Many states are now accepting online testing for Continuing Education credits. Go online today and see if your state is on the list and view selected chapters for FREE. Don't get frustrated trying to find a local class or miss valuable time at work if your state accepts this convenient solution.